AF006627

Advances in Modal Logic

Volume 7

Advances in Modal Logic

Volume 7

Edited by

Carlos Areces and Robert Goldblatt

© Individual author and College Publications 2008. All rights reserved.

ISBN 978-1-904987-68-0

College Publications
Scientific Director: Dov Gabbay
Managing Director: Jane Spurr
Department of Computer Science
King's College London, Strand, London WC2R 2LS, UK

http://www.collegepublications.co.uk

Original cover design by Richard Fraser
Cover produced by orchid creative www.orchidcreative.co.uk
Printed by Lightning Source, Milton Keynes, UK

All rights reserved. No part of this publication may be reproduced, stored in a retrieval system or transmitted in any form, or by any means, electronic, mechanical, photocopying, recording or otherwise without prior permission, in writing, from the publisher.

CONTENTS

Preface

Advances in Modal Logic (AiML) is an initiative founded in 1995 and aimed at presenting an up-to-date picture of the state of the art in modal logic and its many applications. It consists of a conference series together with volumes based on the conferences. The conference is the main international forum at which research on all aspects of modal logic is presented. The first one was held in 1996 in Berlin, Germany, and since then it has been organised biennially, with meetings in 1998 in Uppsala, Sweden; in 2000 in Leipzig, Germany (jointly with ICTL-2000); in 2002 in Toulouse, France; in 2004 in Manchester, UK; and in 2006 in Noosa, Australia.

Information about AiML and related events, including conference proceedings, is available at the website `www.aiml.net`.

The seventh conference in the AiML series was hosted by LORIA, the Laboratoire Lorrain de Recherche en Informatique et ses Applications, in Nancy, France, and held on 9–12 September 2008. This proceedings volume contains invited and contributed papers from the conference. The invited speakers and their subjects were

- Mai Gerhke: *Using duality theory to export methods from modal logic*;
- Guido Governatori: *Labelled modal tableaux*;
- Agi Kurucz: *Axiomatising many-dimensional modal logics*;
- Larry Moss: *Relational syllogistic logics, and other connections between modal logic and natural logic*; and
- Michael Zakharyaschev: *Topology, connectedness, and modal logic.*

The Programme Committee received 42 paper submissions. 17 were selected for this book by a process in which each paper received several independent evaluations. They include papers on the metatheory of a variety of modal logics; on systems for spatial and temporal reasoning and interpreting natural language; on the emerging coalgebraic perspective; and on historical views of the nature of modality. It is intended that all these papers will be made available on the AiML website `www.aiml.net`.

The Steering Committee of AiML for 2006–2008 consisted of Carlos Areces, Patrick Blackburn, Robert Goldblatt, Ian Hodkinson, Mark Reynolds, Renate Schmidt, Nobu-Yuki Suzuki, Yde Venema, and Frank Wolter. The Programme Committee for the meeting was co-chaired by Carlos Areces and Robert Goldblatt. It comprised the Steering Committee together with Alessandro Artale, Alexandru Baltag, Guram Bezhanishvili, Philippe Balbiani, Stephane Demri, Melvin Fitting, Guido Governatori, Silvio Ghilardi, Valentin Goranko, Rajeev Gore, Andreas Herzig, Ramon Jansana, Alexander Kurz, Carsten Lutz, Edwin Mares, Larry Moss, Dirk Pattinson, Ulrike Sattler, Ildikó Sain, Jerry Seligman, Valentin Shehtman, Heinrich Wansing, and Michael Zakharyaschev.

Many other people assisted with the reviewing process, including Hajnal Andreka, Nick Bezhanishvili, Remi Brochenin, James Brotherston, Balder Ten Cate, Roberto Cignoli, Corina Cirstea, Willem Conradie, Giovanna Corsi, Giovanna D'Agostino, Deepak D'Souza, Hans van Ditmarsch, Gaelle Fontaine, Tim French, David Gabelaia, Rajeev Gore, Guillaume Hoffmann, Kentaro Kikuchi, Boris Konev, Roman Kontchakov, Agi Kurucz, Tadeusz Litak, John McCabe-Dansted, Paul McNamara, Richard Mendelsohn, Maja Milicic, Angelo Montanari, Massimo Mugnai, Istvan Nemeti, Alessandra Palmigiano, Andre Platzer, Linda Postniece, Revantha Ramanayake, Mikhail Rybakov, Vladislav Ryzhikov, Katsuhiko Sano, Thomas Schneider, Lutz Schröder, François Schwarzentruber, Pablo Seban, Inanc Seylan, Dmitrij Skvortsov, Thomas Studer, Tomoyuki Suzuki, Dmitry Tishkovsky, Dirk Walther, Florian Widmann, and others.

We would like to thank the members of the Programme Committee and all other reviewers for the time and care that they invested in ensuring that a high standard of scholarship was achieved for the conference and its proceedings. Special thanks go to the authors for their excellent contributions, which amply demonstrate that the field of modal logic continues to thrive.

We thank Patrick Blackburn for organising the conference itself, and Jane Spurr for bringing this volume to publication. Finally, we are pleased to acknowledge the support that the conference received from the following seven organisations: LORIA; INRIA Nancy - Grand Est (Institut National de Recherche en Informatique et en Automatique); Université Henri Poincaré; Université Nancy 2; Communauté Urbaine du Grand Nancy; Conseil Régional Lorraine; Ville de Nancy.

Carlos Areces and Rob Goldblatt

Proof systems for the coalgebraic cover modality

MARTA BÍLKOVÁ, ALESSANDRA PALMIGIANO AND YDE VENEMA

ABSTRACT. We investigate an alternative presentation of classical and positive modal logic where the coalgebraic cover modality is taken as primitive. For each logic, we present a sound and complete Hilbert-style axiomatization. Moreover, we give a two-sided sound and complete sequent calculus for the negation-free language, and for the language with negation we provide a one-sided sequent calculus which is sound, complete and cut-free.

Keywords: modal logic, derivation system, coalgebra, coalgebraic modality, Gentzen calculus, completeness.

1 Introduction

This paper studies some derivation systems for a variant of standard modal logic which is based on the finitary *coalgebraic*, or *cover* modality ∇. This connective ∇ takes a finite *set* α of formulas and returns a single formula $\nabla\alpha$. The semantics of the nabla modality can be explicitly formulated as follows, for an arbitrary Kripke structure $\mathbb{S}$ with accessibility relation R:

$$(1) \quad \mathbb{S}, s \Vdash \nabla\alpha \quad \text{if} \quad \begin{array}{l} \text{for all } a \in \alpha \text{ there is a } t \in R[s] \text{ with } \mathbb{S}, t \Vdash a, \text{ and} \\ \text{for all } t \in R[s] \text{ there is an } a \in \alpha \text{ with } \mathbb{S}, t \Vdash a. \end{array}$$

In short: $\nabla\alpha$ holds at a state s iff the formulas in α and the set $R[s]$ of successors of s 'cover' one another.

Using the standard modal language, ∇ can be seen as a defined operator:

$$(2) \quad \nabla\alpha = \Box(\bigvee \alpha) \wedge \bigwedge \Diamond\alpha,$$

where $\Diamond\alpha$ denotes the set $\{\Diamond a \mid a \in \alpha\}$. But is in fact an easy exercise to prove that with ∇ defined by (1), we have the following semantic equivalences:

$$(3) \quad \begin{array}{lcl} \Diamond\alpha & \equiv & \nabla\{\alpha, \top\} \\ \Box\alpha & \equiv & \nabla\varnothing \vee \nabla\{\alpha\} \end{array}$$

In other words, the standard modalities $\Box$ and $\Diamond$ can be defined in terms of the nabla operator (together with $\vee$ and $\top$). When combined, (2) and (3) show that the language based on the nabla operator is indeed an alternative formulation of standard modal logic.

Readers familiar with classical first-order logic will recognize the quantification pattern underlying (1) and (2) from the theory of Ehrenfeucht-Fraïssé games, Scott sentences, and the like, see [6] for an overview. In modal logic, related ideas made an early appearance in Fine's work on normal forms [3]. As far as we know, however, the first two explicit occurrences of the cover modality as a (primitive) *connective* appeared roughly at the same time, in the work of Barwise & Moss on circularity [1], and that of Janin & Walukiewicz on automata-theoretic approaches towards the modal μ-calculus [7].

Broadly speaking, in the literature one may find two kinds of motivation for a ∇-based approach towards modal logic. To start with, a technical reason is that in some applications, ∇-based modal logic works better because one may almost eliminate conjunctions from the language. This observation, which ultimately goes back to automata-theoretic constructions in [7], has subsequently been used in connection with interpolation and Beth definability properties of modal languages [2, 16], and in order to obtain completeness proofs for modal fixpoint logics [15, 14]

A second and more conceptual reason to prefer the ∇-based perspective on modal logic is that it allows for coalgebraic generalizations. Coalgebra [13] is an emerging mathematical theory of state-based evolving systems. Kripke structures are key examples of coalgebras, and ideas from modal logic have been fruitfully exported to other types of coalgebras, see [17] for an overview. It was Moss' fundamental observation [11] that (1) expresses the semantics of ∇ in terms of some kind of the Egli-Milner *relation lifting* of the satisfaction relation between states and formulas. This insight led Moss to the introduction of *coalgebraic logic*, which is based on a generalization of the cover modality to a coalgebraic modality $\nabla_{\mathcal{T}}$ for coalgebras of type $\mathcal{T}$.

These two ideas can be fruitfully combined. For instance, the coalgebraic perspective of [9] enables many results on fixed point logics and automata to be generalized to a much wider level of generality. Similarly, in [12], an axiomatic approach towards ∇ is combined with a very general lifting construction on Chu spaces to shed some light on the Vietoris construction of Stone spaces.

In this paper we consider various *derivation systems* for ∇-based modal logic. In earlier work [12], the second and third author developed a sound and complete Hilbert-style derivation system for the nabla modality. Here we extend this work in two directions. First, we give an alternative Hilbert-style axiomatization which, besides being more compact, elegant and intrinsic than the previous one, allows for a great generalization to the context of the coalgebraic modal languages defined by Moss. (Such a generalization has indeed been supplied: see Kupke, Kurz & Venema [8].) The main contribution of the present paper, however, concerns a number of Gentzen-style derivation systems. Indeed, we introduce a one-sided Gentzen system (see Definition 15), for an expansion of the Boolean propositional language with the nabla operator which corresponds to the basic modal logic **K** in the language with nabla. This system is sound and complete w.r.t. the class of all Kripke models, and cut-free. Moreover, we present a sound and complete two-sided Gentzen system (see Definition 23) for the positive fragment of the same language. Following the same design criteria that inspired our new Hilbert-style axiomatization, the main feature of both Gentzen systems is that they are generalizable to the coalgebraic setting where $\mathcal{T}$ is an arbitrary weak pullback-preserving Set-endofunctor.

The organization of the paper goes as follows: Section 2 contains preliminaries on the syntax and semantics of nabla, and introduces the technical notion of *slim redistributions* (Definition 5). In Section 3 we present the new Hilbert-style axiomatization C_∇ (Definition 9), recall the previous axiomatization H_∇ and prove soundness and completeness of C_∇. In Section 4 we present the one-sided Gentzen system $G1\nabla$, show its soundness and completeness and compare it with the alternative and equivalent one-sided calculus $GW\nabla$ introduced by Walukiewicz. In Section 5 we present the two-sided calculus $G2\nabla$, prove soundness, completeness and show that it is not cut-free. Section 6 presents some concluding remarks and directions for further research.

2 Preliminaries

Syntax and Semantics Throughout this paper we fix a set Q of propositional variables. In principle, we want to define our language $\mathcal{L}$ as the smallest superset of Q which is closed under Boolean formula constructions, and under applying the finitary nabla modality: if α is a finite set of formulas, then $\nabla\alpha$ is a formula. (In fact, most of our results in some way go through for the infinite version of the cover modality as well, but here we restrict to finitary syntax.)

Before we go into the formal details, we need to discuss two aspects of our approach that depart from standard treatments and which respectively involve a slight modification and a subtle distinction. To start with, when we formulate our axiomatization, it will be convenient to work with arbitrary finite conjunctions and disjunctions, rather than with the binary ones. That is, the slight modification consists in taking $\neg$, $\bigwedge$ and $\bigvee$ as our primitive Boolean operation symbols. We will use $\top$ (and $\bot$) as abbreviations for $\bigwedge\varnothing$ (and $\bigvee\varnothing$) respectively.

Second, to explain and motivate the subtle but crucial distinction that we make, recall from the introduction that one of the goals of Moss' approach [11] was to *generalize* the language and semantics of modal logic from ordinary Kripke structures to coalgebras for an (almost) arbitrary set functor. In order to facilitate this generalization, we will separate two roles of the *power set operation* in our framework, and formalize this by using *distinct* notation. Concretely, given a set X, $\mathcal{P}X$ will denote the power set of X in its standard set-theoretic use, and $\mathcal{P}_\omega$ denotes the finitary variant of $\mathcal{P}$. That is, $\mathcal{P}X$ and $\mathcal{P}_\omega X$ denote the collections of all and of all finite subsets of X, respectively. For instance, we will say that $\bigwedge\varphi$ and $\bigvee\varphi$ are formulas whenever $\varphi \in \mathcal{P}_\omega\mathcal{L}$. In case we are working with the power set as the coalgebraic *functor*, we will use the notation $\mathcal{T}$, or $\mathcal{T}_\omega$ for its finitary version. As an example, we will say that $\nabla\alpha \in \mathcal{L}$ whenever $\alpha \in \mathcal{T}_\omega\mathcal{L}$, and we may represent the accessibility relation R of a Kripke frame (S, R) as the function $S \to \mathcal{T}S$ mapping a state to its set of successors.

This approach, which may seem rather pedantic at first sight, creates the ability to distinguish incidental properties of our set-up (arising from the fact that the coalgebraic functor happens to coincide with the power set functor) from structural/categorical ones (which can be generalized to arbitrary functors). In this way it does not only pave the way for generalizations [8], it has also been very useful to increase our understanding of the concrete case at hand, viz., that of the power set functor. In the sequel, we will as much as possible formulate definitions in terms that either apply to, or else can be generalized to, an arbitrary set functor $\mathcal{T}$.

DEFINITION 1. $\mathcal{L}$ is the smallest set containing all propositional variables which is closed under taking negations (if $a \in \mathcal{L}$ then $\neg a \in \mathcal{L}$), under taking finitary conjunctions and disjunctions (if $\varphi \in \mathcal{P}_\omega\mathcal{L}$ then $\bigvee\varphi, \bigwedge\varphi \in \mathcal{L}$), and under applying the finitary nabla modality (if $\alpha \in \mathcal{T}_\omega\mathcal{L}$ then $\nabla\alpha \in \mathcal{L}$). The negation-free fragment of $\mathcal{L}$ is denoted as $\mathcal{L}^+$. Elements of $\mathcal{L}$ ($\mathcal{L}^+$) are called *formulas* (*positive formulas*, respectively).

CONVENTION 2. In the sequel we will need symbols to refer to formulas ($\mathcal{L}$), sets of formulas ($\mathcal{P}_\omega\mathcal{L}$ and $\mathcal{T}_\omega\mathcal{L}$), and to elements of the sets $\mathcal{T}_\omega\mathcal{P}_\omega\mathcal{L}$ and $\mathcal{P}_\omega\mathcal{T}_\omega\mathcal{L}$. It will be convenient to fix our notation for these objects, and we will do so as indicated by the table below, where on the right side we list the generic symbols that we use to denote objects from the sets on the left side:

$\mathcal{L}$	$a, b, c, \ldots$
$\mathcal{T}_\omega\mathcal{L}$	$\alpha, \beta, \gamma \ldots$
$\mathcal{P}_\omega\mathcal{L}$	$\varphi, \psi, \theta \ldots$
$\mathcal{T}_\omega\mathcal{P}_\omega\mathcal{L}$	$\Phi, \Psi, \Theta \ldots$
$\mathcal{P}_\omega\mathcal{T}_\omega\mathcal{L}$	$A, B, C \ldots$

This language can be interpreted in standard Kripke models $\mathbb{S} = (S, R, V)$, with $R \subseteq S \times S$ and $V : S \to \mathcal{P}(\mathsf{Q})$. We will often think of R coalgebraically, as a map $R : S \to \mathcal{T}(S)$ which maps a state s to the set $R[s]$ of its (immediate) successors. For the inductive definition of the satisfaction relation $\Vdash$, we omit the atomic and Boolean clauses because they are completely standard. For the semantics of ∇, we need the notion of *relation lifting*. There are various ways to lift a relation between two sets to one between the respective power sets; our definition uses the so-called Egli-Milner lifting.

DEFINITION 3. Let $R \subseteq X \times X'$ be a binary relation. Its *(power set) lifting* is defined as the relation $\overline{R} \subseteq \mathcal{P}X \times \mathcal{P}X'$ given by

$$(4) \quad (A, A') \in \overline{R} \quad \text{iff} \quad \forall a \in A\, \exists a' \in A'.\, aRa' \quad \text{and} \quad \forall a' \in A'\, \exists a \in A.\, aRa'.$$

REMARK 4. Modal logicians will recognize in (4) the quantification pattern from the definition of a bisimulation. Indeed, bisimulations between two Kripke models $\mathbb{S}$ and $\mathbb{S}'$ can be characterized as non-empty relations $Z \subseteq S \times S'$ such that $V(s) = V'(s')$ and $(R[s], R'[s']) \in \overline{Z}$ whenever sZs'.

Returning to the semantics of $\mathcal{L}$, given a Kripke model $\mathbb{S} = (S, R, V)$, the clause for ∇ can be very concisely formulated in terms of the lifting $\overline{\Vdash}$ of the satisfaction relation $\Vdash \subseteq S \times \mathcal{L}$:

$$\mathbb{S}, s \Vdash \nabla\alpha \quad \text{iff} \quad R[s] \mathrel{\overline{\Vdash}} \alpha.$$

In words: $\mathbb{S}, s \Vdash \nabla\alpha$ if every $t \in R[s]$ satisfies some $a \in \alpha$, and conversely, every $a \in \alpha$ is satisfied at some successor t of s. Thus $\nabla\alpha$ is indeed equivalent to the formula $\Box \bigvee \alpha \wedge \Diamond \bigwedge \alpha$.

Slim redistributions An important role in the paper is played by the notion of *slim redistribution*. Formulated specifically for the power set functor, a set $\Phi \in \mathcal{P}\mathcal{P}X$ is a slim redistribution of a set $A \in \mathcal{P}\mathcal{P}X$ iff $\bigcup A = \bigcup \Phi$ and $\varphi \cap \alpha \neq \varnothing$ for all $\varphi \in \Phi$ and $\alpha \in A$. Borrowing some intuition from topology, these two conditions tell us that on the one hand every given $\alpha \in A$ is 'covered' by Φ (in the sense that $\alpha \subseteq \bigcup \Phi$) in such a way that every $\varphi \in \Phi$ has nonempty intersection with α (and again, in this relation between every $\alpha \in A$ and Φ, we meet the familiar quantification pattern from the definitions of bisimulation and relation lifting). On the other hand, the requirement that $\bigcup \Phi \subseteq \bigcup A$ is clearly a minimality condition on Φ, that takes care that every such Φ can be effectively constructed from A by scrambling and suitably reorganizing its 'ingredients'.

Our formulation below can be extended to arbitrary functors. It uses the lifted membership relation $\overline{\in}$. In our case, an object $\alpha \in \mathcal{T}X$ is a lifted member of $\Phi \in \mathcal{T}\mathcal{P}X$ if $\alpha \subseteq \bigcup \Phi$ and $\alpha \cap \varphi \neq \varnothing$ for all $\varphi \in \Phi$.

DEFINITION 5. Given a set X, we call an element $\alpha \in \mathcal{T}X$ a *lifted member* of an element $\Phi \in \mathcal{T}\mathcal{P}X$ if $\alpha \mathrel{\overline{\in}} \Phi$, where $\overline{\in} \subseteq \mathcal{T}X \times \mathcal{T}\mathcal{P}X$ denotes the lifted version of the membership relation $\in \subseteq X \times \mathcal{P}X$.

An object $\Phi \in \mathcal{T}\mathcal{P}X$ is a *redistribution* of a set $A \in \mathcal{P}\mathcal{T}X$ if $\alpha \mathrel{\overline{\in}} \Phi$ for all $\alpha \in A$ (hence in particular $\bigcup A \subseteq \bigcup \Phi$). We call such a redistribution *slim* if moreover $\bigcup \Phi = \bigcup A$. The set of all slim redistributions of A is denoted by $SRD(A)$.

Since we will usually be looking at redistributions of finite collections of finite sets of formulas, it is good to note that such redistributions will also consist of *finite* collections of *finite* sets of formulas. That is, if $\mathcal{T} = \mathcal{P}$, then $SRD(A) \subseteq \mathcal{T}_\omega \mathcal{P}_\omega X$ whenever $A \in \mathcal{P}_\omega \mathcal{T}_\omega X$. This is not true for an arbitrary set functor $\mathcal{T}$.

EXAMPLE 6.

1. A key example of a slim redistribution of a set $A \in \mathcal{P}\mathcal{T}\mathcal{L}$ arises semantically. Fix a model $\mathbb{S}$ and a state s in $\mathbb{S}$. Define, for any successor t of s, the set $\varphi_t := \{a \in \bigcup A \mid \mathbb{S}, t \Vdash a\}$. Then consider the set $\Phi_s := \{\varphi_t \mid t \in R[s]\} \in \mathcal{T}\mathcal{P}\mathcal{L}$:

(5) $\mathbb{S}, s \Vdash \bigwedge\{\nabla\alpha \mid \alpha \in A\} \iff \Phi_s \in SRD(A)$.

'only if': to see why Φ_s is a redistribution, take an arbitrary element $\alpha \in A$. In order to prove that α is a lifted member of Φ_s, first observe that every element a of α is true at some successor t of s. Hence, $a \in \varphi_t \subseteq \bigcup \Phi_s$. Second, given an arbitrary element $\varphi \in \Phi_s$, there is a successor t of s such that $\varphi = \varphi_t$. But then since $s \Vdash \nabla\alpha$, some $a \in \alpha$ is true at t, and hence, $a \in \varphi_t$. In other words, $\varphi_t \cap \alpha \neq \varnothing$. This shows that Φ_s is a redistribution of A; but then it is easy to see that as such it is *slim*: each φ_t only takes elements from $\bigcup A$, and so $\bigcup \Phi_s \subseteq \bigcup A$. Conversely, it is easy to show that if Φ_s is a redistribution of A, then $\mathbb{S}, s \Vdash \nabla\alpha$ for every $\alpha \in A$.

2. Given a relation $R \subseteq \prod_{1 \leq i \leq n} \alpha_i$ we will write $R \in \bowtie_{1 \leq i \leq n} \alpha_i$ if R is a *subdirect product* of the relations α_i, that is, if $\pi_i[R] = \alpha_i$ for every i. It is an easy exercise to verify that for each such R, the set $\Big\{\{a_1, \dots, a_n\} \mid (a_1, \dots a_n) \in R\Big\}$ is a slim redistribution of the set $A = \{\alpha_1, \dots, \alpha_n\}$. This shows that in a way, slim redistributions generalize the notion of relation lifting, because in the case that $n = 2$, R is binary, and we have

   (6) $R \in \alpha \bowtie \beta \iff (\alpha, \beta) \in \overline{R}$.

3. Any set $\Phi \in \mathcal{TP}X$ is a redistribution of the empty set $\varnothing$, but the latter only has two slim redistributions: $SRD(\varnothing) = \{\varnothing, \{\varnothing\}\}$.

4. It is easy to see that 'being a slim redistribution of' is a symmetric relation. It is neither reflexive, nor transitive, however. A counterexample to reflexivity is the set $A = \{\{a\}, \{b\}\}$: indeed, $SRD(A) = \{\{a, b\}\}$. By symmetry, this is also a counterexample to transitivity.

5. If A is a set of singletons, say $A = \{\{c\} \mid c \in \varphi\}$, then φ is the only lifted member of A, and $\{\varphi\}$ is the unique slim redistribution of A.

REMARK 7. As mentioned, the notions of lifted membership and slim redistributions can be generalized to a much wider category-theoretic setting. This generalization is based on a standard way to *lift* a relation $R \subseteq S \times S'$ to a relation $\overline{R} \subseteq \mathcal{T}S \times \mathcal{T}S'$, and is associated with a *distributive law*, that is, a natural transformation $\lambda : \mathcal{TP} \to \mathcal{PT}$ given by $\lambda_X(\Phi) := \{\alpha \in \mathcal{PT}X \mid \alpha \,\overline{\in}\, \Phi\}$. We refer to [8] for the details.

3 Hilbert-style axiomatizations

In this section we introduce an equational Hilbert-style presentation for the ∇-modal logic. Because we want to deal simultaneously with the full language $\mathcal{L}$ and its positive fragment $\mathcal{L}^+$, we formulate our systems in terms of *inequalities*, i.e. we will introduce a Hilbert-style presentation of a *2-dimensional* deductive system (cf. [4], 4.1), in which *pairs* of formulas are the basic objects axioms and rules act on:

$(\nabla 1)$	From $\vdash \alpha \overline{\leq} \beta$ infer $\vdash \nabla\alpha \leq \nabla\beta$
$(\nabla 2)$	$\bigwedge\{\nabla\alpha \mid \alpha \in A\} \leq \bigvee\{\nabla\{\bigwedge\varphi \mid \varphi \in \Phi\} \mid \Phi \in SRD(A)\}$
$(\nabla 3)$	$\nabla\{\bigvee\varphi \mid \varphi \in \Phi\} \leq \bigvee\{\nabla\beta \mid \beta \overline{\in} \Phi\}$

Table 1. Axioms and rules of the system C_∇

DEFINITION 8. An *inequality* is nothing but a pair (a, b) of formulas, usually denoted as $a \leq b$. An inequality $a \leq b$ is *valid in a Kripke model* $\mathbb{S}$, notation: $\mathbb{S} \models a \leq b$, if for all states s in $\mathbb{S}$, $s \Vdash a$ implies $s \Vdash b$. An inequality is *valid*, notation: $\models a \leq b$, if it is valid in every Kripke model.

The name of the following axiomatization stems from the fact that it was first presented at the WoLLiC 2007 conference in Rio de Janeiro.

DEFINITION 9 (Carioca Axiomatization). C_∇ is the Hilbert-style derivation system, operating on inequalities, which is given by the axioms and rules of Table 1. C_∇^+ is the version of C_∇ in which the language is restricted to the set $\mathcal{L}^+$ of positive formulas.

Therefore, the only substantial distinction between C_∇ and C_∇^+ concerning derivability lies in the boolean parts of the axiomatizations. We do not specify the purely propositional parts - for C_∇ take any adequate axiomatization of classical propositional logic in a language that includes negation, formulated as a set of inequalities and derivation rules, while for C_∇^+ take the corresponding monotone fragment.

DEFINITION 10. A *derivation* in the system C_∇ is a finite tree labelled with inequalities, such that each leaf is labeled by an axiom of C_∇ and every node that is a parent node is labeled by the conclusion of a derivation rule each premise of which labels one of its children nodes. An inequality $a \leq b$ is a *theorem* of this system, notation: $\vdash_H a \leq b$, if it appears as an item (equivalently, as the last item) of some derivation. Similar definitions apply to the system C_∇^+, with $\vdash_H^+$ denoting theoremhood.

While our notions of derivation and theoremhood are standard, the axioms and rules of the system can do with some explanation. To start with, the reader may be slightly puzzled by our formulation of the derivation rule $(\nabla 1)$, since its premise '$\vdash \alpha \overline{\leq} \beta$' uses syntax that has not been defined as part of the object language. The proper way to read this premise is as follows: 'the relation $Z := \{(a, b) \in \alpha \times \beta \mid \vdash_H a \leq b\}$ is such that $(\alpha, \beta) \in \overline{Z}$', that is, 'for all $a \in \alpha$ there is a $b \in \beta$ such that $\vdash_H a \leq b$, and vice versa'. We choose the presentation in Table 1 because it is shorter and reveals more clearly that the rule is in fact the inequality version of a *congruence* rule. The axioms $(\nabla 2)$ and $(\nabla 3)$ could in fact both be replaced with identities, since in both cases, the reverse inequality of the axiom can be derived as a theorem. What these axioms have in common further is that they can be seen as *distributive laws.* This is the clearest in the case of $(\nabla 3)$, which states that ∇ distributes over some disjunctions. In the case of $(\nabla 2)$ the distribution principle is a bit more involved, but basically, the axiom states that any 'conjunction of nablas' can be replaced with a disjunction of 'nablas of conjunctions'.

REMARK 11. The following Hilbert-style rule:

(∇4) From $\vdash \top \leq \bigvee \varphi$ infer $\vdash \top \leq \bigvee\{\nabla\alpha \mid \alpha \in \mathcal{T}_\omega\varphi\}$

is derivable from (∇0)-(∇3).

Proof. Consider the following derivation tree. Let

$$\Pi \quad = \quad \text{trans}\,\dfrac{\nabla 1\,\dfrac{\top \leq \bigvee\varphi}{\nabla\{\top\} \leq \nabla\{\bigvee\varphi\}} \qquad \nabla\{\bigvee\varphi\} \leq \bigvee B}{\vee\,\dfrac{\nabla\{\top\} \leq \bigvee B}{\nabla\{\top\} \leq \bigvee B \vee \nabla\varnothing}}$$

in the derivation

$$\text{trans}\,\dfrac{\top \leq \nabla\varnothing \vee \nabla\{\top\} \qquad \vee\,\dfrac{\nabla\varnothing \leq \bigvee B \vee \nabla\varnothing \qquad \Pi}{\nabla\varnothing \vee \nabla\{\top\} \leq \bigvee B \vee \nabla\varnothing}}{\top \leq \bigvee B \vee \nabla\varnothing}$$

where $B = \{\nabla\beta \mid \beta \mathrel{\overline{\in}} \{\varphi\}\}$: notice that its leftmost leaf is an instance of the axiom (∇2) and its rightmost leaf is an instance of the axiom (∇3), where $\Phi = \{\{a\} \mid a \in \Phi\}$. Notice also that in the special setting we are in, i.e. dealing with the power set functor, $\beta \mathrel{\overline{\in}} \{\varphi\}$ iff $\beta \in \mathcal{T}\varphi \setminus \varnothing$. Since $\mathcal{T}\varphi = \mathcal{T}_\omega\varphi$ because φ is finite, the conclusion of the derivation tree above can be rewritten as $\top \leq \bigvee\{\nabla\beta \mid \beta \in \mathcal{T}_\omega\varphi\}$. ■

Although the formulation of (∇4) does not use the actual symbol, this rule effectively captures the interaction between the coalgebraic modality and negation. To see why this is so, observe that the conclusion of (∇4) implies that $\neg\nabla\beta \leq \bigvee\{\nabla\alpha \mid \beta \neq \alpha \in \mathcal{T}\varphi\}$.

REMARK 12. The systems C_∇ and C_∇^+ can be seen as streamlined version of the axiom system H_∇ given by the second and third author [12]. This axiomatization has the following set of axioms and rules:

(∇_1) If $\alpha \mathrel{\overline{\leq}} \beta$, then $\nabla\alpha \leq \nabla\beta$
(∇_2) If $\bot \in \alpha$ then $\nabla\alpha = \bot$
(∇_3) $\nabla\alpha \wedge \nabla\beta \leq \bigvee\{\nabla\{a \wedge b \mid (a,b) \in Z\} \mid Z \in \alpha \bowtie \beta\}$
(∇_4) $\nabla\{a \vee b \mid a \in \alpha, b \in \beta\} \cup \{\top\} \leq \nabla\alpha \cup \{\top\} \vee \nabla\beta \cup \{\top\}$
(∇_5) $\top \leq \nabla\varnothing \vee \nabla\{\top\}$
(∇_6) $\nabla\alpha \cup \{a \vee b\} \leq \nabla\alpha \cup \{a\} \vee \nabla\alpha \cup \{b\} \vee \nabla\alpha \cup \{a, b\}$
(∇_7) $\neg\nabla\alpha = \nabla\{\bigwedge\{\neg a \mid a \in \alpha\}, \top\} \vee \bigvee_{a \in \alpha} \nabla\{\neg a\} \vee \nabla\varnothing$

Our new axiomatization generalizes this system, as follows. The axioms (∇_3) and (∇_5) can be seen as instances of our axiom (∇2): take for A the sets $\{\alpha, \beta\}$ and $\varnothing$, respectively. Similarly, the axioms (∇_2), (∇_4) and (∇_6) are all instances of our axiom (∇3): take for Φ the sets $\{\{a\} \mid a \in \alpha\} \cup \{\varnothing\}$, $\{\{a, b\} \mid a \in \alpha \cup \{\top\}, b \in \beta \cup \{\top\}\}$, and $\{\{a\} \mid a \in \alpha\} \cup \{a, b\}$, respectively. Finally, axiom (∇_7) can be derived using the derived rule (∇4). (We'll get back to this in the completeness proof for C_∇.)

Advantages of the Carioca Axiomatization are not only that it shows some of the old axioms to be instances of a more general principle. The main point is that, since in C_∇ the two roles of the powerset as a set-theoretic construction and the powerset as a functor are kept clearly distinct, C_∇ can directly be generalized to an arbitrary set functor, see [8]. As a good way to understand this difference, it can be useful to compare the axiom (∇_7) and the Carioca-style derived rule (∇4): whereas (∇_7) clearly uses the fact that $\alpha \in \mathcal{T}_\omega \mathcal{L}$ is a set of formulas, and thus belongs to $\mathcal{P}_\omega \mathcal{L}$, ($\nabla$4) involves no such confusion.

THEOREM 13 (SOUNDNESS AND COMPLETENESS). *The Carioca axioms are sound and complete. That is, for any pair a, b of formulas:*

(7) $\vdash_H a \leq b$ *iff* $\models a \leq b$.

Similarly, for any pair a, b of positive formulas:

(8) $\vdash_H^+ a \leq b$ *iff* $\models a \leq b$.

Proof. In order to prove *soundness*, that is, the direction from left to right in (7) and (8), we first establish the validity of the axioms, and then show that the rules preserve validity.

Omitting a trivial discussion of the boolean part, we first consider the axiom (∇2). Consider a model $\mathbb{S}$ and a state s such that $\mathbb{S}, s \Vdash \bigwedge\{\nabla\alpha \mid \alpha \in A\}$. We already saw that the set $\Phi_s := \{\varphi_t \mid t \in R[s]\}$, with $\varphi_t := \{a \in \bigcup A \mid \mathbb{S}, t \Vdash a\}$ is a slim redistribution of A. Thus it suffices to show that $\mathbb{S}, s \Vdash \nabla\{\bigwedge\varphi \mid \varphi \in \Phi_s\}$. But this is virtually immediate by the definitions: At an arbitrary successor t of s, the formula $\bigwedge\varphi_t$ holds, and any element of Φ_s is of the form φ_t for some $t \in R[s]$, and so every formula $\bigwedge\varphi_t$ is satisfied at some successor of s.

For axiom (∇3), suppose that $\mathbb{S}, s \Vdash \nabla\{\bigvee\varphi \mid \varphi \in \Phi\}$. For each $t \in R[s]$, define $\beta_t := \{b \in \bigcup\Phi \mid t \Vdash b\}$ and take $\beta := \bigcup\{\beta_t \mid t \in R[s]\}$. We claim that $\beta \overline{\in} \Phi$ and $\mathbb{S}, s \Vdash \nabla\beta$. For the latter claim, it follows from the assumption that each $t \in R[s]$ satisfies the formula $\bigvee\varphi$ for some $\varphi \in \Phi$. Hence for each such t there is some formula $b_t \in \varphi$ with $t \Vdash b_t$. Clearly then, $b_t \in \beta_t \subseteq \beta$, and so indeed every successor of s satisfies some formula in β. Conversely, for each $b \in \beta$, by definition of β there is a $t \in R[s]$ satisfying b.

To see that $\beta \overline{\in} \Phi$, take $b \in \beta$. By definition b belongs to $\bigcup\Phi$, so there is some $\varphi \in \Phi$ such that $b \in \varphi$. Conversely, take an arbitrary $\varphi \in \Phi$. By the assumption $s \Vdash \nabla\{\bigvee\varphi \mid \varphi \in \Phi\}$, there is some $t \in R[s]$ such that $t \Vdash \bigvee\varphi$. That is, some $b \in \varphi$ is satisfied in t. But then this b belongs to β, as an immediate consequence of the definition of β.

Now we show that the derivation rules preserve validity. First we consider the rule ($\nabla 1$). Suppose that $Z \subseteq \mathcal{L} \times \mathcal{L}$ is some relation such that $\mathbb{S} \Vdash a \leq b$ for each $(a, b) \in Z$, and assume that $(\alpha, \beta) \in \overline{Z}$. In order to show that $\mathbb{S} \Vdash \nabla\alpha \leq \nabla\beta$, consider a state s such that $s \Vdash \nabla\alpha$, then it suffices to show that $\mathbb{S}, s \Vdash \nabla\beta$. First take an arbitrary successor t of s. Then from $s \Vdash \nabla\alpha$ we may infer that $t \Vdash a$ for some $a \in A$. Since $(\alpha, \beta) \in \overline{Z}$ there is some $b \in \beta$ with $(a, b) \in Z$, and so from $\mathbb{S} \Vdash a \leq b$ we infer that $t \Vdash b$. For the converse, take an arbitrary element $b \in \beta$. From $(\alpha, \beta) \in \overline{Z}$ we may infer the existence of an $a \in \alpha$ such that $(a, b) \in Z$. But if $a \in \alpha$, there must be a successor t of s where a holds, because $s \Vdash \nabla\alpha$. Then from $\mathbb{S} \Vdash a \leq b$ we may conclude that b holds at this t too. This shows that, indeed, $s \Vdash \nabla\beta$.

Turning to completeness, we first consider the negation-free system C^{+}_{∇}. It was shown in [12] that the (negation-free version of the) system of Remark 12, without the axiom (∇_7) is a complete axiomatization for the valid positive inequalities. But as we mentioned in the same Remark, all the axioms and rules of this system are *instances* of rules and axioms of C^{+}_{∇}, From this completeness (the direction from right to left in (8)) is immediate.

In order to prove completeness of the system C_{∇} with respect to the full language, we follow the same approach. Given the earlier observations it suffices to show the derivability of the axiom (∇_7) in our system, since it is the only axiom or rule that is not an instance of an axiom or rule of C_{∇}.

Fix a set of formulas $\alpha \in T_{\omega}\mathcal{L}$. We will show that

$$(9) \quad \vdash_H \top \leq \nabla\alpha \vee \nabla\{\bigwedge\{\neg a \mid a \in \alpha\}, \top\} \vee \bigvee_{a \in \alpha} \nabla\{\neg a\} \vee \nabla\varnothing$$

We close α under applications of the Boolean connectives, obtaining a set of formulas which is a finite Boolean algebra modulo provable equivalence. Let φ be a set of formulas which contains exactly one representing element for each atom of this Boolean algebra. Then every element $a \in \alpha$ is provable equivalent to a disjunction of formulas in φ, and for all $b \in \varphi$ and $a \in \alpha$ we either have $\vdash_H b \leq a$ or $\vdash_H b \leq \neg a$.

It follows from $\vdash_H \top \leq \bigvee \varphi$, as has been show in derivation of ($\nabla 4$), that $\vdash_H \top \leq \bigvee\{\nabla\beta \mid \beta \subseteq \varphi\}$. So in order to prove that the axiom (∇_7) is derivable in C_{∇}, it suffices to show that for each $\beta \subseteq \varphi$ there is an disjunct d of the formula on the right hand side of (9) such that $\vdash_H \nabla\beta \leq d$. We make a case distinction:

- If $\beta = \varnothing$, simply take $d := \nabla\beta$.
- If $\exists a \in \alpha \, \forall b \in \beta \vdash_H b \leq \neg a$, take $d := \nabla\{\neg a\}$.
- If $\exists b \in \beta \, \forall a \in \alpha \vdash_H b \leq \neg a$, take $d := \nabla\{\bigwedge\{\neg a \mid a \in \alpha\}, \top\}$.
- Otherwise, we have that $\forall b \in \beta \, \exists a \in \alpha. \vdash_H b \leq a$ and $\forall a \in \alpha \, \exists b \in \beta. \vdash_H b \leq a$. In this case, simply apply rule ($\nabla 1$) and obtain $\vdash_H \nabla\beta \leq \nabla\alpha$, so take $d := \nabla\alpha$.

This finishes the proof of the direction from right to left of (7). ∎

REMARK 14. Kupke, Kurz and the third author [8] have generalized Theorem 13 to a much wider setting of set functors that preserve weak pullbacks.

4 One-sided Gentzen calculi

In this section we will introduce the cut free, one-sided sequent proof system $G1\nabla$ for the ∇-presentation of the basic normal modal logic $\mathbf{K}$. $G1\nabla$ arises from the classical propositional left-sided calculus adding one modal rule. We will compare $G1\nabla$ with the alternative one-sided sequent proof system $GW\nabla$ introduced by Janin and Walukiewicz [7], show that they are equivalent, in the sense that the rules of one system are admissible in the other system, and argue that $G1\nabla$ has the advantage of being more suitable for generalizations. The language for this calculus is the restriction $\mathcal{L}^*$ of $\mathcal{L}$ where negations can only be applied to proposition letters: every formula in $\mathcal{L}$ is semantically equivalent to some formula of $\mathcal{L}^*$. To see this, recall that the nabla operator, axiomatized as in the previous section, is interdefinable with normal modal operations $\Box$ and $\Diamond$, and every formula of the basic modal logic $\mathbf{K}$ is equivalent to a formula in which negations can only be applied to proposition letters. Sequents for this calculus are of form $\varphi \Rightarrow \varnothing$, where φ is a finite set of formulas in $\mathcal{L}^*$. We often write a instead of the $\{a\}$ in case of singletons, and φ, ψ instead of $\varphi \cup \psi$.

DEFINITION 15. The sequent calculus $G1\nabla$ consists of the axiom scheme $p, \neg p \Rightarrow \varnothing$ (for p propositional variable) and the following rules:

$$\bigwedge\text{-1}\ \frac{\varphi, \psi \Rightarrow \varnothing}{\varphi, \bigwedge \psi \Rightarrow \varnothing} \qquad \bigvee\text{-1}\ \frac{\{\varphi, a \Rightarrow \varnothing | a \in \psi\}}{\varphi, \bigvee \psi \Rightarrow \varnothing} \qquad \text{weak-1}\ \frac{\varphi \Rightarrow \varnothing}{\varphi, a \Rightarrow \varnothing}$$

$$\nabla\text{-1}\ \frac{\{\varphi_\Phi \Rightarrow \varnothing | \Phi \in SRD(A)\}}{\{\nabla\alpha | \alpha \in A\} \Rightarrow \varnothing}\ \varphi_\Phi \in \Phi$$

The nabla rule is to be read as follows: Given A, if for every $\Phi \in SRD(A)$ there exists some $\varphi_\Phi \in \Phi$ such that $\varphi_\Phi \Rightarrow \varnothing$, then $\{\nabla\alpha | \alpha \in A\} \Rightarrow \varnothing$.

A *derivation* of $\varphi \Rightarrow \varnothing$ in the system $G1\nabla$ is a finite tree, such that each node is labeled by a sequent: the root is labeled by $\varphi \Rightarrow \varnothing$, leaves are labeled by axioms and every node that is a parent node is labeled by the conclusion of a rule each premise of which labels exactly one of its children nodes.

A sequent $\varphi \Rightarrow \varnothing$ is *provable* in $G1\nabla$, notation: $\vdash_{G1\nabla} \varphi \Rightarrow \varnothing$, if there exists a derivation of $\varphi \Rightarrow \varnothing$ in $G1\nabla$.

A sequent $\varphi \Rightarrow \varnothing$ is *valid* in the class K of Kripke structures (notation: $\models_{\mathsf{K}} \varphi \Rightarrow \varnothing$) if φ is not satisfiable in K, i.e. for every model $\mathbb{S} \in \mathsf{K}$ and every state s in $\mathbb{S}$, there exists some $a \in \varphi$ such that $\mathbb{S}, s \not\Vdash a$. Then the next lemma provides the soundness and the semantic invertibility of the nabla rule of $G1\nabla$:

LEMMA 16. *The following are equivalent for every $A \in \mathcal{P}_\omega\mathcal{T}_\omega\mathcal{L}$ and every collection ξ of literals:*

1. $\{\nabla\alpha \mid \alpha \in A\} \cup \xi$ *is satisfiable.*

2. ξ *is satisfiable and for some* $\Phi \in SRD(A)$, φ *is satisfiable for every* $\varphi \in \Phi$.

Proof. Let us show that (1) implies (2): By assumption, ξ is satisfiable and there exists a model $\mathbb{S}$ and a state s in $\mathbb{S}$ such that $\mathbb{S}, s \Vdash \nabla\alpha$ for every $\alpha \in A$. For every $t \in R[s]$, let $\varphi_t = \{a \in \bigcup A \mid \mathbb{S}, s \Vdash a\}$, and let $\Phi_s = \{\varphi_t \mid t \in R[s]\}$. By definition, φ is satisfiable for all $\varphi \in \Phi_s$, and we have already checked that $\Phi_s \in SRD(A)$ (see Example 6.1). Conversely, let us assume that ξ is satisfiable and there exists some $\Phi \in SRD(A)$ such that for every $\varphi \in \Phi$, φ is satisfiable at some state s_φ in some model $\mathbb{S}_\varphi$. Then consider the model $\mathbb{S}$ which consists of the disjoint union of the models $\mathbb{S}_\varphi$ plus one extra point s such that $R[s] = \{s_\varphi \mid \varphi \in \Phi\}$ and $p \in V(s)$ iff $p \in \xi$. It is routine to verify that $\mathbb{S}, s$ satisfies $\{\nabla\alpha \mid \alpha \in A\} \cup \xi$. ■

The following theorem states the soundness and completeness of $G1\nabla$ with respect to K. Since $G1\nabla$ is formulated without the cut rule, once completeness has been established, it immediately follows that the cut rule is redundant.

THEOREM 17 (SOUNDNESS AND COMPLETENESS). *For every* $\mathcal{L}^*$-*sequent* $\varphi \Rightarrow \varnothing$,

$$\vdash_{G1\nabla} \varphi \Rightarrow \varnothing \quad \textit{iff} \quad \models_{\mathsf{K}} \varphi \Rightarrow \varnothing.$$

Proof. The proof of soundness is standard, by induction on the depth of the derivation of $\varphi \Rightarrow \varnothing$. The only case of interest is when the nabla rule is the last rule applied. In this case it follows from the direction (1 $\Rightarrow$ 2) of Lemma 16. As for completeness, it is shown by induction on the number $n(\varphi)$ of connectives in $\{\bigwedge, \bigvee, \nabla\}$ that occur in the elements of φ: if $n(\varphi) = 0$ then φ is a finite collection of literals, and its being non satisfiable implies that $p, \neg p \in \varphi$ for some propositional variable. Then a derivation for $\varphi \Rightarrow \varnothing$ has the axiom $p, \neg p \Rightarrow \varnothing$ at its only leaf followed by applications of the weakening rule to add literals in φ. As for the inductive steps, the only case of interest is when $\varphi = \{\nabla\alpha \mid \alpha \in A\} \cup \xi$ for some finite collection of literals ξ. From the assumptions and direction (2 $\Rightarrow$ 1) of Lemma 16, we get that either ξ is not satisfiable, in which case again we proceed as in the base case, or ξ is satisfiable and for every $\Phi \in SRD(A)$ there exists some φ_Φ such that $\models_{\mathsf{K}} \varphi_\Phi \Rightarrow \varnothing$. Since $n(\varphi_\Phi) < n(\varphi)$, by induction hypothesis $\vdash_{G1\nabla} \varphi_\Phi \Rightarrow \varnothing$ for every $\Phi \in SRD(A)$. Then a derivation for our sequent consists of prolonging all these derivations with an application of the nabla rule, so as to obtain a proof of $\{\nabla\alpha \mid \alpha \in A\} \Rightarrow \varnothing$, followed by applications of weakening to add up the elements in ξ. ■

4.1 Janin and Walukiewicz's tableau

Janin and Walukiewicz [7] introduced a tableau system for the modal μ-calculus based on a language with a family of ∇ modalities. We present here the propositional fragment of their system, with a single ∇ modality as a one-sided sequent proof system $GW\nabla$ for the same language $\mathcal{L}^*$ of $G1\nabla$, prove that it is complete, and relate it to our one sided system $G1\nabla$. Before introducing $GW\nabla$, and as a way of showing its semantic rationale, let us state the following lemma, whose routine proof is omitted:

LEMMA 18. *The following are equivalent for every $A \in \mathcal{P}_\omega \mathcal{T}_\omega \mathcal{L}$, every model $\mathbb{S}$ and every state s of $\mathbb{S}$:*

1. $\mathbb{S}, s \Vdash \{\nabla\alpha \mid \alpha \in A\}$.
2. $\mathbb{S}, s \Vdash \{\nabla\{\bigwedge \varphi_a \mid a \in \alpha\} \mid \alpha \in A\}$, *where for every* $a \in \bigcup A$,

$$(10)\quad \varphi_a = \{a\} \cup \{\bigvee \alpha' \mid \alpha' \in A \text{ and } a \notin \alpha'\}.$$

COROLLARY 19. *The following are equivalent for every $A \in \mathcal{P}_\omega \mathcal{T}_\omega \mathcal{L}$ and every collection ξ of literals:*

1. $\{\nabla\alpha \mid \alpha \in A\} \cup \xi$ *is satisfiable.*
2. ξ *is satisfiable and* $\bigwedge \varphi_a$ *is satisfiable for every* $a \in \bigcup A$.

Proof. ($1 \Rightarrow 2$) follows from Lemma 18. Conversely, assume that for every $a \in \bigcup A$, $\bigwedge \varphi_a$ is satisfiable at some state s_a in some model $\mathbb{S}_a$. Then consider the model $\mathbb{S}$ which consists of the disjoint union of the models $\mathbb{S}_a$ plus one extra point s such that $R[s] = \{s_a \mid a \in \bigcup A\}$ and $s \in V(p)$ iff $p \in \xi$. It is routine to verify that $\mathbb{S}, s$ satisfies $\{\nabla\alpha \mid \alpha \in A\} \cup \xi$. ■

Notice that, given A and $a \in \bigcup A$, $\mathbb{S}, s \Vdash \bigwedge \varphi_a$ iff there exists a choice function $f_a : A \to \bigcup A$ such that $a \in f_a[A] \subseteq \{a \mid \mathbb{S}, s \Vdash a\}$. Therefore, the corollary above can be seen as a reformulation of the satisfiability of a set of nabla formulas (with parameters ranging in A) in terms of the existence of such choice functions for every $a \in \bigcup A$. This corollary semantically motivates the definition of the rule for ∇ in the system $GW\nabla$ that we are about to introduce: indeed it provides its soundness and semantic invertibility.

DEFINITION 20. The propositional fragment of Janin and Walukiewicz's tableaux $GW\nabla$ consists of the axioms, the rules for boolean connectives and the weakening rule that appear in $G1\nabla$, plus the following nabla-rule:

$$W\nabla \; \frac{\varphi_a \Rightarrow \varnothing}{\{\nabla\alpha|\alpha \in A\} \Rightarrow \varnothing} \; a \in \bigcup A$$

where φ_a is defined as in (10). $W\nabla$ is to be read as follows: Given A, if $\varphi_a \Rightarrow \varnothing$ for some $a \in \bigcup A$, then $\{\nabla\alpha|\alpha \in A\} \Rightarrow \varnothing$.

Provability of sequents in $GW\nabla$ (notation: $\vdash_{GW\nabla} \varphi \Rightarrow \varnothing$) is defined analogously to provability in $G1\nabla$. The following theorem states the soundness and completeness of $GW\nabla$ with respect to K. Since $GW\nabla$ is formulated without the cut rule, once completeness has been established, it immediately follows that the cut rule is redundant.

THEOREM 21 (SOUNDNESS AND COMPLETENESS). *For every $\mathcal{L}^*$-sequent $\varphi \Rightarrow \varnothing$,*

$$\vdash_{GW\nabla} \varphi \Rightarrow \varnothing \quad \textit{iff} \quad \models_{\mathsf{K}} \varphi \Rightarrow \varnothing.$$

Proof. It follows the same proof pattern of Proposition 17: The proof of soundness is by induction on the depth of the derivation of $\varphi \Rightarrow \varnothing$. The only case of interest is when the $W\nabla$ is the last rule applied. In this case it follows from direction $(1 \Rightarrow 2)$ of Corollary 19. As for completeness, it is shown by induction on the number $n(\varphi)$ of connectives in $\{\bigwedge, \bigvee, \nabla\}$ that occur in the elements of φ: if $n(\varphi) = 0$ we proceed as in Proposition 17. As for the inductive steps, the only case of interest is when $\varphi = \{\nabla\alpha \mid \alpha \in A\} \cup \xi$ for some finite collection of literals ξ. From the assumptions and the direction $(2 \Rightarrow 1)$ of Corollary 19, we get that either ξ is not satisfiable, in which case again we proceed as in the base case, or ξ is satisfiable and $\bigwedge \varphi_a$ is not satisfiable for some $a \in \bigcup A$. Since $n(\varphi_a) < n(\varphi)$, by induction hypothesis $\vdash_{GW\nabla} \varphi_a \Rightarrow \varnothing$. Then a derivation for our sequent consists of prolonging this derivation with an application of $W\nabla$, so as to obtain a proof of $\{\nabla\alpha \mid \alpha \in A\} \Rightarrow \varnothing$, followed by applications of weakening to add up the elements in ξ. ∎

Now we turn to showing that $G1\nabla$ and $GW\nabla$ are equivalent. From Theorems 17 and 21 it immediately follows that they derive exactly the same sequents. Therefore the rules of one system, being sound, are *admissible* in the other system, which means that if the premises of the application of one rule are provable in one system, then its conclusion is also provable in that system. However what we are going to show is a slightly stronger result, namely that $G1\nabla$ *simulates* $GW\nabla$, i.e. that we can effectively transform a proof of the premises of an application of $W\nabla$ in $G1\nabla$ into a proof of its conclusions. The converse direction is non constructive. We will return to this point in the conclusions. However, we show admissibility of the rule ∇-1 in $GW\nabla$ to clarify how the two rules relate.

LEMMA 22. *$G1\nabla$ and $GW\nabla$ are equivalent.*

Proof. As for showing that $G1\nabla$ simulates $GW\nabla$, assume that $\varphi_a \Rightarrow \varnothing$ is provable in $G1\nabla$ for some $a \in \bigcup A$ and fix $\Phi \in SRD(A)$. We need to show that there exists some $\varphi \in \Phi$ such that $\varphi \Rightarrow \varnothing$ is provable in $G1\nabla$. Since $a \in \bigcup A$, then $a \in \alpha$ for some $\alpha \in A$. Since $\alpha \mathrel{\overline{\in}} \Phi$, then $a \in \varphi$ for some $\varphi \in \Phi$. To finish the proof, let us show that $\varphi \Rightarrow \varnothing$ is provable in $G1\nabla$: Since $\alpha' \mathrel{\overline{\in}} \Phi$ for every $\alpha' \in A$ such that $a \notin \alpha'$, then for every such α' there exists some $a_{\alpha'} \in \alpha'$ such that $a_{\alpha'} \in \varphi$. Let $\varphi' = \{a\} \cup \{a_{\alpha'} \mid \alpha' \in A$ and

$a \notin \alpha'\}$. By construction, $\varphi' \subseteq \varphi$. Moreover, φ' is not satisfiable, for if it was, then so would be $\bigwedge \varphi_a$, against our assumption and the soundness of $G1\nabla$. Hence, by the completeness of $G1\nabla$, $\varphi' \Rightarrow \varnothing$ is provable in $G1\nabla$ and by applying weakening we obtain a proof of $\varphi \Rightarrow \varnothing$.

As for showing that ∇-1 is admissible in $GW\nabla$, assume that for every $\Phi \in SRD(A)$, $\varphi \Rightarrow \varnothing$ is provable in $GW\nabla$ for some $\varphi \in \Phi$ and let us show that $\varphi_a \Rightarrow \varnothing$ is provable in $GW\nabla$ for some $a \in \bigcup A$. Suppose for contradiction that $\bigwedge \varphi_a$ is satisfiable for every $a \in \bigcup A$. Then by Corollary 19, $\bigwedge_{\alpha \in A} \nabla\alpha$ is satisfiable. Then by Lemma 16, there exists some $\Phi \in SRD(A)$ such that $\bigwedge \varphi$ is satisfiable for every $\varphi \in \Phi$, against the assumptions and the soundness of $GW\nabla$. ■

Let us finish this section with some comparing remarks on $GW\nabla$ and $G1\nabla$: we saw that the definition of the rule $W\nabla$ in $GW\nabla$ is grounded in the notion of satisfiability of a set of ∇ formulas, so it more directly reflects its semantics; moreover it has only one premise, so it could be easier to work with in practical situations, e.g. automated reasoning.

On the other hand, the definition of the set φ_a relies on taking disjunctions of sets of type $\alpha \in \mathcal{T}_\omega\mathcal{L}$. This move is certainly legal in the special setting we have adopted in this paper, where $\mathcal{T}$ coincides with the powerset functor, but is no more an option in the general context of a coalgebraic ∇ language based on (almost arbitrary) $\mathcal{T}$. This problem does not occur in the nabla rule of $G1\nabla$, and indeed the system $G1\nabla$ can be imported in a general coalgebraic context.

5 Two-sided Gentzen calculus

One-sided Gentzen calculi are not available for negation-free languages. In this section we focus again on the negation-free fragment of the basic modal logic and introduce a two-sided Gentzen calculus $G2\nabla$ for it. Sequents for this calculus are of form $\varphi \Rightarrow \psi$, φ, ψ being finite sets of $\mathcal{L}^+$ formulas. We will show that $G2\nabla$ is sound, complete but not cut-free.

DEFINITION 23. The sequent calculus $G2\nabla$ consists of the axiom scheme $a \Rightarrow a$ and the following rules:

$$\bigwedge\text{-l}\ \frac{\varphi, \theta \Rightarrow \psi}{\varphi, \bigwedge\theta \Rightarrow \psi} \qquad \bigvee\text{-r}\ \frac{\varphi \Rightarrow \theta, \psi}{\varphi \Rightarrow \bigvee\theta, \psi}$$

$$\bigwedge\text{-r}\ \frac{\{\varphi \Rightarrow a, \psi \mid a \in \theta\}}{\varphi \Rightarrow \bigwedge\theta, \psi} \qquad \bigvee\text{-l}\ \frac{\{\varphi, a \Rightarrow \psi \mid a \in \theta\}}{\varphi, \bigvee\theta \Rightarrow \psi}$$

$$\text{weak-r}\ \frac{\varphi \Rightarrow \psi}{\varphi \Rightarrow \psi, a} \qquad \text{weak-l}\ \frac{\varphi \Rightarrow \psi}{\varphi, a \Rightarrow \psi}$$

$$\text{cut}\ \frac{\varphi \Rightarrow \psi, a \qquad \varphi', a \Rightarrow \psi'}{\varphi, \varphi' \Rightarrow \psi, \psi'}$$

$$\nabla\ \frac{\{\varphi \Rightarrow \theta | (\varphi,\theta) \in \bigcup_{\Phi \in SRD(A)} Y_\Phi\}}{\{\nabla\alpha | \alpha \in A\} \Rightarrow \bigcup_{\Phi \in SRD(A)} \{\nabla\beta | \beta \mathrel{\overline{\in}} \Theta_\Phi\}}\ Y_\Phi \in \Phi \bowtie \Theta_\Phi \text{ for all } \Phi \in SRD(A)$$

The nabla rule is to be read as follows: Given A, if for every $\Phi \in SRD(A)$ there exists some $\Theta_\Phi \in \mathcal{T}_\omega\mathcal{P}_\omega\mathcal{L}$ and some $Y_\Phi \in \Phi \bowtie \Theta_\Phi$ such that $\varphi \Rightarrow \theta$ for every $(\varphi, \theta) \in Y_\Phi$, then the conclusion follows. (For the definition of $\bowtie$, see Example 6(2).)

Provability in $G2\nabla$ (notation: $\vdash_{G2\nabla} \varphi \Rightarrow \psi$) is defined in the usual way. $\varphi \Rightarrow \psi$ is *valid* in the class K of Kripke structures (notation: $\models_\mathsf{K} \varphi \Rightarrow \psi$) if for every model $\mathbb{S} \in \mathsf{K}$ and every state s in $\mathbb{S}$, $\mathbb{S}, s \Vdash \bigwedge \varphi$ implies $\mathbb{S}, s \Vdash \bigvee \psi$. Then the next proposition provides the soundness of $G2\nabla$ w.r.t. the class K of Kripke models:

THEOREM 24 (SOUNDNESS). *For every $\mathcal{L}^+$-sequent $\varphi \Rightarrow \psi$,*

$$\textit{if } \vdash_{G2\nabla} \varphi \Rightarrow \psi \textit{ then } \models_\mathsf{K} \varphi \Rightarrow \psi.$$

Proof. We will focus on showing that the ∇ rule is sound: the proof is then analogous to the previous ones. Let us assume that for every $\Phi \in SRD(A)$ there exists some $\Theta = \Theta_\Phi \in \mathcal{T}_\omega\mathcal{P}_\omega\mathcal{L}$ and some $Y = Y_\Phi \in \Phi \bowtie \Theta$ such that, for every $(\varphi, \theta) \in Y$ and every model $\mathbb{T}$ and state t in $\mathbb{T}$, if $\mathbb{T}, t \Vdash \bigwedge \varphi$ then $\mathbb{T}, t \Vdash \bigvee \theta$; moreover assume that $\mathbb{S}, s \Vdash \bigwedge_{\alpha\in A} \nabla\alpha$ for some model $\mathbb{S}$ and some state s of $\mathbb{S}$. We need to show that there exists some β such that $\beta \mathrel{\overline{\in}} \Theta$ and $\mathbb{S}, s \Vdash \nabla\beta$.

In particular, let $\varphi_{s'} = \{a \in \bigcup A \mid s' \Vdash a\}$ for every $s' \in R[s]$ and let $\Phi_s = \{\varphi'_s \mid s' \in R[s]\}$. Then $\mathbb{S}, s' \Vdash \bigwedge \varphi_{s'}$ and, as it was shown in Example 6, $\Phi \in SRD(A)$. Therefore, by assumptions, there exist Θ and Y as above. Let us take $\beta = \bigcup \Theta \cap \bigcup_{s' \in R[s]} Th(s')$: by definition, for every $b \in \beta$ there is some $\theta \in \Theta$ such that $b \in \theta$. Conversely, fix $\theta \in \Theta$; since $Y \in \Phi \bowtie \Theta$, $(\varphi_{s'}, \theta) \in Y$ for some $s' \in R[s]$; hence $\mathbb{S}, s' \Vdash b$ for some $b \in \theta$. So by definition, $b \in \beta$. This completes the proof that $\beta \mathrel{\overline{\in}} \Theta$. As for showing that $\mathbb{S}, s \Vdash \nabla\beta$: by definition, for every $b \in \beta$ there is some $s' \in R[s]$ such that $s' \Vdash b$. Conversely, fix $s' \in R[s]$; since $Y \in \Phi \bowtie \Theta$, then $(\varphi_{s'}, \theta) \in Y$ for some $\theta \in \Theta$ such that for every every model $\mathbb{T}$ and state t in $\mathbb{T}$, if $\mathbb{T}, t \Vdash \bigwedge \varphi_{s'}$ then $\mathbb{T}, t \Vdash \bigvee \theta$. Since $\mathbb{S}, s' \Vdash \bigwedge \varphi_{s'}$, then there is some $b \in \theta$ such that $\mathbb{S}, s' \Vdash b$. ■

EXAMPLE 25. The following deductions are instances of applications of the nabla rule:

$$\textbf{(i)}\quad \frac{a, b \Rightarrow a \qquad a, b \Rightarrow b}{\nabla\{a\}, \nabla\{b\} \Rightarrow \nabla\{a, b\}}$$

Here $A = \{\{a\}, \{b\}\}$ and $\Phi = \{\{a, b\}\}$ is the only slim redistribution of A. $\Theta_\Phi = \{\{a\}, \{b\}\}$, and $\beta = \{a, b\}$ is the only $\beta \mathrel{\overline{\in}} \Theta_\Phi$.

$$\textbf{(ii)} \quad \frac{\varnothing \Rightarrow \top \qquad \varnothing}{\varnothing \Rightarrow \nabla\varnothing, \nabla\{\top\}}$$

where $A = \varnothing$, $\Phi_1 = \{\varnothing\}$ and $\Phi_2 = \varnothing$ are the only slim redistributions of A, $\Theta_{\Phi_1} = \{\{\top\}\}$ and $\Theta_{\Phi_2} = \varnothing$, and $\beta_1 = \{\top\}$ and $\beta_2 = \varnothing$. We have proved axiom (∇_5).

$$\textbf{(iii)} \quad \frac{\varnothing \Rightarrow \varphi \qquad \varnothing}{\varnothing \Rightarrow \{\nabla\beta | \beta \subseteq \varphi\}}$$

where $A = \varnothing$, $\Phi_1 = \{\varnothing\}$ and $\Phi_2 = \varnothing$ are the only slim redistributions of A, $\Theta_{\Phi_1} = \{\varphi\}$ and $\Theta_{\Phi_2} = \varnothing$, and $\{\beta | \beta \,\overline{\in}\, \{\varphi\}\} = \{\beta | \beta \subseteq \varphi\} \setminus \{\varnothing\}$ and $\{\beta | \beta \,\overline{\in}\, \varnothing\} = \{\varnothing\}$. We have simulated the $(\nabla 4)$ rule.

$$\textbf{(iv)} \quad \frac{\bot \Rightarrow \varnothing}{\nabla\{\bot\} \Rightarrow \varnothing}$$

where $A = \{\{\bot\}\}$, $\Phi = \{\{\bot\}\}$ is the only slim redistribution of A, and $\Theta_\Phi = \{\varnothing\}$. Then there is no $\beta \,\overline{\in}\, \Theta_\Phi$. We have proved an instance of axiom (∇_2).

THEOREM 26 (COMPLETENESS). *$G2\nabla$ is complete w.r.t. the class K of Kripke models.*

Proof. It is enough to show that the Carioca axioms are provable in $G2\nabla$ and that the Carioca rule $(\nabla 1)$ can be simulated in $G2\nabla$. The completeness of $G2\nabla$ then follows from the completeness result for the Carioca axiomatization. Since the rule $(\nabla 4)$ is derivable from the rest of Carioca axiomatization, we do not need to simulate it directly. However, as shown by Example 25(iii), it can be simulated by the ∇ rule which can be useful in further generalizations.

As for $(\nabla 1)$, assume that α and β are such that for every $a \in \alpha$ there exists some $b \in \beta$ such that $a \Rightarrow b$ is provable in $G2\nabla$ and that for every $b \in \beta$ there exists some $a \in \alpha$ such that $a \Rightarrow b$ is provable in $G2\nabla$. We need to show that $\nabla\alpha \Rightarrow \nabla\beta$ is provable in $G2\nabla$. Let $A = \{\alpha\}$ and fix $\Phi \in SRD(A)$. Then it is enough to find some Θ for which there exists some $Y \in \Phi \bowtie \Theta$ such that $\varphi \Rightarrow \theta$ is provable in $G2\nabla$ for every $(\varphi, \theta) \in Y$. Take $\Theta = \{\{b\} \mid b \in \beta\}$. Clearly, if $\beta' \overline{\in} \Theta$, then $\beta' = \beta$. The proof is complete if we show that (a) for every $\varphi \in \Phi$ there is some $\theta \in \Theta$ such that $\varphi \Rightarrow \theta$ is provable in $G2\nabla$ and (b) for every $\theta \in \Theta$ there is some $\varphi \in \Phi$ such that $\varphi \Rightarrow \theta$ is provable in $G2\nabla$. (a): Fix $\varphi \in \Phi$; since $\alpha \overline{\in} \Phi$ then $a \in \varphi$ for some $a \in \alpha$. By assumption, $a \Rightarrow b$ is provable in $G2\nabla$ for some $b \in \beta$. Take $\theta = \{b\}$: by applying weakening, we get the proof of $\varphi \Rightarrow \theta$ we need. (b): Fix $\theta = \{b\} \in \Theta$. By assumption, $a \Rightarrow b$ is provable in $G2\nabla$ for some $a \in \alpha$. Since $\alpha \overline{\in} \Phi$ then $a \in \varphi$ for some $\varphi \in \Phi$. By applying weakening, we get the proof of $\varphi \Rightarrow \theta$ we need.

As for $(\nabla 2)$, we need to find a proof of the sequent $\{\nabla\alpha | \alpha \in A\} \Rightarrow \{\nabla\{\bigwedge\varphi | \varphi \in \Phi\} | \Phi \in SRD(A)\}$. For every $\Phi \in SRD(A)$ take $\Theta_\Phi =$

$\{\{\bigwedge\varphi\}|\varphi \in \Phi\}$ and $Y_\Phi = \{(\varphi, \{\bigwedge\varphi\})|\varphi \in \Phi\}$. Clearly, $Y_\Phi \in \Phi \bowtie \Theta_\Phi$. Then the following is a correct instance of the ∇ rule, whose premises are all provable:

$$\frac{\{\varphi \Rightarrow \bigwedge\varphi|(\varphi, \{\bigwedge\varphi\}) \in \bigcup_{\Phi \in SRD(A)} Y_\Phi\}}{\{\nabla\alpha|\alpha \in A\} \Rightarrow \{\nabla\{\bigwedge\varphi|\varphi \in \Phi\}|\Phi \in SRD(A)\}}$$

indeed for every Φ, if $\beta \overline{\in} \Theta_\Phi$ then $\beta = \{\bigwedge\varphi|\varphi \in \Phi\}$.

As for (∇3), we need to find a proof of the sequent $\nabla\{\bigvee\psi|\psi \in \Psi\} \Rightarrow \{\nabla\beta|\beta\overline{\in}\Psi\}$. Here $A = \{\{\bigvee\psi|\psi \in \Psi\}\}$. For every $\Phi \in SRD(A)$, let $\Theta = \Psi$ and $Y = \{(\{\bigvee\psi\}, \psi)|\psi \in \Psi\}$. Then $Y \in \Phi \bowtie \Theta$, and the following is a correct instance of the ∇ rule whose premises are all provable:

$$\frac{\{\bigvee\psi \Rightarrow \psi|(\{\bigvee\psi\}, \psi) \in \bigcup_{\Phi \in SRD(A)} Y_\Phi\}}{\nabla\{\bigvee\psi|\psi \in \Psi\} \Rightarrow \bigvee\{\nabla\beta|\beta\overline{\in}\Psi\}}$$

■

REMARK 27. $G2\nabla$ is not cut-free.
We will show that our definition of ∇ rule doesn't yield a cut-free system, in particular we show that the following sequent

$$\nabla\{p \vee q\} \Rightarrow \nabla\{p, \top\}, \nabla\{q\}$$

is provable in $G2\nabla$ but not in the system obtained from $G2\nabla$ by removing the cut rule.

$\nabla\{p \vee q\} \Rightarrow \nabla\{p, \top\}, \nabla\{q\}$ *is provable in* $G2\nabla$*:* Let

$$\Pi_1 = \nabla\,\frac{p, q \Rightarrow p, \top \qquad p \Rightarrow p \qquad q \Rightarrow \top}{\nabla\{p, q\} \Rightarrow \nabla\{p, \top\}}$$

$$\Pi_2 = \nabla\,\frac{p \Rightarrow p \qquad p \Rightarrow \top}{\nabla\{p\} \Rightarrow \nabla\{p, \top\}}$$

in the proof

$$\text{cut}\,\dfrac{\text{cut}\,\dfrac{\nabla\,\dfrac{p \vee q \Rightarrow p, q}{\nabla\{p \vee q\} \Rightarrow \nabla\{p, q\}, \nabla\{p\}, \nabla\{q\}} \qquad \Pi_1}{\nabla\{p \vee q\} \Rightarrow \nabla\{p, \top\}, \nabla\{p\}, \nabla\{q\}} \qquad \Pi_2}{\nabla\{p \vee q\} \Rightarrow \nabla\{p, \top\}, \nabla\{q\}}$$

$\nabla\{p \vee q\} \Rightarrow \nabla\{p, \top\}, \nabla\{q\}$ *is not provable in the system obtained by removing the cut rule from* $G2\nabla$*:*

Suppose there was such a proof. Then its last step must be either an application of the weakening, or of the ∇ rule. None of $\nabla\{p \vee q\} \Rightarrow \varnothing$, $\nabla\{p \vee q\} \Rightarrow \nabla\{q\}$, $\nabla\{p \vee q\} \Rightarrow \nabla\{p, \top\}$ is a valid, and hence a provable sequent, thus the last step must be a ∇ inference. Then $A = \{\{p \vee q\}\}$ and the only $\Phi \in SRD(A)$ is again $\{\{p \vee q\}\}$. In that case for some Θ_Φ

(11) $\{\{p, \top\}, \{q\}\} = \{\beta | \beta \mathrel{\overline{\in}} \Theta_\Phi\}$

Notice that (11) means that $\bigcup\{\{p, \top\}, \{q\}\} = \bigcup \Theta_\Phi$. Let us show that there is no Θ_Φ satisfying (11). Let us check all the possibilities for $\theta \in \Theta_\Phi$, i.e. all nonempty subsets of $\{p, q, \top\}$:

Singletons $\{p\}$, $\{q\}$, and $\{\top\}$ are excluded since there is no element both in $\{p, \top\}$ and in $\{q\}$. $\{p, \top\}$ is also excluded since it does not contain q. From the remaining three $\{p, q\}$, $\{\top, q\}$, and $\{p, q, \top\}$, no combination would work as Θ_Φ. Singletons $\{\{p, q\}\}$, $\{\{\top, q\}\}$ wouldn't suffice. For all pairs, the singleton $\{\{p, q, \top\}\}$, and whole $\{\{p, q\}, \{\top, q\}, \{p, q, \top\}\}$, we always have $\{p, q\} \in \Theta_\Phi$ which is not allowed by (11).

This sequent is a key counterexample which shows that in the ∇ rule, the right part of the conclusion is in a sense too robust. Notice, that $\bigcup_{\Phi \in SRD(A)} \{\nabla\beta | \beta \mathrel{\overline{\in}} \Theta_\Phi\}$ is in fact given by a union of *maximal* slim redistributions of Θ_Φ. However, this choice was motivated by semantical soundness of the rule and within our framework it is *the* rule one comes up with. It seems that to obtain a cut-free rule we would need to go much deeper in our structural analysis.

6 Conclusions and further directions

Coalgebraic generalization As we remarked early on, both the new Hilbert-style axiomatization presented here and the Gentzen systems $G1\nabla$ and $G2\nabla$ are designed to keep the roles of $\mathcal{T}$ and $\mathcal{P}$ separated. This paves the way to further generalizations and applications of these proof systems to the context of coalgebraic modal languages associated with arbitrary weak-pullback preserving Set-endofunctors, defined in Moss' style [11]. As a first step in this direction, Kupke, Kurz and the third author [8] showed that the obvious generalization of our system C_∇ to such a general setting is indeed sound and complete in the general case. It would be of interest to see whether such generalizations can be extended to a setting of coalgebraic fixpoint logics (note that the system $GW\nabla$ is part of a tableaux system for the modal μ-calculus.

Complexity In the special setting of this paper, where $A \in \mathcal{PTL}$ is finite, both $G1\nabla$ and $GW\nabla$ produce a PSPACE decision procedure. Indeed, when read backwards, all the rules strip their conclusions of one connective or modal operator. This implies that the length of each branch in a proof search tree is linear in the size of the input sequent and the whole computation can be performed by alternating machines working in linear time. Therefore this procedure is in PSPACE. For a reference to standard decision and proof-search procedures for modal logics see [10, 5].

Refinements The results presented in this paper can be improved further: in particular, we intend to investigate more on a cut-free version of $G2\nabla$, and on a constructive simulation of $G1\nabla$ in $GW\nabla$.

Expanding with the semantic dual of nabla It could be of interest to undertake an analogous independent proof-theoretic study of the coalgebraic operator Δ semantically defined as $\Delta\alpha \equiv \neg\nabla\neg\alpha$, where $\neg\alpha := \{\neg a \mid a \in \alpha\}$. Using the standard modal language, Δ can be seen as a defined operator:

(12) $\Delta\alpha = \Diamond(\bigwedge \alpha) \vee \bigvee \Box\alpha,$

where $\Box\alpha$ denotes the set $\{\Box a \mid a \in \alpha\}$.

Acknowledgements. The research of Marta Bílková has been supported by grant 401/06/0387 of the Grant Agency of the Czech Republic and by grant IAA900090703 of the Grant Agency of Academy of Sciences of the Czech Republic.

The research of Alessandra Palmigiano and Yde Venema has been made possible by VICI grant 639.073.501 of the Netherlands Organization for Scientific Research (NWO).

BIBLIOGRAPHY

[1] J. Barwise and L. Moss. *Vicious Circles*, volume 60 of *CSLI Lecture Notes*. CSLI Publications, 1996.

[2] G. D'Agostino and G. Lenzi. On modal μ-calculus with explicit interpolants. *Journal of applied logic*, 338:256 – 278, 2006.

[3] K. Fine. Normal forms in modal logic. *Notre Dame Journal of Formal Logic*, 16:229–234, 1975.

[4] J. M. Font, R. Jansana and D. Pigozzi. A Survey of Abstract Algebraic Logic. *Studia Logica*, 74:13–97, 2003.

[5] A. Heuerding. *Sequent Calculi for Proof Search in some Modal Logics.* PhD thesis, University of Bern, 1998.

[6] W. Hodges. *Model Theory.* Cambridge University Press, 1993.

[7] D. Janin and I. Walukiewicz. Automata for the modal μ-calculus and related results. In *Proc. MFCS'95*, pages 552–562, Berlin, 1995. Springer. LNCS 969.

[8] C. Kupke, A. Kurz, and Y. Venema. A complete coalgebraic logic, 2008. This volume.

[9] C. Kupke and Y. Venema. Closure properties of coalgebra automata. In *Proceedings of the Twentieth Annual IEEE Symposium on Logic in Computer Science (LICS 2005)*, pages 199–208. IEEE Computer Society Press, 2005.

[10] R. Ladner. The computational complexity of provability in systems of modal propositional logic. *SIAM Journal of Computation*, 6:467–480, 1977.

[11] L. Moss. Coalgebraic logic. *Annals of Pure and Applied Logic*, 96:277–317, 1999. (Erratum published *Ann.P.Appl.Log.* 99:241–259, 1999).

[12] A. Palmigiano and Y. Venema. Nabla algebras and Chu spaces. In *Algebra and Coalgebra in Computer Science (CALCO 2007)*, pages 394–408, Berlin, 2007. Springer-Verlag. LNCS 4624.

[13] J. Rutten. Universal coalgebra: A theory of systems. *Theoretical Computer Science*, 249:3–80, 2000.

[14] L. Santocanale and Y. Venema. Completeness for flat modal fixpoint logics (extended abstract). In N. Dershowitz and A. Voronkov, editors, *LPAR 2007*, pages 499–513, Berlin, 2007. Springer-Verlag. LNCS 4790.

[15] Luigi Santocanale. Completions of μ-algebras. In *Proceedings of the Twentieth Annual IEEE Symposium on Logic in Computer Science (LICS 2005)*, pages 219–228. IEEE Computer Society, 2005.

[16] B. ten Cate, W. Conradie, M. Marx, and Y. Venema. Definitorially complete description logics. In P. Doherty, J. Mylopoulos, and C. Welty, editors, *KR 2006*, pages 79–89. AAAI Press, 2006.

[17] Y. Venema. Algebras and coalgebras. In P. Blackburn, J. van Benthem, and F. Wolter, editors, *Handbook of Modal Logic*, pages 331–426. Elsevier, 2006.

Marta Bílková
Department of Logic, Charles University in Prague &
Institute of Computer Science, Academy of Sciences of the Czech Republic
Pod Vodárenskou věží 2
182 07 Prague 8
Czech Republic
`bilkova@cs.cas.cz`

Alessandra Palmigiano
Institute for Logic, Language and Computation
Universiteit van Amsterdam
Plantage Muidergracht 24
1018 TV Amsterdam
The Netherlands
`apalmigi@science.uva.nl`

Yde Venema
Institute for Logic, Language and Computation
Universiteit van Amsterdam
Plantage Muidergracht 24
1018 TV Amsterdam
The Netherlands
`Y.Venema@uva.nl`

Undecidability for arbitrary public announcement logic

Tim French and Hans van Ditmarsch

ABSTRACT. *Arbitrary public announcement logic* ($APAL$) is an extension of multi-agent epistemic logic that allows agents' knowledge states to be updated by the public announcement of (possibly arbitrary) epistemic formulae. It has been shown to be more expressive than epistemic logic, and a sound and complete axiomatization has been given. Here we address the question of decidability. We present a proof that the satisfiability problem for *arbitrary public announcement logic* ($APAL$) is co-RE complete, via a tiling argument.

Keywords: Epistemic Logic, Public Announcement Logic, Decidability

1 Introduction

Arbitrary announcement logic ($APAL$) is an extension of multi-agent epistemic logics with *public announcements* and *arbitrary announcements*. A public announcement allows the information state of every agent to be updated by publicly informing them that some epistemic formula, ψ is true. An arbitrary announcement is added to the language to allow us to quantify over all possible announcements. This logic is described in detail in [1]. It is shown to be more expressive than normal epistemic logic and a sound and complete axiomatization is given. Furthermore, in [2] a tableau-calculus is presented to determine validity in $APAL$.

While the above results indicate that the set of validities for $APAL$ is recursively enumerable, the full decidability of the satisfiability problem has remained open. Here we show that the satisfiability problem for the logic is undecidable via a tiling argument. This is a surprising result since in [1] it is shown that $APAL$ is bisimulation invariant. Hence every $APAL$ formula is satisfied by a tree-like model, rather than the grid-like models typically required for tiling arguments (see [9] for a detailed analysis of undecidability in extensions of epistemic logic). The undecidability follows from the power of the arbitrary announcement operator. The arbitrary announcement operator, $\diamond\phi$ expresses:

> *"there exists a true formula of epistemic logic, that when publicly announced establishes the truth of ϕ."*

Implicit in this statement is an existential quantification over *all* formulae of epistemic logic, and we show that this expressive power is sufficient to allow us to encode an undecidable tiling problem. This is not an entirely surprising result, despite the many other favorable properties of $APAL$. In [9] a detailed survey is presented of undecidable temporal and epistemic logics, and an analysis is presented of the properties leading to undecidability. The arbitrary announcement operator is transitive in nature and reminiscent of a temporal operator. However, most undecidable logics surveyed in [9] are not bisimulation invariant, indicating a certain uniqueness to this result. Another related result is the undecidability of iterated modal relativization [7]. This logic is shown to be highly undecidable (Σ_1^1-complete), again, by encoding a tiling problem. Other undecidable logics considered in [7] combine *common knowledge* with iterated relativization. 'Relativization' is another term to denote the structural restriction that constitutes the informative effect of an announcement. Iterated relativization is different from arbitrary announcement. The former means that one allows (arbitrary finite length) sequences of model restrictions for a given epistemic formula ('announcement'); note that after announcements of a modal formula, announcing that formula again may still be informative, as in the famous 'Muddy Children Problem' [4]. But the latter means model restriction for *any* epistemic formula. Now the iteration is implicit. It is there because the sequence of two epistemic announcements is again equivalent to an epistemic formula [1].

2 Syntax and semantics

The formulas of $APAL$, $\mathcal{L}_{apal}$ are inductively defined as

$$\phi ::= \; p \mid \neg\phi \mid (\phi \wedge \phi) \mid K_a\phi \mid [\phi]\phi \mid \Box\phi$$

where a is taken from the set of agents A, and p is taken from the set of atomic propositions P. Let $\mathcal{L}_{el}$ be the set of formulas not containing any of the operators $[\phi]$ or $\Box$.

These formulas are interpreted over structures $M = (S, \sim, V)$ where S is a set of worlds, $\sim\colon A \longrightarrow \wp(S \times S)$ assigns a reflexive, transitive and symmetric accessibility relation, $\sim_a$ to each agent a, and $V : P \longrightarrow \wp(S)$ maps each proposition to the set of worlds where it is true.

Let $M = (S, \sim, V)$ and suppose that $s \in S$. The semantics of $APAL$ are given as:

$$\begin{array}{rcl}
M, s \models p & \text{iff} & s \in V(p) \\
M, s \models \neg\phi & \text{iff} & M, s \not\models \phi \\
M, s \models \phi_1 \wedge \phi_2 & \text{iff} & M, s \models \phi_1 \text{ and } M, s \models \phi_2 \\
M, s \models K_a\phi & \text{iff} & \forall t \in S \text{ where } s \sim_a t,\ M, t \models \phi \\
M, s \models [\psi]\phi & \text{iff} & M, s \models \psi \Longrightarrow M^{\psi}, s \models \phi \\
M, s \models \Box\phi & \text{iff} & \forall \psi \in \mathcal{L}_{el},\ M, s \models [\psi]\phi
\end{array}$$

where $M^{\psi} = (S', \sim', V')$ is such that: $S' = \{s \in S \mid M, s \models \psi\}$; for all $a \in A$, $\sim'_a = \sim_a \cap (S' \times S')$; and for all $p \in P$, $V'(p) = V(p) \cap S'$. As usual we take $K_a\phi$ to mean *agent a knows* ϕ, and let $L_a\phi$ abbreviate $\neg K_a\neg\phi$ (*agent a considers* ϕ *possible*).

We say an *APAL* formula ϕ is satisfiable if there exists some model $M = (S, \sim, V)$ and some world $s \in S$ such that $M, s \models \phi$, and if $M, s \models \phi$ for all model-world pairs, M, s, we say ϕ is valid.

Note that when defining the semantics of $\Box\phi$ we restrict the arbitrary announcements to range only over the epistemic formulas (i.e. those in $\mathcal{L}_{el}$). We reason for this is that we obviously cannot allow the arbitrary announcements to range over arbitrary announcements (i.e. formulae of the form $\Box\psi$) as the semantics would then be undefined. Further, we do not let the arbitrary announcements range over announcements (such as $[\psi]\alpha$ where ψ, $\alpha \in \mathcal{L}_{el}$) since such formulas are expressively equivalent to pure epistemic formulae (see [11] for a translation).

The formula $\Box\phi$ expresses the statement *"after publicly announcing any true formula of epistemic logic,* ϕ *must be true."* As we see in its formal semantics above, this statement implicitly quantifies over all true formulae of epistemic logic. For example, suppose ϕ were the formula $K_a p \to K_b p$. The formula $\Box\phi$ is true at some world where p is true, if and only if for every b-related world, u where p is not true, *for every epistemic formula* ψ, there is some a-related world, v, that agrees with u on the interpretation of ψ. (This is because otherwise the announcement of $p \vee \psi$ would be enough to make $K_a p \wedge \neg K_b p$ true). This is a strong property to be able to express. If two sets of worlds cannot be distinguished by any epistemic formula then they are, for the purposes the logic, identical. Given that the epistemic formulae can be arbitrarily large, using this notion of equivalence we are able to encode a grid-like property for finite grids of arbitrary size. This expressivity is exploited to encode an arbitrary tiling problem which is sufficient to show that the satisfiability problem for *APAL* is co-RE complete.

3 Tilings and undecidability

We show the satisfiability problem is undecidable for *APAL* by embedding a tiling problem that is known to be co-RE complete (i.e. equivalent to computing the membership of the complement of any recursively enumerable set).

The tiling problem is as follows:

DEFINITION 1. Let C be a finite set of *colours* and define a C-tile to be a four-tuple over C $\gamma = (\gamma^t, \gamma^r, \gamma^f, \gamma^\ell)$, where the elements are referred to as, respectively, *top, right, floor* and *left.* The tiling problem is, for any given finite set of C-tiles, Γ, determine if there is a function $\lambda : \omega \times \omega \longrightarrow \Gamma$ such that for all $(i, j) \in \omega \times \omega$:

1. $\lambda(i, j)^t = \lambda(i, j + 1)^f$
2. $\lambda(i, j)^r = \lambda(i + 1, j)^\ell$.

The tiling problem has been shown to be co-RE complete by Harel [6] (see [8] for an overview of the application of tiling problems to complexity for modal logics).

4 Encoding the tiling problem

To encode the tiling problem we seek to define a grid like structure in the model M. That is we define a formula *grid* whereby $M, s \models grid$ implies the structure of M is similar to $\omega \times \omega$. To do this we exploit one of the stronger properties of $APAL$: the ability to quantify over *all* epistemic formulas. This allows us to define an equivalence between the worlds in a model and modulo that equivalence, a grid like structure.

Such encodings are rarely elegant and this is no exception. We use the following atoms:

1. We label each world as either *white* (W) or *black* (B) with the understanding that B is an abbreviation for $\neg W$. We intend to label the model in chess-board pattern.
2. We use the set of agents a, b, c, d and t where we suppose that:
   - a and b describe some vertical successor relation (a goes from a black square to a white square, and b goes from a white square to a black square);
   - c and d describe some horizontal successor relation (c goes from a black square to a white square, and d goes from a white square to a black square);
   - the relation for the agent t includes the relations for the agents a, b, c and d.
3. For each tile $\gamma \in \Gamma$ we assign a proposition (also denoted γ) with the understanding that the tiles are mutually exclusive (i.e. $\gamma \rightarrow \bigwedge_{\delta \neq \gamma} \neg\delta$).

Such a structure is represented in Figure 1 (with the assumption that $\sim_t$ is a universal relation over all worlds).

Even though our accessibility relations are equivalence relations, in the multi-agent setting we can enforce directionality by composing equivalence relations for different agents (and grounding them by referring to truths in local or boundary conditions of our structure, such as the actual state, or the top-left state, or ...). More formally, even though $\sim_a$ and $\sim_d$ are equivalence relations, their composition $\sim_a \circ \sim_d$ is not symmetric: we may have that $x \sim_a y \sim_d z$, i.e., $x(\sim_a \circ \sim_d)z$, but not $z(\sim_a \circ \sim_d)x$. Although we were not inspired by this, it deserves mentioning that such emerging asymmetry in multi-agent conditions is used to great effect in the expressivity proofs in Chapter 8 of [11].

The encoding comes in three parts: Firstly, we would like to define t to contain the transitive closure of the other epistemic relations. Next, we define a grid like (or chessboard like) structure over the model. Finally we use this grid-like model to state that the given tiling exists.

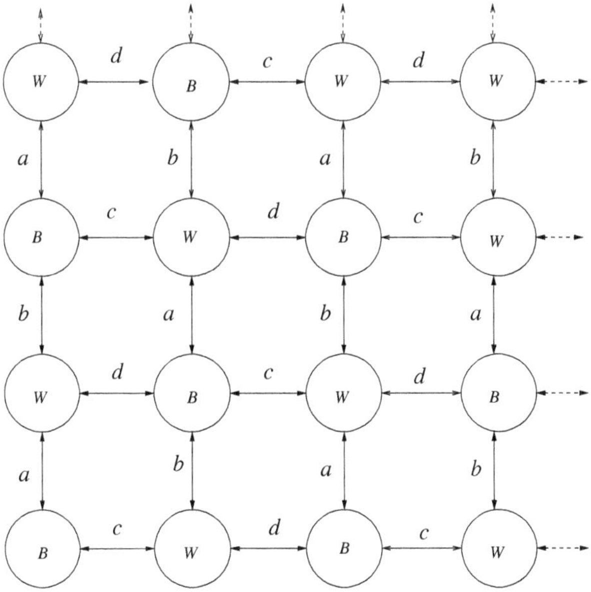

Figure 1. A grid-like model.

First note that these descriptions grossly over-simplify the actual construction. To properly execute these steps we would require a mechanism that allows us to define when one world is equivalent to another. This we do not have. However, the arbitrary announcement mechanism allows us to identify when two sets of worlds cannot be distinguished by any announcement. We will, for the moment ignore these considerations. We will use the term *equivalent* (rather than equal) to describe to worlds that are indistinguishable with respect to epistemic logic, and we will precisely define this notion in the subsequent sections.

4.1 Weak transitive closure

The following formula sets $\sim_t$ to include a weak transitive closure of $\sim_a$, $\sim_b$, $\sim_c$ and $\sim_d$. Particularly at every world w in the model, for each agent $x \in \{a, b, c, d\}$, if there is some world w' of a different colour to w where $w \sim_x w'$, then there is some world u where $w \sim_t u$ and u is equivalent to w'.

$$(1) \quad alt \quad = \quad K_t \left(\begin{array}{l} B \to (L_a W \wedge L_c W) \\ W \to (L_b B \wedge L_d B) \end{array} \right)$$

$$(2) \quad T^* \quad = \quad K_t \Box \bigwedge \left(\begin{array}{l} K_t B \to (K_a B \wedge K_b B \wedge K_c B \wedge K_d B) \\ K_t W \to (K_a W \wedge K_b W \wedge K_c W \wedge K_d W) \end{array} \right).$$

The important part of this formula is the arbitrary announcement ($\Box$) in T^*. This states that no announcement can be made that informs agent t of the colour of the current square, without informing every other agent as well. To the contrary, suppose that this were not true. Specifically suppose the current world is *black*, and agent a considers some *white* world, u, possible where u was demonstrably different to every world t considers possible (say by formula χ_u). Then the public announcement $\chi_u \vee B$ would inform t that the current world is black, but not a.

4.2 Defining a grid

To define a grid-like structure we will require the following properties:

1. Every *black* world has an a-successor that is *white* and a c-successor that is *white*.
2. Every *white* world has a b-successor that is *black* and a d-successor that is *black*.
3. The current world is *black* and both b and d know this.
4. If the current world is *black*:
   - for every *white* world u that is a-reachable from the current world, every *black* world that is d-reachable from u is equivalent to some *black* world that is b-reachable from some *white* world that is c-reachable from the current world.
   - for every *white* world u that is c-reachable from the current world, every *black* world that is b-reachable from u is equivalent to some *black* world that is d-reachable from some *white* world that is a-reachable from the current world.
5. If the current world is *white*:
   - for every *black* world u that is b-reachable from the current world, every *white* world that is c-reachable from u is equivalent to some *white* world that is a-reachable from some *black* world that is d-reachable from the current world.
   - for every *black* world u that is d-reachable from the current world, every *white* world that is a-reachable from u is equivalent to some *white* world that is c-reachable from some *black* world that is b-reachable from the current world.

This is achieved with the following formulas:

$$st = B \wedge K_b B \wedge K_d B \tag{3}$$

$$C1 = B \rightarrow \Box((L_a(W \wedge L_d B)) \rightarrow (K_c(W \rightarrow L_b B))) \tag{4}$$

$$C2 = B \rightarrow \Box((L_c(W \wedge L_b B)) \rightarrow (K_a(W \rightarrow L_d B))) \tag{5}$$

$$C3 = W \rightarrow \Box((L_b(B \wedge L_c W)) \rightarrow (K_d(B \rightarrow L_a W))) \tag{6}$$

$$C4 = W \rightarrow \Box((L_d(B \wedge L_a W)) \rightarrow (K_b(B \rightarrow L_c W))) \tag{7}$$

Again, the arbitrary announcement is used to establish a notion of equivalence between two worlds. The formula *st* simply specifies the state of the initial world (the bottom, left hand corner of the grid which, to extend the chess-board analogy, is black). The other formulae $C1 - C4$ define a weak commutativity property (e.g. every *black* world that is $\sim_a\sim_d$ reachable from the current (*black*) world, is $\sim_c\sim_b$ reachable from the current world).

4.3 The existence of a tiling

Given the previous formulas are sufficient to set up the desired chessboard pattern, it is a simple matter to exploit it to assert the existence of a tiling. Suppose the set Γ is given as above. Let:

$$(8) \qquad blk = B \to \left(\bigwedge_{\gamma \in \Gamma} \left(\gamma \to \bigwedge \begin{bmatrix} K_a(W \to \bigvee_{\gamma^t = \delta^b} \delta)) \\ K_c(W \to \bigvee_{\gamma^r = \delta^\ell} \delta)) \end{bmatrix} \right) \right)$$

$$(9) \qquad wht = W \to \left(\bigwedge_{\gamma \in \Gamma} \left(\gamma \to \bigwedge \begin{bmatrix} K_b(B \to \bigvee_{\gamma^t = \delta^b} \delta)) \\ K_d(B \to \bigvee_{\gamma^r = \delta^\ell} \delta)) \end{bmatrix} \right) \right).$$

The interpretation of these formula is straightforward. Given a tile γ is true at the current state, we assert that the bottom of all successor vertical tiles [1] is the same colour as γ^t. In the case that the current state is *black* the successor vertical states are the a-reachable *white* states, and if the current state is *white* then all successor vertical states are the b-reachable *black* states. A similar characterization exists for the horizontal (*left-right*) correspondence.

Finally we can define the formula:

$$(10) \quad Tile_\Gamma = alt \wedge T^* \wedge st \wedge K_t(C1 \wedge C2 \wedge C3 \wedge C4 \wedge blk \wedge wht).$$

In the following section we show that the existence of a model for this formula is equivalent to the existence of a tiling of the ω-plane for γ.

5 Proof of correctness

In this section we show that the above formula, $Tile_\Gamma$, is satisfiable in $APAL$ if and only if the set of tiles Γ is able to tile the plane $\omega \times \omega$.

We first address the soundness of the construction of the formula $Tiles_\Gamma$.

LEMMA 2. *Given there is a Γ-tiling of the plane, $\lambda : \omega \times \omega \longrightarrow \Gamma$, we may define a model of $APAL$ that satisfies the formula $Tile_\Gamma$.*

Proof. This model is taken directly from the tiling see Figure 2 with the knowledge relation of t being the universal modality. That is we let our model be $M = (S, \sim, V)$ where:

- $S = \omega \times \omega$,

[1] Whilst the existence of multiple vertical successors is not very "grid-like" we will later show this is inconsequential.

- $V(B) = \{(i,j) |\ i+j \text{ is even}\}$ and $V(W) = \{(i,j) |\ i+j \text{ is odd}\}$,
- for each $\gamma \in \Gamma$, $V(\gamma) = \{(i,j) |\ \lambda(i,j) = \gamma\}$,
- $\sim_a$ is the transitive, reflexive and symmetric closure of the relation $\{((i,j),(i,j+1)) |\ (i,j) \in V(B)\}$,
- $\sim_b$ is the transitive, reflexive and symmetric closure of the relation $\{((i,j),(i,j+1)) |\ (i,j) \in V(W)\}$,
- $\sim_c$ is the transitive, reflexive and symmetric closure of the relation $\{((i,j),(i+1,j)) |\ (i,j) \in V(B)\}$,
- $\sim_a$ is the transitive, reflexive and symmetric closure of the relation $\{((i,j),(i+1,j)) |\ (i,j) \in V(W)\}$,
- $\sim_t = \{((i,j),(k,\ell)) |\ i,j,k,\ell \in \omega\}$.

We now show that $M,(0,0) \models Tile_\Gamma$. For the parts of $Tile_\Gamma$ not containing arbitrary announcements this is straightforward. We can see that $M,(0,0) \models alt \wedge st$ by construction and as λ is a tiling it follows that $M,(0,0) \models K_t(blk \wedge wht)$. The remaining formulas T^* and $C1 - C4$ involve arbitrary announcements. Let's first examine T^*. This is equivalent to, for all $i,j \in \omega$,

$$M,(i,j) \models \Box \bigwedge \left(\begin{array}{l} K_tB \rightarrow (K_aB \wedge K_bB \wedge K_cB \wedge K_dB) \\ K_tW \rightarrow (K_aW \wedge K_bW \wedge K_cW \wedge K_dW) \end{array} \right).$$

Suppose that $(i,j) \in V(B)$. If any announcement $[\phi]$ makes K_tB true, it must be that M^ϕ consists only of black worlds, and hence, $M^\phi,(i,j) \models (K_aB \wedge K_bB \wedge K_cB \wedge K_dB)$. A similar argument holds for $(i,j) \in V(W)$.

We will now show that $M,(0,0) \models K_t(C1)$, and the cases for $C2$-$C4$ can be shown similarly. Suppose that $(i,j) \in V(B)$. We must show that for all epistemic ψ where $M,(i,j) \models \psi$, $M^\psi,(i,j) \models L_a(W \wedge L_dB)) \rightarrow K_c(W \rightarrow L_bB)$. Since we are quantifying over all submodels M^ϕ corresponding to epistemic formula, it is sufficient to show that $M',(i,j) \models L_a(W \wedge L_dB)) \rightarrow K_c(W \rightarrow L_bB)$ where $M' = (S', \sim', V')$ is any submodel where $(i,j) \in S'$. In such a case, since $M',(i,j) \models L_a(W \wedge L_dB)$, it follows that $(i,j+1),(i+1,j+1) \in S'$. Also, $(i+1,j)$ is the only one white world c-related to (i,j). If $(i+1,j) \in S'$, then because $(i+1,j+1) \in S'$, we have $M',(i,j) \models K_c(W \rightarrow L_bB)$. If $(i+1,j) \notin S'$ then $M',(i,j) \models K_c(W \rightarrow L_bB)$ as $M',(i,j) \models K_c\neg W$. Thus for every sub-model, M' including (i,j) we have

$$M',(i,j) \models L_a(W \wedge L_dB)) \rightarrow K_c(W \rightarrow L_bB)$$

and thus for every epistemic announcement ψ where $M,(i,j) \models \psi$ we have $M^\psi,(i,j) \models L_a(W \wedge L_dB)) \rightarrow K_c(W \rightarrow L_bB)$. Therefore $M,(0,0) \models K_tC_1$. A similar argument can be applied for $C2$-$C4$ so it follows that given a Γ-tiling exists, we can show, $Tile_\Gamma$ is satisfiable. ∎

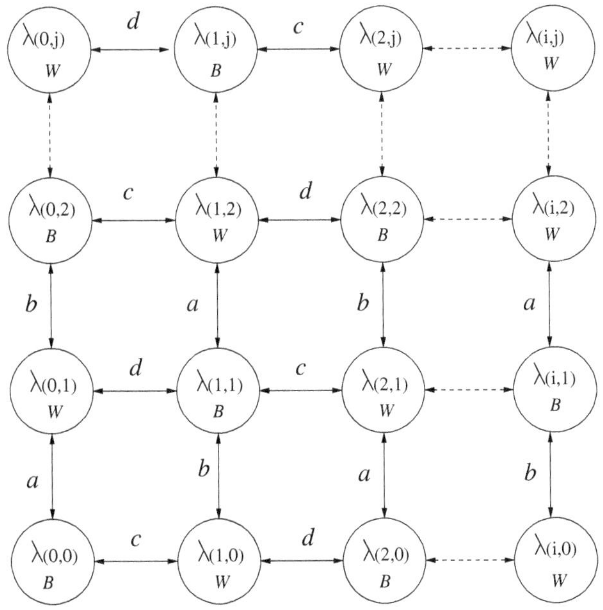

Figure 2. The conversion of a a tiling into a model.

For the completeness argument, we suppose that $M, s \models Tile_\Gamma$. From M we will show that, for each $n \in \omega$ we can construct a tiling, λ_n of an $n \times n$ grid. This is shown to be equivalent to tiling the full $\omega \times \omega$ plane in the following lemma.

LEMMA 3. *If Γ is able to tile an $n \times n$ grid for all $n \in \omega$, then Γ is able to tile the $\omega \times \omega$ plane.*

Proof. Let λ_n be the $n \times n$ tiling, and define λ^* as a tiling of the plane where $\lambda^*(0,0) = \gamma$ for some γ where for some infinite $N_{0,0} \subseteq \omega$, for all $n \in N_{0,0}$ $\lambda_n(0,0) = \gamma$. We then proceed by induction over $\omega \times \omega$ where $(i_1, j_1) \leq (i_2, j_2)$ if and only if $i_1 + j_1 < i_2 + j_2$ or $i_1 + j_1 = i_2 + j_2$ and $i_1 \leq i_2$. For $(i,j) > (0,0)$, we define $\lambda^*(i,j)$ such that

- if $i > 0$ then $\lambda^*(i,j) = \gamma$ where for some infinite $N_{i,j} \subseteq N_{i-1,j+1}$, for all $n \in N_{i,j}$ $\lambda_n(i,j) = \gamma$,

- otherwise (if $i = 0$) we let $\lambda^*(0,j) = \gamma$ where for some infinite $N_{0,j} \subseteq N_{j-1,0}$, for all $n \in N_{0,j}$ $\lambda_n(0,j) = \gamma$

It can be shown that such γ and $N_{i,j}$ can always be found (since there are infinitely many finite tilings and only finitely many tiles, the pigeon hole principle may be applied). Therefore such a λ^* may be defined by induction, (or indeed, Koenig's Lemma). ■

To proceed we require the following definition:

DEFINITION 4. Two worlds are 0-Q-*bisimilar*, iff they satisfy exactly the same set of propositional atoms taken from Q.
For all $n \in \omega$, two worlds, u and v in M, are n-Q-*bisimilar* (written $u \cong_n^Q v$) if and only if:

1. $u \cong_{n-1}^Q v$;

2. for every $x \in \{a, b, c, d, t\}$, for every world w where $u \sim_x w$, there is some world w' where $v \sim_x w'$ and $w \cong_{n-1}^Q w'$; and

3. for every $x \in \{a, b, c, d, t\}$, for every world w where $v \sim_x w$, there is some world w' where $u \sim_x w'$ and $w \cong_{n-1}^Q w'$.

We note for all n and Q, n-Q-bisimilarity is an equivalence relation.

LEMMA 5. *Suppose that the set of propositions, Q, is finite. Then for every n, there is a finite set of $\mathcal{L}_{el}$ formulas $\{\phi_1, \dots, \phi_m\}$ such that for every $u \in S$, there is some $i \leq m$ such that for all $v \in S$, $u \cong_n^Q v$ if and only if $M, v \models \phi_i$.*

Proof. This can be shown by induction. As a base case we take the set of formulas $\phi(Q') = \bigwedge_{x \in Q'} x \wedge \neg \bigvee_{x \in Q \setminus Q'} x$ for all $Q' \subseteq Q$. Clearly, for each $u \in S$, we can let $Q' = \{x | u \in V(x)\}$ and then for all $v \in S$, $M, v \models \phi(Q')$ if and only if $v \cong_0^Q u$.

For the inductive step, suppose that $\{\phi_1, \dots, \phi_m\}$ is a set of formulas such that for every u in S, there is some $i \leq m$ such that for all $v \in S$, $u \cong_n^Q v$ if and only if $M, v \models \phi_i$. For each $u \in S$ let the corresponding formula ϕ_i be denoted ϕ_u^n, and let

$$\phi_u^{n+1} = \phi_u^n \wedge \bigwedge_{x \in A} (succ_x^u \wedge nsucc_x^u)$$

where

$$\begin{aligned} succ_x^u &= \bigwedge \{L_x \phi_v^n | \; v \sim_x u\} \\ nsucc_x^u &= K_x \bigwedge \{\neg \phi_i | \; \forall v \sim_x u, \; M, v \not\models \phi_i\} \end{aligned}$$

for the set of agents, $A = \{a, b, c, d, t\}$.

Then for any $v \in S$ where $M, v \models \phi_u^{n+1}$ we have:

1. $v \cong_n^Q u$ since $M, v \models \phi_u^n$.

2. for every $x \in \{a, b, c, d, t\}$, for every world w where $u \sim_x w$, we have $succ_x^u \rightarrow L_x \phi_w^n$, so $M, v \models L_x \phi_w^n$, and thus there is some world w' where $v \sim_x w'$, $M, w' \models \phi_w^n$, so $w \cong_n^Q w'$.

3. for every $x \in \{a, b, c, d, t\}$, for every world w where $v \sim_x w$, there is some world w' where $u \sim_x w'$ and $w \cong_n^Q w'$. To see this, suppose for contradiction that there was some world w such that $v \sim_x w$ and for every world w' where $u \sim_x w'$ we have $w \not\cong_n^Q w'$. Therefore, we have $nsucc_x^u \to K_x \neg \phi_w^n$ so $M, v \models K_x \neg \phi_w^n$, contradicting $v \sim_x w$.

Therefore $v \cong_{n+1}^Q u$ as required. Conversely, if $v \cong_{n+1}^Q u$, then

1. $M, v \models \phi_u^n$ since $v \cong_n^Q u$.

2. for all $x \in A$, for every world w where $u \sim_x w$ there is some world w' where $v \sim_x w'$ and $w \cong_n^Q w'$ (hence $M, w' \models \phi_w^n$). Therefore $M, v \models succ_x^u$.

3. for every $x \in \{a, b, c, d, t\}$, for every world w where $v \sim_x w$, there is some world w' where $u \sim_x w'$ and $w \cong_n^Q w'$. By the induction hypothesis, for every w where $v \sim_x w$, we have $M, w \models \bigvee_{u \sim_x w'} \phi_{w'}^n$. For all w' where $u \sim_x w'$ we clearly have $\phi_{w'}^n \to \bigwedge\{\neg\phi_i \mid \forall v \sim_x u,\ M, v \not\models \phi_i\}$, so it follows that $M, v \models nsucc_x^u$.

Thus $M, v \models \phi_u^{n+1}$ completing the induction. ■

For the following proofs we define a new operator $\Box_n^Q$, to mean "for all public Q-announcements of depth n". The semantics are given as: $M, u \models \Box_n^Q \phi$ if and only if for all $\mathcal{L}_{el}$ formulae ψ with at most n nestings of knowledge operators and containing only the atoms Q, if $M, u \models \psi$, then $M^\psi, u \models \phi$.

LEMMA 6.

1. *For all n, for all Q, $\Box\phi \to \Box_n^Q \phi$ is a validity.*

2. *For any two worlds u, v where $u \cong_n^Q v$, for any L_{el} formula ϕ of depth at most n and containing only atoms from Q, $M, u \models \phi$ if and only if $M, v \models \phi$.*

3. *For any two worlds u, v where $u \cong_{n+m}^Q v$, for any L_{el} formula ϕ of depth at most n and containing atoms only from Q, $M, u \models \Box_m^Q \phi$ if and only if $M, v \models \Box_m^Q \phi$.*

Proof.

1. Obvious.

2. By induction. Clearly the statement holds for the case $n = 0$. Suppose the statement holds for n. Every L_{el} formula ϕ, of depth $n+1$, can be written as a Boolean combination of atoms and formulas $K_{x_i}\phi_i$ (for $i = 1, \ldots, m$) where ϕ_i is a formula of depth at most n. If $u \cong_{n+1}^Q v$, then for every $u' \sim_x u$ there is some $v' \sim_x v$ where $u' \cong_n^Q v'$, and vice-versa. By the induction hypothesis, $M, u' \models \phi_i$ for all $u' \sim_{x_i} u$ if and only if $M, v' \models \phi_i$ for all $v' \sim_{x_i} v$. It follows that $M, u \models \phi$ if and only if $M, v \models \phi$.

3. To prove this statement we extend the induction above with the case for m. In the case $m = 0$ it is effectively the second part of this lemma. Now suppose the statement holds for a given m (and for all n). Without loss of generality, suppose that $u \cong^Q_{n+m} v$ and $M, u \models \neg\Box^Q_m\phi$ where ϕ is of depth n. Therefore there is some announcement, ψ, of depth m that makes ϕ false. This is equivalent to $M, u \models \psi \wedge \neg\phi^\psi$ where ϕ^ψ is inductively defined by replacing subformulas $K_x(\alpha)$ with $K_x(\psi \to \alpha^\psi)$, and otherwise acting as the identity (see Proposition 4.22 of [11] for a formal description of this translation). Now $\psi \wedge \neg\phi^\psi$ is of depth $n + m$ so the result follows from the second part of this Lemma.

■

We will refer to a formula of depth n, containing only atoms from Q as an n-Q-formula. The above lemmas and definitions will be applied to show how the arbitrary announcements in the formula $Tile_\Gamma$ can allow us to establish that two worlds are n-Q-bisimilar for arbitrary n and arbitrary Q. To this end, for $n \in \omega$ and finite $Q \subset P$, we define the formula $Tile^{(n,Q)}_\Gamma$ to be the formula $Tile_\Gamma$ with every arbitrary announcement $\Box$ replaced by $\Box^Q_n$ (and likewise for subformulas such as $C1^{(n,Q)}$).

There are two types of public announcement in the formula $Tile_\Gamma$. The first appears in the sub-formula T^*, and the following lemma shows how this allows the knowledge relation of agent t to act as a weak kind of transitive closure for the other knowledge relations.

LEMMA 7. *Let $n \geq 1$, $x \in \{a, b, c, d\}$ and $Q \subset P$ be a finite set of propositional atoms including $\{B, W\} \cup \Gamma$. Suppose that $u \in V(B)$ (resp. $u \in V(W)$), $u \sim_x v$ for some $v \in V(W)$ (resp. $v \in V(B)$), $u \cong^Q_n w$, and $M, u \models (alt \wedge T^*)^{(n,Q)}$. Then there is some w' such that $w \sim_t w'$ and $w' \cong^Q_{n-1} v$.*

Proof. Suppose that $M, u \models (alt \wedge T^*)^{(n,Q)}$, $u \cong^Q_n w$, and $u \sim_x v$ where $u \in V(B)$ and $v \in V(W)$. From T^* we have $M, u \models \Box^Q_{n-1}(K_tB \to K_xB)$. As $u \cong^Q_n w$ and $n \geq 1$, by Lemma 6.3 we have $M, w \models \Box^Q_{n-1}(K_tB \to K_xB)$. Now, suppose for contradiction that for all $w' \sim_t w$ we have $w' \not\cong^Q_{n-1} v$. By Lemma 5 there is some formula ϕ^{n-1}_v such that $M, v \models \phi^{n-1}_v$, and for all $w' \sim_t w$ we have $M, w' \models \neg\phi^{n-1}_v$. Therefore we may make the public announcement $\psi = B \vee \phi^{n-1}_v$. Thus $M^\psi, w \models K_tB$, and since $M, w \models \Box^Q_{n-1}(K_tB \to K_xB)$, we have $M^\psi, w \models K_xB$. It follows that $M, w \models \psi \wedge K_x(\psi \to B)$. However, $M, u \models \psi \wedge \neg K_x(\psi \to B)$. Since $\psi = \phi^{n-1}_v \vee B$, ψ has depth $n - 1$ and thus $K_x(\psi \to B)$ has depth n. As $u \cong^Q_n w$, by Lemma 6.2, u and w agree on all formulas of depth n. This contradicts the inference that $M, u \models \neg K_x(\psi \to B)$ and $M, w \models K_x(\psi \to B)$. ■

The other occurrences of arbitrary announcements in the formula $Tile_\Gamma$ appear in the formulae $C1-C4$. These formulas use the arbitrary announcements to establish a weak type of commutativity property, which is essential

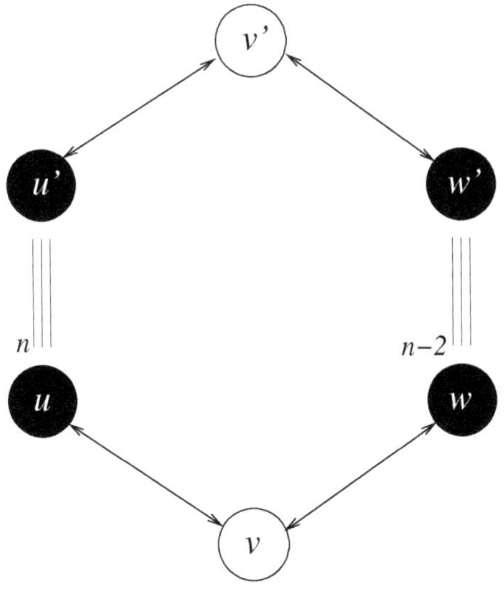

Figure 3. The construction of Lemma 8, inferring the worlds w and w' are $n-2$-bisimilar, given the worlds u and u' are n bisimilar.

in defining a grid. The following lemmas clarify this property and show that it is enforced by the formula $Tile_\Gamma$. The first lemma deals with the black worlds and the second lemma deals (symmetrically) with the white worlds.

LEMMA 8. *Suppose that $u \in V(B)$, $u \sim_a v$, for some $v \in V(W)$. Let $n \geq 2$ and suppose also that $u \cong_n^Q u'$ for some finite $Q \subset P$ including $\{B, W\} \cup \Gamma$, and $u' \sim_c v'$ for some $v' \in V(W)$. Given that $M, u \models (C1 \wedge C2)^{(n,Q)}$:*

1. *for all $w \in V(B)$ where $v \sim_d w$, if there is some $w' \in V(B)$ where $v' \sim_b w'$ then there is some such w' where either $w \cong_{n-2} w'$ or $w' \cong_{n-2}^Q u'$.*

2. *for all $w' \in V(B)$ where $v' \sim_b w'$, if there is some w where $v \sim_d w$, then there is some such w where either $w \cong_{n-2} w'$ or $w \cong_{n-2}^Q u$.*

Proof. We will show case 1, and case 2 can be shown similarly. So given the assumptions of the Lemma, let $w \in V(B)$ be such that $v \sim_d w$. Consider the announcement $\psi = W \vee \phi_u^{n-2} \vee \phi_w^{n-2}$. Since $M, u \models C1^{(n,Q)}$, $M^\psi, u \models L_a(W \wedge L_d B) \to K_c(W \to L_b B)$, and by Lemma 5, $M^\psi, u \models L_a(W \wedge L_d B)$. We may apply modus ponens and Lemma 6 to deduce, $M_{u'}^\psi \models K_c(W \to L_b B)$. Therefore $M_{v'}^\psi \models L_b B$, so there is some $w' \in V(B)$ where $v' \sim_b w'$, and $M, w' \models \psi$. As $w' \in V(B)$, we have either $M, w' \models \phi_u^{n-2}$ and thus $w' \cong_n^Q u'$, or $M, w' \models \phi_w^{n-2}$, and thus $w \cong_{n-2}^Q w'$, (by Lemma 5). This scenario is represented in Figure 3 ■

Notice that the lemma does not perfectly capture the notion of commutativity. Ideally we would like to have:
For all $u, w, u' \in V(B)$ for all $v \in V(W)$ where $u \sim_a v$, $v \sim_d w$ and $u \cong_n^Q u'$, there's some $v' \in V(W)$ and some $w' \in V(B)$ such that $u' \sim_c v'$, $v' \sim_b w'$ and $w' \cong_{n-2}^Q w$, (and vice-versa).
However, we must consider the additional possibility that there's some $v' \in V(W)$ and some $w' \in V(B)$ such that $u' \sim_c v'$, $v' \sim_b w'$ and $w' \cong_{n-2}^Q u'$. In such a case we would have, by the second part of the lemma, that there is some w where $v \sim_d w$ and either $w \cong_{n-2}^Q w'$ or $w \cong_{n-2}^Q u$. In either case, as $w' \cong_{n-2}^Q u'$ and $u \cong_n^Q u'$, we will have $w' \cong_{n-2}^Q u' \cong_{n-2}^Q u \cong_{n-2}^Q w$, which is sufficient for our purposes.

LEMMA 9. *Suppose that $u \in V(W)$, $u \sim_b v$, for some $v \in V(B)$. Suppose also that $u \cong_n^Q u'$ for some finite $Q \subset P$ including $\{B, W\} \cup \Gamma$, and $u' \sim_d v'$ for some $v' \in V(B)$. Given that $M, u \models (C3 \wedge C4)^{(n,Q)}$:*

1. *for all $w \in V(W)$ where $v \sim_c w$, if there is some $w' \in V(W)$ where $v' \sim_a w'$ then there is some such w' where either $w \cong_{n-2}^Q w'$ or $w' \cong_{n-2}^Q u'$.*

2. *for all $w' \in V(W)$ where $v' \sim_a w'$, if there is some w where $v \sim_c w$, then there is some such w where either $w \cong_{n-2}^Q w'$ or $w \cong_{n-2}^Q u$.*

Proof. The proof of this is symmetrical to the proof of Lemma 8 ∎

These lemmas are sufficient to establish a finite grid structure, as depicted in Figure 4. The formulas *blk* and *wht* are then clearly sufficient to encode a finite tiling, so if $M, s \models Tile_\Gamma$ then by Lemma 3 a Γ tiling exists.
Recall

(11) $Tile_\Gamma = alt \wedge T^* \wedge st \wedge K_t(C1 \wedge C2 \wedge C3 \wedge C4 \wedge blk \wedge wht)$.

We give the following Lemma.

LEMMA 10. *Suppose that $M, s \models Tile_\Gamma$. Then for all $n \in \omega$, for $Q = \Gamma \cup \{B, W\}$ we may define a partial function $f : \{0, \ldots, n\}^2 \longrightarrow S$ such that:*

1. *$f(0,0) = s$;*

2. *if $f(i,j) \in V(B)$, then*

   (a) *if $i < n$, there is some $u \in V(W)$ where $f(i+1,j) \cong_{k(i,j)}^Q u$ and $f(i,j) \sim_c u$, and*

   (b) *if $j < n$, there is some $u \in V(W)$ where $f(i,j+1) \cong_{k(i,j)}^Q u$ and $f(i,j) \sim_a u$;*

   (c) *$M, f(i,j) \models blk$*

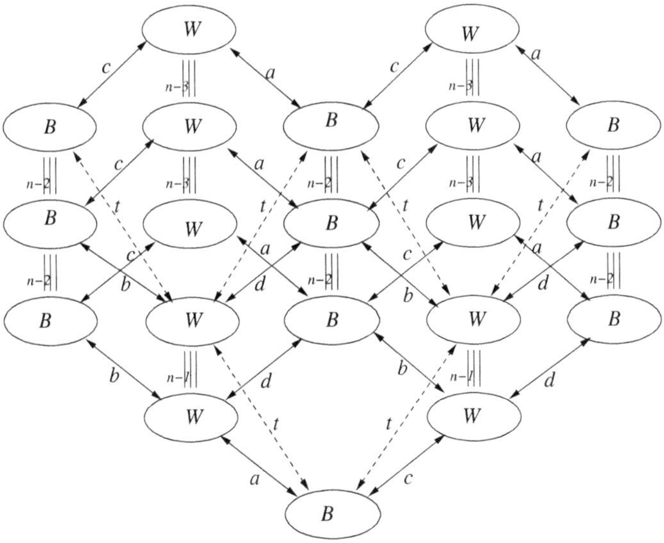

Figure 4. The construction of the finite grid.

3. if $f(i,j) \in V(W)$, *then*

(a) if $i < n$, *there is some* $u \in V(B)$ *where* $f(i+1,j) \cong^{Q}_{k(i,j)} u$ *and* $f(i,j) \sim_d u$, *and*

(b) if $j < n$, *there is some* $u \in V(B)$ *where* $f(i,j+1) \cong^{Q}_{k(i,j)} u$ *and* $f(i,j) \sim_b u$;

(c) $M, f(i,j) \models wht$

where $k(i,j) = 2n + 3 - (i+j)$.

We can show this by construction, applying Lemmas 7, 8 and 9. The function $k(i,j)$ is chosen such that for all $i,j \leq n$, $k(i,j) \geq 3$. This allows the preconditions of the Lemmas 7, 8 and 9 to be met for all $i,j \leq n$. The proof is illustrated in Figure 4. You can view this figure as a cube, cut in half diagonally up from a bottom corner. The base of the shape makes a finite grid. We construct a function, f, mapping $\{0,\ldots,n\}^2$ to the states of the model, such that $f(i,j)$ is $k(i,j)$-Q-bisimilar to the corresponding world at the base of the grid. As we get further from the corner $k(i,j)$ decreases so this correspondence becomes progressively weaker. By the time $i+j > 2n$, $k(i,j) < 3$ so the preconditions for the necessary lemmas is not met. However, by this stage we have already defined an $n \times n$ grid.

Proof. We construct the a function, F, satisfying the stated properties by induction over $i+j$, where $i+j \leq 2n$. For the induction hypothesis we assume for all g,h where $g+h < i+j$, $f(g,h)$ is defined such that:

1. if $f(g,h) \in V(B)$, then
   (a) if $0 < g < n$, then $f(g-1,h) \in V(W)$ and for some $u \in V(B)$ we have $f(g,h) \cong^Q_{k(g,h)} u$ and $f(g-1,h) \sim_d u$, and
   (b) if $0 < h < n$, then $f(g,h-1) \in V(W)$ and for some $u \in V(B)$ we have $f(g,h) \cong^Q_{k(g,h)} u$ and $f(g,h-1) \sim_b u$;
2. if $f(g,h) \in V(W)$, then
   (a) if $0 < g < n$ then $f(g-1,h) \in V(B)$ and for some $u \in V(W)$ we have $f(g,h) \cong^Q_{k(g,h)} u$ and $f(g-1,h) \sim_c u$, and
   (b) if $0 < h < n$ then $f(g,h-1) \in V(B)$ and for some $u \in V(W)$ we have $f(g,h) \cong^Q_{k(g,h)} u$ and $f(g,h-1) \sim_a u$;
3. there is some u where $f(g,h) \cong^Q_{k(g,h)} u$ and

   (12) $M,u \models (alt \wedge T^* \wedge K_t(C1 \wedge C2 \wedge C3 \wedge C4 \wedge blk \wedge wht))^{(k(g,h),Q)}$.

For brevity let,

(13) $Hyp = alt \wedge T^* \wedge K_t(C1 \wedge C2 \wedge C3 \wedge C4 \wedge blk \wedge wht)$.

For the base case of this induction, suppose $M,s \models Tile_\Gamma$, and let $n \in \omega$. We define $f(0,0) = s$. Then

1. $s \cong_{k(0,0)} f_n(0,0)$,
2. $M,s \models alt \wedge T^*$,
3. $M,s \models C1 \wedge C2 \wedge C3 \wedge C4$,

so it clearly satisfies the inductive hypothesis.

For the induction, suppose that the inductive hypothesis holds for the pair i,j. There are three cases to consider, $i=0$, $j=0$ and $i,j \neq 0$.

1. if $i=0$ we may assume $j \neq 0$ (since $f(0,0)$ is already defined). By the induction hypothesis, $f(i,j-1)$ is defined. We suppose, without loss of generality, that $f(i,j-1) \in V(B)$ and the case of $f(i,j-1) \in V(W)$ may be handled similarly. Also by the induction hypothesis, there is some $u \cong^Q_{k(i,j-1)} f(i,j-1)$, where $M,u \models Hyp^{(k(i,j-1),Q)}$. By Lemma 6 we have $M, f(i,j-1) \models L_a W$. Therefore there is some world $v \in V(W)$ where $f(i,j-1) \sim_a v$ and we let $f(i,j) = v$. By Lemma 7 there is some world $w \sim_t u$ such that $w \cong^Q_{k(i,j)} v$ and since $\models Hyp \rightarrow K_t Hyp$, $M,w \models Hyp^{(k(i,j),Q)}$ as required.
2. the case for $j=0$ is symmetric to the case above.

3. if $i, j \neq 0$ suppose, without loss of generality, that $f(i,j) \in V(B)$ (and the case for $f(i-1,j-1) \in V(W)$ is handled similarly). By the induction hypothesis for some $u \cong^Q_{k(i-1,j-1)} f(i-1,j-1)$, we have $M, u \models Hyp^{(k(i-1,j-1),Q)}$, $u \sim_a v$ for some $v \cong^Q_{k(i-1,j)} f(i-1,j) \in V(W)$, and $u \sim_c v'$ for some $v' \cong_{k(i,j-1)} f(i,j-1)$. By Lemma 8, either:

   (a) there is some $w, w' \in V(B)$ such that $v \sim_d w$, $v' \sim_b w'$ and $w' \cong^Q_{k(i,j)} w$. In such a case we let $f(i,j) = w$; or

   (b) there is some $w' \in V(B)$ where $v' \sim_b w'$ and $w' \cong^Q_{k(i,j)} u$. In this case we let $f(i,j) = u$. By Lemma 8 we also have for all $w' \in V(B)$ where $v' \sim_b w'$ there is some $w \in V(B)$ where $v \sim_d w$ and $w \cong^Q_{k(i,j)} u$.

   Also by the induction hypothesis we have $M, v \models Hyp^{(f(i-1,j),Q)}$, so we may apply Lemma 7 to show that there is some world z such that $z \cong_{k(i,j)} f_n(i,j)$ and $v \sim_t z$. As the formula

   (14) $Hyp^{(f(i-1,j),Q)} \rightarrow K_t(Hyp^{(f(i-1,j),Q)})$

   is a tautology of epistemic logic and $M, v \models Hyp^{(f(i-1,j),Q)}$ we have $M, z \models (alt \wedge T^* \wedge K_t(C1 \wedge C2 \wedge C3 \wedge C4 \wedge blk \wedge wht))^{(k(i,j),Q)}$, for some $z \cong_{k(i,j)} f_n(i,j)$, as required.

Therefore, the induction hypothesis holds for the pair (n,n). Thus for all (i,j) where $i + j < 2n$ we have

1. if $f(i,j) \in V(W)$, then

   (a) $f(i+1,j) \in V(B)$ and for some $u \in V(W)$ we have $f(i,j) \cong^Q_{k(i,j)} u$ and $f(i+1,j) \sim_d u$, and

   (b) $f(i,j+1) \in V(B)$ and for some $u \in V(W)$ we have $f(i,j) \cong^Q_{k(i,j)} u$ and $f(i,j+1) \sim_b u$;

   (c) $M, f(i,j) \models wht$

2. if $f(i,j) \in V(B)$, then

   (a) $f(i+1,j) \in V(W)$ and for some $u \in V(B)$ we have $f(i,j) \cong^Q_{k(i,j)} u$ and $f(i+1,j) \sim_c u$, and

   (b) $f(i,j+1) \in V(W)$ and for some $u \in V(B)$ we have $f(i,j) \cong^Q_{k(i,j)} u$ and $f(i,j+1) \sim_a u$;

   (c) $M, f(i,j) \models blk$

so f satisfies the required properties. ■

Note that this lemma only defines an $n \times n$ grid, since as Lemmas 6, 8 and 9 are used the induction and Lemma 6 is only available when $k(i,j) > 2$, and Lemmas 8 and 9 are only available when $k(i,j) > 1$. However, n is chosen arbitrarily. Because $f(0,0) \cong_n^Q f(0,0)$ for all n, we can seed the induction with an arbitrarily large n. This allows us to apply Lemma 3 to define a tiling.

COROLLARY 11. *If $M, u \models Tile_\Gamma$ then a Γ-tiling exists*

Proof. If $M, u \models Tile_\Gamma$, then by Lemma 10, for all n we may define f such that for all i, j where $i + j < 2n$,

1. if $f(i,j) \in V(B)$ then
   (a) if $j < n$, $f(i,j) \sim_a u$ for some $u \cong_{k(i,j+1)}^Q f(i,j+1)$.
   (b) if $i < n$, $f(i,j) \sim_c u$ for some $u \cong_{k(i,j+1)}^Q f(i+1,j)$.
   (c) $M, f(i,j) \models blk$.
2. if $f(i,j) \in V(W)$ then
   (a) if $j < n$, $f(i,j) \sim_b u$ for some $u \cong_{k(i,j+1)}^Q f(i,j+1)$.
   (b) if $i < n$, $f(i,j) \sim_d u$ for some $u \cong_{k(i,j+1)}^Q f(i+1,j)$.
   (c) $M, f(i,j) \models wht$.

Recall the formulas:

$$(15) \qquad blk \quad = \quad B \to \left(\bigwedge_{\gamma \in \Gamma} \left(\gamma \to \bigwedge \left[\begin{array}{l} K_a(W \to \bigvee_{\gamma^t = \delta^b} \delta)) \\ K_c(W \to \bigvee_{\gamma^r = \delta^\ell} \delta)) \end{array} \right] \right) \right)$$

$$(16) \qquad wht \quad = \quad W \to \left(\bigwedge_{\gamma \in \Gamma} \left(\gamma \to \bigwedge \left[\begin{array}{l} K_b(B \to \bigvee_{\gamma^t = \delta^b} \delta)) \\ K_d(B \to \bigvee_{\gamma^r = \delta^\ell} \delta)) \end{array} \right] \right) \right).$$

Applying the semantics of epistemic logic:

1. if $f(i,j) \in V(B)$, and $M, f(i,j) \models \gamma$ then for some $u \cong_{k(i,j+1)}^Q f(i,j+1)$ we have $f(i,j) \sim_a u$ and $u \in V(W)$. Therefore $M, u \models \delta$ for some δ where $\gamma^t = \delta^b$. Thus $M, f(i+1,j) \models \delta$ for some δ where $\gamma^t = \delta^b$. Likewise for some $u \cong_{k(i,j+1)}^Q f(i+1,j)$ we have $f(i,j) \sim_c u$ and $u \in V(W)$, so $M, f(i,j+1) \models \delta$ for some δ where $\gamma^r = \delta^\ell$.

2. if $f(i,j) \in V(W)$, and $M, f(i,j) \models \gamma$ then for some $u \cong_{k(i,j+1)}^Q f(i,j+1)$ we have $f(i,j) \sim_a u$ and $u \in V(B)$. Therefore $M, u \models \delta$ for some δ where $\gamma^t = \delta^b$. Thus $M, f(i+1,j) \models \delta$ for some δ where $\gamma^t = \delta^b$. Likewise for some $u \cong_{k(i,j+1)}^Q f(i+1,j)$ we have $f(i,j) \sim_c u$ and $u \in V(B)$, so $M, f(i,j+1) \models \delta$ for some δ where $\gamma^r = \delta^\ell$.

Therefore f defines a tiling of an $n \times n$ grid with the tiles of Γ. Therefore the formula $Tile_\Gamma$ is satisfiable if and only if Γ can tile the $n \times n$ grid for arbitrary n. Applying Lemma 3 it follows a γ-tiling exists. ■

Thus the satisfiability problem $APAL$ is co-RE hard, as it is able to embed the tiling problem. As we know from [2] that the set of valid formulas for $APAL$ is recursively enumerable, it follows that the satisfiability problem is co-RE complete.

6 Future work

Up to this point arbitrary public announcement logic has shown some promise for practical reasoning applications: it has an axiomatization, a tableau-calculus, it is bisimulation invariant, naturally extends epistemic logic, and model checking is PSPACE-complete [1]. The notion of an arbitrary public announcement is also a natural concept (consider the plea, "*is there anything I can say to make you believe X*"). Therefore, a natural avenue of investigation is to consider whether we may be able to somehow expressively weaken $APAL$ to a decidable logic which also enjoys all these favorable properties.

One area of investigation may be to consider the set of formulae (or more abstractly, model-properties) that announcements may range over. From the encoding $Tile_\Gamma$ we have given, we note that the encoding of the tiling problem for Γ only requires five agents in the language (although it is conceivable a more complex encoding could do with less). Also, from the proof of correctness we have given, we note that the arbitrary announcements do not need to range over all epistemic formulae. The announcements only need to range over all formulae containing the atoms from $\Gamma \cup \{B, W\}$ (or the atoms appearing in the formula). However, the proof does require that the arbitrary announcements range over formulae of unbounded depth, so it may be of interest to consider restrictions where the arbitrary announcements ranged over announcements of a bounded depth (say, formulas of depth at most one).

We also note that generalizing the set of formulae that the arbitrary announcements range over would not affect this undecidability result. For example, if we allowed fixed-point operators to appear in the arbitrary announcements, then the proofs of Lemmas 7 and 8 would remain. However we may consider restrictions of the set of formulae. One natural restriction to consider would be to restrict arbitrary announcements to positive knowledge formulae (formulae where the knowledge operators K_x always appear in the scope of an even number of negations). For such a restricted set of formulae Lemma 5 would not hold so decidability may still be achievable.

An alternative approach is discussed [10]. Here it is suggested that informative events such as announcements, rather than simply being evaluated with respect to the given model, could add additional information (at an atomic level) into the model. This approach is motivated by the observation that while the $APAL$ is bisimulation invariant, it is not the case that two models bisimilar with respect to a subset of atoms, X, agree on all formulas that contain only the atoms X. The suggested *Future event logic* quantifies over refinements of a model, which includes all public announcements of epistemic formulae, but also other public or non-public informative events that can be described as action models [3]. It is shown to be decidable via a reduction to a bisimulation quantified logic [5]. Future event logic is interesting in its own right, and it aims to provide the sort of link between dynamic and temporal epistemics pointed out in [9]

Acknowledgements. The authors would like to thank the anonymous reviewers for their comments and helpful suggestions. Hans van Ditmarsch acknowledges support of the Netherlands Institute of Advanced Study where he was Lorentz Fellow in 2008.

BIBLIOGRAPHY

[1] P. Balbiani, A. Baltag, H.P. van Ditmarsch, A. Herzig, T. Hoshi, and T. De Lima. What can we achieve by arbitrary announcements? A dynamic take on Fitch's knowability. In D. Samet, editor, *Proceedings of TARK XI*, pages 42–51, Louvain-la-Neuve, Belgium, 2007. Presses Universitaires de Louvain.

[2] P. Balbiani, H. van Ditmarsch, A. Herzig, and T. de Lima. A tableau method for public announcement logics. In *Proceedings of the International Conference on Automated Reasoning with Analytic Tableaux and Related Methods*, volume 4548 of *Lecture Notes in Computer Science*, pages 43–59. Springer, 2007.

[3] A. Baltag and L.S. Moss. Logics for epistemic programs. *Synthese*, 139:165–224, 2004. Knowledge, Rationality & Action 1–60.

[4] R. Fagin, J. Halpern, Y. Moses, and M. Vardi. *Reasoning about knowledge.* MIT Press, 1995.

[5] T. French. *Bisimulation quantifiers for modal logic.* PhD thesis, The University of Western Australia, 2006. Available from `http://people.csse.uwa.edu.au/tim/`.

[6] D. Harel. Effective transformations on infinite trees, with applications to high undecidability, dominoes, and fairness. *J. A.C.M.*, 33(1):224–248, 1986.

[7] J.S. Miller and L.S. Moss. The undecidability of iterated modal relativization. *Studia Logica*, 79(3):373–407, 2005.

[8] E. Spaan. *Complexity of Modal Logics.* PhD thesis, Universiteit van Amsterdam, 1993.

[9] J.F.A.K. van Benthem and E. Pacuit. The tree of knowledge in action: towards a common perspective. *Advances in Modal Logic*, 6:87–106, 2006.

[10] H.P. van Ditmarsch and T. French. Simulation and information. (Electronic) Proceedings of LOFT 2008, Amsterdam, `http://www.illc.uva.nl/LOFT2008/listofacceptedpapers.html`, 2008.

[11] H.P. van Ditmarsch, W. van der Hoek, and B.P. Kooi. *Dynamic Epistemic Logic*, volume 337 of *Synthese Library*. Springer, 2007.

Hans van Ditmarsch
Computer Science, University of Otago, New Zealand & IRIT, France
`hans@cs.otago.ac.nz`

Tim French
School of Computer Science and Software Engineering,
The University of Western Australia, Australia
`tim@csse.uwa.edu.au`

Cut-elimination and proof-search for bi-intuitionistic logic using nested sequents

RAJEEV GORÉ, LINDA POSTNIECE AND ALWEN TIU

ABSTRACT. We propose a new sequent calculus for bi-intuitionistic logic which sits somewhere between display calculi and traditional sequent calculi by using nested sequents. Our calculus enjoys a simple (purely syntactic) cut-elimination proof as do display calculi. But it has an easily derivable variant calculus which is amenable to automated proof search as are (some) traditional sequent calculi. We first present the initial calculus and its cut-elimination proof. We then present the derived calculus, and then present a proof-search strategy which allows it to be used for automated proof search. We prove that this search strategy is terminating and complete by showing how it can be used to mimic derivations obtained from an existing calculus GBiInt for bi-intuitionistic logic. As far as we know, our new calculus is the first sequent calculus for bi-intuitionistic logic which uses no semantic additions like labels, which has a purely syntactic cut-elimination proof, and which can be used naturally for backwards proof-search.

Keywords: Bi-intuitionistic logic, display calculi, proof search.

1 Introduction

Bi-intuitionistic logic (BiInt) is obtained by extending intuitionistic logic with a binary connective variously called "subtraction" or "exclusion" or even "co-implication", which we write as $-\!\!<$. Intuitively, the formula $A -\!\!< B$ reads "A excludes B". We assume the reader is familiar with the Kripke semantics for intuitionistic logic using a binary reflexive and transitive relation $\leq$. Then the forcing relation for the exclusion operator is defined as

$$w \Vdash A -\!\!< B \quad \text{iff} \quad \exists v \leq w . v \Vdash A \text{ and } v \nVdash B.$$

In a sequent setting, the introduction rules for exclusion are dual to the introduction rules for implication, e.g., the left-introduction rule $-\!\!<_L$ is dual to the right-introduction rule $\rightarrow_R$ for implication as shown below:

$$\frac{\Gamma, A \vdash B}{\Gamma \vdash A \rightarrow B} \rightarrow_R \qquad\qquad \frac{A \vdash B, \Delta}{A -\!\!< B \vdash \Delta} -\!\!<_L$$

If we remove the implication connective from bi-intuitionistic logic, we obtain a logic called *dual intuitionistic logic*. Both intuitionistic logic and dual intuitionistic logic have simple sequent calculus formulations which enjoy cut elimination. However, the combined logic has surprisingly no known simple sequent calculus formulation which enjoys true cut-elimination.

Rauszer [21] formalised a sequent calculus for BiInt, but it was later found to fail cut elimination (see [5] for a counterexample). Crolard's work [7] is based upon Rauszer's calculus and so also fails cut-elimination. In response, Pinto and Uustalu announced a labelled sequent calculus with cut-elimination for BiInt [28], but have yet to provide a full paper outlining details. Their calculus does not provide a purely syntactic account of BiInt since it uses labelled formulae of the form $x : A$ to capture the semantic notion that "Kripke world x makes formula A true". Independently, a purely syntactic cut-free calculus for (proof-search in) BiInt was given by Postniece (previously Buisman) and Goré [5, 13] by combining a "refutation" calculus and a "provability" calculus. However, their cut-elimination result is obtained indirectly via a semantic argument which shows that the cut-free fragment of their calculus is complete with respect to the Kripke semantics of BiInt. Thus the only truly cut-free calculus for BiInt appears to be the display calculus due to Goré [12] (see [30] for a variation).

Although display calculi were not designed for automated proof-search there is a surprising lack of interest in the study of proof search for display logics: the only exceptions are the works of Wansing [29] and Restall [23]. The main difficulty in using display calculi for proof search are the invertible structural display postulate rules which are at the heart of display calculi. Although these rules guarantee the display property, they allow "pointless" shuffling of structures and easily lead to non-termination of proof search if applied naively. Another issue is the presence of explicit contraction and weakening rules in display calculi which are couched in terms of structures rather than formulae. Replacing these rules with ones based on formulae can break one of the conditions for a display calculus, namely, the (C6/C7) condition that "each rule is closed under simultaneous substitution of arbitrary structures for congruent parameters" [18]. Absorbing them completely to obtain a "contraction-free" calculus is thus not an obvious step.

Here, we present two sequent-like calculi LBiInt_1 and LBiInt_2 for bi-intuitionistic logic which sit somewhere in between display calculi and traditional sequent calculi in terms of cut-elimination and proof-search.

LBiInt_1: The calculus LBiInt_1 shares some features of display calculi, in that it has certain structural rules that allow shuffling of structures in a sequent, akin to the display postulates used in display calculi to display a formula nested in a structure. The syntactic judgments in LBiInt_1 can be seen as a tree of (traditional) sequents, and the structural rules can be used to "display" a sequent by bringing it to the root of an equivalent tree. The logical rules of LBiInt_1 are similar to those in Gentzen's traditional sequent calculus, as they apply only to the topmost sequent in the tree of sequents. The virtue of LBiInt_1 is twofold: its contraction and weakening rules can be restricted to formulae while its purely syntactic cut-elimination proof is simple and very similar to the cut-elimination proof for display calculi.

LBiInt$_2$: The calculus LBiInt$_2$ is a refinement of LBiInt$_1$ and is obtained by absorbing all the structural rules of LBiInt$_1$ into the logical rules. The calculus LBiInt$_2$ is easily shown to be sound, since its rules are derivable in LBiInt$_1$. But from a proof-search perspective, we are able to associate a terminating and systematic backward proof-search strategy for applying the rules of LBiInt$_2$. However, we currently do not have a direct syntactic proof of completeness of LBiInt$_2$ with respect to LBiInt$_1$. Instead, completeness of LBiInt$_2$ is shown by an encoding of the calculus GBiInt [5], which is known to be sound and complete for bi-intuitionistic logic. The translation is natural and shows an interesting duality between GBiInt and LBiInt$_2$. It also gives a first simple proof theoretic account of the proof search strategy associated with GBiInt (which is largely semantically motivated).

Our methodology is to use structures (called nested sequents) which are similar to the structures in display calculi but which are more restricted than those used in display calculi. In particular, not all the display structural connectives used in Goré's calculus [12] are allowed and certain display postulates are missing. The idea is to get as close as possible to sequent calculus, because then we may be able to use the standard saturation techniques for proof search common in sequent calculus. Since our calculi are not display calculi, Belnap's general cut elimination theorem [2] cannot be used directly to prove cut elimination for our calculi. One way of showing cut elimination for LBiInt$_1$ would be an indirect proof via a detour through display calculus. That is, one first designs a corresponding display system for LBiInt$_1$ for which Belnap's cut elimination theorem can be used, e.g., by modifying Goré's calculus to work with a more restricted form of structures, and then showing that the cut free proofs of this display system can be mapped to cut-free proofs of LBiInt$_1$. We show here a simple and direct cut elimination proof, without detour through display calculus and Belnap's theorem, by using a certain proof substitution technique, which is very similar to Belnap's original cut elimination proof. We believe that this cut elimination proof can be extended to other logics which contain pairs of adjoint connectives like classical modal (tense) logics such as $KtS4$ and its cousins. If so, then there is a possibility of obtaining general characterisations of cut admissibility like those of Belnap's for these logics.

Outline of the paper Sections 2-4 present the calculus LBiInt$_1$ and its meta theory via theorems on cut elimination, soundness and completeness. While the structural rules in LBiInt$_1$ are somewhat more restrictive than display calculi, and hence reduce slightly the non-determinism arising from the structural rules of display calculi, they still pose some difficulty in proof search. In Section 5, we present a restricted version of LBiInt$_1$, called LBiInt$_2$, in which all the structural rules are omitted and are instead absorbed into introduction rules. We give a terminating proof search strategy for LBiInt$_2$. The idea behind backward proof search for LBiInt$_2$ is that the introduction rules for implication and subtraction can be used to 'suspend'

Identity and cut:

$$\frac{}{X, A \vdash A, Y}\ id \qquad \frac{X_1 \vdash Y_1, A \quad A, X_2 \vdash Y_2}{X_1, X_2 \vdash Y_1, Y_2}\ cut$$

Structural rules:

$$\frac{X \vdash Y}{X, A \vdash Y}\ w_L \qquad \frac{X \vdash Y}{X \vdash A, Y}\ w_R \qquad \frac{X, A, A \vdash Y}{X, A \vdash Y}\ c_L \qquad \frac{X \vdash A, A, Y}{X \vdash A, Y}\ c_R$$

$$\frac{(X_1 < Y_1), X_2 \vdash Y_2}{X_1, X_2 \vdash Y_1, Y_2}\ s_L \qquad \frac{X_1 \vdash Y_1, (X_2 > Y_2)}{X_1, X_2 \vdash Y_1, Y_2}\ s_R$$

$$\frac{X_2 \vdash Y_2, Y_1}{X_1, (X_2 < Y_2) \vdash Y_1}\ < \qquad \frac{X_1, X_2 \vdash Y_2}{X_1 \vdash Y_1, (X_2 > Y_2)}\ >$$

Logical rules:

$$\frac{X, B_i \vdash Y}{X, B_1 \wedge B_2 \vdash Y}\ \wedge_L\ i \in \{1, 2\} \qquad \frac{X \vdash A, Y \quad X \vdash B, Y}{X \vdash A \wedge B, Y}\ \wedge_R$$

$$\frac{X, A \vdash Y \quad X, B \vdash Y}{X, A \vee B \vdash Y}\ \vee_L \qquad \frac{X \vdash B_i, Y}{X \vdash B_1 \vee B_2, Y}\ \vee_R\ i \in \{1, 2\}$$

$$\frac{X \vdash A, Y \quad X, B \vdash Y}{X, A \to B \vdash Y}\ \to_L \qquad \frac{X, A \vdash B}{X \vdash Y, A \to B}\ \to_R$$

$$\frac{A \vdash B, Y}{X, A \prec B \vdash Y}\ \prec_L \qquad \frac{X \vdash A, Y \quad X, B \vdash Y}{X \vdash A \prec B, Y}\ \prec_R$$

Figure 1. LBiInt$_1$: a sequent calculus for bi-intuitionistic logic

proof search of a (top-level) sequent and to 'restart' it at a later stage. Such restart rules are already known in the literature, but as far as we are aware, our work is the first time they have been given a purely proof-theoretic setting. In the same section we also show that we can encode the sound and complete calculus GBiInt [5] into LBiInt$_2$, thereby giving an indirect proof of the completeness of LBiInt$_2$. Section 6 discusses related and future work. An extended version of the paper with more details will be made available on the web.

2 System LBiInt$_1$

Formulas of bi-intuitionistic logic are given by the following grammar:

$$A := p \mid A \to A \mid A \prec A \mid A \wedge A \mid A \vee A.$$

Negative and positive structures [1] are expressions generated, respectively, from the following grammars:

$$N := \emptyset \mid A \mid (N, N) \mid N < P \qquad\qquad P := \emptyset \mid A \mid (P, P) \mid N > P.$$

The structural (comma) connective "," is associative and commutative and $\emptyset$ is its unit. We always consider structures modulo these equivalences.

A *sequent* is an expression of the form $X \vdash Y$, where X is a negative structure and Y is a positive structure. To reduce parentheses, we assume that the structural connective "," binds tighter than $>$ and $<$. Thus, we write $X, Y > Z$ to mean $(X, Y) > Z$.

A *context* is a structure with a hole, denoted with $Z[]$. We write $Z[X]$ to denote a structure resulting from filling the hole in the context $Z[]$ with the structure X. Note that such a replacement does not always give a legal structure. For example, if $Z[]$ is $[] < p$ and X is $q > r$, then $Z[X] = (q > r) < p$ is not a structure since we have a positive structure to the left of $<$. A k-hole context is a context with k holes. Given a k-hole context $Z[\cdots]$ we write $Z[X^k]$ to stand for the structure obtained from $Z[\cdots]$ by replacing each hole with an occurrence of the structure X. An *anti-positive context* is a context $Z[]$ such that $Z[X]$ is a negative structure for every positive structure X. An *iso-positive context* is a context $Z[]$ such that $Z[X]$ is a positive structure for every positive structure X. Likewise, an *anti-negative context* is a context $Z[]$ such that $Z[X]$ is a positive structure for every negative structure X, and $Z[]$ is an *iso-negative context* if $Z[X]$ is a negative structure for every negative structure X. These definitions extend straightforwardly to multiple-hole contexts.

The structural connective comma "," is a proxy for conjunction (on the left) and disjunction (on the right), while $<$ is a proxy for exclusion and $>$ is a proxy for implication as we shall show later. Note that, unlike display calculi, $<$ appears only on the left of turnstile at the top level while $>$ appears only on the right at the top level, thus there is no overloading of these structural connectives.

Our first sequent system LBiInt_1 for bi-intuitionistic logic is given in Figure 1. The introduction rules for the logical connectives are the standard ones. The logical rules are non-invertible, since they lose structures or formulas going upwards. Since we have contraction and weakening, on both sides of the sequent, it is possible to formulate invertible logical rules by implicit contraction, as we shall see later. LBiInt_1 is very similar to the display calculus for bi-intuitionistic logic of Goré [11], but with some differences:

- Sequents are of a more restricted form than in display calculus. For example, we do not allow sequents of the form $X \vdash A < B$.

---

[1] Only recently, we have realised that this may not be the simplest class of structures needed for our calculus. That is, the structural connective $<$ may not be needed, as we can overload the connective $>$, by interpreting it differently in positive and negative contexts, just as the structural connective ',' can be overloaded to represent both disjunction and conjunction in different contexts. The current notation was chosen to conform with Goré's display system, where $<$ is used as a structural proxy for $-\!\!<$.

- The contraction and the weakening rules are applicable to formulae only, not structures in general like in display calculi. But we shall see that the general contraction and weakening are derivable from the "atomic" ones in LBiInt_1, which is not the case for Goré's system.
- The structural rules s_L and s_R are more general than the display postulates in display logic. These rules are derivable in Goré's system, but one needs to use contraction and weakening on structures.

As a consequence of these differences, cut elimination for LBiInt_1 does not necessarily follow from cut elimination for its display calculus counterpart. However, it may be possible to modify Goré's system in such a way that there is a mapping between the cut free proofs of both LBiInt_1 and the modified system. We leave the details of such a connection to future work.

The following two propositions state the admissibility of the general contraction and weakening rules. These can be proved by using the structural rules s_L, s_R, $>$ and $<$.

PROPOSITION 1. **Admissibility of general contraction.** *The two contraction rules shown below are cut-free admissible in $LBiInt_1$:*

$$\frac{X, Y, Y \vdash Z}{X, Y \vdash Z}\ gc_L \qquad \frac{X \vdash Y, Y, Z}{X \vdash Y, Z}\ gc_R$$

Proof. We prove this simultaneously by induction on the size of Y. We show a derivation of the gc_L rule; the case for gc_R is symmetric. The non-trivial case is when $Y = Y_1 < Y_2$. We show that in this case, the contraction rule can be reduced to contractions on smaller structures, which therefore are admissible by the induction hypothesis:

$$\dfrac{\dfrac{\dfrac{\dfrac{\dfrac{\dfrac{\dfrac{X, (Y_1 < Y_2), (Y_1 < Y_2) \vdash Z}{(Y_1 < Y_2), (Y_1 < Y_2) \vdash X > Z}\ >}{(Y_1 < Y_2), Y_1 \vdash Y_2, (X > Z)}\ s_L}{Y_1, Y_1 \vdash Y_2, Y_2, (X > Z)}\ s_L}{Y_1, Y_1 \vdash Y_2, (X > Z)}\ gc_R}{Y_1 \vdash Y_2, (X > Z)}\ gc_L}{Y_1 < Y_2 \vdash X > Z}\ <}{X, (Y_1 < Y_2) \vdash Z}\ s_R$$

■

PROPOSITION 2. **Admissibility of general weakening.**
The two weakening rules below are cut-free admissible in $LBiInt_1$:

$$\frac{X \vdash Z}{X, Y \vdash Z}\ gw_L \qquad \frac{X \vdash Z}{X \vdash Y, Z}\ gw_R$$

PROPOSITION 3. *The id rule can be restricted to the atomic form:*

$$\frac{}{p \vdash p}\ id$$

So from now on, we assume that all *id* rules are of the atomic form.

3 Cut elimination

Although the proof system LBiInt_1 shares some similarity with traditional Gentzen's systems, cut elimination for LBiInt_1 as presented here follows a different technique from the standard cut elimination technique for sequent calculus. In particular, when the cut formula is not principal in either one of the premises of the cut rule, no cut reductions are required in our cut elimination proof. Instead, the structural rules s_L and s_R allow us to carry the context of one premise of the cut to its other premise resulting in a "proof substitution" akin to the normalisation proofs in natural deduction. Apart from Belnap's cut-elimination proof for display logic, the closest technique we know of is the cut elimination proof for classical logic in a proof system using *deep inference* [3].

For example, suppose π_1 is the cut-free derivation below where the occurrence of p in the root sequent participates in n instances of *id* in the leaves of π_1:

$$\begin{array}{c} \dfrac{}{p \vdash p}\ id \quad \cdots \quad \dfrac{}{p \vdash p}\ id \\ \vdots \\ X_1 \vdash Y_1, p \end{array}$$

Let ξ be the derivation below which ends in an instance of cut on p:

$$\dfrac{\begin{array}{c}\pi_1 \\ X_1 \vdash Y_1, p\end{array} \quad \begin{array}{c}\pi_2 \\ p, X_2 \vdash Y_2\end{array}}{X_1, X_2 \vdash Y_1, Y_2}\ cut$$

Then a cut free derivation for $X_1, X_2 \vdash Y_1, Y_2$ can be obtained by replacing the parametric ancestors of the cut formula p in π_1 with the structure $(X_2 > Y_2)$ and replacing the leaves of π_1, where the cut formula p is used, with the derivation π_2. This cut-free derivation is schematically presented as follows:

$$\begin{array}{c} \dfrac{\begin{array}{c}\pi_2 \\ p, X_2 \vdash Y_2\end{array}}{p \vdash (X_2 > Y_2)} > \quad \cdots \quad \dfrac{\begin{array}{c}\pi_2 \\ p, X_2 \vdash Y_2\end{array}}{p \vdash (X_2 > Y_2)} > \\ \vdots \\ \dfrac{X_1 \vdash Y_1, (X_2 > Y_2)}{X_1, X_2 \vdash Y_1, Y_2}\ s_R \end{array}$$

The reductions for the cases where the cut formula is non-atomic follow essentially the same idea. That is, we substitute the cut formula on one premise of the cut rule with the context of the other premise, and expand this context when the cut formula is used. The only difference is that in the case of non-atomic cut formula, we need to produce extra cuts to make this substitution work. But all the cuts produced are of smaller size, therefore the whole process terminates.

In the following, we write $|A|$ for the *size* of the formula A: the number of logical operators appearing in A. In an instance of a cut rule

$$\frac{X_1 \vdash Y_1, A \quad A, X_2 \vdash Y_2}{X_1, X_2 \vdash Y_1, Y_2}\ cut$$

the formula A is called the *cut formula* of the cut instance. The *cut-rank* of the cut instance is $|A|$. Given a derivation π, we denote with $mc(\pi)$ the maximum of the cut-ranks in π. If there are no cuts in π then $mc(\pi) = 0$.

Lemma 4 states the proof substitutions needed to eliminate atomic cuts.

LEMMA 4. *Suppose $p, X \vdash Y$ is cut-free derivable for some fixed p, X and Y. Then for any k-hole anti-positive context $Z_1[\cdots]$ and any l-hole iso-positive context $Z_2[\cdots]$, if $Z_1[p^k] \vdash Z_2[p^l]$ is cut-free derivable, then $Z_1[(X > Y)^k] \vdash Z_2[(X > Y)^l]$ is cut-free derivable.*

Proof. Let π be a cut-free derivation of $p, X \vdash Y$ and let ξ be a cut-free derivation of $Z_1[p^k] \vdash Z_2[p^l]$. We construct a cut-free derivation ξ' of $Z_1[(X > Y)^k] \vdash Z_2[(X > Y)^l]$ by induction on the height of ξ. Most cases follow straightforwardly from the induction hypothesis. The only non-trivial case is when p is active in the derivation, i.e., when ξ ends with an *id* rule or a contraction rule applied to an occurence of p to be substituted for:

- Suppose ξ is

$$\frac{}{Z_1'[p^k], p \vdash p, Z_2'[p^{l-1}]}\ id$$

  Note that the p immediately to the left of the turnstile cannot be part of the p^k by the restrictions on the context $Z_1[\cdots]$. The derivation ξ' is then constructed as follows, where we use double lines to abbreviate derivations:

$$\dfrac{\dfrac{\overset{\pi}{p, X \vdash Y}}{p \vdash (X > Y)}\ >}{Z_1'[(X > Y)^k], p \vdash (X > Y), Z_2'[(X > Y)^{l-1}]}\ gw_R; gw_L$$

- Suppose ξ is

$$\frac{\overset{\xi_1}{Z_1[p^k] \vdash p, p, Z_2'[p^{l-1}]}}{Z_1[p^k] \vdash p, Z_2'[p^{l-1}]}\ c_R$$

  By induction hypothesis, we have a cut-free derivation ξ_1' of

$$Z_1[(X > Y)^k] \vdash (X > Y), (X > Y), Z_2'[(X > Y)^{l-1}].$$

The derivation ξ' is then constructed as follows:

$$\dfrac{\begin{array}{c}\xi_1' \\ Z_1[(X > Y)^k] \vdash (X > Y), (X > Y), Z_2'[(X > Y)^{l-1}]\end{array}}{Z_1[(X > Y)^k] \vdash (X > Y), Z_2'[(X > Y)^{l-1}]}\ gc_R$$

Note that gw_R and gc_R and gw_L are cut-free derivable in LBiInt_1 by Proposition 1 and Proposition 2. ■

Lemmas 5-9 state the proof substitutions needed for non-atomic cuts.

LEMMA 5. *Let ξ be a derivation of*

$$Z_1[(A_1 \vee A_2)^k] \vdash Z_2[(A_1 \vee A_2)^l]$$

for some k-hole iso-negative context $Z_1[\cdots]$ and l-hole anti-negative context $Z_2[\cdots]$, such that $mc(\xi) < |A_1 \vee A_2|$. Let π_i be a derivation of $X \vdash Y, A_i$, for some $i \in \{1, 2\}$, such that $mc(\pi_i) < |A_1 \vee A_2|$. Then there is a derivation ξ' with $mc(\xi') < |A_1 \vee A_2|$ of

$$Z_1[(X < Y)^k] \vdash Z_2[(X < Y)^l].$$

Proof. By induction on the height of ξ. In the following, we let $A = A_1 \vee A_2$. Most cases follow straightforwardly from the induction hypothesis. The only interesting case is when a left-rule is applied to an occurence of $A_1 \vee A_2$ which is to be replaced by $X < Y$. That is, ξ is

$$\dfrac{\begin{array}{c}\xi_1 \\ Z_1'[A^{k-1}], A_1 \vdash Z_2[A^l]\end{array} \quad \begin{array}{c}\xi_2 \\ Z_1'[A^{k-1}], A_2 \vdash Z_2[A^l]\end{array}}{Z_1'[A^{k-1}], A_1 \vee A_2 \vdash Z_2[A^l]}\ \vee_L$$

By induction hypothesis, we have a derivation ξ_i', for each $i \in \{1, 2\}$, of

$$Z_1'[(X < Y)^{k-1}], A_i \vdash Z_2[(X < Y)^l]$$

with $mc(\xi_i') < |A_1 \vee A_2|$. The derivation ξ' is then constructed as follows:

$$\dfrac{\dfrac{\begin{array}{c}\pi_i \\ X \vdash Y, A_i\end{array}}{X < Y \vdash A_i}\ < \qquad \begin{array}{c}\xi_i' \\ Z_1'[(X < Y)^{k-1}], A_i \vdash Z_2[(X < Y)^l]\end{array}}{Z_1'[(X < Y)^{k-1}], (X < Y) \vdash Z_2[(X < Y)^l]}\ cut$$

■

LEMMA 6. *Let ξ be a derivation of*

$$Z_1[(A_1 \wedge A_2)^k] \vdash Z_2[(A_1 \wedge A_2)^l]$$

for some k-hole iso-negative context $Z_1[\cdots]$ and l-hole anti-negative context $Z_2[\cdots]$ with $mc(\xi) < |A_1 \wedge A_2|$. Let π_1 be a derivation of $X \vdash Y, A_1$ and *let* π_2 *be a derivation of* $X \vdash Y, A_2$ *with* $mc(\pi_1) < |A_1 \wedge A_2|$ *and* $mc(\pi_2) < |A_1 \wedge A_2|$. *Then there is a derivation* ξ' *with* $mc(\xi') < |A_1 \wedge A_2|$ *of*

$$Z_1[(X < Y)^k] \vdash Z_2[(X < Y)^l].$$

Proof. Analogous to the proof of Lemma 5. ∎

LEMMA 7. *Let* ξ *be a derivation of*

$$Z_1[(A \to B)^k] \vdash Z_2[(A \to B)^l]$$

for some k*-hole iso-negative context* $Z_1[\cdots]$ *and* l*-hole anti-negative context* $Z_2[\cdots]$ *with* $mc(\xi) < |A \to B|$. *Let* π *be a derivation of* $X, A \vdash B$ *with* $mc(\pi) < |A \to B|$. *Then there is a derivation* ξ' *with* $mc(\xi') < |A \to B|$ *of*

$$Z_1[X^k] \vdash Z_2[X^l].$$

Proof. By induction on the height of ξ. As in the previous lemmas, the non-trivial case is when ξ ends with $\to_L$ on $A \to B$:

$$\dfrac{\overset{\xi_1}{Z_1'[(A \to B)^{k-1}] \vdash A, Z_2[(A \to B)^l]} \quad \overset{\xi_2}{Z_1'[(A \to B)^{k-1}], B \vdash Z_2[(A \to B)^l]}}{Z_1'[(A \to B)^{k-1}], A \to B \vdash Z_2[(A \to B)^l]} \to_L$$

By induction hypothesis, we have derivations ξ_1' and ξ_2' respectively of the sequents below where $mc(\xi_1') < |A \to B|$ and $mc(\xi_2') < |A \to B|$:

$$Z_1'[X^{k-1}] \vdash A, Z_2[X^l] \qquad\qquad Z_1'[X^{k-1}], B \vdash Z_2[X^l]$$

In the following, we let V_1 denote $Z_1'[X^{k-1}]$ and V_2 denote $Z_2[X^l]$. The derivation ξ' is constructed as follows:

$$\dfrac{\dfrac{\overset{\xi_1'}{V_1 \vdash A, V_2} \quad \dfrac{\overset{\pi}{X, A \vdash B} \quad \overset{\xi_2'}{V_1, B \vdash V_2}}{V_1, A, X \vdash V_2}\, cut}{V_1, V_1, X \vdash V_2, V_2}\, cut}{V_1, X \vdash V_2}\, gc_L; gc_R$$

∎

LEMMA 8. *Let* ξ *be a derivation of*

$$Z_1[(A {-\!\!<} B)^k] \vdash Z_2[(A {-\!\!<} B)^l]$$

for some k*-hole iso-negative context* $Z_1[\cdots]$ *and* l*-hole anti-negative context* $Z_2[\cdots]$ *with* $mc(\xi) < |A {-\!\!<} B|$. *Let* π_1 *be a derivation of* $X \vdash Y, A$ *and let* π_2 *be a derivation of* $X, B \vdash Y$ *with* $mc(\pi_1) < |A {-\!\!<} B|$ *and* $mc(\pi_2) < |A {-\!\!<} B|$. *Then there is a derivation* ξ' *with* $mc(\xi') < |A {-\!\!<} B|$ *of*

$$Z_1[(X < Y)^k] \vdash Z_2[(X < Y)^l].$$

Proof. The non-trivial case is when ξ ends with $-\!\!<_L$ on $A-\!\!<B$:

$$\dfrac{\begin{array}{c}\xi_1\\ A \vdash B, Z_2[(A-\!\!<B)^l]\end{array}}{Z_1'[(A-\!\!<B)^{k-1}], A-\!\!<B \vdash Z_2[(A-\!\!<B)^l]}\ -\!\!<_L$$

By induction hypothesis, we have a derivation ξ_1' of

$$A \vdash B, Z_2[(X < Y)^l]$$

with $mc(\xi_1') < |A-\!\!<B|$. Let V denote the structure $Z_2[(X < Y)^l]$. Then ξ' is constructed as follows:

$$\dfrac{\dfrac{\dfrac{\dfrac{\begin{array}{c}\pi_1\\ X \vdash Y, A\end{array} \quad \dfrac{\begin{array}{c}\xi_1'\\ A \vdash B, V\end{array} \quad \begin{array}{c}\pi_2\\ X, B \vdash Y\end{array}}{A, X \vdash Y, V}\ cut}{X, X \vdash Y, Y, V}\ cut}{\overline{\overline{X \vdash Y, V}}\ gc_L; gc_R}}{X < Y \vdash V}\ <}{Z_1'[(X < Y)^{k-1}], X < Y \vdash V}\ gw_L$$

■

LEMMA 9. *Let ξ be a derivation of $Z_1[A^k] \vdash Z_2[A^l]$ where A is a non-atomic formula, $Z_1[\cdots]$ is a k-hole anti-positive context, $Z_2[\cdots]$ is an l-hole iso-positive context, and $mc(\xi) < |A|$. Let π be a derivation of $A, X \vdash Y$ with $mc(\pi) < |A|$. Then there is a derivation ξ' with $mc(\xi') < |A|$ of $Z_1[(X > Y)^k] \vdash Z_2[(X > Y)^l]$.*

Proof. By induction on the height of ξ and case analysis on A. The non-trivial case is when ξ ends with a right-introduction rule on A. That is, in this case, we have $Z_2[A^l] = (Z_2'[A^{l-1}], A)$ for some iso-positive context $Z_2'[\cdots]$. We distinguish several cases depending on A. We show here the cases where A is either a disjunction $C \vee D$, or an implication $C \to D$.

- Suppose $A = C \vee D$ and ξ is the following derivation:

$$\dfrac{\begin{array}{c}\xi_1\\ Z_1[(C \vee D)^k] \vdash Z_2'[(C \vee D)^{l-1}], C\end{array}}{Z_1[(C \vee D)^k] \vdash Z_2'[(C \vee D)^{l-1}], C \vee D}\ \vee_R$$

By induction hypothesis, we have a derivation ξ_1' of

$$Z_1[(X > Y)^k] \vdash Z_2'[(X > Y)^{l-1}], C$$

such that $mc(\xi_1') < |C \vee D|$. Let $W_1 = Z_1[(X < Y)^k]$ and let $W_2 = Z_2[(X > Y)^{l-1}]$. Applying Lemma 5 to π and ξ_1', we obtain a derivation θ of

$$(W_1 < W_2), X \vdash Y$$

such that $mc(\theta) < |C \vee D|$. The derivation ξ' is then constructed as follows:

$$\dfrac{\dfrac{\begin{array}{c}\xi'_1\\ (W_1 < W_2), X \vdash Y\end{array}}{W_1 < W_2 \vdash X > Y}\,>}{W_1 \vdash W_2, (X > Y)}\,s_L$$

Clearly, $mc(\xi') < |C \vee D|$.

- Suppose $A = C \to D$ and ξ is

$$\dfrac{\begin{array}{c}\xi_1\\ Z_1[(C \to D)^k], C \vdash D\end{array}}{Z_1[(C \to D)^k] \vdash Z'_2[(C \to D)^{l-1}], C \to D}\,\to_R$$

By induction hypothesis, we have a derivation ξ'_1 of

$$Z_1[(X > Y)^k], C \vdash D$$

Then the derivation ξ' is constructed as follows:

$$\dfrac{\begin{array}{c}\theta\\ Z_1[(X > Y)^k], X \vdash Y\end{array}}{Z_1[(X > Y)^k] \vdash Z_2[(X > Y)^{l-1}], (X > Y)}\,s_L$$

where θ is obtained by applying Lemma 7 to π and ξ'_1.

The other cases are treated analogously, using Lemmas 6 and Lemma 8. ■

Finally, cut elimination is proved by simple proof substitutions, the construction of which is given by the preceding lemmas.

THEOREM 10. *If $X \vdash Y$ is LBiInt$_1$-derivable then it is also cut-free derivable.*

Proof. As typical in cut elimination proofs, we remove topmost cuts in succession. Let π be a derivation of LBiInt$_1$ with a topmost cut instance

$$\dfrac{\begin{array}{c}\pi_1\\ X_1 \vdash Y_1, A\end{array} \quad \begin{array}{c}\pi_2\\ X_2, A \vdash Y_2\end{array}}{X_1, X_2 \vdash Y_1, Y_2}\,cut$$

Note that π_1 and π_2 are both cut-free since this is a topmost instance in π. We use induction on the size of A to eliminate this topmost instance of cut.

If A is an atomic formula p then the cut free derivation is constructed as follows where ξ is obtained from applying Lemma 4 to π_1 and π_2:

$$\dfrac{\begin{array}{c}\xi\\ X_1 \vdash Y_1, (X_2 > Y_2)\end{array}}{X_1, X_2 \vdash Y_1, Y_2}\,s_R$$

If A is non-atomic, using Lemma 9 we get the following derivation θ:

$$\dfrac{\begin{array}{c}\xi\\ X_1 \vdash Y_1, (X_2 > Y_2)\end{array}}{X_1, X_2 \vdash Y_1, Y_2}\, s_R$$

We have $mc(\theta) < |A|$ by Lemma 9, therefore by induction hypothesis, we can remove all the cuts in θ to get a cut-free derivation of $X_1, X_2 \vdash Y_1, Y_2$. ∎

4 Soundness and completeness of LBiInt_1

To prove soundness, we first define an interpretation of sequents as formulae as shown below using two extra logical constants $\top$ and $\bot$ which were not part of our original language for formulae. But it is easy to show that the system LBiInt_1 can be extended to cover these constants in the obvious way.

DEFINITION 11. The two mutual-recursively functions τ_N and τ_P respectively translate a negative and positive structure into a BiInt-formula:

$$\begin{array}{lcllcl}
\tau_N(\emptyset) & = & \top & \tau_P(\emptyset) & = & \bot \\
\tau_N(A) & = & A & \tau_P(A) & = & A \\
\tau_N(X, Y) & = & \tau_N(X) \wedge \tau_N(Y) & \tau_P(X, Y) & = & \tau_P(X) \vee \tau_P(Y) \\
\tau_N(X < Y) & = & \tau_N(X) \multimap\!\!< \tau_P(Y) & \tau_P(X > Y) & = & \tau_N(X) \to \tau_P(Y)
\end{array}$$

We assume the reader is familiar with the Kripke semantics [22, 12] for BiInt using a binary reflexive and transitive relation $\leq$, which extends the usual Kripke semantics for Int with an extra clause for exclusion given below:

$$w \Vdash A \multimap\!\!< B \quad \text{iff} \quad \exists v \leq w . v \Vdash A \text{ and } v \nVdash B.$$

THEOREM 12. **Soundness.** *Every LBiInt_1-derivable BiInt formula is valid.*

Proof. We show that for every rule ρ of LBiInt_1

$$\dfrac{X_1 \vdash Y_1 \quad \cdots \quad X_n \vdash Y_n}{X \vdash Y}\, \rho$$

the following holds: if for every $i \in \{1, \ldots, n\}$, the formula $\tau_N(X_i) \to \tau_P(Y_i)$ is valid then the formula $\tau_N(X) \to \tau_P(Y)$ is valid. Since the formula-translation $(\tau_N(X) \wedge A) \to (A \vee \tau_P(Y))$ of the *id* rule is obviously valid, it then follows that every formula derivable in LBiInt_1 is also valid.

For all the rules of LBiInt_1, except $<$ and $\multimap\!\!<_L$, we can show the stronger statement that the following formula is valid:

$$[(\tau_N(X_1) \to \tau_P(Y_1)) \wedge \cdots \wedge (\tau_N(X_n) \to \tau_P(Y_n))] \to (\tau_N(X) \to \tau_P(Y)).$$

Soundness of $<$ and $\multimap\!\!<_L$ are shown in the standard way, by reasoning about the forcing relation $\Vdash$ and the reflexive and transitive relation $\leq$. ∎

$$\{X\} = \{A \mid X = (A, Y) \text{ for some } A \text{ and } Y\}$$

$$\frac{}{X, A \vdash A, Y}\,id \qquad \frac{\{X_1\}, X_2 \vdash Y_2}{X_1 \vdash Y_1, (X_2 > Y_2)} > \{X_1\} \not\subseteq \{X_2\}$$

$$\frac{X_2 \vdash Y_2, \{Y_1\}}{X_1, (X_2 < Y_2) \vdash Y_1} < \{Y_1\} \not\subseteq \{Y_2\}$$

$$\frac{X, B_1 \wedge B_2, B_i \vdash Y}{X, B_1 \wedge B_2 \vdash Y}\wedge_L \qquad \frac{X \vdash A \wedge B, A, Y \quad X \vdash A \wedge B, B, Y}{X \vdash A \wedge B, Y}\wedge_R$$

$$\frac{X, A \vee B, A \vdash Y \quad X, A \vee B, B \vdash Y}{X, A \vee B \vdash Y}\vee_L \qquad \frac{X \vdash B_1 \vee B_2, B_i, Y}{X \vdash B_1 \vee B_2, Y}\vee_R$$

$$\frac{X, A \to B \vdash A, Y \quad X, A \to B, B \vdash Y}{X, A \to B \vdash Y}\to_L \qquad \frac{X \vdash Y, A \to B, B}{X \vdash Y, A \to B}\to_{R1}$$

$$\frac{X, A \prec B, A \vdash Y}{X, A \prec B \vdash Y}\prec_{L1} \qquad \frac{X \vdash A, A \prec B, Y \quad X, B \vdash A \prec B, Y}{X \vdash A \prec B, Y}\prec_R$$

$$\frac{A \vdash B, \{Y\}, (X, A \prec B > Y)}{X, A \prec B \vdash Y}\prec_{L2} \qquad \frac{(X < Y, A \to B), \{X\}, A \vdash B}{X \vdash Y, A \to B}\to_{R2}$$

Figure 2. System LBiInt_2

Completeness is shown by embedding Rauszer's sequent calculus G1 [21] for BiInt into LBiInt_1. The calculus G1 contains the cut rule, and is shown to be complete by Rauszer [21]. The encoding of G1 into LBiInt_1 is obvious since all the rules of G1 are easily derivable from the rules of LBiInt_1.

THEOREM 13. **Completeness.** *Every BiInt-valid formula is $LBiInt_1$-derivable.*

5 Proof search

System LBiInt_1 is not suitable for proof search, since the structural rules s_L, s_R, $<$ and $>$ can easily lead to non-termination if applied naively. In addition, we also have the usual problems with the contraction rules since they can be applied *ad infinitum*. We now present a refined version of LBiInt_1, called LBiInt_2, in which all the structural rules, except for $>$ and $<$, are absorbed into logical rules. The resulting calculus, for the intuitionistic fragment, resembles contraction-free calculi for the traditional Gentzen systems for intuitionistic logic, e.g., the system **G3i** in [26]. The underlying idea behind LBiInt_2 is that the right-introduction rule for $\to$ and the left introduction rule for $\prec$ act as an instruction to store the current state (of proof search), and the rules $>$ and $<$ act as an instruction to restart previously stored computation states.

The inference rules for LBiInt_2 are given in Figure 2 using the notation

$\{X\}$ to denote the *set of formulae* that appear at the top-level of X:

$$\{X\} = \{A \mid X = (A, Y) \text{ for some } A \text{ and } Y\}.$$

Intuitively, the set $\{X\}$ denotes X with all the substructures of the form $Y > Z$ or $Y < Z$ removed. For example, if X is $(A, B, (C > D))$, then $\{X\}$ is the set $\{A, B\}$. The right introduction rule for $\to$ splits into two rules: $\to_{R1}$ and $\to_{R2}$. The $\to_{R1}$ rule is strictly speaking not necessary as it can be derived using $\to_{R2}$ and $<$. However, it is useful in our proof search strategy which relies on a *saturation* process on sequents, as we shall see later. The rule $\to_{R2}$ incorporates some features of the structural rule s_L. The left introduction rule for $-\!\!<$ splits also into two rules with roles symmetric to those for $\to$.

5.1 A terminating proof search strategy

We classify the rules of LBiInt_2 into three groups:

Static Rules: $= \{id, \wedge_L, \wedge_R, \vee_L, \vee_R, \to_L, -\!\!<_R, -\!\!<_{L1}, \to_{R1}\}$;

Jump Rules: $= \{-\!\!<_{L2}, \to_{R2}\}$; and

Return Rules: $= \{<, >\}$.

We call a sequent of static rule applications a *saturation*.

DEFINITION 14. A sequent $X \vdash Y$ is *saturated* iff it satisfies 1-8, and is *strongly saturated* iff it additionally satisfies 9:

1. $\{X\} \cap \{Y\} = \emptyset$
2. If $A \wedge B \in \{X\}$ then $A \in \{X\}$ and $B \in \{X\}$
3. If $A \wedge B \in \{Y\}$ then $A \in \{Y\}$ or $B \in \{Y\}$
4. If $A \vee B \in \{X\}$ then $A \in \{X\}$ or $B \in \{X\}$
5. If $A \vee B \in \{Y\}$ then $A \in \{Y\}$ and $B \in \{Y\}$
6. If $A \to B \in \{X\}$ then $A \in \{Y\}$ or $B \in \{X\}$
7. If $A -\!\!< B \in \{Y\}$ then $A \in \{Y\}$ or $B \in \{X\}$
8. If $A \to B \in \{Y\}$ then $B \in \{Y\}$ If $A -\!\!< B \in \{X\}$ then $A \in \{X\}$
9. If $A \to B \in \{Y\}$ then $A \in \{X\}$ If $A -\!\!< B \in \{X\}$ then $B \in \{Y\}$.

We say that an LBiInt_2 rule ρ is applicable to a sequent $\gamma_0 = X_0 \vdash Y_0$ if for every premise $X_i \vdash Y_i$ of ρ, $\{X_i\} \not\subseteq \{X_0\}$ or $\{Y_i\} \not\subseteq \{Y_0\}$. Thus only jump and return rules are applicable to saturated sequents.

DEFINITION 15 (Proof search strategy).
Function Prove
Input: sequent γ_0
Output: *true* (i.e. γ_0 is derivable) or *false* (i.e. γ_0 is not derivable)

1. If *id* is applicable to γ_0 then return *true*
2. Else if a static rule ρ is applicable to γ_0 then
   (a) Let $\gamma_1, \cdots, \gamma_n$ be the premises of ρ obtained from γ_0
   (b) Return $\bigwedge_{i=1}^{n} Prove(\gamma_i)$
3. Else if $Prove(\gamma_1) = true$ for some premise instance γ_1 obtained from γ_0 by applying $\rho \in \{ -\!\!<_{L2}, \rightarrow_{R2}, <, > \}$ backward then return *true*
4. Else return *false.*

We shall show that the search strategy given in Definition 15 terminates, if given an input sequent with a certain simple structure, which is defined in the following.

DEFINITION 16. A structure is a *flat structure* if it contains no occurrences of the structural connectives $>$ and $<$. We use Γ and Δ to stand for flat structures since flat structures can be viewed as sets of formulae. The set of (positive/negative) *linear structures* is the smallest set of structures that satisfies the following:

1. Every flat structure is a linear structure.
2. If X is a positive linear structure and Γ is a flat structure, then $\Gamma < X$ is a negative linear structure.
3. If X is a negative linear structure and Δ is a flat structure, then $X > \Delta$ is a positive linear structure.
4. If X is a positive (negative) linear structure and Δ is a flat structure, then (X, Δ) is a positive (resp. negative) linear structure.

A sequent $X \vdash Y$ is a *linear sequent* if either X is a flat structure and Y is a positive linear structure, or X is a negative linear structure and Y is a flat structure.

The intuition of Definition 16 is that a linear sequent $X \vdash Y$ can take the form $(X' < Y'), \Gamma \vdash \Delta$, or $\Gamma \vdash \Delta, (X'' > Y'')$, or $\Gamma \vdash \Delta$, where $X' < Y'$ and $X'' > Y''$ store the sequent corresponding to the previous state of computation, and Γ and Δ are sets of formulae.

LEMMA 17. *Let $X \vdash Y$ be a linear sequent. Then for every LBiInt$_2$-derivation π of $X \vdash Y$, every sequent in π is a linear sequent.*

Proof. Given a derivation π of a linear sequent $X \vdash Y$, we show by induction on the length of π that every sequent in π is a linear sequent. This is straightforward by showing that in every rule of LBiInt$_2$, if the conclusion of the rule is a linear sequent, then every premise of the rule is also a linear sequent, which can be verified by inspection of the rules of LBiInt$_2$. ∎

Note that as a consequence of Lemma 17, every sequent that arises during proof search for a linear sequent $X \vdash Y$, using the search procedure given in Definition 15, is a linear sequent.

We now define a translation from linear sequents to linked lists, consisting of nodes that are pairs of sets of formulae, linked by labels marked either R or R^{-1}.

DEFINITION 18.

$$\begin{array}{lcl} list(\Gamma \vdash \Delta) & = & \langle \Gamma, \Delta \rangle \\ list((X' < Y'), \Gamma \vdash \Delta) & = & list(X' \vdash Y') \;\; R \;\; \langle \Gamma, \Delta \rangle \\ list(\Gamma \vdash \Delta, (X'' > Y'')) & = & list(X'' \vdash Y'') \;\; R^{-1} \;\; \langle \Gamma, \Delta \rangle \end{array}$$

We write $length(L)$ to mean the number of nodes in the list L.

COROLLARY 19. *A backward LBiInt$_2$ rule application to a linear sequent $X \vdash Y$ can be viewed as an operation on $list(X \vdash Y)$, where the conclusion (resp. premise) is the list before (resp. after) the operation. The jump rules append a node to the list, and the static rules saturate the end node. The return rules remove a node from the end of the list, and add subformulae to the penultimate node.*

For example, below left is is an instance of $\rightarrow_{R2}$ with the corresponding list structures of the premise and conclusion on the right:

$$\frac{(C < B, A \rightarrow B), C, A \vdash B}{C \vdash B, A \rightarrow B} \rightarrow_{R2} \qquad \frac{\langle \{C\}, \{B, A \rightarrow B\} \rangle \;\; R \;\; \langle \{C, A\}, \{B\} \rangle}{\langle \{C\}, \{B, A \rightarrow B\} \rangle}$$

We now define a metric that we will use in the main termination proof.

DEFINITION 20. The degree of a formula is:

$$\begin{array}{rcl} deg(p) & = & 0 \\ deg(A \wedge B) = deg(A \vee B) & = & max(deg(A), deg(B)) \\ deg(A \rightarrow B) = deg(A -\!\!< B) & = & 1 + max(deg(A), deg(B)). \end{array}$$

The degree of a sequent is:

$$\begin{array}{rcl} deg_L(X \vdash Y) & = & max\{deg(A) \mid A \in \{X\}\} \\ deg_R(X \vdash Y) & = & max\{deg(B) \mid B \in \{Y\}\} \\ deg(X \vdash Y) & = & max(deg_L(X \vdash Y), deg_R(X \vdash Y)). \end{array}$$

Note that only logical connectives contribute to these metrics.

We denote with $sf(A)$ the set of subformulae of A, and

$$sf(\Gamma) = \bigcup_{A \in \Gamma} sf(A)$$

the set of subformulae of Γ. In the following, we assume that the initial input to the search procedure Prove is a linear sequent $\Gamma_0 \vdash \Delta_0$, and we define $m = |sf(\Gamma_0 \cup \Delta_0)|$.

LEMMA 21. *Let $X \vdash Y$ be any sequent encountered during proof search. Using jump rules, $list(X \vdash Y)$ can be extended at most $\mathcal{O}(m^2)$ times.*

Proof. We show that the number of jump rule applications is bounded by $\mathcal{O}(m^2)$.

First, we show that there can be at most m consecutive jumps in the same direction. In the forward case, consider a backwards application of $\rightarrow_{R2}$ with principal formula $A \rightarrow B$. After this application, A will be added to the LHS of the sequent, and remain on the LHS during saturation and forward jumps. Should $A \rightarrow B$ reappear on the RHS, B will be added to the RHS by the $\rightarrow_{R1}$ rule during saturation, so a repeated application of $\rightarrow_{R2}$ to $A \rightarrow B$ will be blocked by the general blocking condition. Thus since the number of $\rightarrow$-formulae is bounded by m and we can only jump on each $\rightarrow$-formula once, there can be at most m consecutive forward jumps. The backward case is symmetric.

We now show that we can switch direction at most m times. Consider a direction switch, e.g., a forward jump using $\rightarrow_{R2}$ followed by a backward jump $-\!\!<_{L2}$ (the other case is symmetric), and any static rule applications in between. Let γ_0 and γ_1 be the conclusion and premise of the $\rightarrow_{R2}$ rule respectively, and let γ_2 and γ_3 be the conclusion and premise of the $-\!\!<_{R2}$ rule respectively, as shown below:

$$\dfrac{\vdots \atop \gamma_3 = C \vdash D, \Delta, ((X < Y, A \rightarrow B), \Gamma, C -\!\!< D > \Delta)}{\gamma_2 \;=\; (X < Y, A \rightarrow B), \Gamma, C -\!\!< D \vdash \Delta}\; -\!\!<_{L2}$$

$$\vdots$$

$$\dfrac{\gamma_1 \;=\; (X < Y, A \rightarrow B), \{X\}, A \vdash B}{\gamma_0 \;=\; X \vdash Y, A \rightarrow B}\; \rightarrow_{R2}$$

$$\vdots$$

Let $d_0 = deg(\gamma_0)$. We will show that $deg(\gamma_3) \leq d_0 - 1$. By inspection of the rules and Definition 20, we have the following:

$$\begin{array}{l}
deg_L(\gamma_1) \leq d_0 \\
deg_R(\gamma_1) \leq d_0 - 1 \\
deg_L(\gamma_2) = deg_L(\gamma_1) \leq d_0 \\
deg_R(\gamma_2) \leq max(deg_L(\gamma_1) - 1, deg_R(\gamma_1)) = d_0 - 1 \\
deg_L(\gamma_3) \leq deg_L(\gamma_2) - 1 = d_0 - 1 \\
deg_R(\gamma_3) \leq deg(\gamma_2) - 1 = d_0 - 1
\end{array}$$

Therefore $deg(\gamma_3) = max(deg_L(\gamma_3), deg_R(\gamma_3)) \leq d_0 - 1$.

After a direction switch, we can again make at most m jumps in one direction. Therefore the total number of jump rule applications is bounded by $\mathcal{O}(m^2)$. ■

LEMMA 22. *Let $X \vdash Y$ be any sequent encountered during proof search. Then the saturation process for $X \vdash Y$ terminates after $\mathcal{O}(m)$ steps.*

Proof. Every application of a static rule adds a subformula of $sf(\Gamma_0 \cup \Delta_0)$ to the sequent. After at most m applications of static rules, the sequent will contain all subformulae of the original sequent, and hence will be saturated. ■

THEOREM 23. *The proof search strategy of Definition 15 terminates.*

Proof. Suppose for a contradiction that the strategy does not terminate. From Lemmas 21 and 22, we can conclude that the only way to get non-termination is for the jump and return rules to repeatedly create and remove nodes.

The length of the list is at least 1 because the first node cannot be removed. We call a node that cannot be removed *stable*. Every time a return rule removes node i from the list, it adds one or more new subformulae of $\Gamma_0 \cup \Delta_0$ to node $i-1$. After at most m such updates, node $i-1$ will contain every subformula, and the return rules will no longer be applicable to node i because their side conditions will not hold. Then node $i-1$ will become stable. Eventually all nodes will become stable, and the return rules will no longer be applicable to the end of the list. Contradiction. ■

5.2 Soundness and completeness of LBiInt$_2$

For soundness of LBiInt$_2$ we show that every LBiInt$_2$ rule is derivable in LBiInt$_1$.

THEOREM 24. **Soundness of LBiInt$_2$.** *If the sequent $X \vdash Y$ is derivable in LBiInt$_2$ then it is also derivable in LBiInt$_1$.*

To prove completeness of LBiInt$_2$, we take a detour through Buisman and Goré's calculus GBiInt [5, 13]. That is, we show that every derivation of a formula in GBiInt can be translated into a derivation of the same formula in LBiInt$_2$. Due to space limits, we give here only the outline of the translation. The full proof is available in the extended version of the paper.

The GBiInt proof system makes use of two forms of sequents:

$$\mathcal{S}\ \Gamma \lhd \Delta\ \mathcal{P} \qquad\qquad \mathcal{S}\ \Gamma \rhd \Delta\ \mathcal{P}.$$

The former denotes a *refutation* of the sequent, whereas the latter denotes its provability. In both sequents, Γ and Δ are sets of formulas, and $\mathcal{S}$ and $\mathcal{P}$ are sets of sets of formulas. $\mathcal{S}$ and $\mathcal{P}$ are called *variables* in [5], and they contain a counter model for "failed" proof search, i.e., a refutation. Proof search in GBiInt proceeds by saturation of sequents, until they become strongly saturated, at which point we can close the search by the following rule:

$$\frac{}{\{\Gamma\}\ \Gamma \lhd \Delta\ \{\Delta\}}\ Ret$$

This rule applies only in case where $\Gamma \vdash \Delta$ is strongly saturated. It basically says that there is no proof for this sequent, since one can construct a counter-model for this sequent. The information about the counter-model is then stored in the variables of the sequents, and passed back to a previous search point. To see how this information is used, consider the following (simplified) right introduction rule for implication in GBiInt (which is also called $\rightarrow_{R2}$):

$$\frac{\mathcal{S}\ \Gamma, A \lhd B\ \mathcal{P} \quad \Gamma \rhd \Delta, A \rightarrow B, P_1 \quad \cdots \quad \Gamma \rhd \Delta, A \rightarrow B, P_n}{\Gamma \rhd \Delta, A \rightarrow B} \rightarrow_{R2}$$

where $\mathcal{P} = \{P_1, \ldots, P_n\}$ and $P_i \not\subseteq \Delta$. Here, in order to simplify presentation, we consider an instance of the GBiInt rule $\rightarrow_{R2}$ where the lower sequent contains no variables. Intuitively, proof search using this rule tries to decompose $A \rightarrow B$ and see if it can find a counter model (the left premise). If the left premise does return some counter-model, i.e., $\mathcal{P}$ is not empty, then continue the search (right-premise) using the extra information gathered from the counter-model. Naturally, the formula $A \rightarrow B$ needs no further decomposition in subsequent proof search.

The translation from GBiInt to LBiInt$_2$ is surprisingly natural and uncovers a nice duality between the two calculi. To illustrate the idea behind the translation, consider the following derivation scheme in GBiInt, where the last rule is $\rightarrow_{R2}$. Let π be a GBiInt derivation:

$$\begin{array}{c} \dfrac{}{\{S_1\}\ S_1 \lhd P_1\ \{P_1\}}\ Ret \quad \cdots \quad \dfrac{}{\{S_n\}\ S_n \lhd P_n\ \{P_n\}}\ Ret \\ \vdots \\ \{S_1, \ldots, S_n\}\ \Gamma, A \lhd B\ \{P_1, \ldots, P_n\} \end{array}$$

with n-instances of Ret in its leaves. Let ξ_i, for each $i \in \{1, \ldots, n\}$, be a GBiInt derivation of $\Gamma \rhd \Delta, A \rightarrow B, P_i$ and let ξ be the GBiInt derivation

$$\frac{\begin{array}{c}\pi \\ \mathcal{S}\ \Gamma, A \lhd B\ \mathcal{P}\end{array} \quad \begin{array}{c}\xi_1 \\ \Gamma \rhd \Delta, A \rightarrow B, P_1\end{array} \quad \cdots \quad \begin{array}{c}\xi_n \\ \Gamma \rhd \Delta, A \rightarrow B, P_n\end{array}}{\Gamma \rhd \Delta, A \rightarrow B} \rightarrow_{R2}$$

where $\mathcal{S} = \{S_1, \ldots, S_n\}$ and $\mathcal{P} = \{P_1, \ldots, P_n\}$. Then we construct an LBiInt$_2$ derivation of $\Gamma \vdash \Delta, A \rightarrow B$ as follows: We first translate each ξ_i into an LBiInt$_2$ derivation ξ_i', and then "plug" in the derivation ξ_i to the i-th instance of Ret in π. This can be done by storing the context $\Gamma < \Delta$ in the left-hand side of the sequents appearing in π, and restoring them at the leaves which are instances of Ret. The following schematic derivation shows this process of storing and restoring of context:

$$\begin{array}{c} \dfrac{\begin{array}{c}\xi_1' \\ \Gamma \vdash \Delta, A \rightarrow B, P_1\end{array}}{(\Gamma < \Delta, A \rightarrow B), S_1 \vdash P_1} < \quad \cdots \quad \dfrac{\begin{array}{c}\xi_n' \\ \Gamma \vdash \Delta, A \rightarrow B, P_n\end{array}}{(\Gamma < \Delta, A \rightarrow B), S_n \vdash P_n} < \\ \vdots \\ \dfrac{(\Gamma < \Delta, A \rightarrow B), \Gamma, A \vdash B}{\Gamma \vdash \Delta, A \rightarrow B} \rightarrow_{R_2} \end{array}$$

Notice that while proof search in GBiInt involves backtracking and passing back information from failed attempts, in LBiInt$_2$ we simply go forward and restart a stored computation state when proof search in the current state (i.e., top level sequent) can no longer progress.

THEOREM 25. *If A is derivable in GBiInt then A is also derivable in* $LBiInt_2$.

COROLLARY 26. $LBiInt_2$ *is sound and complete with respect to BiInt.*

Since procedure Prove from Defn 15 mimics this translation, we have:

THEOREM 27. *Any formula A is* $LBiInt_2$*-derivable if and only if* Prove($\emptyset \vdash A$) *returns true.*

Thus LBiInt_2 gives us a decision procedure for BiInt. The fact that BiInt is decidable is already known:

THEOREM 28. *The decision problem for BiInt is* PSPACE*-complete.*

Proof. We first show that BiInt is in PSPACE. To do this, we can easily extend the polynomial Gödel translation of intuitionistic logic into the modal logic S4 [10], to a translation of BiInt into the tense modal logic KtS4. Since KtS4 is in PSPACE [24], we know that BiInt is also in PSPACE.

We now show that BiInt is PSPACE-hard. To do this, we use the fact that BiInt is an extension of Int, which is PSPACE-complete [25], and hence PSPACE-hard. Therefore BiInt is PSPACE-complete. ■

6 Related and future work

We know of only two other sequent calculi for BiInt, one due to Pinto and Uustalu [28], and the other due to Crolard [7]. But the labelled calculus of Pinto and Uustalu is still not available in full detail, and Crolard's calculus fails cut-elimination for the same reasons as does Rauszer's original calculus [21]. The display calculus for BiInt due to Goré [12] and its more recent extension due to Wansing [30] both have a purely syntactic cut-elimination proof of course. But as stated in our introduction, neither of these is really suitable for proof-search.

Hypersequents have been used for many modal [1], intutionistic and intermediate logics [19, 6]. Similarly, Dosen's "higher level" sequents [8] can cater for many different logics in one (cut-free) setting: for example, both intuitinistic logic, classical logic and modal logics S4 and S5. But we know of no actual work which uses either framework for intuitionistic logics with a "converse" modality like BiInt.

The closest calculus to ours appears to be the sequent calculus for the Lambek-Grishin logic **LG** of Moortgat [20], for which he proves a purely syntactic cut-elimination result. Briefly, the logic **NL** is a substructural intuitionistic logic which has a single non-associative and non-commutative conjunction $\otimes$, a single non-associative and non-commutative disjunction $\oplus$, and two implication connectives $A \to B$ and $B \leftarrow A$. The logic **LG** is an extension of **NL** with two exclusion connectives $A \prec B$ and $B \succ A$, and extra "mixed associativity" principles like $((A \prec B) \otimes C) \to ((A \otimes C) \prec B)$ amongst others. Moortgat uses invertible residuation rules which are similar to our rules $s_L, s_R, <, >$ since both logics permit such "flip-flopping". But the lack of associativity and commutativity of $\oplus$ and $\otimes$ means there is less non-determinism in the calculus in terms of proof-search. Indeed, the decision procedure for **LG** obtained by Moortgat runs in polynomial time while it is known that the decision problem for BiInt is PSPACE-complete.

The main difference between Int and BiInt is of course the exclusion connective $-\!\!<$, whose Kripke semantic clause "looks backward against" the accessibility relation $\leq$. Thus it makes sense to look at sequent-like calculi for the modal companions of BiInt, namely the tense logics Kt and KtS4.

There are many sequent-like calculi for the related normal modal logics Kt, its reflexive and transitive cousin KtS4, or even just good old classical modal logic S5 [27, 17, 1, 14, 15, 16, 4]. In each such calculus there is a rule (or rules) which allow us to "return" to previously seen worlds when the rules are viewed from the perspective of counter-model construction. These calculi can be broadly separated into two groups: those which use two-sided sequents, and those which use a negation-normal-form to make do with right-sided sequents. It should be possible to extend most of the former calculi to handle BiInt, but it is unlikely that the latter can be so extended since BiInt has no negation-normal-form theorem.

On this note, we intend to consider the following ideas for future work. Suppose we posit two additional structural connectives $\circ$ and $\bullet$ with $\circ$ standing as a proxy for $\langle F\rangle/[F]$ on the left/right, and $\bullet$ standing as a proxy for $\langle P\rangle/[P]$ on the left/right. Are the following rules enough to give us a cut-free calculus for tense logic Kt which is also amenable for backward proof-search:

$$\frac{\Gamma, A, \vdash \circ(X > Y), \Delta}{X, [P]\Gamma, \langle P\rangle A \vdash Y, \langle P\rangle\Delta}\ \langle P\rangle_L \qquad \frac{\Gamma, \circ(X < Y) \vdash A, \Delta}{X, [P]\Gamma \vdash Y, [P]A, \langle P\rangle\Delta}\ [P]_R$$

$$\frac{\Gamma, A, \vdash \bullet(X > Y), \Delta}{X, [F]\Gamma, \langle F\rangle A \vdash Y, \langle F\rangle\Delta}\ \langle F\rangle_L \qquad \frac{\Gamma, \bullet(X < Y) \vdash A, \Delta}{X, [F]\Gamma \vdash Y, [F]A, \langle F\rangle\Delta}\ [F]_R$$

As we have noted in the introduction, the calculus LBiInt_1 (and its derivative LBiInt_2) is motivated by display calculi. It can be seen as an attempt to tame proof search in display calculi. In this preliminary work, we have been able to derive more proof-search friendly calculi essentially by constraining the use of the display postulates of display calculi. However, there is still a methodological gap in our results. We have not been able to show a direct relation between LBiInt_2 and LBiInt_1: that is, we still need the help of an "external" calculus GBiInt for our completeness result for LBiInt_2. It is important that this gap be closed in order to generalise these results beyond bi-intuitionistic logic. Our ultimate goal is to obtain a systematic way to "sequentialize" a given display calculus to one with nested sequents, and derive a proof search strategy for the latter.

Acknowledgments. We thank the anonymous referees for their comments on an earlier draft of the paper.

BIBLIOGRAPHY

[1] A. Avron. The method of hypersequents in proof theory of propositional non-classical logics. In C Steinhorn W Hodges, M Hyland and J Truss, editors, *Logic: Foundations to Applications*, pages 1–32. Oxford Science Publications, 1996.

[2] N. Belnap. Display logic. *Journal of Philosophical Logic*, 11:375–417, 1982.

[3] K. Brünnler. Atomic cut elimination for classical logic. In *CSL*, volume LNCS 2803, pages 86–97. Springer, 2003.

[4] K. Brünnler. Deep sequent systems for modal logic. In G. Governatori et al, editor, *Advances in Modal Logic 6*, pages 107–119. College Publications, 2006.

[5] L. Buisman and R. Goré. A cut-free sequent calculus for bi-intuitionistic logic. In N Olivetti, editor, *TABLEAUX*, volume 4548 of *LNCS*, pages 90–106. Springer, 2007.

[6] A. Ciabattoni and M. Ferrari. Hypersequent calculi for some intermediate logics with bounded Kripke models. *Journal of Logic and Computation*, 11(2):283–294, 2001.

[7] T. Crolard. Subtractive logic. *Theoretical Computer Science*, 254(1–2):151–185, 2001.

[8] K. Došen. Sequent-systems for modal logic. *Journal of Symbolic Logic*, 50(1):149–169, 1985.

[9] S. Feferman, editor. *Gödel's Collected Works*. Oxford University Press, Oxford, 1980.

[10] K. Gödel. Eine interpretation des intuitionistischen aussagenkalkuls. *Ergebnisse eines Mathematischen Kolloquiums.*, 4:39–40, 1933. English translation in [9].

[11] R. Goré. Substructural logics on display. *LJIGPL*, 6(3):451–504, 1998.

[12] R. Goré. Dual intuitionistic logic revisited. In Roy Dyckhoff, editor, *Proc. TABLEAUX-2000*, volume LNAI 1847, pages 252–267. Springer, 2000.

[13] R. Goré and L. Postniece. Combining derivations and refutations for cut-free completeness in bi-intuitionistic logic. *Journal of Logic and Computation*. To appear. Available via `http://users.rsise.anu.edu.au/~linda/BiIntLong.pdf`.

[14] A. Heuerding, M. Seyfried, and H. Zimmermann. Efficient loop-check for backward proof search in some non-classical propositional logics. In *Analytic Tableaux and Related Methods*, volume 1071 of *LNAI*, pages 210–225, 1996.

[15] A. Indrzejczak. Multiple sequent calculus for tense logics. International Conference on Temporal Logic, Leipzig 2000. 93–104.

[16] A. Indrzejczak. Cut-free double sequent calculus for S5. *LJIGPL*, 6(3):505–516, 1998.

[17] R. Kashima. Cut-free sequent calculi for some tense logics. *Studia Logica*, 53:119–135, 1994.

[18] M. Kracht. Power and weakness of the modal display calculus. In H. Wansing, editor, *Proof Theory of Modal Logics*, pages 92–121. Kluwer, 1996.

[19] A. Ciabattoni, M. Baaz and C. G. Fermüller. Hypersequent calculi for Gödel logics — a survey. *Journal of Logic and Computation*, 13:835–861, 2003.

[20] M. Moortgat. Symmetries in natural language syntax and semantics: The Lambek-Grishin calculus. In *Logic, Language, Information and Computation: Proc. 14th International Workshop WoLLIC 2007*, LNCS 4576, pages 264–284. Springer, 2007.

[21] C. Rauszer. A formalization of the propositional calculus of H-B logic. *Studia Logica*, 33:23–34, 1974.

[22] C. Rauszer. An algebraic and Kripke-style approach to a certain extension of intuitionistic logic. *Dissertationes Mathematicae*, 168, 1980.

[23] G. Restall. *An Introduction to Substructural Logics*. Routledge, New York, 2000.

[24] E. Spaan. The complexity of propositional tense logics. In M. de Rijke, editor, *Diamonds and Defaults*, pages 287–309. Kluwer Academic Publishers, 1993.

[25] R. Statman. Intuitionistic propositional logic is polynomial-space complete. *Theoretical Computer Science*, 9:67–72, 1979.

[26] A. S. Troelstra and H. Schwichtenberg. *Basic Proof Theory*. Cambridge University Press, 1996.

[27] K. Trzesicki. Gentzen-style axiomatization of tense logic. *Bulleting of the Section of Logic*, 13(2):75–84, 1984.

[28] T. Uustalu and L. Pinto. Days in logic '06 conference abstract. Online at `http://www.mat.uc.pt/~kahle/dl06/tarmo-uustalu.pdf`, accessed on 27th October, 2006.

[29] H. Wansing. Modal tableaux based on residuation. *Journal of Logic and Computation*, 7(6):719–731, 1997.
[30] H. Wansing. Constructive negation, implication and co-implication. *Manuscript*, March:to appear, 2007.

Rajeev Goré
Computer Sciences Laboratory,
The Australian National University
Australia
Rajeev.Gore@anu.edu.au

Linda Postniece
Computer Sciences Laboratory,
The Australian National University
Australia
Linda.Postniece@anu.edu.au

Alwen Tiu
Computer Sciences Laboratory,
The Australian National University
Australia
Alwen.Tiu@anu.edu.au

Valentini's cut-elimination for provability logic resolved

RAJEEV GORÉ AND REVANTHA RAMANAYAKE

ABSTRACT. In 1983, Valentini presented a syntactic proof of cut-elimination for a sequent calculus GLS_V for the provability logic GL where we have added the subscript V for "Valentini". The sequents in GLS_V were built from sets, as opposed to multisets, thus avoiding an explicit contraction rule. From a syntactic point of view, it is more satisfying and formal to explicitly identify the applications of the contraction rule that are 'hidden' in these set-based proofs of cut-elimination. There is often an underlying assumption that the move to a proof of cut-elimination for sequents built from multisets is easy. Recently, however, it has been claimed that Valentini's arguments to eliminate cut do not terminate when applied to a multiset formulation of GLS_V with an explicit rule of contraction. The claim has led to much confusion and various authors have sought new proofs of cut-elimination for GL in a multiset setting.

Here we refute this claim by placing Valentini's arguments in a formal setting and proving cut-elimination for sequents built from multisets. The formal setting is particularly important for sequents built from multisets, in order to accurately account for the interplay between the weakening and contraction rules. Furthermore, Valentini's original proof relies on a novel induction parameter called "width" which is computed 'globally'. It is difficult to verify the correctness of his induction argument based on "width". In our formulation however, verification of the induction argument is straightforward. Finally, the multiset setting also introduces a new complication in the the case of contractions above cut when the cut-formula is boxed. We deal with this using a new transformation based on Valentini's original arguments.

Finally, we show that the algorithm purporting to show the non-termination of Valentini's arguments is not a faithful representation of the original arguments, but is instead a transformation already known to be insufficient.

Keywords: cut elimination, provability logic, Gödel-Löb logic

1 Introduction

The provability logic GL is obtained by adding Löb's axiom $\Box(\Box A \supset A) \supset \Box A$ to the standard Hilbert calculus for propositional normal modal logic K [11]. Interpreting the modal operator $\Box A$ as the provability predicate "A is provable in Peano arithmetic", it can be shown that GL is complete with respect to the formal provability interpretation in Peano arithmetic (see [14]). For an introduction to provability logic see [13].

In 1981, Leivant [5] proposed a syntactic proof of cut-elimination for a sequent calculus for GL. Valentini [16] soon described a counter-example to this proof, proposing a more complicated proof for the sequent calculus GLS_V for GL. The calculus GLS_V is a sequent calculus for classical propositional logic together with a single modal rule GLR. Valentini's proof appears to be the first proof of cut-elimination for a sequent calculus for GL and relies on a complicated transformation justified by a Gentzen-style induction on the degree of the cut-formula and the cut-height, as well as a new induction parameter — the width of a cut-formula. Roughly speaking, the width of a cut-instance is the number of GLR rule instances above that cut which contain a parametric ancestor of the cut-formula in their conclusion. However, Valentini's proof is very brief, informal and difficult to check. For example, he only considers a cut-instance where the cut-formula is left and right principal by the GLR rule (the Sambin Normal Form), noting that "the presence of the new parameter [width] does not affect the [remainder of the standard cut-elimination proof]" [16]. While it is true that the standard transformations appropriately reduce the degree and/or cut-height, he fails to observe that these transformations can sometimes increase the width of lower cuts, casting doubt on the validity of the induction. A careful study of the proposed transformation is required to confirm that the proof is not affected (see Remark 21).

Several other solutions for cut-elimination have been proposed. Borga [2] presented one solution, while Sasaki [12] described a proof for a sequent calculus very similar to GLS_V, relying on cut-elimination for $K4$. Note that only Leivant and Valentini used traditional Gentzen-style methods involving an induction over the degree of the cut-formula and the cut-height.

All four authors used sequents $X \Rightarrow Y$ where X and Y are *sets*, so these calculi did not require a rule of contraction as there is no notion of a set containing an element multiple times (unlike a multiset where the number of occurrences is important). Thus the following instance of the $L\wedge$ rule is legal in GLS_V even though it 'hides' a contraction on $P \wedge Q$:

$$\frac{P \wedge Q, P \Rightarrow R}{P \wedge Q \Rightarrow R}\ L\wedge$$

From a syntactic viewpoint, it is more satisfying and formal to explicitly identify the contractions that are 'hidden' in these set-based proofs of cut-elimination. The appropriate formalisation to understand the reliance on contraction is to use multisets. But then the contraction rules often pose new problems that require attention. For example, Gentzen [4] in his original proof of the *Hauptsatz* for the classical sequent calculus LK, introduced a 'multicut' rule to deal with a complication in the case of contractions above cut. Nevertheless, even that proof is not purely syntactic in the following sense: since multicut is not a rule of the original calculus, the proof has to detour via a conservative extension. Proofs of purely syntactic cut-elimination for LK have subsequently appeared in the literature (see [10],[3],[1] for example). In the case of GL, it turns out that additional complications also arise when formulating Valentini's arguments in a multiset setting, for example, due to the interplay between weakening and contraction rules (see Remark 16).

Thus the move to a proof of cut-elimination for sequents built from multisets is not straightforward. Moen [7] attempted to lift Valentini's set-based arguments to obtain a proof for sequents built from multisets before concluding that this was not possible. Specifically, he presents a concrete derivation ϵ containing cut, and claims that a multiset formulation of Valentini's argument does not terminate when applied to ϵ. Not surprisingly, this claim has ignited the search for new proofs of purely syntactic cut-elimination in a Gentzen-style multiset setting for GL.

In response, Negri [8] and Mints [6] proposed two different solutions. Negri uses a non-standard multiset sequent calculus in which sequents are built from multisets of labelled formulae of the form $x : A$, where A is a traditional formula and x is an explicit name for a Kripke world. She shows that contraction is height-preserving admissible in this calculus and uses this to handle contractions above cut in her cut-elimination argument. In our view, the use of semantic information in the calculus deviates from a purely proof theoretic approach. Mints [6] solves the problem using a sequent calculus similar to the multiset-formulation of GLS_V, but does not state how to handle contractions above cut.

So there are two issues to consider:

1. formalise "width" more precisely to clarify Valentini's original definition, and check whether it is a suitable induction measure, and

2. determine whether Valentini's arguments can be used to obtain a purely syntactic proof of cut-elimination in a *multiset* setting.

Our contribution is as follows: we have successfully translated Valentini's set-based arguments for cut-elimination to a sequent calculus built from multisets. To this end, we have formalised the notion of parametric ancestor and width to correspond intuitively with Valentini's original definition. With this formalisation we show that Valentini's arguments can be applied in the multiset setting, noting that although certain transformations may increase the width of lower cuts, this does not affect the proof. In the case where the last rule in either premise derivation of the cut-rule is a contraction on the cut-formula, we avoid the multicut rule by using von Plato's arguments [10] when the cut-formula is not boxed, and a new argument for the case when the cut-formula is boxed. Thus we obtain a purely syntactic proof of cut-elimination in a multiset setting. We also believe that we have identified a mistake in Moen's claim that Valentini's arguments (in a multiset setting) do not terminate. It appears that Moen has not used a faithful representation of Valentini's arguments for the inductive case, but instead a transformation he titles Val-II(core) that is already known to be insufficient [11]. We discuss this further in Section 5. Of course, the incorrectness of Moen's claim does not imply the correctness of Valentini's arguments in a multiset setting. Indeed the whole point is that complications do arise in the multiset setting, and that these have to be dealt with carefully.

Initial sequents: $A \Rightarrow A$ for each formula A

Logical rules:

$$\frac{X \Rightarrow Y, A}{X, \neg A \Rightarrow Y}\,L\neg \qquad \frac{A, X \Rightarrow Y}{X \Rightarrow Y, \neg A}\,R\neg$$

$$\frac{A_i, X \Rightarrow Y}{A_1 \wedge A_2, X \Rightarrow Y}\,L\wedge \qquad \frac{X \Rightarrow Y, A_1 \qquad X \Rightarrow Y, A_2}{X \Rightarrow Y, A_1 \wedge A_2}\,R\wedge$$

$$\frac{A_1, X \Rightarrow Y \qquad A_2, X \Rightarrow Y}{A_1 \vee A_2, X \Rightarrow Y}\,L\vee \qquad \frac{X \Rightarrow Y, A_i}{X \Rightarrow Y, A_1 \vee A_2}\,R\vee$$

$$\frac{X \Rightarrow Y, A \qquad B, X \Rightarrow Y}{A \supset B, X \Rightarrow Y}\,L\supset \qquad \frac{A, X \Rightarrow Y, B}{X \Rightarrow Y, A \supset B}\,R\supset$$

Modal rule:

$$\frac{\Box X, X, \Box B \Rightarrow B}{\Box X \Rightarrow \Box B}\,GLR$$

Structural rules:

$$\frac{X \Rightarrow Y}{A, X \Rightarrow Y}\,LW \qquad \frac{X \Rightarrow Y}{X \Rightarrow Y, A}\,RW$$

$$\frac{A, A, X \Rightarrow Y}{A, X \Rightarrow Y}\,LC \qquad \frac{X \Rightarrow Y, A, A}{X \Rightarrow Y, A}\,RC$$

Cut-rule:

$$\frac{X \Rightarrow Y, D \qquad D, U \Rightarrow W}{X, U \Rightarrow Y, W}\,cut$$

Table 1. The sequent calculus GLS

Finally, we remind the reader that it is trivial to show that the cut-rule is redundant for both set and multiset sequent calculus formulations for GL by proving that the calculus without the cut-rule is sound and complete for the Kripke semantics of GL. However, our purpose here is to resolve the claim about the failure of *syntactic* cut-elimination based on Valentini's arguments for a sequent calculus built with multisets.

2 Preliminaries

We use the letters $P, Q, \ldots$ to denote propositional variables. Formulae are defined in the usual way in terms of propositional variables, the logical constant $\bot$ and the logical connectives $\wedge$ (conjunction), $\vee$ (disjunction), $\supset$ (implication) and $\Box$ (necessity, or in this context, provability). Formulae are denoted by $A, B, \ldots$. Multisets of formulae are denoted by X, Y, U, V, W, G and also as a list of comma-separated formula enclosed in "$\langle$" and "$\rangle$". A formula A is said to be *boxed* if it is of the form $\Box B$ for some formula B and is *not boxed* otherwise. The relation '$\equiv$' is used to denote syntactic equality. Let X be the multiset $\langle A_1, \ldots, A_n \rangle$. Then we define the multiset $\Box X$ to be $\langle \Box A_1, \ldots, \Box A_n \rangle$. Furthermore $B \in X$ iff $B \equiv A_i$ for some $1 \leq i \leq n$. The negation of $B \in X$ is denoted by $B \notin X$. The notation $(A)^m$ or A^m denotes m comma-separated occurrence of A.

A *sequent* is a tuple (X, Y) of multisets X and Y of formulae and is written $X \Rightarrow Y$. We sometimes use $\mathcal{S}$ or $\mathcal{S}'$ to denote a sequent. The multiset X (Y) is called the antecedent (succedent). The multiset sequent calculus we use here is called GLS (Table 1). For the logical and structural rules in GLS, the multisets X and Y are called the *context*. In the conclusion of each of these rules, the formula occurrence not in the context is called the *principal formula*. This follows standard practice (see [15]). For the GLR rule, each formula in $\Box X, X, \Box B, B$ is called a principal formula. The $\Box B$ in the succedent of the conclusion of the GLR rule is called the *diagonal formula* (and is of course boxed). In the cut-rule, the formula D is the *cut-formula*. A rule with one premise (two premises) is called a unary (binary) rule.

A binary rule where the context in both premises is required to be identical is called an *additive* binary rule (eg: $L\vee, R\wedge$). A binary rule where the context in each premise need not be identical is called a *multiplicative* binary rule (eg: *cut*). The term context-sharing (context-independent) is also used to refer to an additive (multiplicative) rule (see [15]).

Note, we have deleted the initial sequent $\bot \Rightarrow \bot$ and the $\bot$-rule that appears in GLS_V. As [13] observes, it is not necessary to include $\bot$ although its presence can be convenient from a semantic viewpoint. As our concerns here are proof theoretic we shall not require it. We have also replaced the multiplicative $L\supset$ in GLS_V with an additive version. As all the other logical rules in GLS are additive, it seems appropriate to use an additive $L\supset$. We observe that the definitions and proofs in this paper apply, with slight amendment, to a sequent calculus built from multisets that is obtained directly from GLS_V.

A *derivation* (in GLS) is defined recursively with reference to Table 1 as:

(i) an initial sequent $A \Rightarrow A$ for any formula A is a derivation, and

(ii) an application of a logical, modal, structural or cut-rule to derivations concluding its premise(s) is a derivation.

This is the standard definition. Viewing a derivation as a tree, we call the root of the tree the *end-sequent* of the derivation. If there is a derivation with end-sequent $X \Rightarrow Y$ we say that $X \Rightarrow Y$ is *derivable* in GLS. Let $\bigwedge X$ ($\bigvee Y$) denote the conjunction (disjunction) of all formula occurrences in X (Y). Interpreting the sequent $X \Rightarrow Y$ as the formula $\bigwedge X \supset \bigvee Y$, from [11] we see that derivability in GLS is sound and complete wrt GL.

We write $\{\pi\}_1^r/^{\rho}X \Rightarrow Y$ to denote the derivation

$$\frac{\pi_1 \qquad \dots \qquad \pi_r}{X \Rightarrow Y}\,\rho$$

where ρ is an instance of a rule with r premises. We refer to $\pi_1, \dots, \pi_r$ as the *premise derivations* of ρ. If ρ is unary (binary) then $r = 1$ ($r = 2$). Rather than $\{\pi\}_1^1$ and $\{\pi\}_1^2$, we write, respectively, "π_1" and "$\pi_1 \quad \pi_2$".

Let ρ be some rule-occurrence in a derivation τ. Then $\rho(A)$ indicates that the principal formula is A; $\rho^*(X)$ denotes some number (≥ 0) of applications of ρ that make each formula occurrence (including multiple formula occurrences) in the multiset X a principal formula. To identify a rule-occurrence in the text we occasionally use subscripts, eg: GLR_1, cut_0.

A derivation τ is *cut-free* if τ contains no instances of the cut-rule. A cut-instance is said to be *topmost* if its premise derivations are cut-free.

DEFINITION 1 (n-ary GLR rule). Given a derivation τ, an instance ρ of the GLR rule appearing in τ is n-ary if there are exactly $n-1$ GLR rule instances on the path between ρ and the end-sequent of τ.

Let $GLR(n, \tau)$ denote the number of n-ary GLR rules in τ. Next we define the height, cut-height, and degree of a formula in the standard manner.

DEFINITION 2 (height, cut-height, degree). The *height* $s(\tau)$ of a derivation τ is the greatest number of successive applications of rules in it plus one. The *cut-height* h of an instance of the cut-rule with premise derivations τ_1 and τ_2 is $s(\tau_1) + s(\tau_2)$. The *degree* $deg(A)$ of a formula A is defined as the number of symbol occurrences in A from $\{\Box, \neg, \wedge, \vee, \supset\}$ plus one.

3 Generalising the notion of derivation

To formalise the notion of width we need a more general structure than a derivation. The structure we have in mind can be obtained from a derivation τ by deleting a proper subderivation τ' in τ. We call this structure a *stub-derivation*. To emphasise the point of deletion we use the annotation stub.

Formally a stub-derivation (in GLS) is defined recursively with reference to Table 1 as follows:

(i) an initial sequent $A \Rightarrow A$ for any formula A is a stub-derivation, and

(ii) for any sequent $\mathcal{S}$ and stub-derivation π, each of

$$\text{(a) } \mathsf{stub}/\mathcal{S} \qquad \text{(b) } \mathsf{stub} \quad \pi/\mathcal{S} \qquad \text{(c) } \pi \quad \mathsf{stub}/\mathcal{S}$$

is a stub-derivation, and

(iii) an application of a logical, modal, structural or cut-rule to stub-derivations concluding its premise(s) is a stub-derivation.

Viewing a stub-derivation τ as a tree, we call the root of the tree the *end-sequent* of the stub-derivation (denoted $ES(\tau)$). The leaves of the tree are called the *top-sequents*. Clearly a derivation is a stub-derivation in which every top-sequent is an initial sequent. Thus a stub-derivation generalises the notion of a derivation.

We use the term 'stub-instance' to refer to an occurrence of either $\mathsf{stub}/\mathcal{S}$ or $\mathsf{stub} \quad \pi/\mathcal{S}$ or $\pi \quad \mathsf{stub}/\mathcal{S}$. An *sstub-derivation* (read: single stub-derivation) is a stub-derivation containing exactly one stub-instance. We write $d[\mathsf{stub}]$ instead of d, to remind the reader that the structure contains exactly one stub-instance.

Let d' be a derivation with end-sequent $\mathcal{S}'$, let $d[\mathsf{stub}]$ be an sstub-derivation with an occurrence of one of the following:

$$\mathsf{stub}/\mathcal{S} \qquad\qquad \mathsf{stub} \quad \pi/\mathcal{S} \qquad\qquad \pi \quad \mathsf{stub}/\mathcal{S}$$

and suppose that

$$\mathcal{S}'/^{\rho}\mathcal{S} \qquad\qquad \mathcal{S}' \quad ES(\pi)/\mathcal{S} \qquad\qquad ES(\pi) \quad \mathcal{S}'/\mathcal{S}$$

respectively is a legal instance of some logical or structural rule ρ. We say that $d[\mathsf{stub}]$ and d' are *compatible* and write $d[\mathsf{stub}] \mapsfrom d'$ to denote

$$\frac{d'}{\mathcal{S}}\,\rho \qquad\qquad \frac{d' \qquad \pi}{\mathcal{S}}\,\rho \qquad\qquad \frac{\pi \qquad d'}{\mathcal{S}}\,\rho$$

respectively, obtained by "attaching" the tree d' to the tree $d[\mathsf{stub}]$ at the node stub under rule ρ. We refer to ρ as a *binding rule* for $d[\mathsf{stub}]$ and d'.

By permitting formula occurrences in a (stub-)derivation to contain $*$ or $\circ$ decorations, we define an *annotated (stub-)derivation*. The forgetful map $\lfloor\cdot\rfloor$ maps an annotated stub-derivation to the stub-derivation obtained by erasing all $*$ and $\circ$ decorations. Clearly $\lfloor\cdot\rfloor$ maps an annotated derivation to a derivation. A *transformed (stub-)derivation* τ' is a (stub-)derivation that is obtained from some existing (stub-)derivation τ by syntactic transformation. We write $A^{\circ n}$ or A^{*n} to mean n occurrences of the formula A° or A^{*} respectively.

Formally a stub-derivation and an annotated stub-derivation are different structures. Because these structures are very similar, for economy of space we will introduce definitions and prove results for stub-derivations alone and note, whenever applicable, that the definitions and results can be extended to annotated stub-derivations.

EXAMPLE 3. Let us denote the sstub-derivation at below left by $d[\mathsf{stub}]$ and the derivation at below right by d':

$$\dfrac{\dfrac{\mathsf{stub}}{B \Rightarrow A \supset B} \qquad A \supset B \Rightarrow A \supset B}{B \vee (A \supset B) \Rightarrow A \supset B}\,L\vee \qquad\qquad \dfrac{B \Rightarrow B}{A, B \Rightarrow B}\,LW$$

Observe that $d[\mathsf{stub}]$ has a stub-instance of type $\mathsf{stub}/\mathcal{S}$, with $\mathcal{S} \equiv B \Rightarrow A \supset B$, and d' has endsequent $\mathcal{S}' \equiv A, B \Rightarrow B$. Because $\mathcal{S}'/\mathcal{S}$ is an instance of $R{\supset}$, $d[\mathsf{stub}]$ and d' are compatible. The derivation $d[\mathsf{stub}] \mapsfrom d'$ is:

$$\dfrac{\dfrac{\dfrac{B \Rightarrow B}{A, B \Rightarrow B}\,LW}{B \Rightarrow A \supset B}\,R{\supset} \qquad A \supset B \Rightarrow A \supset B}{B \vee (A \supset B) \Rightarrow A \supset B}\,L\vee$$

and the binding rule is $R{\supset}$.

EXAMPLE 4. Let us denote the sstub-derivation at below left by $d[\mathsf{stub}]$ and the derivation at below right by d':

$$\frac{\mathsf{stub} \qquad A \supset B \Rightarrow A \supset B}{B \vee (A \supset B) \Rightarrow A \supset B} \qquad\qquad \dfrac{\dfrac{B \Rightarrow B}{A, B \Rightarrow B}\, LW}{B \Rightarrow A \supset B}\, R{\supset}$$

Observe that $d[\mathsf{stub}]$ has a stub-instance of type $\mathsf{stub} \quad \tau/\mathcal{S}$, with $\mathcal{S} \equiv B \vee (A \supset B) \Rightarrow A \supset B$, and d' has endsequent $\mathcal{S}' \equiv B \Rightarrow A \supset B$.

Since $\mathcal{S}' \quad A \supset B \Rightarrow A \supset B/\mathcal{S}$ is an instance of $L\vee$, $d[\mathsf{stub}]$ and d' are compatible. The derivation $d[\mathsf{stub}] \hookleftarrow d'$ is identical to that obtained in Example 3, although here the binding rule is $L\vee$.

DEFINITION 5. Let τ be a stub-derivation and G a formula multiset. The antecedent operator $\oplus$: stub-derivation $\times$ formula multiset $\mapsto$ stub-derivation is defined as follows:

Case $G = \langle\rangle$: let $\tau \oplus G = \tau$

Case $G \neq \langle\rangle$: define $\tau \oplus G$ recursively on τ as follows

1. initial sequent: $(A \Rightarrow A) \oplus G = (A \Rightarrow A/^{LW^*(G)} A, G \Rightarrow A)$
2. stub-instance:
   (a) $(\mathsf{stub}/X \Rightarrow Y) \oplus G = (\mathsf{stub}/X, G \Rightarrow Y)$
   (b) $(\mathsf{stub} \quad \pi/X \Rightarrow Y) \oplus G = (\mathsf{stub} \quad \pi \oplus G/X, G \Rightarrow Y)$
   (c) $(\pi \quad \mathsf{stub}/X \Rightarrow Y) \oplus G = (\pi \oplus G \quad \mathsf{stub}/X, G \Rightarrow Y)$
3. unary non-GLR rule: $(\pi/X \Rightarrow Y) \oplus G = (\pi \oplus G/X, G \Rightarrow Y)$
4. GLR rule: $(\pi/^{GLR} X \Rightarrow Y) \oplus G = (\pi/^{GLR} X \Rightarrow Y)/^{LW^*(G)} X, G \Rightarrow Y$
5. binary additive rule: $(\pi_1 \; \pi_2/X \Rightarrow Y) \oplus G = (\pi_1 \oplus G \; \pi_2 \oplus G/X, G \Rightarrow Y)$
6. cut-rule: $(\pi_1 \quad \pi_2/^{cut} X \Rightarrow Y) \oplus G = (\pi_1 \oplus G \quad \pi_2/^{cut} X, G \Rightarrow Y)$.

That $\oplus$ maps into the set of stub-derivations follows by inspection of the definition. Notice that the recursion terminates at an initial sequent, stub-instance or a GLR rule. The operator $\oplus$ will bind stronger that $\hookleftarrow$.

LEMMA 6. *If d is a stub-derivation and G is a formula multiset, then $d \oplus G$ is a stub-derivation. Furthermore, if d is in fact an sstub-derivation $d[\mathsf{stub}]$, then $d[\mathsf{stub}] \oplus G$ is an sstub-derivation.*

Proof. The result follows immediately from Definition 5. ■

EXAMPLE 7. Refer to the sstub-derivation $d[\mathsf{stub}]$ in Example 3. If G is a non-empty formula multiset, then $d[\mathsf{stub}] \oplus G$ is the stub-derivation:

$$\dfrac{\dfrac{\mathsf{stub}}{B, G \Rightarrow A \supset B} \qquad \dfrac{A \supset B \Rightarrow A \supset B}{A \supset B, G \Rightarrow A \supset B}\, LW^*(G)}{B \vee (A \supset B), G \Rightarrow A \supset B}\, L\vee$$

Form of annotated derivation δ	$\Phi_{\Box B}[\delta]$
$(\Box B)^* \Rightarrow \Box B$	$(\Box B)^\circ \Rightarrow \Box B$
$\dfrac{\dfrac{\{\pi\}_1^r}{G, (\Box B)^{n-1} \Rightarrow H}}{G, (\Box B)^{*n} \Rightarrow H}\, LW(\Box B)$	$\dfrac{\Phi_{\Box B}\left[\dfrac{\{\pi\}_1^r}{G, (\Box B)^{*n-1} \Rightarrow H}\right]}{G, (\Box B)^\circ, (\Box B)^{*n-1} \Rightarrow H}\, LW(\Box B)$
$\dfrac{\dfrac{\{\pi\}_1^r}{G, (\Box B)^{n+1} \Rightarrow H}}{G, (\Box B)^{*n} \Rightarrow H}\, LC(\Box B)$	$\dfrac{\Phi_{\Box B}\left[\dfrac{\{\pi\}_1^r}{G, (\Box B)^{*n+1} \Rightarrow H}\right]}{G, (\Box B)^{*n} \Rightarrow H}\, LC(\Box B)$
$\dfrac{\dfrac{\{\pi\}_1^r}{G', (\Box B)^{n} \Rightarrow H'}}{G, (\Box B)^{*n} \Rightarrow H}\, \rho \neq GLR$	$\dfrac{\Phi_{\Box B}\left[\dfrac{\{\pi\}_1^r}{G', (\Box B)^{*n} \Rightarrow H'}\right]}{G, (\Box B)^{*n} \Rightarrow H}\, \rho$
$\dfrac{\dfrac{\{\pi\}_1^r}{\Box G, G, (\Box B)^n, B^n, \Box A \Rightarrow A}}{\Box G, (\Box B)^{*n} \Rightarrow \Box A}\, \mathrm{GLR}$	$\dfrac{\dfrac{\{\pi\}_1^r}{\Box G, G, (\Box B)^n, B^n, \Box A \Rightarrow A}}{\Box G, (\Box B)^{\circ n} \Rightarrow \Box A}\, \mathrm{GLR}$
$\dfrac{\dfrac{\{\pi\}_1^r}{G', (\Box B)^n \Rightarrow H'} \quad \dfrac{\{\pi'\}_1^s}{G'', (\Box B)^n \Rightarrow H''}}{G, (\Box B)^{*n} \Rightarrow H}\, \rho \neq cut$	$\dfrac{\Phi_{\Box B}\left[\dfrac{\{\pi\}_1^r}{G', (\Box B)^{*n} \Rightarrow H'}\right] \quad \Phi_{\Box B}\left[\dfrac{\{\pi'\}_1^s}{G'', (\Box B)^{*n} \Rightarrow H''}\right]}{G, (\Box B)^{*n} \Rightarrow H}\, \rho$
antecedent of $ES(\delta)$ does not contain a $(\Box B)^*$ formula occurrence	δ

Table 2. Definition of $\Phi_{\Box B}$. Multisets G and $\Box G$ contain no occurrences of annotated formulae. The function is defined on the class of cut-free annotated derivations.

By observation, we can confirm that $d[\mathsf{stub}] \oplus G$ is a sstub-derivation as predicted by Lemma 6. Notice that $d[\mathsf{stub}] \oplus G$ and d' (from Example 3) are not compatible, because there is no logical or structural inference rule that can take us from the premise sequent $A, B \Rightarrow B$ to the conclusion sequent $B, G \Rightarrow A \supset B$.

Definition 5 can be extended in the obvious way to apply to annotated stub-derivations. Then it is easily verified that Lemma 6 holds under the substitution of "annotated (s)stub-derivation" for "(s)stub-derivation" in the statement of the lemma.

Cut-elimination often involves tracing the "parametric ancestors" of the cut-formula. The following definition uses the symbols $\circ$ and $*$ as annotations to help trace the parametric ancestors.

DEFINITION 8. ($f_C[\cdot]$: annotated derivation wrt C).
Let τ be a cut-free derivation with endsequent $X \Rightarrow Y$, and C a formula.

1. if C is not boxed then let $f_C[\tau] = \tau$.

2. if C is boxed ($C \equiv \Box B$) and $\Box B \notin X$ then let $f_{\Box B}[\tau] = \tau$.

3. if C is boxed ($C \equiv \Box B$) and $\Box B \in X$. Then τ must be a derivation of the form $\Box B \Rightarrow \Box B$ or $\{\pi\}_1^r / X', \Box B \Rightarrow Y$.

   Set $f_{\Box B}[\tau]$ as $\Phi_{\Box B}[(\Box B)^* \Rightarrow \Box B)]$ or $\Phi_{\Box B}[\{\pi\}_1^r / X', (\Box B)^* \Rightarrow Y]$ respectively, where $\Phi_{\Box B}$ is defined on the class of cut-free annotated derivations as shown in Table 2.

Observe that the annotation operator $f_C[\cdot]$ is a total function mapping derivations to annotated derivations.

REMARK 9. Let τ be a derivation with endsequent $X \Rightarrow Y$. If $\Box B \in X$ then the formula occurrences $(\Box B)^\circ$ and $(\Box B)^*$ in $f_{\Box B}[\tau]$ are each called a *parametric ancestor* of the formula occurrence $\Box B \in X$ in the endsequent. Intuitively, the annotation $\circ$ denotes the final parametric ancestor when tracing ancestors upwards. That is, the $\Box B$ is introduced at that point.

DEFINITION 10. Define $\partial^\circ(B, \tau)$ for a formula B and an annotated derivation τ, as the number of instances of the GLR rule in τ whose conclusion contains an occurrence of the annotated formula B° in the antecedent.

LEMMA 11. *Let $d[\mathsf{stub}]$ be an annotated sstub-derivation and G a formula multiset. Then*

(a) $\partial^\circ(B, d[\mathsf{stub}] \oplus G) = \partial^\circ(B, d[\mathsf{stub}])$

(b) Let d' be a derivation such that $d[\mathsf{stub}]$ and d' are compatible. Then $\partial^\circ(B, d[\mathsf{stub}] \hookleftarrow d') = \partial^\circ(B, d[\mathsf{stub}]) + \partial^\circ(B, d')$.

Proof.

(a) Because $\partial^\circ(B, \cdot)$ counts the number of instances of the GLR rule with conclusion sequents containing the formula occurrence B°, the result is an immediate consequence of the fact that $\oplus$ does not introduce formulae into the conclusion sequent of an instance of the GLR rule (see Definition 5(4)).

(b) By the definition of compatibility, the binding rule for $d[\mathsf{stub}]$ and d' cannot be GLR. Thus if an instance ρ of the GLR rule appears in $d[\mathsf{stub}] \hookleftarrow d'$, then ρ must appear in one of $d[\mathsf{stub}]$ or d'. Also, if an instance ρ of the GLR rule appears in either $d[\mathsf{stub}]$ or d', then ρ must appear in $d[\mathsf{stub}] \hookleftarrow d'$. The result follows immediately.

■

REMARK 12. Lemma 11(a) holds even if G contains decorated formulae.

DEFINITION 13 (width). Let cut_0 be a topmost cut as shown below:

$$\dfrac{\dfrac{\{\pi\}_1^r}{X \Rightarrow Y, B}\,\rho \qquad \dfrac{\{\sigma\}_1^s}{B, U \Rightarrow W}}{X, U \Rightarrow Y, W}\,cut_0$$

Then, the width of cut_0 is defined as:

$$width(cut_0) = \begin{cases} \partial^\circ(B, f_B[\pi_1]) & \text{if } \rho = GLR \text{ (so } \{\pi\}_1^r = \pi_1) \\ GLR(2, \{\pi\}_1^r / X \Rightarrow Y, B) & \text{otherwise} \end{cases}$$

REMARK 14.

(i) The width has been defined only for a topmost cut as this context is sufficient for our purposes.

(ii) $width(cut_0)$ is independent of the right premise derivation of cut_0.

EXAMPLE 15. Let us calculate $width(cut_0)$ in the following:

$$\dfrac{\dfrac{\dfrac{\dfrac{\dfrac{\{\pi\}_1^r}{\Box C, C, \Box\Box B, \Box B, \Box B \Rightarrow B}}{\Box C, \Box\Box B \Rightarrow \Box B}\,GLR \qquad \dfrac{\dfrac{\{\sigma\}_1^s}{\Box D \Rightarrow \Box B}}{\Box D, \Box\Box B \Rightarrow \Box B}\,LW}{\Box C \vee \Box D, \Box\Box B \Rightarrow \Box B}\,L\vee}{\Box(\Box C \vee \Box D), \Box C \vee \Box D, \Box\Box B \Rightarrow \Box B}\,LW}{\Box(\Box C \vee \Box D) \Rightarrow \Box\Box B}\,GLR \qquad \Box\Box B, U \Rightarrow W}{\Box(\Box C \vee \Box D), U \Rightarrow W}\,cut_0$$

Writing the left premise derivation of cut_0 as $\mu / \Box(\Box C \vee \Box D) \Rightarrow \Box\Box B$, we get $width(cut_0) = \partial^\circ(\Box\Box B, f_{\Box\Box B}[\mu])$ where $f_{\Box\Box B}[\mu]$ is

$$\dfrac{\dfrac{\dfrac{\dfrac{\{\pi\}_1^r}{\Box C, C, \Box\Box B, \Box B, \Box B \Rightarrow B}}{\Box C, (\Box\Box B)^\circ \Rightarrow \Box B}\,GLR \qquad \dfrac{\dfrac{\{\sigma\}_1^s}{\Box D \Rightarrow \Box B}}{\Box D, (\Box\Box B)^\circ \Rightarrow \Box B}\,LW}{\Box C \vee \Box D, (\Box\Box B)^* \Rightarrow \Box B}\,L\vee}{\Box(\Box C \vee \Box D), \Box C \vee \Box D, (\Box\Box B)^* \Rightarrow \Box B}\,LW$$

Because $f_{\Box\Box B}[\mu]$ contains only one GLR rule whose conclusion contains the formula occurrence $(\Box\Box B)^\circ$ in its antecedent, we have $width(cut_0) = 1$.

REMARK 16. Let μ be the left premise derivation of cut_0 from Definition 13. Valentini [16, pg 473] defines the width as the cardinality of $GLR^{(2)}$, where $GLR^{(2)}$ in our notation is the set of all instances ρ of GLR such that:

(a) ρ is a 2-ary GLR rule in μ, and

(b) B is the diagonal formula of every 1-ary GLR rule in μ below ρ, and

(c) B is not introduced by weakening below ρ.

Applying Valentini's original definition to the following derivation in GLS

$$\dfrac{\dfrac{\dfrac{\dfrac{\dfrac{\{\pi\}_1^r}{\Box\Box X, \Box X, \Box X, X, \Box\Box C, \Box C, \Box C \Rightarrow C}}{\Box\Box X, \Box X, \Box\Box C \Rightarrow \Box C}\,GLR}{\Box\Box X, \Box X, \Box\Box C, \Box\Box C \Rightarrow \Box C}\,LW(\Box\Box C)}{\Box\Box X, \Box X, \Box\Box C \Rightarrow \Box C}\,LC(\Box\Box C)}{\Box\Box X \Rightarrow \Box\Box C}\,GLR \quad \Box\Box C, U \Rightarrow W}{\Box\Box X, U \Rightarrow W}\,cut_0$$

we compute the width of cut_0 as 0 (due to condition *(c)*). Using the definition in this paper we have $width(cut_0) = 1$. Our definition considers the interplay of the weakening and contraction rules, and is required to obtain the cut-elimination result for GLS. In GLS_V however, there are no contraction rules so Valentini's original definition suffices.

Thus Moen is certainly justified in asking whether Valentini's arguments can be lifted to multiset-based sequents. However, we will see that Moen's claims about failure of cut-elimination in the new setting are incorrect.

4 Cut-elimination for GLS

We have the following decomposition lemma.

LEMMA 17. *Let τ be a cut-free derivation of the form $\{\pi\}_1^r/^{\rho}X, \Box B \Rightarrow Y$ and suppose that $\partial^\circ(\Box B, f_{\Box B}[\tau]) > 0$. If*

(i) $\rho = GLR$ then $f_{\Box B}[\tau] = \{\pi\}_1^r/^{GLR}X, (\Box B)^\circ \Rightarrow Y$.

(ii) $\rho \neq GLR$ then we can write the annotated derivation $f_{\Box B}[\tau]$ in the form $d[\mathsf{stub}] \leftarrow\!\!\dashv d'$ such that

$$\partial^\circ(\Box B, d[\mathsf{stub}] \leftarrow\!\!\dashv d') = \partial^\circ(\Box B, d[\mathsf{stub}]) + \partial^\circ(\Box B, d').$$

Furthermore, denote the endsequent of d' as $U \Rightarrow W$. Then for any multiset M, and any derivation d'' with endsequent $U, M \Rightarrow W$, we have that $d[\mathsf{stub}] \oplus M$ and d'' are compatible.

Proof. If $\rho = GLR$ then it follows immediately from Definition 8 that $f_{\Box B}[\tau] = \{\pi\}_1^r/^{GLR}X, (\Box B)^\circ \Rightarrow Y$.

Now suppose that $\rho \neq GLR$. Because $\partial^\circ(\Box B, f_{\Box B}[\tau]) > 0$, we can write $f_{\Box B}[\tau]$ in the form below, where $n \geq 1$, and $\Box G$ contains no annotated formulae, and $\eta/X, (\Box B)^* \Rightarrow Y$ is an annotated sstub-derivation possibly containing branches:

$$\dfrac{\dfrac{\{\pi'\}_1^s}{\Box G, G, (\Box B)^n, B^n, \Box A \Rightarrow A}}{\Box G, (\Box B)^{\circ n} \Rightarrow \Box A}\,GLR_1$$
$$\eta$$
$$X, (\Box B)^* \Rightarrow Y$$

We can identify the annotated derivation $f_{\Box B}[\tau]$ with $d[\mathsf{stub}] \leftarrow\!\!| \; d'$ where d' (below left) is an annotated derivation and $d[\mathsf{stub}]$ (below right) is an annotated sstub-derivation.

$$\dfrac{\dfrac{\{\pi'\}_1^s}{\Box G, G, (\Box B)^n, B^n, \Box A \Rightarrow A}}{\Box G, (\Box B)^{\circ n} \Rightarrow \Box A}\,GLR \qquad\qquad \begin{array}{c}\mathsf{stub}\\ \eta \\ X, (\Box B)^* \Rightarrow Y\end{array}$$

From Lemma 11(b) we have

$$\partial^\circ(\Box B, f_{\Box B}[\tau]) = \partial^\circ(\Box B, d[\mathsf{stub}] \leftarrow\!\!| \; d') = \partial^\circ(\Box B, d[\mathsf{stub}]) + \partial^\circ(\Box B, d').$$

Write the endsequent of d' as $U \Rightarrow W$. Observe that GLR_1 must be a 1-ary GLR rule in $f_{\Box B}[\tau]$. If this were not the case, the antecedent of the conclusion of GLR_1 could not contain occurrences of B°. Thus the path (through η) between the leaf stub in $d[\mathsf{stub}]$ and the endsequent $X, (\Box B)^* \Rightarrow Y$ of $d[\mathsf{stub}]$ contains no GLR rule instances. From Definition 5 and the compatibility of $d[\mathsf{stub}]$ and d', for any multiset M and any derivation d'' with endsequent $U, M \Rightarrow W$, it follows that $d[\mathsf{stub}] \oplus M$ and d'' are compatible. ■

DEFINITION 18 (rank of a cut). For a topmost cut cut_0 the *rank* $rk(cut_0)$ is the triple (d, n, h) where d is the degree of the cut-formula, n is the width of cut_0, and h is the cut-height of cut_0.

LEMMA 19. *Let τ be the following derivation where cut_0 is a topmost cut:*

$$\dfrac{\dfrac{\dfrac{\{\pi\}_1^r}{\Box X, X, \Box B \Rightarrow B}}{\Box X \Rightarrow \Box B}\,GLR \qquad \dfrac{\{\sigma\}_1^s}{\Box B, U \Rightarrow W}}{\Box X, U \Rightarrow W}\,cut_0$$

and suppose ($\star$): for any derivation δ, every topmost cut in δ with rank $<$ $rk(cut_0)$ is eliminable.

Then there is a transformed cut-free derivation τ' of $X, \Box X \Rightarrow B$.

Proof. Let μ denote the subderivation $\{\pi\}_1^r / \Box X, X, \Box B \Rightarrow B$ of τ.

Case width($\mathbf{cut_0}$) $= \mathbf{0}$: Hence $\partial^\circ(\Box B, f_{\Box B}[\mu]) = 0$. Then the annotated derivation $f_{\Box B}(\mu)$ must have final parametric ancestors of the form $(\Box B)^\circ \Rightarrow \Box B$ or $X' \Rightarrow Y' /^{LW(\Box B)} X', (\Box B)^\circ \Rightarrow Y'$ only.

Let $\Box B^{(*|\circ)}$ stand for an annotated occurrence of $\Box B$ where the annotation is not known. Consider the substitution $(f_{\Box B}[\mu])\{\Box B^{(*|\circ)} := \Box X\}$ obtained by replacing every occurrence

1. of $(\Box B)^*$ with $\Box X$, and
2. of $(\Box B)^\circ \Rightarrow \Box B$ with a derivation of $\Box X \Rightarrow \Box B$ (the left premise derivation of cut_0), and
3. of $\dfrac{X' \Rightarrow Y'}{X', (\Box B)^\circ \Rightarrow Y'}\,LW(\Box B)$ with $\dfrac{X' \Rightarrow Y'}{X', \Box X \Rightarrow Y'}\,LW^*(\Box X)$

As the endsequent of $f_{\Box B}[\mu]$ was $\Box X, X, (\Box B)^* \Rightarrow B$ we have that $(f_{\Box B}[\mu])\{\Box B^{(*|\circ)} := \Box X\}$ is a cut-free derivation of $\Box X, X, \Box X \Rightarrow B$. Applying repeated left contraction gives a cut-free derivation of $\Box X, X \Rightarrow B$.

Case width(cut$_0$) > 0 : Hence $\partial^\circ(\Box B, f_{\Box B}[\mu]) > 0$. First suppose that the last rule in μ is GLR. Then μ must be of the form:

$$\dfrac{\dfrac{\{\pi'\}_1^s}{\Box\Box X', \Box X', \Box X', X', \Box\Box A, \Box A, \Box A \Rightarrow A}}{\Box\Box X', \Box X', \Box\Box A \Rightarrow \Box A}\; GLR$$

where $X = \Box X'$ and $B \equiv \Box A$.

Then the following is a derivation of $\Box X, X \Rightarrow B$, with $deg(cut_1) = deg(cut_0)$ and $width(cut_1) = 0 < width(cut_0)$:

$$\dfrac{\dfrac{\dfrac{\dfrac{\dfrac{\Box A \Rightarrow \Box A}{\Box A, A, \Box\Box A \Rightarrow \Box A}\; LW^*(A, \Box\Box A)}{\Box A \Rightarrow \Box\Box A}\; GLR \qquad \dfrac{\{\pi'\}_1^s}{\Box\Box X', \Box X', \Box X', X', \Box\Box A, \Box A, \Box A \Rightarrow A}}{\Box A, \Box\Box X', \Box X', \Box X', X', \Box A, \Box A \Rightarrow A}\; cut_1}{\Box\Box X', \Box X', \Box X', X', \Box A \Rightarrow A}\; LC^*(\Box A)}{\Box\Box X', \Box X' \Rightarrow \Box A}\; GLR$$

The required derivation is obtained by using ($\star$) to eliminate cut_1.

If the last rule in μ is not GLR, by Lemma 17 we can write $f_{\Box B}[\mu]$ as $d[\mathsf{stub}] \leftarrow\!\!\dashv d'$, where d' and $d[\mathsf{stub}]$ are respectively:

$$\dfrac{\dfrac{\{\pi'\}_1^t}{\Box G, G, (\Box B)^n, B^n, \Box A \Rightarrow A}}{\Box G, (\Box B)^{\circ n} \Rightarrow \Box A}\; GLR \qquad\qquad \begin{matrix}\mathsf{stub}\\ \eta \\ \Box X, X, (\Box B)^* \Rightarrow B\end{matrix}$$

where $n \geq 1$, and $\Box G$ does not contain annotated formulae, and

$$\partial^\circ(\Box B, d[\mathsf{stub}] \leftarrow\!\!\dashv d') = \partial^\circ(\Box B, d[\mathsf{stub}]) + \partial^\circ(\Box B, d').$$

Let d'' be the annotated derivation

$$\dfrac{\Box A \Rightarrow \Box A}{A, \Box A, \Box G, (\Box B)^{\circ n} \Rightarrow \Box A}\; LW^*(A, \Box G, (\Box B)^n)$$

Then $d[\mathsf{stub}] \oplus \langle A, \Box A\rangle$ and d'' are compatible (Lemma 17). Note that $\partial^\circ(\Box B, d') = 1$ and $\partial^\circ(\Box B, d'') = 0$. Let Λ_{11} be the derivation:

$$\dfrac{\dfrac{\dfrac{\lfloor d[\mathsf{stub}] \oplus \langle A, \Box A\rangle \leftarrow\!\!\dashv d''\rfloor}{\Box A, \Box X \Rightarrow \Box B}\; GLR \qquad \dfrac{\{\pi\}_1^r}{\Box X, X, \Box B \Rightarrow B}}{\Box A, \Box X, \Box X, X \Rightarrow B}\; cut_1}{\Box A, \Box X, X \Rightarrow B}\; LC^*$$

and Λ_{12} the derivation

$$\dfrac{\dfrac{\dfrac{\lfloor d[\mathsf{stub}] \oplus \langle A, \Box A\rangle \leftarrow\!\!\dashv d''\rfloor}{\Box A, \Box X \Rightarrow \Box B}\; GLR \qquad \dfrac{\{\pi'\}_1^t}{\Box G, G, (\Box B)^n, B^n, \Box A \Rightarrow A}}{\Box A, \Box A, \Box X, \Box G, G, (\Box B)^{n-1}, B^n \Rightarrow A}\; cut_2}{\Box A, \Box X, \Box G, G, (\Box B)^{n-1}, B^n \Rightarrow A}\; LC$$

Consider the derivation Λ_1:

$$\dfrac{\dfrac{\dfrac{\Lambda_{11} \qquad \Lambda_{12}}{\Box A, \Box X, X, \Box A, \Box X, \Box G, G, (\Box B)^{n-1}, B^{n-1} \Rightarrow A}\ cut_3}{\Box A, \Box X, X, \Box G, G, (\Box B)^{n-1}, B^{n-1} \Rightarrow A}\ LC^*}{\Box X, \Box G, (\Box B)^{n-1} \Rightarrow \Box A}\ GLR$$

For $i \in \{1, 2\}$, observe that $deg(cut_i) = deg(cut_0)$. Furthermore,

$$\begin{aligned}
width(cut_i) &= \partial^\circ(\Box B, f_{\Box B}(\lfloor d[\mathsf{stub}] \oplus \langle A, \Box A\rangle \hookleftarrow d''\rfloor)) && \text{Def. of width}\\
&= \partial^\circ(\Box B, d[\mathsf{stub}] \oplus \langle A, \Box A\rangle \hookleftarrow d'') && \text{By inspection}\\
&= \partial^\circ(\Box B, d \oplus \langle A, \Box A\rangle[\mathsf{stub}]) + \partial^\circ(\Box B, d'') && \text{Lemma 11(b)}\\
&< \partial^\circ(\Box B, d \oplus \langle A, \Box A\rangle[\mathsf{stub}]) + \partial^\circ(\Box B, d') &&\\
&= \partial^\circ(\Box B, d[\mathsf{stub}]) + \partial^\circ(\Box B, d') && \text{Lemma 11(a)}\\
&= \partial^\circ(\Box B, d[\mathsf{stub}] \hookleftarrow d') && \text{Lemma 11(b)}\\
&= width(cut_0)
\end{aligned}$$

Because $deg(cut_i) = deg(cut_0)$ and the premises of both cut_1 and cut_2 are cut-free, by appealing twice to $(\star)$ we can in turn eliminate cut_1 and cut_2. In the resulting derivation, since $deg(cut_3) < deg(cut_0)$ we can eliminate cut_3 by $(\star)$. We thus obtain a cut-free derivation Λ_2 of $\Box X, \Box G, (\Box B)^{n-1} \Rightarrow \Box A$.

Let Λ_3 be the annotated derivation

$$\dfrac{\Lambda_2}{\Box X, \Box G, (\Box B)^n \Rightarrow \Box A}\ LW(\Box B)$$

Clearly $\partial^\circ(\Box B, \Lambda_3) = 0$. Furthermore, by Lemma 17, $d[\mathsf{stub}] \oplus \Box X$ and Λ_3 are compatible. Recall that $\lfloor\cdot\rfloor$ is the forgetful map. The endsequent of $\lfloor(d[\mathsf{stub}] \oplus X) \hookleftarrow \Lambda_3\rfloor$ is thus $\Box X, \Box X, X, \Box B \Rightarrow B$. Now consider the derivation

$$\dfrac{\dfrac{\dfrac{\lfloor(d[\mathsf{stub}] \oplus \Box X) \hookleftarrow \Lambda_3\rfloor}{\Box B, \Box X, X \Rightarrow B}\ LC^*(\Box X)}{\Box X \Rightarrow \Box B}\ GLR \qquad \dfrac{\{\pi\}_1^r}{\Box X, X, \Box B \Rightarrow B}}{\dfrac{X, \Box X, \Box X \Rightarrow B}{X, \Box X \Rightarrow B}\ LC^*(\Box X)}\ cut_4(\Box B)$$

By a similar calculation to the above we obtain $width(cut_4) < width(cut_0)$. Because $deg(cut_4) = deg(cut_0)$ and the premises of cut_4 are cut-free, appealing to $(\star)$ we can eliminate cut_4. We thus obtain a cut-free derivation of $X, \Box X \Rightarrow B$ as required. ■

Without loss of generality it suffices to consider a derivation concluded by a cut-rule with cut-free premise derivations.

THEOREM 20 (Cut-elimination). *Let τ be a derivation concluded by an instance cut_0 of the cut-rule with cut-free premise derivations. Then there is a transformed cut-free derivation τ' with identical endsequent.*

Proof. Induction on the rank (d, n, h) of cut_0 under the standard lexicographic ordering. We say that the cut-formula is *left principal* if an occurrence of the cut-formula in the succedent of the left premise is a principal formula. The term *right principal* is defined analogously. This follows standard practice.

1 Cut with an initial sequent as premise. This is the base case. The transformations are standard (see [9],[15]).

2 Cut with neither premise an initial sequent.

(a) Cut-formula is left and right principal.

First suppose that the cut-formula is boxed. There are five possibilities:

(i) the cut-formula is left and right principal by the GLR rule. The derivation must then be in SNF:

$$\dfrac{\dfrac{\dfrac{\{\pi\}_1^r}{\Box X, X, \Box B \Rightarrow B}}{\Box X \Rightarrow \Box B}\,GLR \qquad \dfrac{\dfrac{\{\sigma\}_1^s}{\Box B, \Box U, B, U, \Box C \Rightarrow C}}{\Box B, \Box U \Rightarrow \Box C}\,GLR}{\Box X, \Box U \Rightarrow \Box C}\,cut_0$$

The induction hypothesis implies that for any derivation δ, any topmost cut in δ with rank $< rank(cut_0)$ is eliminable. This is precisely condition $(\star)$ in Lemma 19. Hence we can obtain a cut-free derivation of $\Box X, X \Rightarrow B$. Consider the derivation

$$\dfrac{\dfrac{\Box X, X \Rightarrow B \qquad \dfrac{\dfrac{\dfrac{\{\pi\}_1^r}{\Box X, X, \Box B \Rightarrow B}}{\Box X \Rightarrow \Box B}\,GLR \qquad \dfrac{\{\sigma\}_1^s}{\Box B, \Box U, B, U, \Box C \Rightarrow C}}{\Box X, \Box U, B, U, \Box C \Rightarrow C}\,cut_1}{\Box X, X, \Box X, \Box U, U, \Box C \Rightarrow C}\,cut_2}{\Box X, \Box U \Rightarrow \Box C}\,GLR$$

Observe that $rk(cut_1) = (d, n, h-1)$. By the induction hypothesis we can eliminate cut_1. In the resulting derivation, since $deg(cut_2) < d$, the result follows from another application of the induction hypothesis.

(ii) the cut-formula $\Box B$ is left principal by the GLR rule and right principal by $LC(\Box B)$.

Then τ is as below where both premises of cut_0 are cut-free and $m \geq 0$:

$$\dfrac{\dfrac{\dfrac{\{\pi\}_1^r}{\Box X, X, \Box B \Rightarrow B}}{\Box X \Rightarrow \Box B}\,GLR \qquad \dfrac{\dfrac{\{\sigma\}_1^s}{(\Box B)^{m+2}, U \Rightarrow W}\,\rho}{\Box B, U \Rightarrow W}\,LC^{m+1}(\Box B)}{\Box X, U \Rightarrow W}\,cut_0$$

In general ρ need not be the GLR rule. However if $\rho \neq GLR$ then either (1) $\rho = LW(\Box B)$ and we delete ρ and the $LC(\Box B)$ rule that follows, or (2) $\Box B$ is not principal by ρ.

In the former case the result is immediate. In the latter case the result is obtained by applying ρ *after* cut_0 and invoking the induction hypothesis. Observe that this is possible even if ρ is a binary rule.

If $\rho = GLR$ it follows that $U \equiv \Box V$ and $W \equiv \Box C$ for some multiset V and formula C, and $s = 1$ and $\sigma_1 \equiv \{\sigma'\}_1^{s'}/(\Box B)^{m+2}, B^{m+2}, \Box V, V, \Box C \Rightarrow C$. Thus τ must be of the form

$$\dfrac{\dfrac{\dfrac{\{\pi\}_1^r}{\Box X, X, \Box B \Rightarrow B}}{\Box X \Rightarrow \Box B}\,GLR \qquad \dfrac{\dfrac{\dfrac{\{\sigma'\}_1^{s'}}{(\Box B)^{m+2}, B^{m+2}, \Box V, V, \Box C \Rightarrow C}}{(\Box B)^{m+2}, \Box V \Rightarrow \Box C}\,\rho = GLR}{\Box B, \Box V \Rightarrow \Box C}\,LC^{m+1}(\Box B)}{\Box X, \Box V \Rightarrow \Box C}\,cut_0$$

A derivation of $\Box X, X \Rightarrow B$ is obtained as in (i) using Lemma 19. Consider the derivation:

$$\dfrac{\dfrac{\dfrac{\Box X, X \Rightarrow B \qquad \dfrac{\dfrac{\Box X \Rightarrow \Box B \qquad \dfrac{\dfrac{\{\sigma'\}_1^{s'}}{(\Box B)^{m+2}, B^{m+2}, \Box V, V, \Box C \Rightarrow C}}{\Box B, B^{m+2}, \Box V, V, \Box C \Rightarrow C}\,LC^{m+1}(\Box B)}{\Box X, B^{m+2}, \Box V, V, \Box C \Rightarrow C}\,cut_1}{\Box X, B, \Box V, V, \Box C \Rightarrow C}\,LC^{m+1}(B)}{\Box X, X, \Box X, \Box V, V, \Box C \Rightarrow C}\,cut_2}{\Box X, X, \Box V, V, \Box C \Rightarrow C}\,LC^*}{\Box X, \Box V \Rightarrow \Box C}\,GLR$$

Now cut_1 has identical degree and width compared to cut_0, and smaller cut-height. Hence, we can eliminate cut_1 using the induction hypothesis. In the resulting derivation $deg(cut_2) < d$ so the result follows from the induction hypothesis.

(iii) the cut-formula $\Box B$ is left principal by $RC(\Box B)$ and right principal by the GLR rule.

Then τ has the following form where both premises of cut_0 are cut-free:

$$\dfrac{\dfrac{\dfrac{\{\pi\}_1^r}{X \Rightarrow Y, \Box B, \Box B}}{X \Rightarrow Y, \Box B}\,RC_1 \qquad \dfrac{\sigma}{\Box B, \Box U \Rightarrow \Box C}\,GLR}{X, \Box U \Rightarrow Y, \Box C}\,cut_0$$

Because the conclusion of (any) GLR rule has a exactly one formula in the succedent, it follows that at least one of the $\Box B$ formula occurrences in the succedent of the premise of RC_1 can be traced upwards in $\{\pi\}_1^r$ to $RW(\Box B)$ rule application(s). In particular, when tracing upwards, it is impossible to encounter a GLR rule application *before* encountering a $RW(\Box B)$ rule application. Deleting these $RW(\Box B)$ rule applications and the RC_1 contraction rule certainly preserves the derivation structure because all the binary rules excluding the cut-rule are additive. This new derivation contains a single instance of cut with identical degree of cut-formula and reduced cut-height compared to cut_0. Furthermore, observe that it must be the case that the width is $\leq n$. The result follows from the induction hypothesis.

If the calculus uses multiplicative binary rules instead, the result still holds, although the transformations are slightly more complicated.

In each instance, the proof can be formalised using an annotation function similar in structure to $f_{\Box B}$. We omit the details.

(iv) the cut-formula $\Box B$ is left and right principal by $RC(\Box B)$ and $LC(\Box B)$ respectively.

A combination of the strategies in (ii) and (iii) suffice.

(v) the cut-formula $\Box B$ is either left or right principal by $RW(\Box B)$ or $LW(\Box B)$ respectively.

Trivial.

When the cut-formula is not boxed and the cut-formula is left and right principal by the respective left and right introduction rule the transformations are standard (see [9],[15] for example) — derivation τ is transformed to a derivation τ' containing cuts $\{cut_i\}_{i\geq 1}$ on strictly smaller cut-formulae (i.e. $deg(cut_i) < d$ for $i \geq 1$).

If the cut-formula is right principal by $LC(B)$ then τ has the form below where B is principal by ρ:

$$\dfrac{\dfrac{\{\pi\}_1^r}{X \Rightarrow Y, B}\rho \qquad \dfrac{\dfrac{\{\sigma\}_1^s}{B, B, U \Rightarrow W}}{B, U \Rightarrow W} LC(B)}{X, U \Rightarrow Y, W} cut_0$$

We must have $\rho \neq GLR$. This is the well-known case of contractions above cut that occurs in cut-elimination for LK.

There are several proofs of cut-elimination avoiding Gentzen's multicut rule, for classical sequent calculi, appearing in the literature (for example, see [10],[3],[1]). We adapt the transformations proposed by von Plato [10] for the classical calculus $G0c$. Although all the binary rules in $G0c$ are multiplicative, and all the binary rules in GLS are additive, the same argument can be lifted here.

That argument in [10] relies on invertibility of all logical rules in $G0c$. Invertibility is not required to be height-preserving. A similar result holds for GLS too. We omit the details.

The cases corresponding to (iii)-(v) can be dealt with similarly.

(b) Cut-formula is left principal only.

(c) Cut-formula neither left nor right principal.

We analyse the last inference rule in the *right (left)* premise derivation of cut_0. The standard transformations suffice here (see [9],[15] for example). In particular, observe that for any instance cut_1 of the cut-rule appearing in a transformed derivation, it must be the case that $width(cut_1) \leq n$. ■

REMARK 21. In general, it is possible for the width of lower cuts to increase under the cut-elimination transformations. For example, consider some transformation which reduces some topmost cut instance cut_b (for "before") to the derivation below containing the cut instance cut_a (for "after") where $\{\pi\}_1^r$ and $\{\sigma\}_1^s$ need not be cut-free:

$$\dfrac{\{\pi\}_1^r \qquad \{\sigma\}_1^s}{G \Rightarrow H} cut_a$$

The cut-elimination transformations which ultimately turn cut_a into a topmost cut may produce a derivation where $width(cut_a) > width(cut_b)$. However, we have seen in the proof of Lemma 19 that $width(cut_4)$ does not increase despite any reductions above it. This is because the cut_4 in that proof is 'shielded' by the GLR instance concluding Λ_1. This shielding is crucial for the success of the proof.

5 Moen's Val-II(core) is not Valentini's reduction

We have carefully examined Moen's slides titled "The proposed algorithms for eliminating cuts in the provability calculus GLS do not terminate" [7].

Moen sets out to reduce a cut in SNF using the transformation he titles Val-II(core). Moen claims that Val-II(core) is the "...core of Valentini's reduction" [7]. Yet Val-II(core) does not appear in [16]. However it appears in [11, page 322] with the comment "this reduction is not sufficient".

Thus Moen is incorrect in claiming that he has demonstrated that Valentini's algorithm does not terminate — Moen is using the wrong algorithm. In fact, for his concrete derivation ϵ, the width of the cut-formula is 1 so the reduction is immediate. Applying the base case transformations, and then the classical transformations, we obtained a cut-free derivation of the end-sequent of ϵ.

6 Conclusion

We have resolved the issue surrounding the use of Valentini's arguments for cut-elimination in a multiset setting for GL. In order to formally define the measure width, we formalised the notion of 'tracing up' a derivation (i.e. identifying the parametric ancestors) via a constructive function. This constructive function can be used to aid the formalisation of various other notions in proof theory.

BIBLIOGRAPHY

[1] K. Bimbó. $LE^t_{\rightarrow}$, $LR^{\circ}_{\underset{\sim}{\wedge}}$, LK and cut-free proofs. *Journal of Philosophical Logic*, 36:557–570, 2007.

[2] M. Borga. On Some Proof Theoretical Properties of the Modal Logic GL. *Studia Logica*, 42:453–459, 1983.

[3] M. Borisavljević, K. Došen and Z. Petrić. On permuting cut with contraction. *Math. Struct. in Comp. Science*, 10:99–136, 2000.

[4] G. Gentzen. The Collected Papers of Gerhard Gentzen, ed. M. Szabo.

[5] D. Leivant. On the Proof Theory of the Modal Logic for Arithmetic Provability. *Journal of Symbolic Logic*, 46:531–538, 1981.

[6] G. Mints. Cut elimination for provability logic. Collegium Logicum 2005.

[7] A. Moen. The proposed algorithms for eliminating cuts in the provability calculus GLS do not terminate. *NWPT 2001*, Norwegian Computing Center, 2001-12-10. http://publ.nr.no/3411

[8] S. Negri. Proof Analysis in Modal Logic. *Journal of Philosophical Logic*, 34:507–544, 2005.

[9] S. Negri and J. von Plato. *Structural Proof Theory*. CUP, 2001.

[10] J. von Plato. A proof of Gentzen's *Hauptsatz* without multicut. *Archive of Mathematical Logic*, 40:9–18, 2001.

[11] G. Sambin and S. Valentini. The Modal Logic of Provability. The Sequential Approach. *Journal of Philosophical Logic*, 11:311–342, 1982.
[12] K. Sasaki. Löb's Axiom and Cut-elimination Theorem. *Journal of the Nanzan Academic Society Math. Sci. and Information Engineering*, 1:91–98, 2001.
[13] V. Švejdar. On Provability Logic. *Nordic Journal of Philosophical Logic*, 4:95–116, 2000.
[14] R.M. Solovay. Provability Interpretations of Modal Logic. *Israel Journal of Mathematics*, 25:287–304, 1976.
[15] A.S. Troesltra and H. Schwichtenberg. *Basic Proof Theory.* CUP, 2000.
[16] S. Valentini. The Modal Logic of Provability: Cut-elimination. *Journal of Philosophical Logic*, 12:471–476, 1983.

Rajeev Goré
Computer Sciences Laboratory,
The Australian National University
Australia
Rajeev.Gore@anu.edu.au

Revantha Ramanayake
Computer Sciences Laboratory,
The Australian National University
Australia
revantha@rsise.anu.edu.au

Labelled modal tableaux

Guido Governatori

ABSTRACT. Labelled tableaux are extensions of semantic tableaux with annotations (labels, indices) whose main function is to enrich the modal object language with semantic elements. This paper consists of three parts. In the first part we consider some options for labels: simple constant labels vs labels with free variables, logic depended inference rules vs labels manipulation based on a label algebra. In the second and third part we concentrate on a particular labelled tableaux system called KEM using free variable and a specialised label algebra. Specifically in the second part we show how labelled tableaux (KEM) can account for different types of logics (e.g., non-normal modal logics and conditional logics). In the third and final part we investigate the relative complexity of labelled tableaux systems and we show that the uses of KEM's label algebra can lead to speed up on proofs.

Keywords: labelled tableaux, non-normal modal logic, conditional logic, relative complexity

1 Introduction

Since the seminal work by Fitch [12] labels have been widely used in modal logic to simulate possible world semantics in the proof theory to improve, simplify and speed up proofs. Usually the main function of labels is to "import" or simulate semantic structures in the object language. Accordingly, in semantic based proof methods (cf., among others, [32]), labels represent possible worlds and accessibility relations (using sequences of atomic labels) in Kripke models.

Semantic tableaux (cf., [35]) is one of the most common form of semantic based proof procedures and, we believe, it offers the best format for the use of labels. The basic idea is to supplement the object language with a label language and a label algebra. The basic entities of labelled deductions are labelled formulas, i.e., expressions of the form $A : x$, where x is a label drawn form the label language, and A (the declarative unit) is a well-formed formula of the logic at hand (cf., [14]). Intuitively the meaning of a labelled formula such as $A : x$ is that the declarative unit (A) is true at the world(s) denoted be the label x.

The structure of the paper is as follows: in Section 2 we introduce the formalism and we discuss some options to combine labels and tableaux. In Section 3 we examine how to extend labelled tableaux to cover other logics having possible world semantics, namely non-normal modal logic (modal

 © 2008, the author(s).

logic where necessitation does not hold) and conditional logics. Finally in Section 4 we investigate the relative complexity of two labelled tableaux systems presented in Section 2 and we discuss the general issue for the methodology to compare this kind of proof systems.

2 Labelled modal tableaux

As is well known semantic tableaux calculus is a refutation proof method[1]. Therefore a proof of A is a failed attempt to provide a model for $\neg A$. A tableaux for a formula A is a (binary) tree whose root is $A : i_0$ where i_0 is the initial label, and the nodes are derived from previous nodes according to the inference rules of the system. A branch is *closed* iff it contains a pair of complementary formulas (the notion of complementary formulas may vary from system to system), otherwise it is *open*; a tree is *closed* iff every branch in it is closed. Finally a proof of A is a closed tree with root $\neg A : i_0$. A tree is *complete* iff every rule that can be applied has been applied.

A labelled modal tableaux systems is defined by the structure of labels and the inference rules for analysing the formulas. In most systems (new) labelled formulas are generated from previous formulas using inference rules that closely resemble the semantic evaluation of the premises. Given the semantic conditions, it is indeed possible that the conclusion of a premise holds in a set of possible worlds instead of a single worlds; for example, just consider the semantic clause for $\Box A$ which requires A to be true in all worlds accessible from the world where $\Box A$ is currently evaluated. We have two alternative ways for representing such conclusions using labels:

1. we can use ground labels and generate all possible/relevant instances of such worlds;

2. we can use a label with a free-variable, where the variable is intended to range over such worlds.

The second issue we have to tackle is how to represent the structure of the model. Different modal logics determine different structures on Kripke frames (or better, possible world frames in the general case). Again we have multiple options.

1. We can define logic dependant inference rules assigning formulas to existing labels or generating new labels and formulas (see Section 2.2 for examples).

2. We can use a single logic neutral inference rule for a modal operator, make use of an explicit representation of the relevant semantics structures (e.g., the accessibility relation) and, then use an external mechanism to resolve and compute the semantic structure and propagate the formulas to the labels accordingly. For example for a transitive

[1]For a comprehensive account of tableaux for modal logic we refer the reader to [17].

logic one can have the following rules (see, among others [5, 16])

$$\frac{\Diamond A : w}{\begin{array}{l} A : w' \\ wRw' \end{array}} w' \text{ new on the branch} \qquad \frac{wRw', w'Rw''}{wRw''}$$

3. For the last alternative, the option we are going to investigate in the rest of the paper, we follow the previous case in so far as each modal operator has a single inference rule common to all logics and the various logics are differentiate by logic specific operations that manipulates the labels. In other words every logic has its own label 'algebra'.

In the rest of the section we are going to introduce a two label tableaux systems, both using free variables, the first SST adopts the first strategy, and the second KEM adopts the third strategy.

2.1 Label formalism

In this section we present the KEM label formalism. The formalism will also be used for SST. In fact most of the differences between the formalisms of the two systems are just notational ones, and the differences that are not notational are not relevant for the present investigation.

KEM has two basic kinds of atomic labels: variables and constants. Formally, let $\Phi_C = \{w_1, w_2, \dots\}$ and $\Phi_V = \{W_1, W_2, \dots\}$ be two arbitrary sets of *atomic labels*: the set of *constant world-symbols* (or simply *constants*) and the set of *variable world-symbols* (or simply *variables*). A *label* is then an element of the set of labels $\Im$ defined as follows:

DEFINITION 1. $\Im = \bigcup_{1 \leq p} \Im_p$ where $\Im_p$ is:

$$\begin{aligned} \Im_1 &= \Phi_C \cup \Phi_V \\ \Im_2 &= \Im_1 \times \Phi_C \\ \Im_{n+1} &= \Im_1 \times \Im_n, \ n > 1. \end{aligned}$$

Thus, a label i is either a variable or a constant or a "structured" sequence of atomic labels. For a structured label $i = (k', k)$ we have the following cases: (i) k' is an atomic world-symbol and (ii) $k \in \Phi_C$ or $k = (m', m)$ where (m', m) is a label. As we have alluded to in the previous section, we may think of constant and variable world-symbols as denoting respectively worlds and sets of worlds in a standard Kripke setting. A label of the form (k', k) is called a "world-path". For instance, the label (W_1, w_1) represents a path from w_1 to the set W_1 of worlds accessible from w_1; $(w_2, (W_1, w_1))$ represents a path which takes us to a world w_2 accessible by any world accessible from w_1 (i.e., accessible by the sub-path (W_1, w_1)) according to the appropriate accessibility relation. Thus a label of the form (k', k) is "structurally" designed to record information about the accessibility relation when we move from a label (a world or a set of worlds) to another label. We define the length of a label i, $\ell(i)$, as the number of atomic labels in i. From now on we shall use $i, j, k, \ldots$ to denote arbitrary labels.

DEFINITION 2. For a label $i = (j, k)$, we shall call j the *head* and k the *body* of i, and denote them by $h(i)$ and $b(i)$ respectively.

The notions of body and head are obviously recursive (they can be defined as projection functions), and allow us to identify any sub-label of a given label; thus, if $b(i)$ denotes the body of i, then $b(b(i))$ will denote the body of $b(i)$, $b(b(b(i)))$ will denote the body of $b(b(i))$, and so on. We call each of $b(i)$, $b(b(i))$, etc., a *segment* of i. Let $s(i)$ denote any segment of i (obviously, by definition every segment $s(i)$ of a label i is a label); $h(s(i))$ will denote the head of $s(i)$. With $s^n(i)$ we will denote the segment of i of length n, i.e., $s^n(i) = s(i)$ such that $\ell(s(i)) = n$. We shall use $h^n(i)$ as an abbreviation for $h(s^n(i))$. A label is *restricted* if its head is a constant, and *unrestricted* otherwise.

DEFINITION 3. For any label $i, \ell(i) \geq n$, we define the *countersegment-n* of i, as follows:

$$c^n(i) = h(i) \times (\cdots \times (h^k(i) \times (\cdots \times (h^{n+1}(i), w_0)))) \text{ for } n < k < \ell(i)$$

where w_0 is a dummy label, i.e., a label not appearing in i (the context in which such a notion is applied will tell us what w_0 stands for).

If $n = \ell(i)$ we have that $c^n(i) = w_0$, and $s^n(i) = i$.

EXAMPLE 4. If $i = (w_4, (W_3, (w_3, (W_2, w_1))))$, then $\ell(i) = 5$, $h^3(i) = w_3$, $s^3(i) = (w_3, (W_2, w_1))$, and its countersegment-3 is $c^3(i) = (w_4, (W_3, w_0))$; intuitively $c^n(i)$, is what remains of i after deleting $s^n(i)$.

To clarify the notion of countersegment, which will be used frequently in this work, we present, in the following table the list of the segments of i in the left-hand column and the relative countersegments in the right-hand column.

$s^1(i) = w_1$	$c^1(i) = (w_4, (W_3, (w_3, (W_2, w_0))))$
$s^2(i) = (W_2, w_1)$	$c^2(i) = (w_4, (W_3, (w_3, w_0)))$
$s^3(i) = (w_3, (W_2, w_1))$	$c^3(i) = (w_4, (W_3, w_0))$
$s^4(i) = (W_3, (w_3, (W_2, w_1)))$	$c^4(i) = (w_4, w_0)$
$s^5(i) = i$	$c^5(i) = w_0$

The dummy label w_0 is considered as an atomic label, and it is used to encapsulated complex labels into an atomic one.

We are now ready to introduce the two labelled tableaux systems.

2.2 Single step tableaux (SST)

Single Step Tableaux [27] originate from and add modularity to Fitting's prefix tableaux [13]. The free-variable version we shall focus on here has been proposed by Beckert and Goré [7].

The basic idea of SST is that (modal) formulas are used to move the evaluation point to the "neighbourhood" of the labels they are associated with, that is, each time we are allowed to move only one step apart. In other words the information that can be extracted from a formula is propagated only to the labels that the current label extends immediately or an are an immediate extension of the current label.

SST has the following inference rules. For the presentation of the inference rules of SST, and subsequently of KEM we shall assume familiarity with Smullyan-Fitting α, β, ν, π unifying notation [13]. For the propositional part we give only the rules for $\wedge$.

(α-rules)
$$\frac{A \wedge B : i}{\begin{array}{c} A : i \\ B : i \end{array}}$$

(β-rules)
$$\frac{\neg(A \wedge B) : i}{\neg A : i \quad | \quad \neg B : i}$$

(π-rules)
$$\frac{\Diamond A : i}{A : (w_{\lceil \pi \rceil}, i)} \qquad \frac{\neg\Box A : i}{\neg A : (w_{\lceil \pi \rceil}, i)}$$

where $\lceil \cdot \rceil$ is an arbitrary but fixed bijection from the set of formulas to N

(ν_D-rules)
$$\frac{\Box A : i}{A : (W_n, i)} \qquad \frac{\neg\Diamond A : i}{\neg A : (W_n, i)}$$

(ν_T-rules)
$$\frac{\Box A : i}{A : i} \qquad \frac{\neg\Diamond A : i}{A : i}$$

(ν_4-rules)
$$\frac{\Box A : i}{\Box A : (W_n, i)} \qquad \frac{\neg\Diamond A : i}{\neg\Diamond A : (W_n, i)}$$

(ν_{4^r}-rules)
$$\frac{\Box A : i}{\Box A : b(i)} \qquad \frac{\neg\Diamond A : i}{\neg\Diamond A : b(i)}$$

(ν_B-rules)
$$\frac{\Box A : i}{A : b(i)} \qquad \frac{\neg\Diamond A : i}{\neg A : b(i)}$$

(ν_5-rules)
$$\frac{\Box A : i}{\Box A : (W_n, h^1(i))} \qquad \frac{\neg\Diamond A : i}{\neg\Diamond A : (W_n, h^1(i))}$$

In the above rules W_n must a be a new label, i.e., a label that does not previously occur in the tableaux. The ν_D-rules are also known as the ν_K-rules. SST has an additional mechanism to keep track of which labels are denoting, and the mechanism essentially differentiates between serial and non-serial logics (see [7] for the details).

The α-, π-, and ν_D-rules are common to KEM and SST and the β-rules are the usual branching rules of tableau methods. The ν_T-rules are the rules specific to reflexive logics; the ν_4-rules for the transitive logics; the ν_{4^r}-rules and ν_5-rules for Euclidean logics; and finally the ν_B-rules are the specific rules for symmetric logics. The intuition behind these logic is that we 'move' a formula to a labels that is one single step (using the accessibility relation) from the labels the formula in the antecedent is associated to. Thus for the ν_B-rules the idea is that symmetry allows us to travel backward in the accessibility relation. For the full list of characterisation of the fifteen basic modal logics see [7].

The tableaux system for a logic is given by a combination of the above rules. For example SST for S4 has the following (modal) rules: π, ν_D, ν_T and ν_4; and the symmetric and serial logic DB is characterised by the rules π, ν_D and ν_B. The main consequence of n sets of ν-rule is that every time we have a formula of type ν we have to introduce n new labelled formulas.

We say that two labelled formulas $A : i$ and $B : j$ are complementary in SST when $B = \neg A$ and there exists a substitution ρ which is a unifier of i and j.

Let L be one of the fifteen basic modal logics. With $\vdash_{\mathrm{SST(L)}} A$ we mean that there is a close SST-tree for $\neg A : w_1$; or, in other words, that SST proves that A is a theorem of L.

THEOREM 5. $\vdash_{\mathrm{SST(L)}} A$ *iff* $\models_{\mathsf{L}} A$.

For the proof and for detailed accounts of SST see [7, 29].

2.3 KEM

KEM (see [18, 1, 20]) is a labelled analytic proof system based on a combination of tableau and natural deduction inference rules which allows for a suitably restricted ("analytic") application of the cut rule and a specialised, yet modular, unification mechanism for the labels.

Unifications

In the course of proofs labels are manipulated in a way closely related to the semantics of the logic under analysis. Labels are compared and matched using a specialised logic dependent unification mechanism. The notion of two labels i and j being unifiable means that the intersection of their denotations is not empty and that we can "move" to such a set of worlds through the path corresponding to the result of the unification of the two labels.

The definition of the unification appropriate for the various logics (or logic unification) is carried out in several steps with the help of several auxiliary notions of unification.

First we have to provide the foundation of our unification (σ-unification). The basic unification is defined, as usual, in terms of a substitution, then we use the basic unification to define the unifications corresponding to the various modal axioms (axiom unifications); in the same way a modal logic is obtained by combining several axioms we combine the axiom unifications in combined unification. Finally we apply, in a recursive way, the combined unification to define the unification for the logic (logic unification).

Before presenting the formal machinery for the various unifications we have to give the notation used for them. Let L be a modal logic, and $\mathsf{A}_1, \ldots, \mathsf{A}_n$ be the axioms of L. With σ^{A_i} we denote the unification for the axiom A_i; with $\sigma^{\mathsf{A}_1 \ldots \mathsf{A}_n}$ the unification obtained from the combination of the σ^{A_i}-unifications; and with σ_{L} the unification for the logic L. Given two labels i and j and a unification σ^* we shall use $[i,j]\sigma^*$ to denote both the result of the σ^*-unification of i and j, and the fact that i and j σ^*-unify.

DEFINITION 6. A substitution is a mapping $\rho : \Im_1 \to \Im_1$ such that

$$\rho(i) = \begin{cases} i & i \in \Phi_C \\ j & \text{otherwise} \end{cases}$$

Accordingly we have that two atomic ("world") labels i and j σ-unify iff there is a substitution ρ such that $\rho(i) = \rho(j)$. The notion of σ-unification (or label unification) is extended to the case of composite labels (path labels) as follows:

DEFINITION 7. Let $i, j \in \Im$

$$[i,j]\sigma = k \text{ iff } \quad \exists \rho : \quad \begin{aligned} &h(k) = \rho(h(i)) = \rho(h(j)) \text{ and} \\ &b(k) = [b(i), b(j)]\sigma \end{aligned}$$

Clearly σ is symmetric, i.e., $[i,j]\sigma$ iff $[j,i]\sigma$. Moreover this definition offers a flexible and powerful mechanism: in Section 3.1 we show that different classes of modal logics (in particular classes of non-normal modal logics such as regular and monotonic modal logics) are determined by conditions on the underlying substitution but the axiom unifications can be left unchanged. At the same time it allows for an independent computation of the elements of the result of the unification, and variables can be freely renamed without affecting the result of a unification.

Let A be a modal axiom. In general the "axiom" unification can be described as follows:

$$[i,j]\sigma^{\mathsf{A}} \iff [f_A(i), g_A(j)]\sigma \text{ and } C^A$$

where f_A and g_A are given logic-dependent functions from labels to labels and C^A is a set of constraints.

We now give the axiom unifications for the axioms for the fifteen basic normal modal logics[2].

DEFINITION 8. Let $i, j \in \Im$

$$[i,j]\sigma^{\mathsf{K}} = [i,j]\sigma \quad \begin{aligned} &\text{if at least one of } i \text{ and } j \text{ is restricted, and} \\ &\forall n \leq \ell(i), [s^n(i), s^n(j)]\sigma^{\mathsf{K}} \end{aligned}$$

[2]For the full details and explanations of the unifications, see [15, 20].

DEFINITION 9. Let $i, j \in \Im$. $[i,j]\sigma^{\mathsf{D}} = [i,j]\sigma$

EXAMPLE 10. For the difference between σ^{K} and σ^{D}, let us consider first the labels

$$i = (w_3, (W_1, w_1)) \qquad\qquad j = (W_2, (w_2, w_1))$$

Obviously i and j σ^{K}- and σ^{D}-unify on $(w_3, (w_2, w_1))$ with the substitution

$$\rho : \; W_1 \mapsto w_2, \quad W_2 \mapsto w_3$$

On the other hand the labels

$$i = (w_2, (W_1, w_1)) \qquad\qquad j = (W_2, (W_1, w_1))$$

σ^{D}- but not σ^{K}-unify. This is due to the fact that both $s^2(t)$ and $s^2(s)$ are variables, while in the definition of σ^{K} it is required that at least one of them is a constant. The reason for this condition on σ^{K} is that the interpretation of W_1 is the set of worlds accessible from w_1, but such a set may be empty so the denotation of W_1 would be empty; this is not the case with σ^{D} since the corresponding accessibility relation is serial, so W_1 cannot be empty.

To simplify the remaining definition of axiom unifications we introduce the following notation: given two labels i and j (of different length), we use $\bar{\imath}$ to denote the longest of the two and $\bar{\jmath}$ for the shortest of the two.

DEFINITION 11. Let $i, j \in \Im$

$$[i,j]\sigma^{\mathsf{T}} = [s^{\ell(\bar{\jmath})}(i), j]\sigma \text{ if } \forall n \geq \ell(\bar{\jmath}), [h^n(\bar{\imath}), h(\bar{\jmath}))]\sigma = [h(\bar{\imath}), h(\bar{\jmath})]\sigma$$

DEFINITION 12. Let $i, j \in \Im$

$$[i,j]\sigma^{\mathsf{4}} = c^{\ell(\bar{\jmath})}(\bar{\imath}) \text{ if } h(\bar{\jmath}) \in \Phi_V \text{ and } w_0 = [\bar{\jmath}, s^{\ell(\bar{\jmath})}(\bar{\imath})]\sigma$$

DEFINITION 13. Let $i, j \in \Im$

$$[i,j]\sigma^{\mathsf{5}} = \begin{cases} ([h(t), h(s)]\sigma; c^1(s^2(t))) & \text{if } \ell(t) > 2, \ell(s) > 1, h(t) \in \Phi_V, \text{ or} \\ & h(t) = h(s) \in \Phi_C \\ [t,s]\sigma & \text{if } \ell(t) = \ell(s) = 2 \\ ([t, h(s)]\sigma; c^1(s^2(s))) & \text{if } \ell(s) > 2, \ell(t) > 1, h(s) \in \Phi_V, \text{ or} \\ & h(t) = h(s) \in \Phi_C \end{cases}$$

where $w_0 = [s^1(t), s^1(s)]\sigma$.

For the unification for axiom B or σ^{B}-unification we first have to introduce some auxiliary definitions and constructions. For a label i and an integer n we define the sets of restricted and unrestricted segments longer that n.

$$\Phi_C^{i,n} = \{s^m(i) : m > n \text{ and } h^m(i) \in \Phi_C\}$$
$$\Phi_V^{i,n} = \{s^m(i) : m > n \text{ and } h^m(i) \in \Phi_V\}$$

We can give now the key notion to be used in the definition of the σ^{B}-unification.

DEFINITION 14. Given a label i and an integer n, i has the *bmorphism property for* n iff there is a morphism $\eta : \Phi_C^{i,n} \mapsto \Phi_V^{i,n}$ such that

1. η is injective, and
2. if $\eta(s^k(i)) = s^l(i)$, then $k < l$.

We are now ready to give the definition of σ^{B}-unification.

DEFINITION 15. Let $i, j \in \Im$

$$[i,j]\sigma^{\mathsf{B}} = [s^{\ell(\bar{j})}(\bar{\imath}), \bar{j}]\sigma \text{ iff } (1)\ \ell(\bar{\imath}) - \ell(\bar{j}) = 2n, (n > 0), \text{ and}$$
$$(2)\ \bar{\imath} \text{ has the bmorphism property for } \ell(\bar{j}).$$

According to the above definition labels like

$$(W_1, (w_2, w_1)) \qquad\qquad w_1$$

provide a simple instance of this unification. Intuitively W_1 denotes the set of worlds accessible from w_2, but, since w_2 is accessible from w_1, so, by symmetry, w_1 is one of the world accessible from w_2.

The key idea of σ^{B}-unification is to match world symbols laying an even number of steps apart, where the number of steps is determined by the sequences of variable and constants. In the above example the head of the first label is a variable we can go back by two steps. In general every constant must be compensated for by a variable following it.

EXAMPLE 16. Let us consider the labels

$$(1) \qquad i = (W_3, (W_2, (w_3, (W_1, (w_2, w_1))))) \qquad\qquad j = (W_4, w_1)$$

The labels i and j σ^{B}-unify since the difference of the lengths of the two labels is even ($\ell(i) = 6$ and $\ell(j) = 2$); moreover $s^2(i) = (w_2, w_1)$ and j σ-unify, and the restricted segment $s^4(i)$ can be mapped to the unrestricted segment $s^5(i)$.

In similar way the labels

$$(2) \qquad i = (W_3, (w_3, (W_2, (w_2, w_1)))) \qquad\qquad j = w_1$$

σ^{B}-unify, with the injective morphism η that maps $s^2(i)$ to $s^3(i)$ and $s^4(i)$ to $s^5(i)$. On the other hand the labels

$$(3) \qquad i = (w_3, (W_1, (w_2, w_1))) \qquad\qquad j = (W_2, w_1)$$

do not σ^{B}-unify since there is no (injective) morphism that satisfies condition (2) of Definition 14.

Before introducing the main unification, the unification for the various logics at hand we introduce the combined unification (or $\sigma^{\mathsf{A}_1 \ldots \mathsf{A}_n}$-unification), where $\mathsf{A}_1 \ldots \mathsf{A}_n$ is the list of axioms defining a logic L.

DEFINITION 17. Let $i, j \in \Im$

$$[i,j]\sigma^{\mathsf{A}_1\ldots\mathsf{A}_n} = \begin{cases} [i,j]\sigma^{\mathsf{A}_1} & \text{or} \\ \vdots \\ [i,j]\sigma^{\mathsf{A}_n} \end{cases}$$

Finally we are ready to give the main unification for a logic L (or σ_{L}), where the logic is defined by axioms $\mathsf{A}_1 \ldots \mathsf{A}_n$. This unification which will be used within the inference rules.

DEFINITION 18. Let $i, j \in \Im$

$$[i,j]\sigma_{\mathsf{L}} = \begin{cases} [i,j]\sigma^{\mathsf{A}_1\ldots\mathsf{A}_n} & \text{or} \\ [c^n(i), c^m(j)]\sigma^{\mathsf{A}_1\ldots\mathsf{A}_n} & \exists n, m : 1 \le n \le \ell(i), 1 \le m \le \ell(j) \end{cases}$$

where $w_0 = [s^n(i), s^m(j)]\sigma_{\mathsf{L}}$.

Notice that σ_{L} has a recursive definition.

EXAMPLE 19. Let us consider the labels

$$(4) \qquad i = (W_2, (w_2, w_1)) \qquad\qquad j = (W_3, (w_3, w_1))$$

The two labels do not σ^{B}-unify, they have the same length and do not σ^{D}-unify: the segments of length 2 are restricted and they have different heads thus there is no substitution ρ such that $\rho(w_2) = \rho(w_3)$. However, the two labels σ_{DB}-unify. We use the following decompositions

$$c^1(i) = (W_2, (w_2, w_0)) \qquad s^1(i) = w_1$$
$$c^3(j) = w_0 \qquad s^3(j) = (W_3, (w_2, w_1)).$$

It is easy to see that $[c^1(i), c^3(j)]\sigma^{\mathsf{B}} = w_0$ and $w_0 = [s^1(i), s^3(j)]\sigma^{\mathsf{B}} = w_1$.

When we consider the interpretation of the labels in (4) we have that i, intuitively, denotes the set of worlds accessible from w_2 which is accessible from w_1 and, similarly, the interpretation of j is the set of worlds accessible from w_3 which, in turn, is accessible from w_1. Since the accessibility relation is symmetric, w_1 belongs to both interpretations; thus the denotations of i and j have a non empty intersection and thus they labels hold unify. The σ_{DB}-unification takes care of cases like this.

Inference rules

For the propositional part of KEM we exemplify only the rules for conjunction.

$$(\alpha\text{-rules}) \qquad \frac{A \wedge B : i}{\begin{array}{c} A : i \\ B : i \end{array}}$$

The α-rules are just the familiar linear branch-expansion rules of the tableau method.

(β-rules)
$$\frac{\neg(A \wedge B) : i \quad A : j}{\neg B : [i,j]\sigma_{\mathsf{L}}} \qquad \frac{\neg(A \wedge B) : i \quad B : j}{\neg A : [i,j]\sigma_{\mathsf{L}}}$$

The β-rules are nothing but natural inference patterns such as Modus Ponens, Modus Tollens and Disjunctive syllogism generalised to the modal case. To apply such rules it is required that the labels of the premises unify and the label of the conclusion is the result of their unification.

(π-rules)
$$\frac{\Diamond A : i}{A : (w_n, i)} \qquad \frac{\neg\Box A : i}{\neg A : (w_n, i)}$$

where w_n is new, that is, it does not occur in the tree.

(ν-rules)
$$\frac{\Box A : i}{A : (W_n, i)} \qquad \frac{\neg\Diamond A : i}{\neg A : (W_n, i)}$$

where W_n is new.

ν- and π- rules allow us to expand labels according to the intended semantics, where, with "new" we mean that the label does not occur previously in the tree.

(PB)
$$\frac{}{A : i \quad | \quad \neg A : i}$$

PB (the "Principle of Bivalence") represents the semantic counterpart of the cut rule of the sequent calculus (intuitive meaning: a formula A is either true or false in any given world). PB is a zero-premise inference rule, so in its unrestricted version it can be applied whenever we like. However, we impose a restriction on its application. Then PB can be only applied w.r.t. immediate sub-formulas of unanalysed β-formulas, that is β formulas for which we have no immediate sub-formulas with the appropriate labels in the branch (tree).

(PNC)
$$\frac{A : i \quad \neg A : j}{\times}[i,j]\sigma_{\mathsf{L}}$$

The rule PNC (*Principle of Non-Contradiction*) states that two labelled formulas are σ_{L}-complementary when the two formulas are complementary and their labels σ_{L}-unify.

Let L be one of the fifteen basic modal logics. With $\vdash_{\text{KEM}(\mathsf{L})} A$ we mean that there is a close KEM-tree for $\neg A : w_1$; or, in other words, that SST proves that A is a theorem of L.

THEOREM 20. [15, 20] $\vdash_{\text{KEM}(\mathsf{L})} A$ *iff* $\models_{\mathsf{L}} A$.

3 Beyond basic modal logics

3.1 Non-normal modal logics

Normal modal logics are extensions of classical propositional logic with axiom K (i.e., $\Box(A \to B) \to (\Box A \to \Box B)$) and the necessitation rule (i.e., $\vdash A / \vdash \Box A$). However, weaker extensions are possible, when we consider the following rules to extend classical propositional logic:

(RE) $$\frac{\vdash A \leftrightarrow B}{\vdash \Box A \leftrightarrow \Box B}$$

(RK) $$\frac{\vdash (A_1 \wedge \cdots \wedge A_n) \to A}{\vdash (\Box A_1 \wedge \cdots \wedge \Box A_n) \to \Box A}\; n \geq 0$$

and, in particular, we shall consider

(Nec) $$\frac{\vdash A}{\vdash \Box A}\; (\mathrm{RK}, n = 0)$$

(RM) $$\frac{\vdash A \to B}{\vdash \Box A \to \Box B}\; (\mathrm{RK}, n = 1)$$

(RR) $$\frac{\vdash (A \wedge B) \to C}{\vdash (\Box A \wedge \Box B) \to \Box C}\; (\mathrm{RK}, n = 2)$$

We can now classify modal logics according to their deductive power.

DEFINITION 21. A modal logic L is *classical* iff it is closed under RE; *monotonic* iff it is closed under RM; emphregular iff it is closed under RR; *normal* iff it is closed under RK.

According to [9] the smallest classical logic is called E, the smallest regular logic R, the smallest monotonic logic M, and the smallest normal logic K.

The semantics of non-normal modal logic is given in terms of neighbourhood semantics. A model is a structure

$$\mathcal{M} = \langle W, N, v \rangle$$

where W is a set of possible worlds, N is a function from W to 2^{2^W} and v is an evaluation function: $v : \mathit{WFF} \times W \mapsto \{T, F\}$, where WFF is the set of well-formed formulas.

Before providing the evaluation clauses for the formulas we need to define the notion of truth set.

DEFINITION 22. Let $\mathcal{M}$ be a model and A be a formula. The truth set of A wrt to $\mathcal{M}$, $\|A\|^{\mathcal{M}}$ is thus defined: $\|A\|^{\mathcal{M}} = \{w \in W : v(A, w) = T\}$.

The evaluation clauses for atomic and boolean formulas are as usual while those for modal operators are given below.

DEFINITION 23. Let w be a world in $\mathcal{M} = \langle W, N, v \rangle$:

1. $w \models \Box A \iff \|A\|^{\mathcal{M}} \in N_w$;

2. $w \models \Diamond A \iff W - \|A\|^{\mathcal{M}} \notin N_w$.

It is natural to add some conditions on the function N in neighbourhood models. The conditions relevant for the present work are given in the following definition.

DEFINITION 24. Let $\mathcal{M}$ be a model. For every world $w \in W$ and every proposition A, and B.

(m) If $\|A\| \cap \|B\| \subseteq N_w$, then $\|A\| \in N_w$ and $\|B\| \in N_w$;

(c) If $\|A\| \in N_w$ and $\|B\| \in N_w$, then $\|A\| \cap \|B\| \in N_w$;

(n) $W \in N_w$.

According as the function N in a neighbourhood model satisfies condition (m), (c), or (n), we shall say that the model is *supplemented*, is *closed under intersections*, or *contains the unit*. When a model is both supplemented and closed under intersections then we shall call it a *quasi-filter*; when a quasi-filter contains the unit it is a *filter*.

The conditions determining the minimal non-normal modal logics are as follows:

1. E is characterised by the class of neighbourhood models;

2. M is characterised by the class of supplemented models;

3. R is characterised by the class of quasi-filters;

4. K is characterised by the class of filters.

From now on we shall use $\models_{\mathsf{L}} A$ to denote that A is valid in the class of model characterising L.

KEM for non-normal modal logic

Here we illustrate how to modify KEM to capture monotonic and regular modal logics. The required modifications involve the definition of substitution, Definition 6.

We shall denote the constants occurring in labels obtained as the result of an unification with $*$, and we shall denote the set of such constants by Φ^*_C.

It is worth noting that the variables can be mapped on more than a label in the course of a proof; imposing restrictions on the number of labels a variable can be mapped to in the course of a proof makes us able to characterise the classes of modal logics at hand. More precisely the world substitutions for the classes of logics under analysis are:

Monotonic logics

$$\rho^M : \Phi_V \mapsto \Im_{\text{branch}} \qquad \text{injective}$$
$$\mathbf{1}_{\Phi_C^*}$$

The condition for monotonic logics states that a variable can be mapped to a unique label in a branch of a KEM-proof, while constants are mapped to themselves only if they are the result of a unification. It is worth noting that it is possible to map a variable to different labels if they occur in distinct branches.

Regular logics

$$\rho^R : \Phi_V \mapsto \Im$$
$$\mathbf{1}_{\Phi_C^*}$$

For regular logics the restriction on variables is released, while that on constants still obtains.

THEOREM 25. [21] *Let* L *be either* M *or* R. $\vdash_{\text{KEM(L)}} A$ *iff* $\models_{\mathsf{L}} A$.

Unfortunately at the moment it is not know whether it is possible to capture classical modal logics by imposing similar restrictions to the substitution functions.

More on non-normal modal logics

Jones and Pörn [24, 25] defined a non-normal multi-modal deontic logic where, semantically, the set of worlds accessible from a given world w is partitioned into ideal and sub-ideal worlds: an accessible world is ideal if all obligations in w are respected and sub-ideal if some of the obligations in w are violated. In addition each world is either an ideal or a sub-ideal version of itself.

First of all atomic labels are indexed with either d, s or nothing. Thus, for example, (w_2^d, w_1) means that w_2 is an ideal version of w_1; (W_1^s, w_1) denotes the set of all subideal versions of w_1, and (w_2, w_1) says that we do not know if w_2 is an ideal or subideal version of w_1.

To accommodate the above conditions it is possible to define new inference rules operating on labels instead of the declarative units of labelled formulas.

$$\text{Exc}\frac{\begin{array}{c}A:(W^s,i)\\ A:(W^d,j)\end{array}}{A:(W,[i,j]\sigma_{\mathsf{JP}})} \qquad \text{LPNC}\frac{\begin{array}{c}A:i^s\\ B:i^d\end{array}}{\times} \qquad \text{LPB}\frac{}{A:i^s \mid B:i^d}\, i \text{ restricted}$$

The above three rules give us that the set of worlds accessible from a given is a partition. Exec tells us that if A holds in all ideal versions of a world as well as in all subidal versions then it holds in all accessible worlds. LPNC says that it is not possible to have a world that is at the same time an ideal and a subideal ideal version of another world. Finally, the meaning of LPB is that the classes in a partition are mutually exclusive and so we can create two mutually exclusive branches for our tableaux tree.

$$\text{Ref}\frac{\begin{array}{c}\Box A : i \\ \neg A : j\end{array}}{\begin{array}{c}\Box A : k^s \\ \neg A : k^s\end{array}}[i, j]\sigma_{\mathsf{JP}} = k$$

Finally the Ref rules (reflexivity rules) allows us on detection of a violation (i.e, we both have an obligation and the negation of the content of the obligations with two labels that unify) to determine when a world is a subideal version of itself. For a full account, we refer to [19, 4].

Labelled tableaux systems using the explicit representation of the accessibility relation have been proposed for non-normal modal logic by [16].

3.2 Conditional logics

Conditional logics are extensions of classical logic with a binary intensional operators, >, meant to represent hypothetical, conditional or counterfactual reasoning [26, 31]. Different possible world semantics have been put forward for conditional logics: the system of sphere semantics [26] and the selection function semantics [31]. If one wants to use labels to mimic the semantics of conditional logics, one has to chose the most appropriate semantics. Most of the (labelled) tableaux systems for conditional logics assume the selection function semantics, where a model $\mathcal{M}$ is a structure $\langle W, f, v\rangle$, where W is a set of possible worlds, f is a selection function which picks for every formula A a subset $f(A, w)$ of W for each world $w \in W$, and v is a valuation function assigning to every formula A and $w \in W$ a boolean value. We refer to $f(A, w)$ as the set of A-worlds relative to w. The valuation condition for a conditional formula $A > B$ is as follows:

$$w \models A > B \text{ iff } f(A, w) \subseteq \|B\|$$

Chellas [8] proposed the reading of a conditional operators as a parametrised modal operator, that is $A > B$ can be understood as $[A]B$. Based on this reading it is possible to consider conditional logics as a type of multi-modal logic and to use the idea of having different types of labels for each modal operator. Accordingly, to cope with conditional logics, we extend the label formalism, and atomic labels can be indexed by formulas. Hence we can have labels as (W_1^A, w_1), intuitively denoting the set of A-worlds relative to the world denoted by w_1 and (w_2^B, w_1) for a possible word in $f(B, w_1)$.

In general the unification for a conditional logic has the following structure:

$$[i^Y, j^Z]\sigma_> = [i, j]\sigma$$

and for each 'turning point' $(i'^{Y'}, j'^{Z'})$ one of the following conditions (for normal conditional logics) (i) $Y' \equiv Z'$ or (ii) $Y' \equiv \top$ and $h(i') \in \Phi_V$ or (iii)

For the inference rules we have to consider that now formulas can occur both in the declarative part of a labelled formula but also as index of a label. Thus the notation $A@X : i^Y$ means that X is either Y or C. Based on the intuition described so fare the inference rules for $>$ are as follows:

$$\frac{A > B@X : i^Y}{B : (W_n^A, i^Y)} \qquad \frac{\begin{array}{c}A > B@X : i^Y \\ A@X' : j^A\end{array}}{A : (c^{\ell(j)-1})^A} \qquad \frac{\neg(A > B))X@i^Y}{\neg B : (w_n^A, i^Y)}$$

The presence of two rules for the case of a positive conditional is due to the fact that positive conditional behaves both as α and β formulas (according to Smullyan classification). Notice that the β-version can be problematic in some conditional logic.

In [2] we discuss the issues of the design of a labelled tableaux for conditional logic (some of them are related to the fact that one has to begin a new tableaux to check the equivalence of two formulas when computing a unification), and in [3] we provided a sound and complete system for the flat fragment of some particular logics. Pozzato and co-workers [33] used and extended the ideas of [2, 3] to provide sequents and tableaux systems for a larger class of conditional or conditional like logics. Priest [34], on the other hand presented tableaux for conditional logics using the propagation of fomulas based on the representation of the semantic structure in first-order logic.

4 Relative complexity: the beauty of symmetry

In the last few years several comparisons (competitions) of theorem provers for modal logic have been held (cf. [6, 28, 30]) and experimental research has been carried out (cf. [23, 22]). Despite the potential interest for eventual applications, we believe that this kind of research provided little or no insight on better theoretical architectures for modal theorem provers. Very often the overall performance is heavily influenced by external factors, for example, language specific optimisations of the implementation.

In this section we compare SST and KEM from a theoretical perspective. This means that we do not consider implementation issues, but only logical ones; moreover we are not interested in the propositional features and in the interaction of modal operators and propositional connectives, but only in the modal characteristics.

To prove that a proof system $\mathcal{A}$ is essentially better than a proof system $\mathcal{B}$ we have to exhibit at least one formula (or a class of formulas) for which $\mathcal{A}$ is better than $\mathcal{B}$, and for all formulas $\mathcal{A}$ is not essentially worse than $\mathcal{B}$.[3] There are many distinct modal logics, and it is possible that results in a logic do not apply to a different logic. Moreover $\mathcal{A}$ may cover some modal logics which are not covered by $\mathcal{B}$ and the other way around. However, any general purpose modal theorem prover should cover the basic fifteen normal modal logics. Among them, some offer too simple modal structures while other lend themselves to specialised optimisation procedures (in particular the logics

[3] We shall give a precise definition of what "better" and "worse" mean in this context in Section 4.2.

with a finite number of distinct modalities). In both cases, these logics do not provide the best scenario to really test the theoretical architecture behind a modal theorem prover. Therefore we have to identify a modal logic with the following properties:

1. it is one of the basic fifteen normal modal logics;

2. the proof procedures are modular for both systems, that is, they are the combination of the proof procedures of the single components of the logic; and

3. there are no specialised proof procedures.

The basic normal modal logic DB satisfies the criteria listed above to be a representative candidate to test the capability of a theorem prover for modal logic. Moreover, due to some well-known difficulties [13], symmetric logics lie outside most of the current modal theorem prover methods, though they play an important role in different applications areas.

4.1 The complexity of KEM unifications for DB

To provide a comparison of the two methods at hand first we have to study the complexity of the KEM unification procedure. We start by showing that the unification of two world symbols can be computed in constant time.

LEMMA 26. *The σ-unification of two world symbols w and w' can be computed in constant time.*

Proof. It is immediate to see that the unification of two world symbols requires at most three steps, and thus it has constant complexity. ■

As we have seen the unification of two world symbols is just the first basic step of the unification. The next step is the σ-unification of two labels; in this case, we can prove that its complexity is linear in the length of the two labels.

LEMMA 27. *The σ-unification of two labels i and j can be computed in linear time.*

Proof. All we have to do is to see whether the word symbols in the two labels stepwise unify.

Thus at the end we have to verify n unifications of world symbols, but from Lemma 26, we know that the unification of world symbols can be computed in constant time. Therefore the σ-unification of two labels can be computed in linear time. ■

The next unification we have to examine is the unification for the axiom B.

LEMMA 28. *The σ^{B}-unification of two labels i and j can be computed in linear time.*

Proof. The computation of the σ^{B}-unification of two labels i and j can be reduced to three sub-problems. (1) To compute the lengths of the two labels and to determine whether the difference is even. This obviously can be computed in linear time: all we have to do is to scan sequentially the two labels. (2) To determine whether an injective morphism from $\Phi_C^{\bar{\imath},\ell(\bar{\jmath})}$ to $\Phi_V^{\bar{\imath},\ell(\bar{\jmath})}$ exists. It is to implement a linear time algorithm that scans the labels and increments or decrements a counter to verify that there is such an injective morphism. (3) To compute $[s^{\ell(\bar{\jmath})}(\bar{\imath}), \bar{\jmath}]\sigma$: by Lemma 27 the σ-unification of two labels has linear complexity. Therefore the complexity of σ^{B} is linear. ■

Unfortunately we cannot prove such good complexity results for σ_{DB}; however, for special labels we can prove the following result.

LEMMA 29. *The σ_{DB}-unification of two labels i and j such that $\ell(i) = 1$ can be computed in quadratic time.*

Proof. For a label j of length n there are n distinct segments and n distinct countersegments, namely

$$
\begin{aligned}
c^n(j) &= w_0 & s^n(j) &= j\\
c^{n-1}(j) &= (h^n(j), w_0) & s^{n-1}(j) &= b(j)\\
c^{n-2}(j) &= (h^n(j), (h^{n-1}(j), w_0)) & s^{n-2}(j) &= b(b(j))\\
&\vdots & &\vdots
\end{aligned}
$$

Now we have to see whether i either σ^{B}- or σ-unifies with the countersegments and whether i σ_{DB}-unifies with the segments. Thus we have to compute $2n$ linear unifications and n σ_{DB}-unifications. Let us examine the first of these, i.e., $[s^{n-1}(j), i]\sigma_{\mathsf{DB}}$. This time the length of $s^{n-1}(j)$ is $n-1$, and thus we have $n-1$ ways to split it in segments and countersegments. That is:[4]

$$
\begin{aligned}
c^{n-1}(c^{n-1}(j)) &= w_0 & s^{n-1}(j) &= b(j)\\
c^{n-2}(c^{n-1}(j)) &= (h^{n-1}(j), w_0) & s^{n-2}(j) &= b(b(j))\\
&\vdots & &\vdots
\end{aligned}
$$

A close inspection shows that only the countersegments are different from the previous step. Therefore we can repeat this process for all the segments of j, and each time we can replace the σ_{DB} unification for the appropriate segment of length m, with $2m$ linear unifications. Hence, at the end, the number of linear unifications we have to compute is

$$2 \sum_{n=1}^{n=\ell(j)} n = O(n^2)$$

which shows that the σ_{DB}-unification for the case at hand is quadratic. ■

[4]Notice that for $m \leq n$ $s^m(s^n(i)) = s^m(i)$.

4.2 KEM vs. SST

So far the standard way to compare the relative complexity of two proof systems was given by the notion of p-simulation.

DEFINITION 30. A proof system $\mathcal{A}$ p-simulates a proof system $\mathcal{B}$ iff there is a function g, computable in polynomial time, which maps derivations in $\mathcal{B}$ for any given formula ϕ, to derivations in $\mathcal{A}$ for ϕ (cf. [10]).

The main problem with p-simulation is that it considers only proofs, i.e., closed trees in tableaux terminology, and it says nothing about open trees. While this notion is fully appropriate for semi-decidable logics and non deterministic proof systems, it does not offer a good measure to compare tableaux-like proof-systems for decidable modal propositional logics. The main point is that this notion does not contemplate proof-procedures. Modal tableaux proof-procedures, in effect, are systematic searches for models that make the initial formula true with respect to the initial world. In this perspective modal tableaux can show that a formula is not a theorem by showing that the negation of the formula is satisfiable. However, to show that a formula is satisfiable we have to complete its tree. In general, to complete a tree we have to explore the whole search space generated by the formula.

Therefore, to obviate the above problem, we propose a stepwise simulation. Here the main idea is that a proof system $\mathcal{A}$ stepwise simulates a proof system $\mathcal{B}$ iff $\mathcal{A}$ does not perform any inference steps for which no corresponding inference steps exist in $\mathcal{B}$.

DEFINITION 31. A proof system $\mathcal{A}$ p-search-simulates a proof system $\mathcal{B}$ iff there is a polynomial function g such that for any formula ϕ, g maps derivations (trees) from ϕ in $\mathcal{A}$ to derivations (trees) from ϕ in $\mathcal{B}$ (cf. [11]).

Note that a stepwise simulation is independent of whether the considered derivations (trees) are proofs or not.

We are now ready to present the main result of the paper. To prove it we have to identify a formula (or a class of formulas) whose complete KEM-tree is polynomial in the size of the formula while the complete SST-tree is exponential in the size of the formula. Surprisingly the formula is extremely simple, namely:

$$(5) \quad p \rightarrow (\Box\Diamond)^n p$$

As we shall see (5) involves only one propositional linear step and there are no interaction between propositional connectives and modal operators. Therefore the discriminant is only the way the two proof systems deal with modal operators.

THEOREM 32. *The length of the complete proof of* $p \rightarrow (\Box\Diamond)^n p$ *in* KEM *is* $O(n^2)$.

Proof.

$$
\begin{array}{rl}
1. & \neg(p \rightarrow (\Box\Diamond)^n p) : w_1 \\
2. & p : w_1 \\
3. & \neg(\Box\Diamond)^n p : w_1 \\
4. & \neg\Diamond(\Box\Diamond)^{n-1} p : (w_2, w_1) \\
5. & \neg(\Box\Diamond)^{n-1} p : (W_1, (w_2, w_1)) \\
 & \vdots \\
2n+3. & \neg p : (W_n, (w_{n+1}, (\ldots, (W_1, (w_2, w_1)) \ldots)))
\end{array}
$$

The initial formula, i.e., $\neg(p \rightarrow (\Box\Diamond)^n p) : w_1$, is of type α, then we expand the tree with two nodes both labelled with w_1: the first of such nodes (2) consists of p which is atomic and does not need further investigations; the second node (3) contains a formula of type π labelled with w_1. From (3) we obtain (4), which is of type ν. Applying the ν-rule on it, we get (5). We repeat the above steps $n-1$ times, for a total of $2n+3$ steps (nodes).

At this point we have two complementary formulas, the formulas in (2) and $(2n+3)$. We have to verify whether the two labels σ_{DB}-unify.

From Lemma 29 we know that the complexity of the instance of σ_{DB}-unification at hand is quadratic in the length of the labels involved, which in turn, is linear in the size of the formula. Therefore the complexity of the complete KEM-proof of $p \rightarrow (\Box\Diamond)^n p$ is $2n + 3 + O(n^2) = O(n^2)$. ■

THEOREM 33. *The length of the complete proof of* $p \rightarrow (\Box\Diamond)^n p$ *in* SST *is* $O(2^{n+1})$.

Proof.

$$
\begin{array}{rl}
1. & \neg(p \rightarrow (\Box\Diamond)^n p) : w_1 \\
2. & p : w_1 \\
3. & \neg(\Box\Diamond)^n p : w_1 \\
4. & \neg\Diamond(\Box\Diamond)^{n-1} p : (w_2, w_1) \\
5. & \neg(\Box\Diamond)^{n-1} p : (W_1, (w_2, w_1)) \\
6. & \neg(\Box\Diamond)^{n-1} p : w_1 \\
7. & \neg\Diamond(\Box\Diamond)^{n-2} p : (w_3, (W_1, (w_2, w_1))) \\
8. & \neg\Diamond(\Box\Diamond)^{n-2} p : (w_3, w_1) \\
9. & \neg(\Box\Diamond)^{n-2} p : (W_2, (w_3, (W_1, (w_2, w_1)))) \\
10. & \neg(\Box\Diamond)^{n-2} p : (w_2, w_1) \\
11. & \neg(\Box\Diamond)^{n-2} p : (W_2, (w_3, w_1)) \\
12. & \neg(\Box\Diamond)^{n-2} p : w_1 \\
 & \vdots
\end{array}
$$

The formula we start with $(\neg(p \rightarrow (\Box\Diamond)^n p) : w_1)$ is of type α, and then we obtain two formulas $p : w_1$ and $\neg(\Box\Diamond)^n p : w_1$. At this point we have an atomic formula and a formula of type π. We apply the π-rule on it deriving $\neg\Diamond(\Box\Diamond)^{n-1} p : (w_2, w_1)$. Now we have a formula of type ν, and we have to apply both the ν-rule for D and B, thus we have to produce the formulas $\neg(\Box\Diamond)^{n-1} p : w_1$ and $\neg(\Box\Diamond)^{n-1} p : (W_1, (w_2, w_1))$. These last two formulas are of type π, and from them we obtain $\neg\Diamond(\Box\Diamond)^{n-2}p : (w_3, w_1)$ and $\neg\Diamond(\Box\Diamond)^{n-2}p : (w_3, (W_1, (w_2, w_1)))$; both formulas produce two new formulas. It is then clear that each formula of type ν produces two new formulas of less complexity, showing thus a geometrical progression; it is then immediate to see that the formula determining the number of steps is

$$
\begin{aligned}
2\sum_{m=1}^{n} 2^{m-1} + 2^m &= 2\left(\frac{2^{(n-1)+1} - 1}{2-1}\right) + 2^n \\
&= 2(2^n - 1) + 2^n \\
&= 2^{n+1} + 2^n - 2
\end{aligned}
$$

thus the complexity of the complete proof of $p \to (\Box\Diamond)^n p$ in SST is $O(2^{n+1})$. ■

It is true that there are shorter proofs for (5) in SST. However, if instead of (5) we consider the formula

(6) $\quad p \to (\Box\Diamond)^n q$

which is not a theorem of DB, then the search space for it is $O(2^{n+1})$, since (6) has the same modal structure as (5) and we have to explore the whole search space before we can conclude that its negation has a model. This is the reason why when we compare proof systems using p-search-simulation we have to consider exhaustive proof-search procedures and worst-case scenarios.

THEOREM 34. SST *cannot p-search-simulate* KEM.

Proof. From Theorem 32 and Theorem 33 it follows that SST cannot p-simulate KEM since the complexity of $p \to (\Box\Diamond)^n p$ is $O(2^{n+1})$ for SST, while for KEM it is $O(n^2)$. ■

Let us now examine the question whether KEM p-search-simulates SST or whether the two systems cannot p-search-simulate each other. To show that a system $\mathcal{A}$ p-search-simulates a system $\mathcal{B}$ we have to define a polynomial procedure that transforms a tree for ϕ in $\mathcal{A}$ in a tree for ϕ in $\mathcal{B}$ for any formula ϕ.

LEMMA 35. *The rule* ν_B *is a derived rule in* KEM, *and it can be derived in polynomial time.*

Proof.

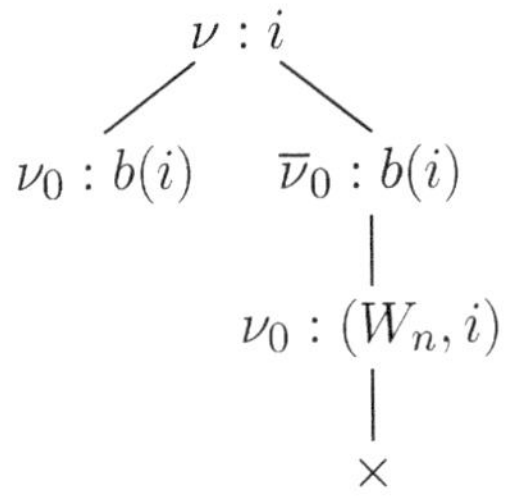

We apply PB with respect to ν_0, and with label $b(i)$; in the right branch we apply the ν rule and we obtain $\nu_0 : (w_n, i)$, but $[b(i), (W_n, i)]\sigma_{\mathsf{DB}}$, and thus the branch is closed. In particular it is possible to show that the labels involved σ^{B}-unify, and we have seen (Lemma 28) that the σ^{B}-unification can be computed in linear time. Therefore the derivation of ν_B has linear complexity. ■

Lemma 35 allows us to define a proof-search in KEM where we use both the new derived ν-rule and the original ν-rules of KEM, and the unification is restricted to σ. It is immediate to see that this proof procedure corresponds to SST, and the components involved have linear complexity, we have thus proved the following theorem.

THEOREM 36. KEM *p-search-simulates* SST.

Notice that the results above extends immediately to the ground version of SST [27, 29] as well as to Fitting's prefix tableaux [13].

5 Conclusions

Labels can be a very powerful tool for the design of (semantic based) deductive systems. In this paper we have seen how labels can be used to create tableaux system for a variety of logics amenable of possible world semantics. In addition we have shown that the use of free-variable labels with particular logic dependant label algebra can speed up the complexity of modal tableaux.

Acknowledgements. This work was supported by Australia Research Council under Discovery Project No. DP0452628 on "Combining modal logic for dynamic and multi-agents systems".

National ICT Australia is funded by the Australian Government's Department of Communications, Information Technology and the Arts and the Australian Research Council through Backing Australia's Ability and the ICT Centre of Excellence program.

BIBLIOGRAPHY

[1] A. Artosi, P. Benassi, G. Governatori, and A. Rotolo. Shakespearian modal logic: A labelled treatment of modal identity. In M. Kracht, M. de Rijke, H. Wansing, and M. Zakharyaschev, editors, *Advances in Modal Logic. Volume 1*, pages 1–21. CSLI Publications, 1998.

[2] A. Artosi and G. Governatori. A tableaux methodology for deontic conditional logics. In *ΔEON'98*, pages 65–81, Bologna, 1998.

[3] A. Artosi, G. Governatori, and A. Rotolo. Labelled tableaux for non-monotonic reasoning: Cumulative consequence relations. *Journal of Logic and Computation*, 12(6):1027–1060, 2002.

[4] A. Artosi, G. Governatori, and G. Sartor. Towards a computational treatment of deontic defeasibility. In M. Brown and J. Carmo, editors, *Deontic Logic Agency and Normative Systems*, pages 27–46. Springer-Verlag, 1996.

[5] M. Baldoni, L. Giordano, and A. Martelli. A tableau for multimodal logics and some (un)decidability results. In Harrie C. M. de Swart, editor, *TABLEAUX'98*, LNCS 1397, pages 44–59. Springer, 1998.

[6] P. Balsiger and A. Heuerding, *Comparison of theorem provers for modal logics – introduction and summary.* In H. C. M. de Swart, editor, *TABLEAUX '98*, LNCS 1397, pages 25–26, Springer-Verlag, 1998.
[7] B. Beckert and R. Goré. Free variable tableaux for propositional modal logics. *Studia Logica*, 69(1):59–96, 2001.
[8] B.F. Chellas. Basic conditional logic. *Journal of Philosophical Logic*, 4:133–153, 1975.
[9] B.F. Chellas. *Modal Logic, An Introduction.* Cambridge University Press, 1980.
[10] S.A. Cook and R.A. Reckhow. The relative efficiency of propositional proof systems. *Journal of Symbolic Logic*, 44:36–50, 1979.
[11] H. de Nivelle, R. Schmidt, and U. Hustadt. Resolution-based methods for modal logics. *Logic Journal of IGPL*, 8:265–292, 2000.
[12] F.B. Fitch. Tree proofs in modal logic. *Journal of Symbolic Logic*, 31:152, 1966.
[13] M. Fitting. *Proof Methods for Modal and Intuitionistic Logics.* Reidel, 1983.
[14] D.M. Gabbay. *Labelled Deductive System.* Oxford University Press, 1996.
[15] D.M. Gabbay and G. Governatori. Fibred modal tableaux. In D. Basin, M. D'Agostino, D.M. Gabbay, S. Matthews, and L. Viganó, editors, *Labelled Deduction*, pages 163–194. Kluwer, 2000.
[16] R. Girle. *Modal Logic and Philosophy.* Acumen, 2000.
[17] R. Goré. Tableau methods for modal and temporal logics. In M. D'Agostino, D.M¿ Gabbay, R. Heinle, and J. Posegga, editors, *Handbook of Tableaux Methods*, pages 297–396. Kluwer, 1999.
[18] G. Governatori. Labelled tableaux for multi-modal logics. In P. Baumgartner, R. Hähnle, and J. Posegga, editors, *TABLEAUX'95*, LNAI 918, pages 79–94, Springer-Verlag, 1995.
[19] G. Governatori. Labelling ideality and subideality. In D.M. Gabbay and H.J. Ohlbach, editors, *Practical Reasoning*, LNAI 1085, pages 291–304, Springer-Verlag, 1996.
[20] G. Governatori. *Un modello formale per il ragionamento giuridico.* PhD thesis, CIRFID, University of Bologna, Bologna, 1997.
[21] G. Governatori and A. Luppi. Labelled tableaux for non-normal modal logics. In E. Lamma and P. Mello, editors, *AI*IA 99*, LNAI 1792, pages 119–130, Springer-Verlag, 2000.
[22] I. Horrocks, P.F. Patel-Schneider, and R. Sebastiani. An analysis of empirical testing for modal decision procedures. *Logic Journal of IGPL*, 8:293–323, 2000.
[23] U. Hustadt and R. Schmidt. On evaluating decision procedures for modal logic. In *IJCAI'97*, pages 202–207, 1997.
[24] A.J.I. Jones and I. Pörn. Ideality, sub-ideality and deontic logic. *Synthese*, 65:275–290, 1985.
[25] A.J.I. Jones and I. Pörn. "Ought" and "Must". *Synthese*, 66:89–93, 1986.
[26] D. Lewis. *Counterfactuals.* Basil Blackwell, 1986.
[27] F. Massacci. Strongly analytic tableaux for normal modal logic. In A. Bundy, editor, *CADE-12*, LNAI 814, pages 723–737, Springer-Verlag, 1994.
[28] F. Massacci. Design and results of the Tableaux-99 non-classical (modal) systems comparison. In N.V. Murray, editor, *TABLEAUX '99*, LNCS 1617, pages 14–18, Springer-Verlag, 1999.
[29] F. Massacci. Single step tableaux for modal logic. *Journal of Automated Reasoning*, 24(3):319–364, 2000.
[30] F. Massacci and F.M. Donini. Design and results of tancs-2000 non-classical (modal) systems comparison. In R. Dyckhoff, editor, *TABLEAUX 2000*, LNCS 1847 pages 52–56, Springer-Verlag, 2000.
[31] Donald Nute. *Topics in Conditional Logic.* Reidel, Dordrecht, 1980.
[32] H.J. Ohlbach. Semantic based translation methods for modal logics. *Journal of Logic and Computation*, 1:691–746, 1991.
[33] G.L. Pozzato. *Proof Methods for Conditional and Preferential Logics.* PhD thesis, Università degli Studi di Torino, 2007.
[34] G. Priest. *An Introduction to Non-Classical Logic.* Cambridge University Press, 2001.

[35] R.M. Smullyan. *First-Order Logic*. Springer-Verlag, Berlin, 1968.

Guido Governatori
Queensland Research Laboratory,
National ICT Australia,
Brisbane, Queensland, Australia
guido.governatori@nicta.com.au

Many-valued hybrid logic

Jens Hansen, Thomas Bolander and Torben Braüner

ABSTRACT. In this paper we define a many-valued semantics for hybrid logic and we give a sound and complete tableau system which is proof-theoretically well-behaved, in particular, it gives rise to a decision procedure for the logic. This shows that many-valued hybrid logics is a natural enterprise and opens up the way for future applications.

Keywords: Modal logic, hybrid logic, many-valued logic, tableau systems.

1 Introduction

Classical hybrid logic is obtained by adding to ordinary, classical modal logic further expressive power in the form of a second sort of propositional symbols called nominals, and moreover, by adding so-called satisfaction operators. A nominal is assumed to be true at exactly one world, so a nominal can be considered the name of a world. Thus, in hybrid logic a name is a particular sort of propositional symbol whereas in first-order logic it is an argument to a predicate. If i is a nominal and ϕ is an arbitrary formula, then a new formula $@_i\phi$ called a satisfaction statement can be formed. The part $@_i$ of $@_i\phi$ is called a satisfaction operator. The satisfaction statement $@_i\phi$ expresses that the formula ϕ is true at one particular world, namely the world at which the nominal i is true. Hybrid logic is proof-theoretically well-behaved, which is documented in the forthcoming book [7]. Hybrid-logical proof-theory includes a long line of work on tableau systems for hybrid logic, see [1, 2, 6, 4, 15, 3].

Now, classical hybrid logic can be viewed as a combination of two logics, namely classical, two-valued logic (where the standard propositional connectives are interpreted in terms of the truth-values true and false) and hybrid modal logic (where modal operators, nominals, and satisfaction operators are interpreted in terms of a set of possible worlds equipped with an accessibility relation). The present paper concerns many-valued hybrid logic, that is, hybrid logic where the two-valued logic basis has been generalized to a many-valued logic basis. To be more precise, we shall define a many-valued semantics for hybrid logic, and we shall give a tableau system that is sound and complete with respect to the semantics. Not only is the many-valued semantics a generalization of the two-valued semantics, but if we chose a two-valued version of the many-valued tableau system, then modulo minor reformulations and the deletion of superfluous rules, the tableau system obtained is identical to an already known tableau systems for hybrid logic.

Our many-valued semantics is a hybridized version of a many-valued semantics for modal logic given in the papers [11, 12, 13]. A notable feature of this semantics is that it allows the accessibility relation as well as formulas to take on many truth-values (in other many-valued modal logics it is only formulas that can take on many truth-values).

A leading idea behind our work is that we distinguish between the way of reasoning and what the reasoning is about, and in accordance with this idea, we generalize the way of reasoning from two-valued logic to many-valued logic such that we reason in a many-valued way about time, space, knowledge, states in a computer, or whatever the subject-matter is. Given our distinction between the way of reasoning and what the reasoning is about, we take it that the concerns of hybrid logic basically are orthogonal to as whether the logic basis is two-valued or many-valued. Thus, it is expectable that the already known proof-theoretically well-behaved tableau systems for two-valued hybrid logic can be generalized to proof-theoretically well-behaved tableau systems for many-valued hybrid logic. Accordingly, if we define a many-valued hybrid logic and give a tableau system that satisfies standard proof-theoretic requirements (it is cut-free, it satisfies a version of the subformula property, and it gives rise to a decision procedure), then we learn more about hybrid logic and we provide more evidence that hybrid logic and hybrid-logical proof-theory is a natural enterprise.

This paper is structured as follows. In the second section of the paper we define the many-valued semantics for hybrid logic and we make some remarks on the relation to intuitionistic hybrid logic. In the third section we introduce a tableau system, in the fourth section we prove termination, and in the fifth section we prove completeness.

2 A many-valued hybrid logic language

In this section a Many-Valued Hybrid Logic language (denoted by **MVHL**) is presented and a semantics for the language is given. We have included global modalities, one reason being that they are used in our motivation for our choice of semantics for the nominals, but our termination and completeness proofs later in the paper do not include global modalities. In the following let $\mathcal{T}$ denote a fixed finite Heyting algebra. That is, $\mathcal{T}$ is a finite lattice such that for all a and b in $\mathcal{T}$ there is a greatest element x of $\mathcal{T}$ satisfying $a \wedge x \leq b$. The element x is called the *relative pseudo-complement* of a with respect to b (denoted $a \Rightarrow b$). To avoid notational ambiguity in relation to the syntax of our hybrid logic, we will in the following use the symbol $\Rightarrow$ for relative pseudo-complement, and $\sqcup$ and $\sqcap$ for meet and join, respectively. The largest and smallest elements of $\mathcal{T}$ are denoted $\top$ and $\bot$, respectively. The elements of the Heyting algebra $\mathcal{T}$ are going to be used as truth values for our many-valued logic. Thus, in the following, we will often refer to the elements of $\mathcal{T}$ as *truth values*.[1]

[1]In order to give reasonable semantics for $\wedge$ and $\vee$ a Lattice structure is needed. A complete Lattice would be enough if the accessibility relation was only allowed to have two values, but since we also allows for the accessibility relation to take values in $\mathcal{T}$, the

2.1 Syntax for MVHL

Let a countable infinite set of propositional variables PROP and a countable infinite set of nominals NOM be given. In addition to the usual connectives of propositional model logic, we include the global modalities E and A, and for every $i \in \mathsf{NOM}$, a satisfaction operator $@_i$.

DEFINITION 1 (**MVHL**-formulas). The set of **MVHL**-formulas is given by the following grammar:

$$\varphi ::= p \mid a \mid i \mid (\psi_1 \wedge \psi_2) \mid (\psi_1 \vee \psi_2) \mid (\psi_1 \to \psi_2) \mid \Box\psi \mid \Diamond\psi \mid @_i\psi \mid E\psi \mid A\psi,$$

where $p \in \mathsf{PROP}$, $a \in \mathcal{T}$, and $i \in \mathsf{NOM}$.

In general we will use i, j, k and so on for nominals and a, b, c for elements of $\mathcal{T}$.

2.2 Semantics for MVHL

The semantics for **MVHL** is a Kripke semantics in which the accessibility relation is allowed to take values in $\mathcal{T}$. This is inspired by [13]. A model $\mathcal{M}$ is a tuple $\mathcal{M} = \langle W, R, \mathbf{n}, \nu \rangle$, where W is the set of worlds, and R a mapping $R : W \times W \to \mathcal{T}$ called the accessibility relation. $\mathbf{n}$ is a function interpreting the nominals, i.e. $\mathbf{n} : \mathsf{NOM} \to W$. Finally the valuation $\nu : W \times \mathsf{PROP} \to \mathcal{T}$ assigns truth values to the propositional variables at each world.

Now given a model $\mathcal{M} = \langle W, R, \mathbf{n}, \nu \rangle$, we can extend the valuation ν to all formulas in the following inductive way, where $w \in W$:

$$
\begin{aligned}
\nu(w, a) &:= a \quad \text{for } a \in \mathcal{T} \\
\nu(w, i) &:= \begin{cases} \top & \text{, if } \mathbf{n}(i) = w \\ \bot & \text{, else} \end{cases} \\
\nu(w, \varphi \wedge \psi) &:= \nu(w, \varphi) \sqcap \nu(w, \psi) \\
\nu(w, \varphi \vee \psi) &:= \nu(w, \varphi) \sqcup \nu(w, \psi) \\
\nu(w, \varphi \to \psi) &:= \nu(w, \varphi) \Rightarrow \nu(w, \psi) \\
\nu(w, \Box\varphi) &:= \bigsqcap \{R(w, v) \Rightarrow \nu(v, \varphi) \mid v \in W\} \\
\nu(w, \Diamond\varphi) &:= \bigsqcup \{R(w, v) \sqcap \nu(v, \varphi) \mid v \in W\} \\
\nu(w, @_i\varphi) &:= \nu(\mathbf{n}(i), \varphi) \\
\nu(w, A\varphi) &:= \bigsqcap \{\nu(v, \varphi) \mid v \in W\} \\
\nu(w, E\varphi) &:= \bigsqcup \{\nu(v, \varphi) \mid v \in W\}
\end{aligned}
$$

The semantics chosen for the hybrid logical constructions is discussed in the following. The semantics for $@_i\varphi$ is obvious, its truth value is simply the truth value of φ at the world i denotes. The semantics chosen for the

structure of a Heyting algebra is needed. For further discussions of the choice of a finite Heyting algebra as the set of truth values see [12, 13].

global modalities A and E reflect the fact that these modalities are simply the global versions of the modalities $\Box$ and $\Diamond$. The choice of semantics for nominals is less obvious. In this paper we have chosen to assign each nominal i the value $\top$ in exactly one world, and $\bot$ in all other worlds. This is in agreement with the the standard semantics for hybrid logic in which a nominal "points to a unique world". It would probably also be possible to allow nominals to take values outside the set $\{\top, \bot\}$, but at least a nominal should receive the value $\top$ in one and only one world in order for the semantics to be in accordance with classical, two-valued, hybrid logic (and for nominals to be semantically different from ordinary propositional symbols). Our decision of making the semantics of nominals two-valued rests primarily on the fact that it allows us to preserve the following well-known logical equivalence from classical, two-valued, hybrid logic:

$$@_i\varphi \leftrightarrow E(i \wedge \varphi)$$
$$@_i\varphi \leftrightarrow A(i \rightarrow \varphi)$$

With the chosen semantics, these equivalences also hold in MVHL:

$$\nu(w, @_i\varphi) = \nu(\mathbf{n}(i), \varphi) = \bigsqcup\{\nu(v, i) \sqcap \nu(v, \varphi) \mid v \in W\} = \nu(w, E(i \wedge \varphi))$$
$$\nu(w, @_i\varphi) = \nu(\mathbf{n}(i), \varphi) = \bigsqcap\{\nu(v, i) \Rightarrow \nu(v, \varphi) \mid v \in W\} = \nu(w, A(i \rightarrow \varphi)).$$

Here we have been using that the following holds in a Heyting algebra: $\top \sqcap a = a$, $\bot \sqcap a = \bot$, $a \sqcup \bot = a$, $\top \Rightarrow a = a$ and $\bot \Rightarrow a = \top$. Another pleasant property resulting from the choice of semantics for nominals is the following:

$$\nu(w, @_i\Diamond j) = \nu(\mathbf{n}(i), \Diamond j) = \bigsqcup\{R(\mathbf{n}(i), v) \sqcap \nu(v, j) \mid v \in W\} = R(\mathbf{n}(i), \mathbf{n}(j)).$$

This identity expresses that the reachability of the world denoted by j from the world denoted by i is described by the formula $@_i\Diamond j$. This property also holds in classical hybrid logic. Identity between worlds denoted by nominals can also be expressed as usual, since we have:

$$\nu(w, @_ij) = \top \text{ iff } \mathbf{n}(i) = \mathbf{n}(j).$$

2.3 The relation to intuitionistic hybrid logic

As pointed out in the paper [12], there is a close relation between the many-valued modal logic given in that paper and intuitionistic modal logic. We shall in this subsection consider the relation between many-valued hybrid logic and a variant of the intuitionistic hybrid logic given in the paper [9] (which in turn is a hybridization of an intuitionistic modal logic introduced in a tense-logical version in [10]). In the present subsection we do not assume that a finite Heyting algebra has been fixed in advance, so the only atomic formulas we consider are ordinary propositional symbols, nominals, and the symbol $\bot$. We first define an appropriate notion of an intuitionistic model, which can be seen as a restricted variant of the notion of a model given in [9][2].

DEFINITION 2. A *restricted model* for intuitionistic hybrid logic is a tuple

$$(W, \leq, D, \{R_w\}_{w\in W}, \{\nu_w\}_{w\in W})$$

where

1. W is a non-empty finite set partially ordered by $\leq$;
2. D is a non-empty set;
3. for each w, R_w is a binary relation on D such that $w \leq v$ implies $R_w \subseteq R_v$; and
4. for each w, ν_w is a function that to each ordinary propositional symbol p assigns a subset of D such that $w \leq v$ implies $\nu_w(p) \subseteq \nu_v(p)$.

The elements of the set W are states of knowledge and for any such state w, the relation R_w is the set of known relationships between possible worlds and the set $\nu_w(p)$ is the set of possible worlds at which p is known to be true. Note that the definition requires that the epistemic partial order $\leq$ preserves these kinds of knowledge, that is, if an advance to a greater state of knowledge is made, then what is known is preserved.

Given a restricted model $\mathfrak{M} = (W, \leq, D, \{R_w\}_{w\in W}, \{\nu_w\}_{w\in W})$, an *assignment* is a function $\mathbf{n}$ that to each nominal assigns an element of D. The relation $\mathfrak{M}, \mathbf{n}, w, d \models \phi$ is defined by induction, where w is an element of W, $\mathbf{n}$ is an assignment, d is an element of D, and ϕ is a formula.

$$\begin{array}{rll}
\mathfrak{M}, \mathbf{n}, w, d \models p & \text{iff} & d \in \nu_w(p) \\
\mathfrak{M}, \mathbf{n}, w, d \models i & \text{iff} & d = \mathbf{n}(i) \\
\mathfrak{M}, \mathbf{n}, w, d \models \phi \wedge \psi & \text{iff} & \mathfrak{M}, \mathbf{n}, w, d \models \phi \text{ and } \mathfrak{M}, \mathbf{n}, w, d \models \psi \\
\mathfrak{M}, \mathbf{n}, w, d \models \phi \vee \psi & \text{iff} & \mathfrak{M}, \mathbf{n}, w, d \models \phi \text{ or } \mathfrak{M}, \mathbf{n}, w, d \models \psi \\
\mathfrak{M}, \mathbf{n}, w, d \models \phi \rightarrow \psi & \text{iff} & \text{for all } v \geq w, \\
 & & \mathfrak{M}, \mathbf{n}, v, d \models \phi \text{ implies } \mathfrak{M}, \mathbf{n}, v, d \models \psi \\
\mathfrak{M}, \mathbf{n}, w, d \models \bot & \text{iff} & \text{falsum} \\
\mathfrak{M}, \mathbf{n}, w, d \models \Box\phi & \text{iff} & \text{for all } v \geq w, \text{ for all } e \in D, \\
 & & dR_v e \text{ implies } \mathfrak{M}, \mathbf{n}, v, e \models \phi \\
\mathfrak{M}, \mathbf{n}, w, d \models \Diamond\phi & \text{iff} & \text{for some } e \in D, dR_w e \text{ and } \mathfrak{M}, \mathbf{n}, w, e \models \phi \\
\mathfrak{M}, \mathbf{n}, w, d \models @_i\phi & \text{iff} & \mathfrak{M}, \mathbf{n}, w, \mathbf{n}(i) \models \phi \\
\mathfrak{M}, \mathbf{n}, w, d \models A\phi & \text{iff} & \text{for all } v \geq w, \text{ for all } e \in D, \mathfrak{M}, \mathbf{n}, v, e \models \phi \\
\mathfrak{M}, \mathbf{n}, w, d \models E\phi & \text{iff} & \text{for some } e \in D, \mathfrak{M}, \mathbf{n}, w, e \models \phi
\end{array}$$

[2]Compare to Definition 2, p. 237, of the paper [9]. The differences are the following: i) In [9] the set W need not be finite. ii) Instead of D there is a family $\{D_w\}_{w\in W}$ of non-empty sets such that $w \leq v$ implies $D_w \subseteq D_v$, R_w is a binary relation on D_w, and $\nu_w(p)$ is a subset of D_w. iii) There is a family $\{\sim_w\}_{w\in W}$ where $\sim_w$ is an equivalence relation on D_w such that $w \leq v$ implies $\sim_w \subseteq \sim_v$ and such that if $d \sim_w d'$, $e \sim_w e'$, and $dR_w e$, then $d' R_w e'$, and similarly, if $d \sim_w d'$ and $d \in \nu_w(p)$, then $d' \in \nu_w(p)$. The equivalence relations are used for the interpretation of nominals. Such a model for intuitionistic hybrid logic corresponds to a standard model for intuitionistic first-order logic with equality where equality is interpreted using the equivalence relations, cf. [16].

This semantics can be looked upon in two different ways: As indicated above, it can be seen as a restricted variant of the semantics given in [9], but it can also be seen as a hybridized version of a semantics given in the paper [12]. In the latter paper, the epistemic worlds of the semantics are thought of as experts and the epistemic partial order is thought of as a relation of dominance between experts: One expert dominates another one if whatever the first expert says is true is also said to be true by the second expert.

As pointed out in [12], the intuitionistic semantics for modal logic is in a certain sense equivalent to the many-valued semantics. This also holds in the hybrid-logical case. In what follows, we outline this equivalence. It can be shown that given a restricted model $\mathfrak{M} = (W, \leq, D, \{R_w\}_{w \in W}, \{\nu_w\}_{w \in W})$, cf. Definition 2, and an assignment $\mathbf{n}$, the $\leq$-closed subsets of W ordered by $\subseteq$ constitute a finite Heyting algebra, and moreover, a many-valued model $(D, R^*, \mathbf{n}, \nu^*)$ can be defined by letting

- $R^*(d, e) = \{w \in W \mid dR_w e\}$ and
- $\nu^*(d, p) = \{w \in W \mid d \in \nu_w(p)\}$.

By a straightforward extension of the corresponding proof in [12], it can be proved that for any formula ϕ, it is the case that $\nu^*(d, \phi) = \{w \in W \mid \mathfrak{M}, \mathbf{n}, w, d \models \phi\}$. Conversely, given a finite Heyting algebra $\mathcal{T}$ and a many-valued model $(D, R, \mathbf{n}, \nu)$, a restricted model $\mathfrak{M} = (W, \subseteq, D, \{R^*_w\}_{w \in W}, \{\nu^*_w\}_{w \in W})$ can be defined by letting

- $W = \{w \mid w$ is a proper prime filter in $\mathcal{T}\}$,
- $dR^*_w e$ if and only if $R(d, e) \in w$, and
- $d \in \nu^*_w(p)$ if and only if $\nu(d, p) \in w$.

Details can be found in the paper [12]. Again, by a straightforward extension of the corresponding proof in that paper, it can be proved that for any formula ϕ, it is the case that $\mathfrak{M}, \mathbf{n}, w, d \models \phi$ if and only if $\nu(d, \phi) \in w$.

Thus, in the above sense the intuitionistic semantics for hybrid logic is equivalent to the many-valued semantics for hybrid logic. It is an interesting question whether there is such an equivalence if instead of the restricted models of Definition 2 one considers the more general models for intuitionistic hybrid logic given in the paper [9][3]. We shall leave this to further work.

3 A tableau calculus for MVHL

In the following we will present a tableau calculus for MVHL. The basic notions for tableaux are defined as usual (see e.g. [14]). The formulas

[3]As indicated in the previous footnote, in the intuitionistic semantics of [9], nominals are interpreted using a family $\{\sim_w\}_{w \in W}$ of equivalence relations, not identity. This seems to imply that in an equivalent many-valued semantics, nominals should be allowed to take on arbitrary truth-values, not just top and bottom.

occurring in our tableaux will all be of the form $@_i(a \to \varphi)$ or $@_i(\varphi \to a)$ prefixed either a T or an F, where $i \in \mathsf{NOM}$ and $a \in \mathcal{T}$. That is, the formulas occurring in our tableaux will be *signed formulas* of hybrid logic. A signed formula of the form $T@_i(a \to \varphi)$ is used to express that the formula $a \to \varphi$ is true at i, that is, receives the value $\top$ at i. If $\nu(\mathbf{n}(i), a \to \varphi) = \top$ then, by definition of ν, $a \Rightarrow \nu(\mathbf{n}(i), \varphi) = \top$. By definition of relative pseudo-complement we then get that $\top$ is the greatest element of $\mathcal{T}$ satisfying $a \wedge \top \leq \nu(\mathbf{n}(i), \varphi)$. In other words, we simply have $a \leq \nu(\mathbf{n}(i), \varphi)$. Thus what is expressed by a formula $T@_i(a \to \varphi)$ is that the truth value of φ at i is greater than or equal to a. Symmetrically, a signed formula of the formula $T@_i(\varphi \to a)$ expresses that the truth value of φ at i is less than or equal to a. Dually, a signed formula of the form $F@_i(a \to \varphi)$ ($F@_i(\varphi \to a)$) expresses that the truth value of φ at i is *not* greater than or equal to (less than or equal to) a.

The tableau rules are divided into four classes; Branch Closing Rules, Non-modal Rules, Modal Rules and Hybrid Rules. The Branch Closing Rules and Propositional Rules are direct translations of Fitting's corresponding rules for the pure modal case [13].

Branch closing rules:

A tableau branch Θ is said to be *closed* if one of the following holds:

1. $T@_i(a \to b) \in \Theta$, for some a, b with $a \not\leq b$.
2. $F@_i(a \to b) \in \Theta$, for some a, b with $a \leq b$, $a \neq \bot$, and $b \neq \top$.
3. $F@_i(\bot \to \varphi) \in \Theta$, for some formula φ.
4. $F@_i(\varphi \to \top) \in \Theta$, for some formula φ.
5. $T@_i(b \to \varphi), F@_i(a \to \varphi) \in \Theta$, for some a, b with $a \leq b$.
6. $T@_j(a \to i), F@_i(b \to j) \in \Theta$, for some $a, b \neq \bot$.
7. $T@_i(i \to a) \in \Theta$, for some nominal i and truth value a with $a \neq \top$.

The two last conditions, 6 and 7, have no counterpart in Fitting's system, but are required in ours to deal with the semantics chosen for nominals. Note that if a formula $F@_i(a \to i)$ with $a \neq \top$ occurs on a branch then the branch can also be closed: In case $a = \bot$, condition 3 immediately implies closure. If $a \neq \bot$ then using the reversal rule ($\mathbf{F} \geq$) (see below), we can add a formula $T@_i(i \to b)$ to the branch, where b is one of the maximal members of $\mathcal{T}$ not above a. Because b is not above a, b cannot be $\top$. Thus condition 7 implies closure.

Non-modal rules:

The tableau rules for the propositional connectives and the rules capturing the properties of the Heyting algebra are given in Figure 1 and Figure 2, respectively. The rules of Figure 2 are called *reversal rules*, as in [13]. The

$$\frac{T@_i(a \to (\varphi \wedge \psi))}{\begin{array}{c}T@_i(a \to \varphi)\\ T@_i(a \to \psi)\end{array}}\ (\mathbf{T}\wedge)^1 \qquad \frac{F@_i(a \to (\varphi \wedge \psi))}{F@_i(a \to \varphi) \mid F@_i(a \to \psi)}\ (\mathbf{F}\wedge)^1$$

$$\frac{T@_i((\varphi \vee \psi) \to a)}{\begin{array}{c}T@_i(\varphi \to a)\\ T@_i(\psi \to a)\end{array}}\ (\mathbf{T}\vee)^2 \qquad \frac{F@_i((\varphi \vee \psi) \to a)}{F@_i(\varphi \to a) \mid F@_i(\psi \to a)}\ (\mathbf{F}\vee)^2$$

$$\frac{F@_i(a \to (\varphi \to \psi))}{\begin{array}{c|c|c}T@_i(b_1 \to \varphi) & \cdots & T@_i(b_n \to \varphi)\\ F@_i(b_1 \to \psi) & \cdots & F@_i(b_n \to \psi)\end{array}}\ (\mathbf{F}\!\to)^3 \qquad \frac{T@_i(a \to (\varphi \to \psi))}{F@_i(b \to \varphi) \mid T@_i(b \to \psi)}\ (\mathbf{T}\!\to)^4$$

[1] Where $a \neq \bot$.
[2] Where $a \neq \top$.
[3] Where $a \neq \bot$ and $b_1, \ldots, b_n$ are all the members of $\mathcal{T}$ with $b_i \leq a$ except $\bot$.
[4] Where $a \neq \bot$ and b is any member of $\mathcal{T}$ with $b \leq a$ except $\bot$.

Figure 1. Propositional Rules for **MVHL**.

reversal rules together with the closure rules ensure that no formula can be assigned more than one truth value (relative to a given world and a given branch).

Modal rules:

These modal rules, presented in Figure 3, differ from the ones of Fitting and heavily employs the hybrid logic machinery. Note that the tableau rules contain formulas of the form $T@_i(a \leftrightarrow \Diamond j)$. Such formulas are simply used as shorthand notation for the occurrence of both the formulas $T@_i(a \to \Diamond j)$ and $T@_i(\Diamond j \to a)$. In each of the rules of our calculus, the leftmost premise is called the *principal premise*. If α is a signed formula on one of the forms $T@_i(a \to \varphi)$, $T@_i(\varphi \to a)$, $F@_i(a \to \varphi)$ or $F@_i(\varphi \to a)$, we call φ the *body* of α and i its *prefix*. If α and β are two signed formulas such that the body of α is a subformula of the body of β, then α is said to be a *quasi-subformula* of β.

Hybrid rules:

These hybrid rules, presented in Figure 4, are inspired by the standard rules from classical hybrid logic (see [1, 6, 4]). Note that for the **(NOM)** rule, two versions are needed. Furthermore a new rule is needed due to the fact that we are in a many-valued setting, this is the rule **(NOM EQ)**, which ensures our semantic definition of nominals as being $\top$ in exactly one world.

A *tableau proof* of a formula ϕ is a closed tableau with root $F@_i(\top \to \phi)$, where i is an arbitrary nominal not occurring in ϕ. The intuition here is

$$\frac{F@_i(a \to \varphi)}{T@_i(\varphi \to b_1) \mid \cdots \mid T@_i(\varphi \to b_n)}\ (\mathbf{F}\geq)^{1,2} \qquad \frac{T@_i(a \to \varphi)}{F@_i(\varphi \to b)}\ (\mathbf{T}\geq)^{1,3}$$

$$\frac{F@_i(\varphi \to a)}{T@_i(b_1 \to \varphi) \mid \cdots \mid T@_i(b_n \to \varphi)}\ (\mathbf{F}\leq)^{1,4} \qquad \frac{T@_i(\varphi \to a)}{F@_i(b \to \varphi)}\ (\mathbf{T}\leq)^{1,5}$$

[1] φ is a formula other than a propositional constant from $\mathcal{T}$.
[2] Where $b_1, ..., b_n$ are all maximal members of $\mathcal{T}$ with $a \not\leq b_i$ and $a \neq \bot$.
[3] Where b is any maximal member of $\mathcal{T}$ with $a \not\leq b$ and $a \neq \bot$.
[4] Where $b_1, ..., b_n$ are all minimal members of $\mathcal{T}$ with $b_i \not\leq a$ and $a \neq \top$.
[5] Where b is any minimal member of $\mathcal{T}$ with $b \not\leq a$ and $a \neq \top$.

Figure 2. Reversal Rules for **MVHL**.

$$\frac{F@_i(a \to \Box\varphi)}{\begin{array}{c|c|c} T@_i(b_1 \leftrightarrow \Diamond j) & \cdots & T@_i(b_n \leftrightarrow \Diamond j) \\ F@_j((a \sqcap b_1) \to \varphi) & \cdots & F@_j((a \sqcap b_n) \to \varphi) \end{array}}\ (\mathbf{F}\Box)^{1}$$

$$\frac{T@_i(a \to \Box\varphi) \qquad T@_i(b \to \Diamond j)}{T@_j((a \sqcap b) \to \varphi)}\ (\mathbf{T}\Box)$$

$$\frac{F@_i(\Diamond\varphi \to a)}{\begin{array}{c|c|c} T@_i(b_1 \leftrightarrow \Diamond j) & \cdots & T@_i(b_n \leftrightarrow \Diamond j) \\ F@_j(\varphi \to (b_1 \Rightarrow a)) & \cdots & F@_j(\varphi \to (b_n \Rightarrow a)) \end{array}}\ (\mathbf{F}\Diamond)^{1,2}$$

$$\frac{T@_i(\Diamond\varphi \to a) \qquad T@_i(b \to \Diamond j)}{T@_j(\varphi \to (b \Rightarrow a))}\ (\mathbf{T}\Diamond)^{2}$$

$$\frac{F@_i(E\varphi \to a)}{F@_j(\varphi \to a)}\ (\mathbf{FE})^{3} \qquad \frac{T@_i(E\varphi \to a)}{T@_j(\varphi \to a)}\ (\mathbf{TE})^{4}$$

$$\frac{T@_i(a \to A\varphi)}{T@_j(a \to \varphi)}\ (\mathbf{TA})^{4} \qquad \frac{F@_i(a \to A\varphi)}{F@_j(a \to \varphi)}\ (\mathbf{FA})^{3}$$

[1] Where $\mathcal{T} = \{b_1, ..., b_n\}$ and j is a nominal new to the branch.
[2] Where the principal premise is a quasi-subformula of the root formula.
[3] Where j is a nominal new to the branch.
[4] Where j is a nominal already occurring on the branch.

Figure 3. Modal Rules for **MVHL**.

$$\frac{T@_i(@_j\varphi \to a)}{T@_j(\varphi \to a)}\ (@_L) \qquad \frac{T@_i(a \to @_j\varphi)}{T@_j(a \to \varphi)}\ (@_R)$$

$$\frac{F@_i\varphi \qquad T@_i(a \to j)}{F@_j\varphi}\ (\textbf{F-NOM})^{1,2} \qquad \frac{T@_i\varphi \qquad T@_i(a \to j)}{T@_j\varphi}\ (\textbf{T-NOM})^{1,2}$$

$$\frac{T@_k(\Diamond i \to b) \qquad T@_i(a \to j)}{T@_k(\Diamond j \to b)}\ (\textbf{BRIDGE}_L)^1$$

$$\frac{T@_k(b \to \Diamond i) \qquad T@_i(a \to j)}{T@_k(b \to \Diamond j)}\ (\textbf{BRIDGE}_R)^1$$

$$\frac{T@_i(\top \to j) \qquad T@_j(\top \to k)}{T@_i(\top \to k)}\ (\textbf{TRANS})$$

$$\frac{T@_i(a \to j)}{T@_i(\top \to j)}\ (\textbf{NOM EQ})^1$$

[1] Where $a \neq \bot$.
[2] Where the principal premise is a quasi-subformula of the root formula.

Figure 4. Hybrid Rules for **MVHL**.

that the root formula $F@_i(\top \to \phi)$ asserts that ϕ does *not* have the value $\top$, and if the tableau closes, this assertion is refuted. If i is a nominal occurring in the root formula of a tableau then i is called a *root nominal* of the tableau. Other nominals occurring on the tableau are called *non-root nominals*.

4 Termination

The tableau calculus presented above is not terminating. This is due to the rules (**TA**) and (**FA**) for the global modality A. If the rules for the global modalities—(**FE**), (**TE**), (**TA**) and (**FA**)—are all removed, we obtain a tableau calculus for the many-valued hybrid logic with these modalities removed. We will refer to this calculus as the *basic calculus*, and refer to its tableaux as *basic tableaux*. In the following we will prove that the basic calculus terminates. The proof closely follows the method introduced in [4].

If α and β are signed formulas on a tableau branch, then β is said to be *produced* by α if β is one of the conclusions of a rule application with principal premise α. The signed formula β is said to be *indirectly produced* by α if there exists a sequence of signed formulas $\alpha, \alpha_1, \alpha_2, \ldots, \alpha_n, \beta$ in which each formula is produced by its predecessor. We now have the following result.

LEMMA 3 (**Quasi-subformula Property**). *Let $\mathcal{T}$ be a basic tableau. For any signed formula α occurring on $\mathcal{T}$, one of the following holds:*

1. *α is a quasi-subformula of the root formula of $\mathcal{T}$.*
2. *α is a formula of one of the forms $T@_i(a \rightarrow \Diamond j)$, $T@_i(\Diamond j \rightarrow a)$, $F@_i(a \rightarrow \Diamond j)$ or $F@_i(\Diamond j \rightarrow a)$, for which one of the following holds:*
   - *(a) j is a root nominal.*
   - *(b) α is indirectly produced by ($\mathbf{F}\Box$) or ($\mathbf{F}\Diamond$) by a number of applications of the reversal rules.*

Proof. The proof goes by induction on the construction of $\mathcal{T}$. In the basic case α is just the root formula, which of course is of type 1. Now assume that α have been introduced by one of the propositional rules. These rules does not take premises of type 2 and thus by induction they must be of type 1. But then the conclusions produced by these rules must also be of type 1, thus α must be of type 1. If α have been produced by once of the reversal rules by a formula of type 1, then α will also by of type 1 and if α is produced by a formula of type 2, α is also of type 2. Now the modal rules. If α have been produced by the rule ($\mathbf{T}\Box$) then the principal premise can not be a formula of type 2 and thus by induction it must be of type 1. But then so is α. Similar for the rule ($\mathbf{T}\Diamond$) where the side condition insures that the principal premise is of type 1. If α is introduced by on of the rules ($\mathbf{F}\Box$) or ($\mathbf{F}\Diamond$) again the premise must be of type 1. These rules produce two formulas, the first one is by definition of type 2b and the second must be of type 1 since the premise is. Thus in this case α is either of type 1 or type 2b. Finally for the hybrid rules. In the rules (**TRANS**), (**NOM EQ**), ($@_L$) or ($@_R$) the premises can not be of type 2 and thus by induction they must be of type 1. But then the conclusions will also be of type 1. Now if the rule used is (**T-NOM**) or (**F-NOM**) then the side condition insures that the principal premise are of type 1. But then the conclusion will also be of type 1. Now assume that one of the rules (**BRIDGE**$_L$) or (**BRIDGE**$_R$) have been applied to produce α. Then the non-principal premise can not be of type 1 and thus must be of type 2, which implies that j is a root nominal. Thus the conclusion α must be of type 2a. This completes the proof. ∎

Note that in the basic calculus the only rules that can introduce new nominals to a tableau are ($\mathbf{F}\Box$) and ($\mathbf{F}\Diamond$).

DEFINITION 4. Let Θ be a branch of a basic tableau. If a nominal j has been introduced to the branch by applying either ($\mathbf{F}\Box$) or ($\mathbf{F}\Diamond$) to a premise with prefix i then we say that j is *generated* by i on Θ, and we write $i \prec_\Theta j$.

LEMMA 5. *Let Θ be a branch of a basic tableau. The graph $G = (N^\Theta, \prec_\Theta)$, where N^Θ is the set of nominals occurring on Θ, is a finite set of well-founded, finitely branching trees.*

Proof. That G is wellfounded follows from the observation that if $i \prec_\Theta j$, then the first occurrence of i on Θ is before the first occurrence of j. That G is finitely branching is shown as follows. For any given nominal i the number of nominals j satisfying $i \prec_\Theta j$ is bounded by the number of applications of (**F**$\Box$) and (**F**$\Diamond$) to premises of the form $F@_i(a \to \Box\varphi)$ and $F@_i(\Diamond\varphi \to a)$. So to prove that G is finitely branching, we only need to prove that for any given i the number of such premises is finite. However, this follows immediately from the fact that all such premises must be quasi-subformulas of the root formula (cf. Lemma 3 and the condition on applications of (**F**$\Diamond$)). What is left is to prove that G is a finite set of trees. This follows from the fact that each nominal in N^Θ can be generated by at most one other nominal, and the fact that each nominal in N^Θ must have one of the finitely many root nominals of Θ as an ancestor. ■

LEMMA 6. *Let Θ be a branch of a basic tableau. Then Θ is infinite if and only if there exists an infinite chain of nominals*

$$i_1 \prec_\Theta i_2 \prec_\Theta i_3 \prec_\Theta \cdots.$$

Proof. The 'if' direction is trivial. To prove the 'only if' direction, let Θ be any infinite tableau branch. Θ must contain infinitely many distinct nominals, since it follows immediately from Lemma 3 that a tableau with finitely many nominals can only contain finitely many distinct formulas. This implies that the graph $G = (N^\Theta, \prec_\Theta)$ defined as in Lemma 5 must be infinite. Since by Lemma 5, G is a finite set of wellfounded, finitely branching trees, G must then contain an infinite path $(i_1, i_2, i_3, \ldots)$. Thus we get an infinite chain $i_1 \prec_\Theta i_2 \prec_\Theta i_3 \prec_\Theta \cdots$. ■

DEFINITION 7. Let Θ be a branch of a basic tableau, and let i be a nominal occurring on Θ. We define $m_\Theta(i)$ to be the maximal length of any formula with prefix i occurring on Θ.

LEMMA 8 (**Decreasing length).** *Let Θ be a branch of a basic tableau. If $i \prec_\Theta j$ then $m_\Theta(i) > m_\Theta(j)$.*

Proof. For any signed formula α, we will use $|\alpha|$ to denote the length of α. Assume $i \prec_\Theta j$. Let α be a signed formula satisfying: 1) α has maximal length among the formulas on Θ with prefix j; 2) α is the earliest occurring formula on Θ with this property. We need to prove $m_\Theta(i) > |\alpha|$. The formula α can not have been introduced on Θ by applying any of the propositional rules (Figure 1), since this would contradict maximality of α. It can not have been directly produced by any of the reversal rules (Figure 2) either, since this would contradict the choice of α as the earliest possible on Θ of maximal length with prefix j. By the same argument, α can not have been directly produced by any of the rules (**BRIDGE**$_L$), (**BRIDGE**$_R$), (**TRANS**) or (**NOM EQ**). Assume now α has been introduced by applying ($@_L$) or ($@_R$) to a premise of the form $T@_k(@_j\varphi \to a)$

or $T@_k(a \to @_j\varphi)$. By Lemma 3, the premise must be a quasi-subformula of the root formula. Thus j must be a root nominal. However, this is a contradiction, since by assumption j is generated by i, and can thus not be a root nominal. Thus neither ($@_L$) nor ($@_R$) can have been the rule producing α. Now assume that α has been produced by an application of either (**F-NOM**) or (**T-NOM**). Since α has index j, the non-principal premise used in this rule application must have the form $T@_i(a \to j)$. By Lemma 3, this premise must be a quasi-subformula of the root formula, and thus j is again a root nominal, which is a contradiction. Thus α can not have been produced by (**F-NOM**) or (**T-NOM**) either. Thus α must have been introduced by one of the rules (**F**$\Box$), (**T**$\Box$), (**F**$\Diamond$) or (**T**$\Diamond$). Consider first the case of the (**F**$\Box$) and (**F**$\Diamond$) rules. If an instance of one of these produced α, then this instance must have been applied to a premise β with prefix i, since we have assumed $i \prec_\Theta j$ and by Lemma 5 there cannot be an $i' \neq i$ satisfying $i' \prec_\Theta j$. (Note that if α is of the form $T@_j(b \to \Diamond k)$ or $T@_j(\Diamond k \to b)$ produced by a formula $F@_j(a \to \Box\varphi)$ or $F@_j(\Diamond\varphi \to a)$, this would lead to a contradiction with the assumption that α has maximal length with prefix j and is the earliest occurring formula with this property.) Since the rules in question always produce conclusions that are shorter than their premises, β must be longer than α. Since β is a formula with prefix i we then get:

(1) $\quad m_\Theta(i) \geq |\beta| > |\alpha|\,,$

as required. Now consider finally the case where α has been produced by either (**T**$\Box$) or (**T**$\Diamond$). Then α has been produced by a rule instance with non-principal premise of the form $T@_k(b \to \Diamond j)$. Since j is not a root nominal, this premise can not be a quasi-subformula of the root formula. Neither can it be of the tybe ($2a$) mentioned in lemma 3. It must thus be of type ($2b$), that is, it must be indirectly produced by formulas of the form $T@_k(b_m \to \Diamond j')$ or $T@_k(\Diamond j' \to b_m)$ obtained as conclusion by applications of (**F**$\Box$) or (**F**$\Diamond$). Since only reversal rules have been applied in the indirect production from these conclusions, we must have $j = j'$ and thus $k \prec_\Theta j$. Since we already have $i \prec_\Theta j$ we get $k = i$, using Lemma 5. We can conclude that the non-principal premise of the rule instance producing α must have the form $T@_i(b \to \Diamond j)$, and thus the principal premise must be a formula β with index i. Since the rules in question always produce conclusions that are shorter than their premises, β must be longer than α. Since β is a formula with prefix i we then again get the sequence of inequalities (1), as required. ■

We can now finally prove termination of the basic calculus.

THEOREM 9 (**Termination of the basic calculus**). *Any tableau in the basic calculus is finite.*

Proof. Assume there exists an infinite basic tableau. Then it must have an infinite branch Θ. By Lemma 6, there exists an infinite chain

$$i_1 \prec_\Theta i_2 \prec_\Theta i_3 \prec_\Theta \cdots .$$

Now by Lemma 8 we have

$$m_\Theta(i_1) > m_\Theta(i_2) > m_\Theta(i_3) > \cdots$$

which is a contradiction, since $m_\Theta(i)$ is a non-negative number for any nominal i. ■

5 Completeness of the basic calculus

In this section we prove completeness of the basic calculus, that is, the calculus without the global modalities. In this connection we remark that we have proved completeness for a calculus including the global modalities similar to the calculus of the present paper. Let Θ be an open saturated branch in the tableau calculus. We will use this branch to construct a model $\mathcal{M}_\Theta = \langle W_\Theta, R_\Theta, \mathbf{n}_\Theta, \nu_\Theta \rangle$. The set of worlds, W_Θ is simply defined to be the set of nominals occurring on Θ. The definition of the other elements of the model requires a bit more work. First we define the mapping $\mathbf{n}_\Theta$.

Fix a choice function σ that for any given set of nominals on Θ returns one of these nominals. We now define the mapping $\mathbf{n}_\Theta$ in the following way:

$$\mathbf{n}_\Theta(i) = \begin{cases} \sigma\{j \mid T@_i(\top \to j) \in \Theta\} & \text{if } \{j \mid T@_i(\top \to j) \in \Theta\} \neq \emptyset \\ i & \text{otherwise.} \end{cases}$$

A nominal i is called an *urfather* on Θ if $i = \mathbf{n}_\Theta(j)$ for some nominal j.

LEMMA 10. *Let Θ be a saturated tableau branch. Then we have the following properties:*

1. *If $T@_i\varphi \in \Theta$ is a quasi-subformula of the root formula then $T@_{\mathbf{n}_\Theta(i)}\varphi \in \Theta$. Similarly, if $F@_i\varphi \in \Theta$ is a quasi-subformula of the root formula then $F@_{\mathbf{n}_\Theta(i)}\varphi \in \Theta$.*
2. *If $T@_i(\top \to j) \in \Theta$ then $\mathbf{n}_\Theta(i) = \mathbf{n}_\Theta(j)$.*
3. *If i is an urfather on Θ then $\mathbf{n}_\Theta(i) = i$.*

Proof. First we prove (i). Assume $T@_i\varphi \in \Theta$ is a quasi-subformula of the root formula. If $\mathbf{n}_\Theta(i) = i$ then there is nothing to prove. So assume $\mathbf{n}_\Theta(i) = \sigma\{j \mid T@_i(\top \to j) \in \Theta\}$. Then $T@_i(\top \to \mathbf{n}_\Theta(i)) \in \Theta$, and by applying (**T-NOM**) to premises $T@_i\varphi$ and $T@_i(\top \to \mathbf{n}_\Theta(i))$ we get $T@_{\mathbf{n}_\Theta(i)}\varphi$, as needed. The case of $F@_i\varphi \in \Theta$ is proved similarly, using (**F-NOM**) instead of (**T-NOM**). We now prove (ii). Assume $T@_i(\top \to j) \in \Theta$. To prove $\mathbf{n}_\Theta(i) = \mathbf{n}_\Theta(j)$ it suffices to prove that for all nominals k, $T@_i(\top \to k) \in \Theta \Leftrightarrow T@_j(\top \to k) \in \Theta$. So let k be an arbitrary nominal. If $T@_i(\top \to k) \in \Theta$ then we can apply (**T-NOM**) (since $T@_i(\top \to k)$ is a quasi-subformula of the root formula by Lemma 3) to premises $T@_i(\top \to k)$ and $T@_i(\top \to j)$ to obtain the conclusion $T@_j(\top \to k)$, as required. If conversely $T@_j(\top \to k) \in \Theta$ then we can apply (**TRANS**) to premises $T@_i(\top \to j)$ and $T@_j(\top \to k)$ to obtain the conclusion $T@_i(\top \to k)$, as required. We finally prove (iii). Assume i is an urfather. Then $i = \mathbf{n}_\Theta(j)$ for some j. If $j = i$ we are done. Otherwise we have $i = \mathbf{n}_\Theta(j) = \sigma\{k \mid T@_j(\top \to k) \in \Theta\}$ and thus $T@_j(\top \to i) \in \Theta$. This implies $i = \mathbf{n}_\Theta(j) = \mathbf{n}_\Theta(i)$, using item (ii). ■

We now turn to the definition of ν_Θ. As in [13] we will not define a particular valuation ν of the propositional variables occuring on the branch, but only show that any valuation assigning values between a certain lower and upper bound (both given by the branch Θ) will do. Let us first define these bounds.

DEFINITION 11. For a formula φ in the language of MVHL and a nominal i, define:

$$bound^{\Theta,i}(\varphi) = \bigsqcap\{a \mid T@_i(\varphi \to a) \in \Theta\}$$

$$bound_{\Theta,i}(\varphi) = \bigsqcup\{a \mid T@_i(a \to \varphi) \in \Theta\}$$

The intuition is that $bound^{\Theta,i}(\varphi)$ is an upper bound for the truth value of φ at the world i decided by the branch Θ and $bound_{\Theta,i}(\varphi)$ is a lower bound for this truth value.

The following lemma corresponds to Lemma 6.4 of [13] and can be proved in the same way. It ensures that we can actually always chose a value between the lower and the upper bounds.

LEMMA 12. *For all i on Θ and all formulas φ of MVHL*

$$bound_{\Theta,i}(\varphi) \leq bound^{\Theta,i}(\varphi).$$

Later we will show that any valuation assigning a value to p between $bound_{\Theta,i}(p)$ and $bound^{\Theta,i}(p)$ at the world $\mathbf{n}_\Theta(i)$ will do for the truth value of p at this world.

The following lemma corresponds to Proposition 6.5 in [13] and is proven in the same way.

LEMMA 13. *Let φ be any formula in the MVHL language other than a propositional constant from $\mathcal{T}$, and let $a \in \mathcal{T}$, then:*

- *(i) If $T@_i(a \to \varphi) \in \Theta$, then $a \leq bound_{\Theta,i}(\varphi)$.*
- *(ii) If $T@_i(\varphi \to a) \in \Theta$, then $bound^{\Theta,i}(\varphi) \leq a$.*
- *(iii) If $F@_i(a \to \varphi) \in \Theta$, then $a \not\leq bound^{\Theta,i}(\varphi)$.*
- *(iv) If $F@_i(\varphi \to a) \in \Theta$, then $bound_{\Theta,i}(\varphi) \not\leq a$.*

The accessibility relation R_Θ is defined as follows:

$$R_\Theta(i,j) = \bigsqcup\{b \mid T@_i(b \to \Diamond k) \in \Theta, \mathbf{n}_\Theta(k) = j\}.$$

We have the following result, which we are going to use in proving completeness.

LEMMA 14. *If* $T@_i(c \leftrightarrow \Diamond j) \in \Theta$ *then* $R_\Theta(i, \mathbf{n}_\Theta(j)) = c$.

Proof. We will prove $R_\Theta(i, \mathbf{n}_\Theta(j)) \geq c$ and $R_\Theta(i, \mathbf{n}_\Theta(j)) \leq c$. First we prove $R_\Theta(i, \mathbf{n}_\Theta(j)) \geq c$. Since $T@_i(c \leftrightarrow \Diamond j) \in \Theta$ we have $T@_i(c \rightarrow \Diamond j) \in \Theta$, and thus

$$\begin{aligned} R_\Theta(i, \mathbf{n}_\Theta(j)) &= \bigsqcup\{b \mid T@_i(b \rightarrow \Diamond k) \in \Theta, \mathbf{n}_\Theta(k) = \mathbf{n}_\Theta(j)\} \\ &\geq \bigsqcup\{b \mid T@_i(b \rightarrow \Diamond j) \in \Theta\} \\ &\geq c. \end{aligned}$$

We now prove $R_\Theta(i, \mathbf{n}_\Theta(j)) \leq c$. By definition of $\mathbf{n}_\Theta$ we have either $\mathbf{n}_\Theta(j) = j$ or $T@_j(\top \rightarrow \mathbf{n}_\Theta(j)) \in \Theta$. If $T@_j(\top \rightarrow \mathbf{n}_\Theta(j)) \in \Theta$ then since $T@_i(\Diamond j \rightarrow c) \in \Theta$ we get $T@_i(\Diamond \mathbf{n}_\Theta(j) \rightarrow c) \in \Theta$, using ($\mathbf{BRIDGE}_L$). If $\mathbf{n}_\Theta(j) = j$ we obviously also have $T@_i(\Diamond \mathbf{n}_\Theta(j) \rightarrow c) \in \Theta$. Applying Lemma 13 (ii) we then get $bound^{\Theta,i}(\Diamond \mathbf{n}_\Theta(j)) \leq c$. Thus

$$\begin{aligned} R_\Theta(i, \mathbf{n}_\Theta(j)) &= \bigsqcup\{b \mid T@_i(b \rightarrow \Diamond k) \in \Theta, \mathbf{n}_\Theta(k) = \mathbf{n}_\Theta(j)\} \\ &\leq \bigsqcup\{b \mid T@_i(b \rightarrow \Diamond \mathbf{n}_\Theta(j)) \in \Theta\} \quad (\text{using } (\mathbf{BRIDGE}_R)) \\ &= bound_{\Theta,i}(\Diamond \mathbf{n}_\Theta(j)) \\ &\leq bound^{\Theta,i}(\Diamond \mathbf{n}_\Theta(j)) \quad (\text{using Lemma 12}) \\ &\leq c, \end{aligned}$$

as required. ■

The theorem we need for completeness now may be stated in the following way:

THEOREM 15. *Let* ν *be a valuation such that for all propositional variables* p *and all urfather nominals* i

$$bound_{\Theta,i}(p) \leq \nu(i,p) \leq bound^{\Theta,i}(p).$$

Then for all subformulas φ *of the body of root formula of* Θ

$$bound_{\Theta,i}(\varphi) \leq \nu(i,\varphi) \leq bound^{\Theta,i}(\varphi).$$

Proof. By induction on φ. The base cases are where φ is a propositional variable p, a value $c \in \mathcal{T}$ or a nominal j. The case where φ is p follows directly by the assumption. The case where φ is c is easy: First note that for any truth values a, b, if $T@_i(a \rightarrow b) \in \Theta$ then $a \leq b$. This follows from closure rule 1 presented in Section 3. Thus we get:

$$\begin{aligned} bound_{\Theta,i}(c) &= \bigsqcup\{a \mid T@_i(a \rightarrow c) \in \Theta\} \\ &\leq c \\ &\leq \bigsqcap\{a \mid T@_i(c \rightarrow a) \in \Theta\} \\ &= bound^{\Theta,i}(c). \end{aligned}$$

Now assume φ is a nominal j. By definition of ν, $\nu(i,j)$ is $\top$ if $\mathbf{n}_\Theta(j) = i$ and $\bot$ otherwise. Assume first $\mathbf{n}_\Theta(j) = i$. Then $\nu(i,j)$ is $\top$, so trivially we have $bound_{\Theta,i}(j) \leq \nu(i,j)$. We thus only need to prove $\nu(i,j) \leq bound^{\Theta,i}(j)$, that is, we need to prove $\top = bound^{\Theta,i}(j) = \bigsqcap\{a \mid T@_i(j \to a) \in \Theta\}$. This amounts to showing that, for all $a \in \mathcal{T}$, $T@_i(j \to a) \in \Theta$ implies $a = \top$. Assume towards a contradiction that, for some a, $T@_i(j \to a) \in \Theta$ and $a \neq \top$. Since we have assumed $\mathbf{n}_\Theta(j) = i$, by definition of $\mathbf{n}_\Theta$ we get that either $j = i$ or $T@_j(\top \to i) \in \Theta$. If $j = i$ then we have that Θ contains a formula of the form $T@_i(i \to a)$ where $a \neq \top$. This immediately contradicts closure rule 7. Assume instead $T@_j(\top \to i) \in \Theta$. Since we also have $T@_i(j \to a) \in \Theta$ where $a \neq \top$, we can apply ($\mathbf{T} \leq$) to conclude that that Θ must contain a formula of the form $F@_i(t \to j)$ where t is some truth value different from $\bot$. Since Θ then contains both $T@_j(\top \to i)$ and $F@_i(t \to j)$ where $t \neq \bot$, we get a contradiction by closure rule 6. Assume now $\mathbf{n}_\Theta(j) \neq i$. Then $\nu(i,j) = \bot$, and the inequality $\nu(i,j) \leq bound^{\Theta,i}(j)$ thus holds trivially. To prove the other inequality, $bound_{\Theta,i}(j) \leq \nu(i,j)$, we need to show that if $T@_i(a \to j) \in \Theta$ then $a = \bot$. Thus assume toward a contradiction that $T@_i(a \to j) \in \Theta$ and $a \neq \bot$. Then rule (**NOM EQ**) implies $T@_i(\top \to j) \in \Theta$. Thus, by item 2 of Lemma 10, we get $\mathbf{n}_\Theta(i) = \mathbf{n}_\Theta(j)$. Since i is assumed to be an urfather, item 3 of Lemma 10 implies $\mathbf{n}_\Theta(i) = i$. Thus we get $\mathbf{n}_\Theta(j) = \mathbf{n}_\Theta(i) = i$, contradiction the assumption.

Now for the induction step. First the case where φ is $@_j\psi$: Note that $\nu(i, @_j\psi) = \nu(\mathbf{n}_\Theta(j), \psi)$ and by induction hypothesis, since $\mathbf{n}_\Theta(j)$ is an urfather,

$$bound_{\Theta,\mathbf{n}_\Theta(j)}(\psi) \leq \nu(\mathbf{n}_\Theta(j), \psi) \leq bound^{\Theta,\mathbf{n}_\Theta(j)}(\psi).$$

Now by the rule ($@_R$), if $T@_i(a \to @_j\psi) \in \Theta$ then $T@_j(a \to \psi) \in \Theta$, for all $a \in \mathcal{T}$. Thus we get that

$$\begin{aligned}
bound_{\Theta,i}(@_j\psi) &= \bigsqcup\{a \mid T@_i(a \to @_j\psi) \in \Theta\} \\
&\leq \bigsqcup\{a \mid T@_j(a \to \psi) \in \Theta\} \\
&\leq \bigsqcup\{a \mid T@_{\mathbf{n}_\Theta(j)}(a \to \psi) \in \Theta\} \quad \text{(using 1 of Lemma 10)} \\
&= bound_{\Theta,\mathbf{n}_\Theta(j)}(\psi) \\
&\leq \nu(\mathbf{n}_\Theta(j), \psi) \\
&= \nu(i, @_j\psi).
\end{aligned}$$

Similar by the ($@_L$) rule, $T@_i(@_j\psi \to a) \in \Theta$ implies that $T@_j(\psi \to a) \in$

Θ, for all $a \in \mathcal{T}$. Hence

$$\begin{aligned}
\nu(i, @_j\psi) &= \nu(\mathbf{n}_\Theta(j), \psi) \\
&\leq bound^{\Theta, \mathbf{n}_\Theta(j)}(\psi) \\
&= \bigsqcap\{a \mid T@_{\mathbf{n}_\Theta(j)}(\psi \to a) \in \Theta\} \\
&\leq \bigsqcap\{a \mid T@_j(\psi \to a) \in \Theta\} \quad \text{(using 1 of Lemma 10)} \\
&\leq \bigsqcap\{a \mid T@_i(@_j\psi \to a) \in \Theta\} \\
&= bound^{\Theta,i}(@_j\psi),
\end{aligned}$$

and the @-case is done.

In case φ is $\Diamond\psi$, we need to prove that

$$bound_{\Theta,i}(\Diamond\psi) \leq \nu(i, \Diamond\psi) \leq bound^{\Theta,i}(\Diamond\psi),$$

which by definition amounts to showing that

$$\begin{aligned}
\bigsqcup\{a \mid T@_i(a \to \Diamond\psi) \in \Theta\} &\leq \bigsqcup\{R_\Theta(i,j) \sqcap \nu(j,\psi) \mid j \in \Theta\} \\
&\leq \bigsqcap\{a \mid T@_i(\Diamond\psi \to a) \in \Theta\}.
\end{aligned}$$

Proving the first inequality amounts to showing that if $T@_i(a \to \Diamond\psi) \in \Theta$ then

$$a \leq \bigsqcup\{R_\Theta(i,j) \sqcap \nu(j,\psi) \mid j \in \Theta\}.$$

To prove this assume toward a contradiction that

$$T@_i(a \to \Diamond\psi) \in \Theta \text{ and } a \not\leq \bigsqcup\{R_\Theta(i,j) \sqcap \nu(j,\psi) \mid j \in \Theta\},$$

for an $a \in \mathcal{T}$. Then choose a $b \in \mathcal{T}$ such that $b \geq \bigsqcup\{R_\Theta(i,j) \sqcap \nu(j,\psi) \mid j \in \Theta\}$ and b is a maximal member of $\mathcal{T}$ with $a \not\leq b$. Then by the reversal rule $(\mathbf{T}\geq)$, $F@_i(\Diamond\psi \to b) \in \Theta$. Then using the $(\mathbf{F}\Diamond)$ rule there is a $c \in \mathcal{T}$ and a $j \in \Theta$ such that $T@_i(c \leftrightarrow \Diamond j) \in \Theta$ and $F@_j(\varphi \to (c \Rightarrow b)) \in \Theta$. Since $T@_i(c \leftrightarrow \Diamond j) \in \Theta$, Lemma 14 implies $R_\Theta(i, \mathbf{n}_\Theta(j)) = c$. Applying 1 of Lemma 10 to the formula $F@_j(\varphi \to (c \Rightarrow b)) \in \Theta$ we get $F@_{\mathbf{n}_\Theta(j)}(\varphi \to (c \Rightarrow b)) \in \Theta$. Now (iv) of Lemma 13 implies $bound_{\Theta,\mathbf{n}_\Theta(j)}(\psi) \not\leq c \Rightarrow b$. This further implies that $(bound_{\Theta,\mathbf{n}_\Theta(j)}(\psi) \sqcap c) \not\leq b$. But by the induction hypothesis $bound_{\Theta,\mathbf{n}_\Theta(j)}(\psi) \leq \nu(\mathbf{n}_\Theta(j), \psi)$ and thus

$$\begin{aligned}
bound_{\Theta,\mathbf{n}_\Theta(j)}(\psi) \sqcap c &= bound_{\Theta,\mathbf{n}_\Theta(j)}(\psi) \sqcap R_\Theta(i, \mathbf{n}_\Theta(j)) \\
&\leq \nu(\mathbf{n}_\Theta(j), \psi) \sqcap R_\Theta(i, \mathbf{n}_\Theta(j)) \\
&\leq \bigsqcup\{R_\Theta(i, \mathbf{n}_\Theta(j)) \sqcap \nu(\mathbf{n}_\Theta(j), \psi) \mid j \in \Theta\} \\
&\leq \bigsqcup\{R_\Theta(i,j) \sqcap \nu(j,\psi) \mid j \in \Theta\} \leq b,
\end{aligned}$$

which of course is a contradiction.

In order to prove that

$$\bigsqcup\{R_\Theta(i,j) \sqcap \nu(j,\psi) \mid j \in \Theta\} \leq \bigsqcap\{a \mid T@_i(\Diamond\psi \to a) \in \Theta\},$$

we must show that if $T@_i(\Diamond\psi \to a) \in \Theta$, then $R_\Theta(i,j) \sqcap \nu(j,\psi) \leq a$ for all $j \in \Theta$. Thus assume that $T@_i(\Diamond\psi \to a) \in \Theta$ and that $R_\Theta(i,j) \neq \bot$ (or else it's trivial) for an arbitrary $j \in \Theta$. Since $R_\Theta(i,j) \neq \bot$, the definition of R implies that j must be an urfather. Furthermore,

$$R_\Theta(i,j) = \bigsqcup\{b \mid T@_i(b \to \Diamond k) \in \Theta, \mathbf{n}_\Theta(k) = j\}.$$

Let b and k be chosen arbitrarily such that $T@_i(b \to \Diamond k) \in \Theta$ and $\mathbf{n}_\Theta(k) = j$. Then by the $(\mathbf{T}\Diamond)$ rule, $T@_k(\psi \to (b \Rightarrow a)) \in \Theta$. Using 1 of Lemma 10 we get $T@_{\mathbf{n}_\Theta(k)}(\psi \to (b \Rightarrow a)) \in \Theta$, that is, $T@_j(\psi \to (b \Rightarrow a)) \in \Theta$. Now, by induction hypothesis, since j is an urfather,

$$\nu(j,\psi) \leq \mathit{bound}^{\Theta,j}(\psi) \leq b \Rightarrow a.$$

Since k and b were chosen arbitrarily with $T@_i(b \to \Diamond k) \in \Theta$ and $\mathbf{n}_\Theta(k) = j$, we get

$$\nu(j,\psi) \leq \bigsqcap\{b \Rightarrow a \mid T@_i(b \to \Diamond k) \in \Theta, \mathbf{n}_\Theta(k) = j\}.$$

We now get

$$\begin{aligned} R_\Theta(i,j) \sqcap \nu(j,\psi) &\leq \bigsqcup\{b \mid T@_i(b \to \Diamond k) \in \Theta, \mathbf{n}_\Theta(k) = j\} \\ &\quad \sqcap \bigsqcap\{b \Rightarrow a \mid T@_i(b \to \Diamond k) \in \Theta, \mathbf{n}_\Theta(k) = j\} \\ &\leq \bigsqcup\{b \sqcap (b \Rightarrow a) \mid T@_i(b \to \Diamond k) \in \Theta, \mathbf{n}_\Theta(k) = j\} \\ &\leq \bigsqcup\{a \mid T@_i(b \to \Diamond k) \in \Theta, \mathbf{n}_\Theta(k) = j\} \\ &\leq a. \end{aligned}$$

Because $j \in \Theta$ was arbitrary it follows that it holds for all $j \in \Theta$ and the proof of this case is completed.

In case φ is $\Box\psi$, we need to prove that

$$\begin{aligned} \bigsqcup\{a \mid T@_i(a \to \Box\psi) \in \Theta\} &\leq \bigsqcap\{R_\Theta(i,j) \Rightarrow \nu(j,\psi) \mid j \in \Theta\} \\ &\leq \bigsqcap\{a \mid T@_i(\Box\psi \to a) \in \Theta\}. \end{aligned}$$

To prove the first inequality we need to prove that if $j \in \Theta$, then

$$a \leq R_\Theta(i,j) \Rightarrow \nu(j,\psi), \tag{2}$$

for all $a \in \mathcal{T}$ with $T@_i(a \to \Box\psi) \in \Theta$. So let $a \in \mathcal{T}$ be given arbitrarily such that $T@_i(a \to \Box\psi) \in \Theta$. Note that (2) is equivalent to

$$a \sqcap R_\Theta(i,j) \leq \nu(j,\psi).$$

By definition of R_Θ we have

$$R_\Theta(i,j) = \bigsqcup\{b \mid T@_i(b \to \Diamond k) \in \Theta, \mathbf{n}_\Theta(k) = j\}.$$

Let b and k be chosen arbitrarily such that $T@_i(b \to \Diamond k) \in \Theta$ and $\mathbf{n}_\Theta(k) = j$. Then by the **(T□)**-rule it follows that $T@_k((a \sqcap b) \to \psi) \in \Theta$. By 1 of Lemma 10 this implies $T@_j((a \sqcap b) \to \psi) \in \Theta$. Thus we get $bound_{\Theta,j}(\psi) \geq (a \sqcap b)$. Since b and k were chosen arbitrarily with the properties $T@_i(b \to \Diamond k) \in \Theta$ and $\mathbf{n}_\Theta(k) = j$ we then get

$$bound_{\Theta,j}(\psi) \geq \bigsqcup\{a \sqcap b \mid T@_i(b \to \Diamond k) \in \Theta, \mathbf{n}_\Theta(k) = j\}.$$

Using this inequality and the induction hypothesis we now get

$$\begin{aligned} a \sqcap R_\Theta(i,j) &= a \sqcap \bigsqcup\{b \mid T@_i(b \to \Diamond k) \in \Theta, \mathbf{n}_\Theta(k) = j\} \\ &= \{a \sqcap b \mid T@_i(b \to \Diamond k) \in \Theta, \mathbf{n}_\Theta(k) = j\} \\ &\leq bound_{\Theta,j}(\psi) \leq \nu(j,\psi). \end{aligned}$$

Since a was arbitrary this holds for all $a \in \mathcal{T}$ and the inequality have been proven.

To show the other inequality we need to show that

$$\text{if } T@_i(\Box\psi \to a) \in \Theta \text{ then } \bigsqcap\{R_\Theta(i,j) \Rightarrow \nu(j,\psi) \mid j \in \Theta\} \leq a.$$

If $a = \top$ then this is trivial. Thus assume towards a contradiction that there is an $a \neq \top$ with $T@_i(\Box\psi \to a) \in \Theta$ and $\bigsqcap\{R_\Theta(i,j) \Rightarrow \nu(j,\psi) \mid j \in \Theta\} \not\leq a$. Now let $b \leq \bigsqcap\{R_\Theta(i,j) \Rightarrow \nu(j,\psi) \mid j \in \Theta\}$ be a minimal member of $\mathcal{T}$ such that $b \not\leq a$. Then by the reversal rule **(T≤)**, $F@_i(b \to \Box\psi) \in \Theta$. Hence by the **(F□)**-rule there is a nominal $k \in \Theta$ and a $c \in \mathcal{T}$ such that $T@_i(c \leftrightarrow \Diamond k) \in \Theta$ and $F@_k((b \sqcap c) \to \psi) \in \Theta$. From the first it follows that $R_\Theta(i, \mathbf{n}_\Theta(k)) = c$, using Lemma 14. From the second it follows that $F@_{\mathbf{n}_\Theta(k)}((b \sqcap c) \to \psi) \in \Theta$, using 1 of Lemma 10, and thus, by *(iii)* of Lemma 13, $b \sqcap c \not\leq bound^{\Theta,\mathbf{n}_\Theta(k)}(\psi)$. But then from the induction hypothesis it follows that

$$b \sqcap c \not\leq \nu(\mathbf{n}_\Theta(k), \psi) \leq bound^{\Theta,\mathbf{n}_\Theta(k)}(\psi).$$

Hence

$$b \not\leq c \Rightarrow \nu(\mathbf{n}_\Theta(k), \psi) = R_\Theta(i, \mathbf{n}_\Theta(k)) \Rightarrow \nu(\mathbf{n}_\Theta(k), \psi).$$

But by the assumption on b we also have that

$$b \leq \bigsqcap\{R_\Theta(i,j) \Rightarrow \nu(j,\psi) \mid j \in \Theta\} \leq R_\Theta(i, \mathbf{n}_\Theta(k)) \Rightarrow \nu(\mathbf{n}_\Theta(k), \psi),$$

and a contradiction have been reached. This concludes the □ case and thus the entire proof of the theorem. ■

Now completeness can easily be proven, in the following sense.

THEOREM 16. *If there is no tableau proof of the formula* φ, *then there is a model* $\mathcal{M} = \langle W, R, \mathbf{n}, \nu \rangle$ *and a* $w \in W$ *such that* $\nu(w, \varphi) \neq \top$.

Proof. Assume that there is no tableau proof of the formula φ. Then there is an saturated tableau with a open branch Θ starting with the formula $F@_i(\top \to \varphi)$ for a nominal i not in φ. By item 1 of Lemma 10 it follows that also $F@_{\mathbf{n}_\Theta(i)}(\top \to \varphi) \in \Theta$.

The model $\mathcal{M}_\Theta = \langle W_\Theta, R_\Theta, \mathbf{n}_\Theta, \nu_\Theta \rangle$ can now be constructed such that ν_Θ satisfies the assumption of Theorem 15. Since $F@_{\mathbf{n}_\Theta(i)}(\top \to \varphi) \in \Theta$ it follows by Lemma 13 that $\top \not\leq \mathit{bound}^{\Theta, \mathbf{n}_\Theta(i)}(\varphi)$. But by Theorem 15, since φ is a subformula of the root formula and $\mathbf{n}_\Theta(i)$ is an urfather, we know that $\nu_\Theta(\mathbf{n}_\Theta(i), \varphi) \leq \mathit{bound}^{\Theta, \mathbf{n}_\Theta(i)}(\varphi)$ and it thus follows that $\top \not\leq \nu_\Theta(\mathbf{n}_\Theta(i), \varphi)$ and the proof is completed. ■

Acknowledgments. Thanks to the anonymous reviewers for useful comments and corrections. The second and third authors are partially funded by the Danish Natural Science Research Council (the HyLoMOL project).

BIBLIOGRAPHY

[1] P. Blackburn. Internalizing labelled deduction. *Journal of Logic and Computation*, 10:137–168, 2000.

[2] P. Blackburn and M. Marx. Tableaux for quantified hybrid logic. In U. Egly and C. Fermüller, editors, *Automated Reasoning with Analytic Tableaux and Related Methods, TABLEAUX 2002*, volume 2381 of *Lecture Notes in Artificial Intelligence*, pages 38–52. Springer-Verlag, 2002.

[3] T. Bolander and P. Blackburn. Terminating tableau calculi for hybrid logics extending K. In Carlos Areces and Stéphane Demri, editors, *Workshop Proceedings of Methods for Modalities 5*, pages 157–175. École Normale Supérieure de Cachan, France, 2007.

[4] T. Bolander and P. Blackburn. Termination for hybrid tableaus. *Journal of Logic and Computation*, 17(3):517–554, 2007.

[5] T. Bolander and T. Braüner. Two tableau-based decision procedures for hybrid logic. In H. Schlingloff, editor, *4th Workshop "Methods for Modalities" (M4M), Informatik-Bericht Nr. 194*, pages 79–96. Humboldt-Universität zu Berlin, 2005.

[6] T. Bolander and T. Braüner. Tableau-based decision procedures for hybrid logic. *Journal of Logic and Computation*, 16:737–763, 2006. Revised and extended version of [5].

[7] T. Braüner. *Hybrid Logic and Its Proof-Theory.* 2008. To appear.

[8] T. Braüner and V. de Paiva. Towards constructive hybrid logic (extended abstract). In C. Areces and P. Blackburn, editors, *Workshop Proceedings of Methods for Modalities 3*, 2003. 15 pages.

[9] T. Braüner and V. de Paiva. Intuitionistic hybrid logic. *Journal of Applied Logic*, 4:231–255, 2006. Revised and extended version of [8].

[10] W. B. Ewald. Intuitionistic tense and modal logic. *Journal of Symbolic Logic*, 51:166–179, 1986.

[11] M. Fitting. Many-valued modal logics. *Fundamenta Informaticae*, 15:235–254, 1992.

[12] M. Fitting. Many-valued modal logics, II. *Fundamenta Informaticae*, 17:55–73, 1992.

[13] M. Fitting. Tableaus for many-valued modal logic. *Studia Logica*, 55:63–87, 1995.

[14] M. Fitting. *Proof methods for modal and intuitionistic logics*, volume 169 of *Synthese Library*. D. Reidel Publishing Co., Dordrecht, 1983.

[15] J.U. Hansen. A tableau system for a first-order hybrid logic. In J. Villadsen, T. Bolander, and T. Braüner, editors, *Proceedings of the International Workshop on Hybrid Logic 2007*, pages 32–40. *19th European Summerschool in Logic, Language and Information*, 2007.

[16] A. Troelstra and D. van Dalen. *Constructivism in Mathematics: An Introduction.* North-Holland, 1988.

Jens Ulrik Hansen
Programming, Logic and Intelligent Systems Research Group and
Science Studies Research Group
Roskilde University, P.O. Box 260, DK-4000 Roskilde,
Denmark
jensuh@ruc.dk

Thomas Bolander
Informatics and Mathematical Modelling
Richard Petersens Plads, Building 322
Technical University of Denmark, DK-2800 Lyngby,
Denmark
tb@imm.dtu.dk

Torben Braüner
Programming, Logic and Intelligent Systems Research Group
Roskilde University, P.O. Box 260, DK-4000 Roskilde,
Denmark
torben@ruc.dk

Properties of logics of individual and group agency

ANDREAS HERZIG AND FRANÇOIS SCHWARZENTRUBER

ABSTRACT. We provide proof-theoretic results about deliberative STIT logic. First we present STIT logic for individual agents without time, where the problem of satisfiability has recently been shown to be NEXPTIME-complete in the general case. Then we study STIT logic for groups of agents. We prove that satisfiability of STIT formulas involving groups of agents is undecidable by reducing the problem of satisfiability of a formula of the product logic $S5^n$ to group STIT satisfiability problem. We also prove that group STIT is not finitely axiomatizable.

Keywords: logics of agency, deliberative STIT, joint action, decidability, axiomatizability, complexity

1 Introduction

While logics of programs and actions such as PDL allow to reason about the relation between an action and its effects, so-called logics of agency are about the relation between an agent and the effects of his actions. The latter are relevant in game theory, theoretical computer science and philosophy of action.

In game theory, Pauly's coalition logic (CL) allows to reason about the capabilities of coalitions [14]. It provides for expressions of the kind "coalition J can make ϕ true at the next time point".

In theoretical computer science, Alternating-time Temporal Logics ATL and ATL* were introduced by Henzinger et al. in order to reason about distributed processes [2]. The formula $\langle\langle J\rangle\rangle\phi$ reads "coalition J has a strategy such that ϕ holds", where ϕ is a formula of linear temporal logic (that has to satisfy some restrictions in the case of ATL). Goranko showed that CL is nothing but a fragment of ATL (which in turn is a fragment of ATL*), by identifying the CL formula $[J]\phi$ with the ATL formula $\langle\langle J\rangle\rangle X\phi$ [9], where X is the temporal 'next' operator.

In philosophy of action constructions of the form $[i\ stit : \phi]$ were introduced by Belnap et col. [4], read "agent i sees to it that ϕ" or "i brings it about that ϕ". In this paper, we focus on the basic version that is called Chellas STIT [6] (thus baptized by [10]), noted $[i\ cstit : \phi]$ in the literature. (The original operator defined by Chellas is nevertheless notably different since it does not come with the principle of independence of agents that plays a central role in STIT theory.) The Chellas STIT was extended to group agency in [4, Section 10.C] and [11, Section 2.4]. For a set of agents J, the formula $[J\ cstit : \phi]$ reads "group J sees to it that ϕ". We here write $[J]\phi$ instead of $[J\ stit : \phi]$. These logics moreover have a modal operator of historical necessity $\Box$. Recently Broersen et al. showed that ATL can be embedded into the logic of the Chellas STIT, by identifying $\langle\langle J\rangle\rangle X\phi$ with $\neg\Box\neg[J]X\phi$ [5]. This highlights that the modal operators of CL and ATL are nothing but fusions of three modal operators. STIT-logics are therefore the most general formal framework for agency, allowing not only to reason about what agents *can do*, but also about what they *do*.

While it is known that the satisfiability problem is PSPACE-complete for coalition logic CL [14], EXPTIME-complete for ATL [19], and 2EXPTIME-complete for ATL* [15], only little is known about the mathematical properties of STIT logic. Up to now the only known results were restricted to the individual case: Wölfl gave an axiomatization [21], and Xu established axiomatization and decidability in [22] and [4, Chapter 17]. Wansing gave a complete tableaux calculus, but didn't prove termination [20]. In previous work we showed NEXPTIME completeness of the satisfiability problem [3].

The present paper investigates decidability and axiomatizability results for STIT logic without temporal operators. It is organized as follows. In Section 2 we recall the known results about individual STIT. In Section 3 we recall the definition of group STIT, providing also for an alternative semantics and a normal form which is built by rewriting all groups that are different from the 'grand coalition' (alias the set of all agents), to what we call 'anti-individuals' (alias complements of singleton groups). Building on results about the product Logic $S5^n$ that we recall in Section 4, we show in Section 5 that group STIT is undecidable, and in Section 6 that there is no finite axiomatization for it.

2 Individual STIT

In this section, we present the logic of agentive sentences of the form 'individual i sees to it that ϕ'.

The language $\mathcal{L}_{STIT_n}$ of $STIT_n$ logic is built from a countably infinite set of atomic propositions ATM and a finite set of agents $AGT = \{1, \ldots, n\}$. It is defined by the following BNF:

$$\phi \ ::= \ p \mid \neg\phi \mid (\phi \wedge \phi) \mid [i]\phi \mid \Box\phi$$

where p ranges over ATM and i ranges over AGT. $[i]\phi$ is read "agent i sees to it that ϕ (whatever the other agents do)", and $\Box\phi$ is read "ϕ is historically necessary", or "ϕ holds whatever all the agents choose to do".

We use the usual dual operators: $\Diamond\phi$ abbreviates $\neg\Box\neg\phi$, and $\langle i\rangle\phi$ abbreviates $[i]\phi$.

REMARK 1. In the STIT literature the formula $[i]\phi$ is usually written $[i\ cstit : \phi]$, where '*cstit*' stands for 'Chellas STIT'. The formula $\Box\phi$ is sometimes written *Sett* ϕ, where '*Sett*' stands for 'settled'.

REMARK 2. Other STIT operators exist: the so-called deliberative STIT operator $[i\ dstit : \phi]$ can be defined as an abbreviation of $[i]\phi \wedge \neg\Box\phi$; the so-called achievement STIT operator $[i\ astit : \phi]$ is more complex and will not be considered here.

2.1 Semantics of STIT_n

We present two semantics for $\mathcal{L}_{\mathsf{STIT}_n}$. The first one is the original one in terms of Belnap's branching-time models [4], while the second one does not mention time and is closer to standard presentations of Kripke models.

BT+AC models

Semantics is given to formulas of $\mathcal{L}_{\mathsf{STIT}_n}$ in terms of a branching-time (BT) structure augmented by an agent choice (AC) function.

DEFINITION 3 (BT structure). A *BT structure* is of the form $\langle M, < \rangle$, where M is a nonempty set of moments, and $<$ is a partial order on M (transitive and anti-symmetric) that is tree-like: for any m_1, m_2 and m_3 in M, if $m_1 < m_3$ and $m_2 < m_3$, then either $m_1 = m_2$ or $m_1 < m_2$ or $m_2 < m_1$.

A maximal set of linearly ordered moments from M is a *history*. When $m \in h$ we say that moment m is *on* the history h. $Hist$ is the set of all histories.

$$H_m = \{h \mid h \in Hist, m \in h\}$$

is the set of histories passing through m. A *moment-history pair* is a couple m/h, consisting of a moment m and a history h from H_m (i.e., a history and a moment in that history).

BT+AC models are BT structures augmented by agents' choices (AC) and a valuation.

DEFINITION 4 (BT+AC model). A *BT+AC model* is a tuple $\mathcal{M} = \langle M, <, Choice, V \rangle$, where:

- $\langle M, < \rangle$ is a BT structure;
- $Choice : AGT \times M \to 2^{2^{Hist}}$ is a function mapping each agent and each moment m into a partition of H_m such that for all m and all mappings $s_m : AGT \longrightarrow 2^{H_m}$ such that $s_m(i) \in Choice(i, m)$, we have $\bigcap_{i \in AGT} s_m(i) \neq \emptyset$;
- V is valuation function $V : ATM \to 2^{M \times Hist}$.

In terms of game theory, each mapping $s_m : AGT \to 2^{H_m}$ such that $s_m(i) \in Choice(i, m)$ for all i is a *strategy profile* at m. We write $Choice_i^m$ instead of $Choice(i, m)$. Each equivalence class belonging to $Choice_i^m$ can be thought of as a choice that is available to agent i at m: when $h, h' \in Choice_i^m$ then agent i's current choice at the moment-history pair m/h

cannot distinguish between h and h'. Given a moment m, we can view $Choice_i^m$ as a mapping from H_m to 2^{H_m} by defining:

$$Choice_i^m(h) = \{h' \in H_m \mid \text{there is } Q \in Choice_i^m \text{ and } h, h' \in Q\}$$

Thus $Choice_i^m(h)$ returns the particular choice from $Choice_i^m$ containing h, or in other words, the particular action performed by i at the moment-history pair m/h: i's current choice at m/h forces the possible histories to be among $Choice_i^m(h)$.

We call the constraint of nonempty intersection of all possible simultaneous choices of agents at m the *independence constraint.*[1]

A formula is evaluated with respect to a model and a moment-history pair:

$$\begin{array}{lll} \mathcal{M}, m/h \models p & \text{iff} & m/h \in V(p), p \in ATM \\ \mathcal{M}, m/h \models \neg\phi & \text{iff} & \mathcal{M}, m/h \not\models \phi \\ \mathcal{M}, m/h \models \phi \wedge \psi & \text{iff} & \mathcal{M}, m/h \models \phi \text{ and } \mathcal{M}, m/h \models \psi \\ \mathcal{M}, m/h \models \Box\phi & \text{iff} & \mathcal{M}, m/h' \models \phi \text{ for all } h' \in H_m \\ \mathcal{M}, m/h \models [i]\phi & \text{iff} & \mathcal{M}, m/h' \models \phi \text{ for all } h' \in Choice_i^m(h) \end{array}$$

Validity in BT+AC models is defined as truth at every moment-history pair of every BT+AC model. A formula ϕ is satisfiable in BT+AC models if $\neg\phi$ is not valid in BT+AC models.

Kripke models

We now present an alternative semantics for $\mathcal{L}_{\mathsf{STIT}_n}$-formulas that is closer to that of standard modal logics, and was proposed in [13].

DEFINITION 5 (Kripke model). A Kripke model for the logic STIT_n is a tuple $\mathcal{W} = \langle W, R, V \rangle$ where:

- W is a nonempty set;
- R is a mapping associating to every $i \in AGT$ an equivalence relation R_i on W such that for all $(w_1, \ldots, w_n) \in W^n$, $\bigcap_{i \in AGT} R_i(w_i) \neq \emptyset$;
- V is a valuation function $V : ATM \to 2^W$.

Intuitively, R_i is nothing more than the equivalence relation corresponding to the partition $Choice_i^m$. The condition on R corresponds to the independence constraint of BT+AC models.

REMARK 6. Our Kripke models here correspond to the class of point-generated models of the semantics proposed in [3]. There, the independence constraint is formulated in a slightly different way (just because the models might not be point-generated).

A formula is evaluated as usual with respect to a model and a point.

[1]There are other constraints relating the BT structure and the choice function, such as "no choice between undivided histories". They are not relevant here because we do not have temporal operators in our language, and we therefore omit them.

$\mathcal{W}, w \models p$	iff	$w \in V(p)$, for p $\in ATM$
$\mathcal{W}, w \models \neg\phi$	iff	$\mathcal{W}, w \not\models \phi$
$\mathcal{W}, w \models \phi \wedge \psi$	iff	$\mathcal{W}, w \models \phi$ and $\mathcal{W}, w \models \psi$
$\mathcal{W}, w \models \Box\phi$	iff	$\mathcal{W}, w' \models \phi$ for all $w' \in W$
$\mathcal{W}, w \models [i]\phi$	iff	$\mathcal{W}, w' \models \phi$ for all $w' \in R_i(h)$

Validity and satisfiability in Kripke models are defined as usual.

THEOREM 7. *A $STIT_n$-formula is satisfiable in BT+AC models iff it is satisfiable in Kripke models.*

Proof. The proof is done by transforming a given BT+AC model into a Kripke model and vice versa. It is a particular case of the proof of Theorem 11 in Section 3.1. ■

2.2 Axiomatization, decidability and complexity of STIT_n

Xu gave the following axioms:

S5($\Box$) the axiom schemas of S5 for $\Box$;

S5(i) the axiom schemas of S5 for every $[i]$, for every $i \in AGT$;

($\Box \to i$) $\Box\phi \to [i]\phi$, for every $i \in AGT$;

(AIA$_n$) $(\Diamond[1]\phi_1 \wedge \ldots \wedge \Diamond[n]\phi_n) \to \Diamond([1]\phi_1 \wedge \ldots \wedge [n]\phi_n)$.

(AIA$_n$) is called the axiom schema for independence of agents. Xu's axiomatics has the standard inference rules of modus ponens and necessitation for $\Box$. From the latter necessitation rules for every $[i]$ follow by axiom ($\Box \to i$).

From Xu's completeness theorem [4, Chapter 4] and Theorem 7 we get:

THEOREM 8. *[4, Chapter 17] A formula ϕ of $\mathcal{L}_{STIT_n}$ is valid in Kripke models iff ϕ is provable from the schemas S5($\Box$), S5(i), ($\Box \to i$), and (AIA$_n$) by the rules of modus ponens and $\Box$-necessitation.*

REMARK 9. An alternative axiomatization is given in [3], where (AIA$_n$) is replaced by the simpler axiom schema $\Diamond\phi \to \langle k\rangle \bigwedge\{\langle l\rangle\phi \mid 1 \leq l \leq n \text{ and } l \neq k\}$.

It is also shown there that $\Diamond\phi$ can be viewed as an abbreviation of $\langle i\rangle\langle j\rangle\phi$, for some arbitrary i and j.

The complexity of the satisfiability problem for STIT_n-formulas depends of the number of agents.

THEOREM 10. *[3] The problem of deciding satisfiability of a formula of $\mathcal{L}_{STIT_n}$ is NP-complete if $n = 1$, and it is NEXPTIME-complete if $n \geq 2$.*

3 Group STIT

In this section, we extend the individual STIT to group STIT: we study the logic of agentive sentences of the form 'group J sees to it that ϕ'. Now modal operators have as arguments coalitions $J \subseteq AGT$.

Just as in the individual case, the language $\mathcal{L}_{\mathsf{STIT}_n^G}$ of STIT_n^G logic is built from a countable set of atomic propositions ATM and a finite set of agents $AGT = \{1, \ldots, n\}$. But now the modal operators have sets of agents as arguments, and $\mathcal{L}_{\mathsf{STIT}_n^G}$ is defined by the following BNF:

$$\phi \ ::= \ p \mid \neg\phi \mid (\phi \wedge \phi) \mid [J]\phi$$

where p ranges over ATM and J ranges over 2^{AGT}.

The language of individual STIT becomes a fragment of that of group STIT if we identify $[\{i\}]$ with $[i]$. The semantics of group STIT will guarantee that the $\mathcal{L}_{\mathsf{STIT}_n^G}$ formula $[\emptyset]\phi$ has the same interpretation as the $\mathcal{L}_{\mathsf{STIT}_n}$ formula $\Box\phi$.

3.1 Semantics of STIT_n^G

Again, we present the semantics both in terms of branching-time models as defined by Horty and Belnap [10] and Belnap and Perloff [4, Chapter 10], and in terms of Kripke models [13]. We prove that both classes of models have the same logic. The latter will be useful to establish the relationship with product logics.

BT+AC models

Models for STIT_n^G are the same those of the logic STIT_n, i.e. BT+AC models satisfying the independence constraint. The only thing we have to do is to extend the definition of *Choice* in order to interpret group agency.

Horty defines in [11] the notion of collective choice. He first introduces action selection functions s_m from AGT into 2^{H_m} such that for each $m \in M$ and $a \in AGT$, $s_m(a) \in Choice_a^m$. So a selection function s_m selects a particular action for each agent at m. Then for a given m,

$$Select_m = \{s_m : AGT \to 2^{H_m} \mid s_m(a) \in Choice_a^m, \text{for all } a \in AGT\}$$

is the set of all such selection functions. This allows to extend the definition of *Choice*. A collective choice for a nonempty group of agents $\emptyset \subsetneq J \subseteq AGT$ at moment m is defined as:

$$Choice_J^m = \{\bigcap_{j \in J} s_m(j) \mid s_m \in Select_m\}$$

For $J = \emptyset$ we define $Choice_\emptyset^m = \{H_m\}$. We can check that every $Choice_J^m$ is a partition of H_m.

As before,

$$Choice_J^m(h) = \{h' \in H_m \mid \text{ there is } Q \in Choice_J^m \text{ and } h, h' \in Q\}$$

is the particular choice from $Choice_J^m$ containing h, or in other words, the particular joint action performed by coalition J at the moment-history pair m/h. And as before, formulas are interpreted with respect to a model and a moment-history pair:

$$\mathcal{M}, m/h \models [J]\phi \quad \text{iff} \quad \mathcal{M}, m/h' \models \phi \text{ for all } h' \in Choice_J^m(h).$$

Observe that the $\mathcal{L}_{\mathsf{STIT}_n^G}$-formula $[\emptyset]\phi$ is true at m/h if and only if the $\mathcal{L}_{\mathsf{STIT}_n}$-formula $\Box\phi$ is true at m/h.

Validity and satisfiability are defined as before.

Kripke models

Kripke models for STIT_n^G are the same as for STIT_n. Just as we defined $Choice_J^m$ from $Choice_i^m$ in the last section, we here define R_J from the R_is.

Let $\mathcal{W} = \langle W, R, V\rangle$ be a Kripke model for STIT_n. For all nonempty $J \subseteq AGT$, we define

$$R_J = \bigcap_{i \in J} R_i$$

and $R_\emptyset = W \times W$.

A formula is evaluated as usual with respect to a model and a world:

$$\mathcal{W}, w \models [J]\phi \quad \text{iff} \quad \mathcal{W}, w' \models \phi \text{ for all } w' \in R_J(w)$$

THEOREM 11. *A STIT_n^G-formula is satisfiable in BT+AC models iff it is satisfiable in Kripke models.*

Proof. $\boxed{\Rightarrow}$ Let $\mathcal{M}' = \langle M', <, Choice, V'\rangle$ be a BT+AC model such that $\mathcal{M}', m_0/h_0 \models \phi$ for some a moment-history pair m_0/h_0. We define the tuple $\mathcal{W} = \langle W, R, V\rangle$ as follows:

- $W = H_{m_0}$;
- $R_i = \{(h, h') \mid \text{there exists } Q \in Choice_i^{m_0} \text{ such that } h, h' \in Q\}$;
- $V(p)$ is the set of histories $h \in H_{m_0}$ such that $m_0/h \in V'(p)$.

We can check that $\mathcal{W}$ is a Kripke model. We can prove by induction on ψ that for all formulas ψ and for all $h \in H_{m_0}$, $\mathcal{W}, h \models \psi$ iff $\mathcal{M}', m_0/h \models \psi$. Hence, $\mathcal{W}, h_0 \models \phi$.

$\boxed{\Leftarrow}$ Let $\mathcal{W} = \langle W, R, V\rangle$ be a Kripke model such that $\mathcal{W}, w_0 \models \phi$ for some world $w_0 \in W$. We define the BT+AC model $\mathcal{M}' = \langle M', <, Choice, V'\rangle$ as follows:

- $M' = \{m_0\} \cup W$ for some $m_0 \notin W$;
- $< = \{m_0\} \times W$ (and thus $Hist = H_{m_0} = \{\{m_0, w\} \mid w \in W\}$);
- $Choice_i^{m_0} = \{\{\{m_0\} \times R_i(w)\} \mid w \in W\}$, and $Choice_i^w = \{\{h\}, h \in H_w\}$ for every $w \in W$;

- $V'(p)$ is the set of moment-history pairs $m_0/\{m_0, w\}$ such that $w \in V(p)$.

We can check that $\mathcal{M}'$ is a BT+AC model. We can also prove by induction on ψ that for all STIT_n formulas ψ and worlds $w \in W$, $\mathcal{M}', m_0/\{m_0, w\} \models \psi$ iff $\mathcal{W}, w \models \psi$. Hence $\mathcal{M}', m_0/\{m_0, w_0\} \models \phi$. ∎

3.2 Normal form for STIT_n^G formulas

We now show that every formula of $\mathcal{L}_{\mathsf{STIT}_n^G}$ is equivalent to a formula where only the 'grand coalition' AGT and 'anti-individuals' occur, where the latter are complements of singleton groups.

LEMMA 12. *Let $\mathcal{W} = \langle W, R, V\rangle$ be a Kripke model. For all $(w_1, \ldots, w_k) \in W^k$ and $J_1, \ldots, J_k \in 2^{AGT}$ such that $j \neq l$ implies $J_j \cap J_l = \emptyset$ we have: $\bigcap_{j \in \{1 \ldots k\}} R_{J_j}(w_j) \neq \emptyset$.*

Proof. This follows from the independence constraint, which says that

$$\bigcap_{i \in AGT} R_i(w_i) \neq \emptyset \text{ for all } (w_1, \ldots, w_n) \in W^n.$$

∎

The following theorem holds for any J_1 and J_2 (that are not necessarily disjoint).

THEOREM 13. *Let $J_1, J_2 \subseteq AGT$. We have:*

$$\models_{STIT_n^G} [J_1 \cap J_2]\phi \leftrightarrow [J_1][J_2]\phi$$

Proof. Let $\mathcal{M} = \langle W, R, V\rangle$ be a Kripke model. We are going to prove that $R_{J_1 \cap J_2} = R_{J_1} \circ R_{J_2}$.

$\boxed{\supseteq}$ As $J_1 \cap J_2 \subseteq J_1$, we have $R_{J_1} \subseteq R_{J_1 \cap J_2}$ by definition of the relation R_J. Likewise, $R_{J_2} \subseteq R_{J_1 \cap J_2}$. As $R_{J_1 \cap J_2}$ is transitive, we have $R_{J_1} \circ R_{J_2} \subseteq R_{J_1 \cap J_2}$.

$\boxed{\subseteq}$ Let $w, w' \in W$ such that $(w, w') \in R_{J_1 \cap J_2}$. We are going to prove that $R_{J_1}(w) \cap R_{J_1}(w') \neq \emptyset$, i.e. that W contains a point u such that $(w, u) \in R_{J_1}$ and $(u, w') \in R_{J_2}$ (from which it immediately follows that $R_{J_1 \cap J_2} \subseteq R_{J_1} \circ R_{J_2}$).

First, we have

$$R_{J_1}(w) \cap R_{J_1}(w') = R_{J_1 \cap J_2}(w) \cap R_{J_1 \setminus J_1 \cap J_2}(w) \cap R_{J_1 \cap J_2}(w') \cap R_{J_2 \setminus J_1 \cap J_2}(w')$$

by the above Lemma 12. Then, as $R_{J_1 \cap J_2}(w) = R_{J_1 \cap J_2}(w')$, we have

$$R_{J_1}(w) \cap R_{J_1}(w') = R_{J_1 \cap J_2}(w) \cap R_{J_1 \setminus J_1 \cap J_2}(w) \cap R_{J_2 \setminus J_1 \cap J_2}(w').$$

As $J_1 \cap J_2$, $J_1 \setminus J_1 \cap J_2$, and $J_2 \setminus J_1 \cap J_2$ are pairwise disjoint, we have $R_{J_1}(w) \cap R_{J_1}(w') \neq \emptyset$ again by the above Lemma 12. ∎

THEOREM 14. *Let $J \subsetneq AGT$ such that $AGT \setminus J = \{j_1, \ldots, j_r\}$, and let $\bar{j_i} = AGT \setminus \{j_i\}$. Then*

$$\models_{STIT_n^G} [J]\phi \leftrightarrow [\bar{j_1}] \ldots [\bar{j_r}]\phi$$

Proof. By induction on r, with base case $r = 1$ and using Theorem 13 for the induction step. ■

It follows that every STIT_n^G-formula can be written only using the grand coalition $[AGT]$ and anti-individuals $[\bar{i}]$s. From now on, we consider that a STIT_n^G-formula contains only such operators.

4 The product logic $\mathsf{S5}^n$

In this part, we briefly recall the product logic $\mathsf{S5}^n$. The reader is referred to [7] for more details.

Just as $\mathcal{L}_{\mathsf{STIT}_n}$, the language of $\mathsf{S5}^n$ logic is built from a countably infinite set of atomic propositions ATM and a set of parameters $\{1, \ldots, n\}$. It is defined by the following BNF:

$$\phi \ ::= \ p \mid \neg\phi \mid (\phi \wedge \phi) \mid \Box_i\phi$$

where p ranges over ATM and i ranges over $\{1, \ldots, n\}$.

4.1 Semantics of $\mathsf{S5}^n$

A Kripke model for $\mathsf{S5}^n$ is a cartesian product.

DEFINITION 15 ($\mathsf{S5}^n$ model). A $\mathsf{S5}^n$ model is a tuple $\mathcal{X} = (X, R, V)$ where:

- $X = X_1 \times \ldots \times X_n$ for some nonempty sets $X_1, \ldots, X_n$;
- R is a mapping associating to every $i \in \{1, \ldots, n\}$ the equivalence relation

$$R_i = \{\langle (x_1, \ldots, x_n), (y_1, \ldots, y_n)\rangle \in X^2 \mid x_j = y_j \text{ for all } j \neq i\}$$

- $V : ATM \to 2^X$.

Note that the usual $\mathsf{S5}^n$ models are more generally products of equivalence relations. Our $\mathsf{S5}^n$ models here are the subclass of point-generated models (that suffice for the characterization of $\mathsf{S5}^n$).

Definitions of truth conditions, validity and satisfiability are as usual.

4.2 A nonstandard axiomatics for $\mathsf{S5}^n$

We first recall the definition of finite axiomatizability of [7, Chapter 1].

DEFINITION 16 (finite axiomatizability). A logic L is finitely axiomatizable if there is a finite set Ax of formula schemas such that $\phi \in L$ iff there is a sequence $(\phi_1, \ldots, \phi_k)$ of formulas such that for $1 \leq i \leq k$, one of the following holds:

- ϕ_i is a tautology of classical proposition logic or an instance of an axiom in Ax;

- ϕ_i is obtained by necessitation from ϕ_j, where $j < i$;
- ϕ_i is obtained by modus ponens from ϕ_j and ϕ_k, where $j, k < i$;
- $\phi_k = \phi$.

THEOREM 17. *[7, Theorem 8.2] The logic S5^n is not finitely axiomatizable for $n \geq 3$.*

While $\mathsf{S5}^n$ can thus not be axiomatized in the standard way, there exists an axiomatization by means of a nonstandard rule.

THEOREM 18. *[18] S5^n is axiomatized by the following axiom schemas:*

- *$S5(\Box_i)$: the axiom schemas for S5, for every modal operator $\Box_i$*
- $\vdash \Box_i \Box_j \phi \leftrightarrow \Box_j \Box_i \phi$
- *Modus Ponens rule:*
$$\frac{\vdash \phi \qquad \vdash \phi \rightarrow \psi}{\vdash \psi}$$
- *Necessitation rule:*
$$\frac{\vdash \phi}{\vdash \Box_i \phi}$$
- *Rectangle Rule:*
$$\frac{\vdash (p \wedge \tau(\neg\phi \wedge p)) \rightarrow \phi}{\vdash \phi} \text{ if } p \text{ does not occur in } \phi$$
where $\tau(\chi) = \Box_1 \ldots \Box_n[(\bigwedge_{i \in \{1,\ldots,n\}} \Diamond_1 \ldots \Diamond_{i-1} \Diamond_{i+1} \ldots \Diamond_n \chi) \rightarrow \chi]$.

It is the Rectangle Rule which is nonstandard.

REMARK 19. The axiomatics in [18] does not have all the axioms of $S5(\Box_i)$. These are nevertheless valid in $\mathsf{S5}^n$ models and we have chosen to add them explicitly.

4.3 Undecidability of the satisfiability problem for $\mathbf{S5}^n$-formulas

While satisfiability of a formula of $\mathsf{S5}^n$ is decidable for $n = 2$, things get worse beyond.

THEOREM 20. *[18, Theorem 8.6] The problem of satisfiability of a formula of S5^n is undecidable for $n \geq 3$.*

5 Group STIT satisfiability is undecidable

We are going to map the problem of satisfiability in $\mathsf{S5}^n$ to the problem of satisfiability in STIT_n^G. The range of our mapping is the fragment of $\mathcal{L}_{\mathsf{STIT}_n^G}$ formulas where only the 'grand coalition' and 'anti-individuals' occur, i.e. the set of groups J such that either $J = AGT$, or $J = AGT \setminus \{i\}$ for some $i \in AGT$. We note $\bar{i}$ such sets. As satisfiability is undecidable for $\mathsf{S5}^n$, satisfiability in STIT_n^G cannot be decidable either.

The $\mathcal{L}_{\mathsf{S5}^n}$ formula $\Box_i \phi$ will be mapped to the $\mathcal{L}_{\mathsf{STIT}_n^G}$ formula $[\bar{i}]\phi$. For the ease of exposition, we identify these two kinds of formulas from now on, and suppose that formulas $\Box_i \phi$ are part of the STIT_n^G language.

Let $atm(\phi)$ be the set of all atomic propositions occurring in ϕ.

THEOREM 21. *For any $\phi \in \mathcal{L}_{S5^n}$, the following are equivalent:*

1. *ϕ is $S5^n$-satisfiable;*
2. *ϕ is satisfiable in a $STIT_n^G$ model where $R_{AGT} = id_W$;*
3. *$[\emptyset](\bigwedge_{p \in atm(\phi)} [AGT]p \leftrightarrow p) \wedge \phi$ is $STIT_n^G$-satisfiable.*[2]

Proof. $\boxed{1. \Rightarrow 2.}$ Let $\mathcal{X} = \langle X, R, V\rangle$ be an $\mathsf{S5}^n$ model and let $x_0 \in X$ be a point such that $\mathcal{X}, x_0 \models \phi$. We define a triple $\mathcal{W}' = \langle W', R', V'\rangle$ as follows:

- $W' = X$;
- R' is a mapping associating to every $i \in AGT$ the equivalence relation

$$R'_i = \{\langle (x_1, \ldots, x_n), (y_1, \ldots, y_n)\rangle \in W'^2 \mid x_i = y_i\};$$

- $V' = V$.

We can check that for all $(w_1, \ldots, w_n) \in W'^n, \bigcap_{i \in AGT} R_i(w_i) \neq \emptyset$. Thus, $\mathcal{W}'$ is a STIT_n^G-Kripke model as defined in Section 3.1. We can see that

$$\begin{aligned} R'_{\bar{i}} &= \textstyle\bigcap_{j \in \bar{i}} R'_j \text{ (by definition, cf. Section 3.1)} \\ &= \textstyle\bigcap_{j \in \bar{i}} \{\langle (x_1, \ldots, x_n), (y_1, \ldots, y_n)\rangle \in W'^2 \mid x_j = y_j\} \\ &= \{\langle (x_1, \ldots, x_n), (y_1, \ldots, y_n)\rangle \in W'^2 \mid x_j = y_j \text{ for all } j \neq i\} \\ &= R_i \end{aligned}$$

and that

$$\begin{aligned} R'_{AGT} &= \textstyle\bigcap_{j \in AGT} R'_j \text{ (by definition, cf. Section 3.1)} \\ &= \textstyle\bigcap_{j \in AGT} \{\langle (x_1, \ldots, x_n), (y_1, \ldots, y_n)\rangle \in W'^2 \mid x_j = y_j\} \\ &= \{\langle (x_1, \ldots, x_n), (x_1, \ldots, x_n)\rangle \in W'^2\} \\ &= id_{W'} \end{aligned}$$

We can check by induction on ϕ that $\mathcal{X}, z \models \phi$ iff $\mathcal{W}', z \models \phi$ for all $z \in W$.

$\boxed{2. \Rightarrow 3.}$ Let $\mathcal{W} = \langle W, R, V\rangle$ be a STIT_n^G Kripke model such that $R_{AGT} = id_W$, and let $w_0 \in W$ be a world s.t. $\mathcal{W}, w_0 \models \phi$. As $R_{AGT} = id_W$, we have $\mathcal{W}, w_0 \models [\emptyset](\bigwedge_{p \in atm(\phi)} [AGT]p \leftrightarrow p)$. Thus $\mathcal{W}, w_0 \models [\emptyset](\bigwedge_{p \in atm(\phi)} [AGT]p \leftrightarrow p) \wedge \phi$.

[2]Remember that $[\emptyset]$ abbreviates $[\bar{1}] \ldots [\bar{n}]$.

3. $\Rightarrow$ 2. Let $\mathcal{W}' = \langle W', R', V' \rangle$ be a STIT_n^G Kripke model and let $w_0' \in W'$ be a world such that $\mathcal{W}', w_0' \models [\emptyset](\bigwedge_{p \in atm(\phi)} [AGT]p \leftrightarrow p) \wedge \phi$. We prove that there exists a STIT_n^G Kripke model $\mathcal{W} = \langle W, R, V \rangle$ with $R_{AGT} = id_W$ and a point $w_0 \in W$ such that $\mathcal{W}, w_0 \models \phi$. Let $\mathcal{W} = \langle W, R, V \rangle$ where:

- $W = \{R'_{AGT}(x) \mid x \in W'\}$;
- $R_i = \{(R'_{AGT}(x), R'_{AGT}(y)) \mid (x, y) \in R'_i\}$;
- $V(p) = \{U \in W \mid U \subseteq V'(p)\}$.

Notice that $R_J = \{(R'_{AGT}(x), R'_{AGT}(y)) \mid (x, y) \in R'_J\}$, and that $R_{AGT} = id_W$. We can prove by structural induction that for all $w \in W'$ and for all subformulas ψ of ϕ:

$$\mathcal{W}', w \models \psi \text{ iff } \mathcal{W}, R'_{AGT}(w) \models \psi$$

Indeed:

$$\begin{array}{lll}
\mathcal{W}', w \models p & \text{iff} & \mathcal{W}', w \models [AGT]p \\
 & & (\text{because } \mathcal{W}', w \models [\emptyset](\bigwedge_{p \in atm(\phi)} [AGT]p \leftrightarrow p)) \\
 & \text{iff} & \mathcal{W}', w' \models p \text{ for all } y \in R'_{AGT}(z) \\
 & \text{iff} & w' \in V'(p) \text{ for all } w' \in R'_{AGT}(w) \\
 & \text{iff} & R'_{AGT}(w) \subseteq V'(p) \\
 & \text{iff} & R'_{AGT}(w) \in V(p) \\
 & \text{iff} & \mathcal{W}, R'_{AGT}(w) \models p
\end{array}$$

$$\begin{array}{lll}
\mathcal{W}', w \models [\bar{i}]\psi & \text{iff} & \mathcal{W}', w' \models \psi \text{ for all } w' \in R'_{\bar{i}}(w) \\
 & \text{iff} & \mathcal{W}, R'_{AGT}(w') \models \psi \text{ for all } w' \in R'_{\bar{i}}(w) \\
 & \text{iff} & \mathcal{W}, R'_{AGT}(w') \models \psi \\
 & & \text{for all } R'_{AGT}(w') \in R_{\bar{i}}(R'_{AGT}(w)) \\
 & \text{iff} & \mathcal{W}, R'_{AGT}(w) \models [\bar{i}]\psi
\end{array}$$

2. $\Rightarrow$ 1. Let $\mathcal{W} = \langle W, R, V \rangle$ be a STIT_n^G Kripke model with $R_{AGT} = id_W$, and let $w_0 \in W$ be a world such that $\mathcal{W}, w_0 \models \phi$. From $\mathcal{W}$ we define a $\mathsf{S5}^n$ model $\mathcal{X}' = \langle X', R', V' \rangle$ as follows:

- $X' = X_1 \times \ldots \times X_n$ where for all $i \in AGT$, $X_i = \{R_i(w) \mid w \in W\}$;
- R' is a mapping associating to every $i \in AGT$ the equivalence relation

$$R'_i = \{((x_1, \ldots, x_n), (y_1, \ldots, y_n)) \in X'^2 \mid x_j = y_j \text{ for all } j \neq i\};$$

- $V'(p) = \{(x_1, \ldots, x_n) \mid \bigcap_{i \in AGT} x_i \in V(p)\}$ (identifying $\bigcap_{i \in AGT} x_i = \{y\}$ and y).

We can check that

$$\mathcal{X}', (R_1(w), \ldots, R_n(w)) \models \phi \text{ iff } \mathcal{W}, w \models \phi$$

for all $w \in W$. Indeed:

$$\begin{aligned}\mathcal{X}', (R_1(w), \ldots, R_n(w)) \models p \text{ iff } & (R_1(w), \ldots, R_n(w)) \in V'(p)\\ \text{iff } & \textstyle\bigcap_{i \in AGT} R_i(w) \in V(p)\\ \text{iff } & w \in V(p)\\ & (\text{notice that } \textstyle\bigcap_{i \in AGT} R_i(w) = \{w\})\\ \text{iff } & \mathcal{W}, w \models p\end{aligned}$$

and

$$\mathcal{X}', (R_1(w), \ldots, R_n(w)) \models [\bar{i}]\psi$$

iff $\mathcal{X}', (R_1(w), \ldots, R_{i-1}(w), R_i(w'), R_{i+1}(w), \ldots R_n(w)) \models \psi$ for all $w' \in W$

iff $\mathcal{X}', (R_1(w''), \ldots, R_i(w''), \ldots R_n(w'')) \models \psi$ where $R_1(w) \cap \ldots \cap R_i(w') \cap \ldots \cap R_n(w) = \{w''\}$ for all $w' \in W$

iff $\mathcal{X}', (R_1(w''), \ldots, R_i(w''), \ldots R_n(w'')) \models \psi$ for all $w'' \in R_{\bar{i}}(w)$

iff $\mathcal{W}, w'' \models \psi$ for all $w'' \in R_{\bar{i}}(w)$

iff $\mathcal{W}, w \models [\bar{i}]\psi$

■

THEOREM 22. *The problem of satisfiability of a formula of* $STIT_n^G$ *is undecidable for* $n \geq 3$.

Proof. By Theorem 20 and 21. ■

6 Group STIT is not finitely axiomatizable

THEOREM 23. *There is no finite axiomatization of logic* $STIT_n^G$ *if* $n \geq 3$.

Proof. Suppose for a contradiction that STIT_n^G is finitely axiomatizable, i.e. that there exists a finite set of axioms Ax such that for every STIT_n^G-formula ϕ, we have $\models_{\mathsf{STIT}_n^G} \phi$iff there is a deduction of ϕ from (instances of) Ax using Modus Ponens and Necessitation. Let us define an axiomatics Ax' obtained from Ax by removing all $[AGT]$ operators. We are going to prove that for all formulas $\phi \in \mathcal{L}_{\mathsf{S5}^n}$, $\models_{\mathsf{S5}}^n \phi$ iff there is a deduction of ϕ from (instances of) Ax using Modus Ponens and Necessitation. Hence, $\mathsf{S5}^n$ would be finitely axiomatizable and there is a contradiction.

Let us prove first that $\vdash_{Ax'} \phi$ implies $\models_{\mathsf{S5}}^n \phi$. We do so by proving that each instance of Ax' is valid in $\mathsf{S5}^n$. Let us consider an instance ψ' of an axiom of Ax'. ψ' is obtained from an instance ψ of Ax by removing all $[AGT]$ operators. We have $\models_{\mathsf{STIT}_n^G} \psi$. Therefore, ψ is valid in the class of STIT^G_n-models where $R_{AGT} = id_W$. Hence, ψ' is valid in the class of STIT^G_n-models where $R_{AGT} = id_W$. It follows that $\models^n_{\mathsf{S5}} \phi$.

Here is an outline of the $\boxed{\Leftarrow}$-sense of the proof. First, for all $\mathsf{S5}^n$-formulas ϕ,

$$\begin{array}{lll} \models^n_{\mathsf{S5}} \phi & \text{iff} & \models_{\mathsf{STIT}^G_n} [\emptyset](\bigwedge_{p \in atm(\phi)} [AGT]p \leftrightarrow p) \rightarrow \phi \\ & \text{iff} & \vdash_{Ax} [\emptyset](\bigwedge_{p \in atm(\phi)} [AGT]p \leftrightarrow p) \rightarrow \phi \\ & \text{implies (1)} & \vdash_{Ax,[AGT]\psi \leftrightarrow \psi} \phi \\ & \text{implies (2)} & \vdash_{Ax'} \phi \end{array}$$

It remains to prove (1) and (2).

As to (1), it suffices to prove the following:

$$\vdash_{Ax} [\emptyset](\bigwedge_{p \in atm(\phi)} [AGT]p \leftrightarrow p) \rightarrow \phi \text{ implies } \vdash_{Ax,[AGT]\psi \leftrightarrow \psi} \phi$$

This can be established using necessitation and principles of classical propositional logic. Basically the proof goes as follows:

$$\dfrac{\dfrac{\ldots\text{(necessitation and principles of classical propositional logic)}}{\vdash_{Ax,[AGT]\psi \leftrightarrow \psi} [\emptyset](\bigwedge_{p \in atm(\phi)} [AGT]p \leftrightarrow p)} \quad \dfrac{\text{(by hypothesis)}}{\vdash_{Ax,[AGT]\psi \leftrightarrow \psi} [\emptyset](\bigwedge_{p \in atm(\phi)} [AGT]p \leftrightarrow p) \rightarrow \phi}}{\vdash_{Ax,[AGT]\psi \leftrightarrow \psi} \phi}$$

As to (2), suppose $Ax + \psi \leftrightarrow [AGT]\psi$ is the axiom system obtained from Ax by adding the schema $\psi \leftrightarrow [AGT]\psi$. Then we can prove that $\vdash_{Ax + \psi \leftrightarrow [AGT]\psi} \phi$ implies $\vdash_{Ax'} \phi$.

The proof of that goes as follows. Assume that $\vdash_{Ax,[AGT]\psi \leftrightarrow \psi} \phi$. There exists a proof of ϕ, that is to say a sequence $(\phi_1, \ldots, \phi_k)$ such that for $1 \leq i \leq k$, one of the following holds:

- ϕ_i is a tautology, an instance of an axiom in Ax or an instance of $[AGT]\psi \leftrightarrow \psi$;
- ϕ_i is obtained by necessitation from ϕ_j where $j < i$;
- ϕ_i is obtained by modus ponens from ϕ_j and ϕ_k where $j, k < i$;
- $\phi_k = \phi$.

Now, we construct $(\phi'_1, \ldots, \phi'_n)$ where ϕ'_i is ϕ_i in which we have removed all $[AGT]$ operators. The reader can check that $(\phi'_1, \ldots, \phi'_n)$ is a proof of ϕ.

This concludes the proof. ∎

7 Discussion

Now, we are going to propose a generalization of these results, and try to classify some more fragments of STIT_n^G. First, for a given a family $\mathcal{C} \subseteq 2^{AGT}$ of subsets of 2^{AGT} we define the language $\mathcal{L}_{\mathsf{STIT}[\mathcal{C}]}$ by the following BNF:

$$\phi ::= p \mid \neg\phi \mid (\phi \wedge \phi) \mid [J]\phi$$

where p ranges over ATM and J ranges over the set of coalitions $\mathcal{C}$.

Let us call $\mathsf{STIT}[\mathcal{C}]$ the fragment of STIT_n^G where formulas are in $\mathcal{L}_{\mathsf{STIT}[\mathcal{C}]}$. Thus, $\mathsf{STIT}_n^G = \mathsf{STIT}[2^{2^{AGT}}]$ and $\mathsf{STIT}_n = \mathsf{STIT}[\{\emptyset, \{1\}, \ldots, \{n\}\}]$.

We have the following result:

PROPOSITION 24. *Let $\mathcal{C} \subseteq 2^{AGT}$. If $\mathcal{C}$ has a linear structure, then the problem of satisfiability of a formula in $\mathsf{STIT}[\mathcal{C}]$ is NP-complete.*

Proof. We can prove that if a formula is satisfiable, then it is so in a polynomial-sized model. The proof is based on a selection-of-points argument. More details can be found in [16]. ■

We conjecture the following result (which would cover Theorems 10 and 22).

CONJECTURE 25. Given $\mathcal{C} \subseteq 2^{AGT}$, the problem of satisfiability of a formula in $\mathsf{STIT}[\mathcal{C}]$ is:

1. undecidable if there are $J_1, J_2, J_3 \in \mathcal{C}$ such that J_1, J_2, J_3, $J_1 \cap J_2$ and $J_2 \cap J_3$, $J_1 \cap J_3$ are distinct;

2. NEXPTIME-complete if there is no $J_1, J_2, J_3 \in \mathcal{C}$ such that J_1, J_2, J_3, $J_1 \cap J_2$, $J_2 \cap J_3$ and $J_1 \cap J_3$ are distinct, but there exists $J_1, J_2 \in \mathcal{C}$ such that J_1, J_2, $J_1 \cap J_2$ are distinct;

We therefore conjecture, e.g., that: the problem of satisfiability of a formula in $\mathsf{STIT}[\mathcal{C}]$ is undecidable if $\mathcal{C} = \{\{1,3,4\}, \{1,3,5\}, \{4,5\}\}$, that it is NEXPTIME-complete if $\mathcal{C} = \{\{1,3,4\}, \{1,3,5\}, \{1\}\}$, and that it is NP-complete if $\mathcal{C} = \{\emptyset, \{1\}, \{1,2\}, \ldots, \{1,2,\ldots,n\}\}$.

Finally, we also conjecture that a nonstandard axiomatization of STIT_n^G can be obtained from that of $\mathsf{S5}^n$.

8 Conclusion

The paper contains mathematical results for deliberative STIT logic, both in its individual and group version: while the fragment STIT_n allowing to reason only about individual agency is decidable in nondeterministic exponential time (NEXPTIME), the entire logic STIT_n^G (allowing for joint agency) is undecidable and cannot be finitely axiomatized. The result for STIT_n was established in [3], while the general result for STIT_n^G is new.

The results for STIT_n^G apply a fortiori to extensions of STIT_n^G with the temporal 'next' operator. Given these rather negative results, it is interesting to look for decidable fragments of STIT_n^G and its temporal extensions. One of these fragments is Pauly's coalition logic, whose satisfiability problem is decidable in polynomial space (PSPACE-complete). As we said in the introduction, the CL and ATL formula $\langle\langle J\rangle\rangle X\phi$ corresponds to STIT_n^G's $\neg\Box\neg[J]X\phi$: in CL, the three modal operators $\Box$, $[J]$ and X are fused into a single operator. The latter is non-normal: it does not satisfy the K-axiom of standard modal logics. In recent work we have investigated a non-normal modal logic between coalition logic and STIT_n^G where $\Box$ and $[J]$ are fused, while X is the standard temporal 'next' [8]. We called the resulting logic CL^* because it extends CL in the same way as ATL^* extends ATL. We have shown that contrarily to ATL^*, the extension CL^* provides is for free: CL^* has the same complexity as CL. We have also argued that the epistemic extension of CL^* is more powerful than that of CL: contrarily to the latter, CL^* allows to reason about the agents' power, i.e. about agents' knowledge of the right action to choose in order to achieve something. In other words, in the epistemic extension of CL^* we can say that an agent 'knows how to play'. Logics having such expressive power have attracted a lot of attention recently [17, 12, 1].

BIBLIOGRAPHY

[1] Thomas Ågotnes. Action and knowledge in alternating-time temporal logic. *Synthese*, 149(2):377–409, 2006.

[2] Rajeev Alur, Thomas A. Henzinger, and Orna Kupferman. Alternating-time temporal logic. In *Proceedings of the 38th IEEE Symposium on Foundations of Computer Science*, Florida, October 1997.

[3] Philippe Balbiani, Andreas Herzig, and Nicolas Troquard. Alternative axiomatics and complexity of deliberative STIT theories. *Journal of Philosophical Logic*, 2008. to appear.

[4] Nuel Belnap, Michael Perloff, and Ming Xu. *Facing the Future: Agents and Choices in Our Indeterminist World.* Oxford University Press, Oxford, 2001.

[5] Jan Broersen, Andreas Herzig, and Nicolas Troquard. Embedding Alternating-time Temporal Logic in strategic STIT logic of agency. *Journal of Logic and Computation*, 16(5):559–578, 2006.

[6] Brian F. Chellas. Time and modality in the logic of agency. *Studia Logica*, 51(3/4):485–518, 1992.

[7] Dov M. Gabbay, Agnes Kurucz, Frank Wolter, and Michael Zakharyaschev. *Many-Dimensional Modal Logics: Theory and Applications.* Number 148 in Studies in Logic and the Foundations of Mathematics. Elsevier, North-Holland, 2003.

[8] Olivier Gasquet, Andreas Herzig, and François Schwarzentruber. The complexity of reasoning about ability and power. submitted, 2008.

[9] Valentin Goranko. Coalition games and alternating temporal logics. In *TARK'01: Proceedings of the 8th Conference on Theoretical Aspects of Rationality and Knowledge*, pages 259–272, San Francisco, CA, USA, 2001. Morgan Kaufmann Publishers Inc.

[10] John Horty and Nuel Belnap. The deliberative stit: a study of action, omission, ability and obligation. *Journal of Philosophical Logic*, 24(6):583–644, 1995.

[11] John F. Horty. *Agency and Deontic Logic.* Oxford University Press, 2001.

[12] Wojtec Jamroga and Wiebe van der Hoek. Agents that know how to play. *Fundamenta Informaticae*, 63(2-3):185–219, 2004.

[13] Barteld Kooi and Allard Tamminga. Moral conflicts between groups of agents. *Journal of Philosophical Logic*, 37(1):1–21, 2008.

[14] Marc Pauly. A modal logic for coalitional power in games. *Journal of Logic and Computation*, 12(1):149–166, 2002.

[15] Sven Schewe. ATL* satisfiability is 2EXPTIME-complete. In *35th International Colloquium on Automata, Languages and Programming (ICALP 2008)*, 2008.

[16] François Schwarzentruber. $S5_2$ plus inclusion axioms is NP-complete. http://www.irit.fr/~Francois.Schwarzentruber/documents/S52andimplication.pdf, 2008.

[17] Wiebe van der Hoek and Mike Wooldridge. Cooperation, knowledge, and time: Alternating-time temporal epistemic logic and its applications. *Studia Logica*, 75:125–157, 2003.

[18] Yde Venema. Rectangular games. *The Journal of Symbolic Logic*, 63(4), December 1998.

[19] Dirk Walther, Carsten Lutz, Frank Wolter, and Michael Wookdridge. ATL satisfiability is indeed exptime-complete. *Journal of Logic and Computation*, 16:765–787, 2006.

[20] Heinrich Wansing. Tableaux for multi-agent deliberative-STIT logic. In Guido Governatori, Ian Hodkinson, and Yde Venema, editors, *Advances in Modal Logic, Volume 6*, pages 503–520. King's College Publications, 2006.

[21] Stefan Wölfl. Propositional Q-logic. *Journal of Philosophical Logic*, 31:387–414, 2002.

[22] Ming Xu. Axioms for deliberative STIT. *Journal of Philosophical Logic*, 27:505–552, 1998.

Andreas Herzig
IRIT-CNRS,
118 Route de Narbonne, F-31062 Toulouse Cedex 9,
France.
herzig@irit.fr

François Schwarzentruber
IRIT-Université Paul Sabatier,
118 Route de Narbonne, F-31062 Toulouse Cedex 9,
France.
schwarze@irit.fr

Topology, connectedness, and modal logic

ROMAN KONTCHAKOV, IAN PRATT-HARTMANN, FRANK WOLTER AND MICHAEL ZAKHARYASCHEV

ABSTRACT. This paper presents a survey of topological spatial logics, taking as its point of departure the interpretation of the modal logic $\mathcal{S}4$ due to McKinsey and Tarski. We consider the effect of extending this logic with the means to represent topological *connectedness*, focusing principally on the issue of computational complexity. In particular, we draw attention to the special problems which arise when the logics are interpreted not over *arbitrary* topological spaces, but over (low-dimensional) *Euclidean* spaces.

Keywords: Spatial logic, modal logic, topology, connectedness.

1 Introduction: spatial logic and modal logic

In their seminal paper *The algebra of topology* [40], McKinsey and Tarski sought to provide 'an algebraic apparatus adequate for the treatment of portions of point-set topology.' In doing so, they created—*en passant*—a topological framework for the semantics of the modal logic $\mathcal{S}4$, exploiting the striking similarity between Gödel's [24] and Orlov's [43] axioms for 'provability' logic and Kuratowski's axioms for topological spaces. In this framework, proposition letters are interpreted as subsets of a topological space, Boolean connectives as set-theoretic operations on these sets, and the modal box as the topological interior operator. As McKinsey and Tarski showed, a modal formula φ is an $\mathcal{S}4$-validity if and only if, in any interpretation over a topological space T, φ denotes the whole of T. In fact, they showed more. Suppose we are interested not in topological spaces *in general*, but rather in some *specific* dense-in-itself, separable metric space—for example $\mathbb{R}^2$ or $\mathbb{R}^3$. For any such space T, a modal formula φ is an $\mathcal{S}4$-validity if and only if, in any interpretation over T, φ denotes the whole of T. In other words: $\mathcal{S}4$ is the logic of any dense-in-itself, separable metric space.

This situation invites generalization. By a *spatial logic*, let us understand any formal language interpreted over some class of geometrical structures, taken in the most general sense. That is: the variables of this language range over collections of figures in the relevant structures; and its non-logical primitives denote properties and operations defined over those figures. What makes a spatial logic a *logic* is that it has a regimented syntax and formal semantics interpreting it; what makes it *spatial* is that the operative notion of logical consequence is made to depend on the specifically geometrical features of the chosen interpretation. Thus, $\mathcal{S}4$ is a spatial logic whose (propositional) variables range over arbitrary subsets of a topological space in some given class.

Spatial logics, thus understood, have a long pedigree, tracing their origins back to the axiomatic tradition in geometry, which reached its zenith in Hilbert's *Grundlagen der Geometrie* [30]. Strikingly, Hilbert's axiomatization is couched in (lightly mathematicized) idiomatic German: notwithstanding its evident rigour, no attempt is made to articulate the implicit logical syntax or operative inference procedure. This feature prompted a further stage of formalization in another of Tarski's most significant papers: *What is elementary geometry?* [59]. Tarski's geometrical axioms are couched in a first-order language whose variables range over points in the standard model of Euclidean space, and whose non-logical predicates represent notions defined in terms of the metric structure of that space. Again, Tarski showed that the consequences of his axioms coincide with the true statements of that model. Of course, the real achievement here was not simply to shoe-horn Hilbert's perfectly good mathematics into the regimented syntax of a formal language, but rather, to ask what happens when that syntax is restricted. For Tarski showed that his elementary geometry is, on the one hand, *decidable*—there is an algorithmic procedure for determining the truth of any of its formulas—and, on the other, sufficiently expressive that it comes close (in a sense which Tarski was able to make precise) to fixing the familiar model of the plane as $\mathbb{R}^2$. On a practical level, Tarski's work has found application in spatial databases (see, e.g., [34]); from a theoretical point of view, we have the beginnings of one of the central themes in computational logic—the trade-off between expressive power and computational complexity. We remark that the precise complexity of Tarski's geometry (or $\mathrm{Th}(\mathbb{R}, +, \times, \leq)$) seems to be still unknown, with the current lower bound being NExpTime [22] and the upper bound ExpSpace [4].

A quite distinct intellectual tradition has also contributed to recent interest in spatial logics, however: Whitehead's theory of *extensive connection*, which appeared in its most complete form in his *Process and Reality* [65]. Whitehead's goal was to develop a purely *region-based* theory of space, whose sole geometrical primitive was the relation he called *connection*, but which is now (to avoid confusion with established terminology) generally referred to as *contact*. Roughly: two regions contact each other just in case they overlap or touch. Whitehead put forward a collection of postulates governing this relation, and gave reconstructions of various geometrical notions in terms of it. A similar—and in many ways more satisfactory—region-based theory of space was proposed at the same time by de Laguna [15]. In both cases the motivation was essentially metaphysical: to provide a spatial ontology whose basic entities are closer to the data of spatial experience than is the standard Cartesian model of space as $\mathbb{R}^3$. Paradoxically, perhaps, the methodology they employed was resolutely empiricist: the proposed system of postulates and definitions was to be evaluated by its conformity to (pre-theoretic spatial intuition and) spatial experience. The ensuing lack of any formal semantics for the languages in question impeded their mathematical development, despite sporadic revivals in the following decades [10, 11, 7].

Interest in region-based, qualitative spatial logics of this kind was rekindled, however, in the early 1990s, within Artificial Intelligence. (See the recent handbook chapters [49, 12] for comprehensive surveys.) The impetus for this development was the conviction that effective reasoning—geometrical or otherwise—depends on selecting a language with the appropriate representational resources: too little expressive power, and it cannot represent the information required; too much, and the reasoner is overwhelmed by the computational complexity of determining entailments within it. Hence the focus on languages whose variables range over spatial regions: while spatial regions *can* be modelled as sets of points, and so quantified over in second-order logic, a first-order logic whose object-level variables range over regions is, from the point of view of expressive economy, a preferable alternative. And once a region-based domain of quantification has been adopted, the focus on qualitative geometrical primitives follows naturally, since so many salient properties and relations involving regions are qualitative in character.

First-order qualitative theories of space, however, are generally undecidable or even non-recursively enumerable [28, 18, 13], a result which extends to some spatial logics based on the two-variable fragment of first-order logic [39]. Hence, attention has shifted to quantifier-free constraint systems such as 9-intersections or $\mathcal{RCC}$-8 [20, 46]. Intriguingly, research on such systems has led to a renewed and systematic investigation of spatial formalisms within the algebraic framework of Tarski [58, 40]. For spatial relations such as 'contact' or 'part of' form a natural subject for relation algebra; see the surveys [6, 61] and references therein. Furthermore, it turns out that many spatial constraint systems designed in AI can be regarded as natural fragments of $\mathcal{S}4$ augmented with the universal modality and known as $\mathcal{S}4_u$. Thus, the modal logic $\mathcal{S}4_u$ finds itself at a crossroads of different traditions and disciplines related to spatial logics. (See [62, 25] for a broader discussion of modal logics of space.)

Few practical problems in spatial reasoning are purely topological in character, of course; and this has recently prompted several extensions of $\mathcal{S}4_u$ with metric primitives (e.g., [35, 68, 54]). Yet, even from a topological point of view, $\mathcal{S}4$ and its near-relation $\mathcal{S}4_u$ can seem frustratingly inexpressive: for example, very few theorems from standard textbooks on topology can be formulated within them! Perhaps the most glaring expressive defect of these languages is their inability to express the property of *connectedness*—a concept of central theoretical and practical importance. To date, only sporadic attempts have been made to interpret $\mathcal{S}4_u$ over connected spaces, or to augment it with a primitive predicate expressing the property connectedness [9, 53, 66, 44].

The present paper has two main aims, therefore. The first is to present a survey of topological spatial logics, taking $\mathcal{S}4$ as its starting-point. The second is to investigate in detail the extension of these logics with the means to represent topological connectedness (and related notions), focusing principally on issues of computational complexity. A surprising discovery here was how the innocuous-looking *connectedness predicate* can increase complexity from NP to PSPACE, EXPTIME and, if component counting is allowed, to NEXPTIME. In particular, we draw the reader's attention to the special difficulties that arise when these logics are interpreted not over *arbitrary* topological spaces, but over (low-dimensional) *Euclidean* spaces. We also point out the sensitivity of such logics to the geometrical entities—polygons, disc-homeomorphs, *etc.*—over which their variables are taken to range.

2 $\mathcal{S}4_u$ over connected topological spaces

We begin by briefly reviewing the topological semantics for $\mathcal{S}4$, due to McKinsey and Tarski [40]. With a view to the ensuing generalizations, we present the language in unfamiliar guise. Specifically, we re-write the proposition letters as individual variables, the propositional connectives $\wedge$ and $\neg$ as the function-symbols $\cap$ and $^-$, respectively, and the modal box $\Box$ as the function-symbol $^\circ$. In this way, familiar modal formulas become *terms*. Formally, let $\mathcal{V} = \{v_i \mid i < \omega\}$ be a set of variables. Then the $\mathcal{S}4$-*terms* are given by

$$\tau \quad ::= \quad v_i \quad | \quad \overline{\tau} \quad | \quad \tau_1 \cap \tau_2 \quad | \quad \tau^\circ.$$

We abbreviate $\overline{(\overline{\tau}^\circ)}$ by τ^-, $\overline{(\overline{\tau_1} \cap \overline{\tau_2})}$ by $\tau_1 \cup \tau_2$, $v_0 \cap \overline{v_0}$ by $\mathbf{0}$, and $\overline{\mathbf{0}}$ by $\mathbf{1}$.

In the sequel, we assume familiarity with basic general topology. If the topological space T is clear from context, and $X \subseteq T$, we denote the complement of X by $\overline{X}$, the topological closure of X by X^-, and the topological interior of X by X°. (The overloading of symbols here is deliberate.) We follow common practice in identifying topological spaces with their carrier sets, taking the topology to be implicit; in addition, we assume that topological spaces are non-empty. In this context, define a *topological frame* to be a pair $(T, \boldsymbol{S})$, where T is a topological space, and $\boldsymbol{S} \subseteq 2^T$ is a non-empty set of its subsets. A *topological model over* $(T, \boldsymbol{S})$ is a triple $\mathfrak{M} = (T, \boldsymbol{S}, \cdot^{\mathfrak{M}})$, where $\cdot^{\mathfrak{M}}$ is a map from $\mathcal{V}$ to $\boldsymbol{S}$. (The modifier 'topological' will generally be omitted in the sequel.) The *extension* $\tau^{\mathfrak{M}}$ of a term τ in a model $\mathfrak{M}$ is defined inductively by the equations:

$$(\overline{\tau})^{\mathfrak{M}} = \overline{(\tau^{\mathfrak{M}})}, \qquad (\tau_1 \cap \tau_2)^{\mathfrak{M}} = \tau_1^{\mathfrak{M}} \cap \tau_2^{\mathfrak{M}}, \qquad (\tau^\circ)^{\mathfrak{M}} = (\tau^{\mathfrak{M}})^\circ.$$

On the above semantics, variables are constrained to range over certain subsets of the underlying space, as specified by $\boldsymbol{S}$. We refer to the elements of $\boldsymbol{S}$ as *regions*. There is no formal requirement for $\boldsymbol{S}$ to be closed under the term-forming operations of our language (in the present case, $^-$, $\cap$ and $^\circ$). In the special case where every subset of T counts as a region—that is, where $\boldsymbol{S} = 2^T$—we identify the topological frame with the underlying topological space, and simply speak of a model $\mathfrak{M} = (T, \cdot^{\mathfrak{M}})$ over T.

Recall that a topological space is called an *Aleksandrov space* if arbitrary (not only finite) intersections of open sets are open. Aleksandrov spaces can be characterized in terms of pairs the form $F = (W, R)$, where $W \neq \emptyset$ and

R is a transitive and reflexive relation (i.e., a *quasi-order*) on W. Every such pair—or *Kripke frame*—F induces the interior operator $\cdot^{\circ}_F$ on W:

$$X^{\circ}_F = \{x \in X \mid \forall y \in W\ (xRy \rightarrow y \in X)\}, \quad \text{for every } X \subseteq W.$$

In other words, the open sets of the topological space $T_F = W$ induced by F are the *upward closed* (or R-*closed*) subsets of W. It is well-known (see, e.g., [8]) that T_F is an Aleksandrov space and, conversely, every Aleksandrov space is induced by a quasi-order. Topological models over Aleksandrov spaces will be called *Aleksandrov models.*

We are now in a position to characterize the 'logic' of the term-language $\mathcal{S}4$ and its complexity.

THEOREM 1 ([40, 36]). (i) *Let τ be an $\mathcal{S}4$-term and T' a dense-in-itself separable space (e.g., $T' = \mathbb{R}^n$, for some $n \geq 1$). The following conditions are equivalent:*

- $\tau^{\mathfrak{M}} = T$ *for every model* $\mathfrak{M} = (T, \cdot^{\mathfrak{M}})$;
- $\tau^{\mathfrak{M}} = T'$ *for every model* $\mathfrak{M} = (T', \cdot^{\mathfrak{M}})$ *over* T';
- $\tau^{\mathfrak{M}} = T$ *for every (finite) Aleksandrov model* $\mathfrak{M} = (T, \cdot^{\mathfrak{M}})$.

(ii) *The problem of deciding, given an $\mathcal{S}4$-term τ, whether $\tau^{\mathfrak{M}} = T$ for all models $\mathfrak{M} = (T, \cdot^{\mathfrak{M}})$, is* PSPACE*-complete.*

With these resources at our disposal, we present our first topological logic, known under the name of $\mathcal{S}4_u$. (We remark that the original formulation of $\mathcal{S}4_u$ in [26], like those of $\mathcal{S}4$ in [43, 37, 24], made no reference to spatial logic or topology.) The language $\mathcal{S}4_u$ is the set of formulas given by:

$$\varphi \quad ::= \quad \tau_1 = \tau_2 \quad | \quad \neg\varphi \quad | \quad \varphi_1 \wedge \varphi_2,$$

where τ_1 and τ_2 range over $\mathcal{S}4$-terms. We employ the Boolean connectives $\vee$ and $\rightarrow$ as abbreviations in the standard way, additionally writing $\tau_1 \subseteq \tau_2$ for $\tau_1 \cap \overline{\tau_2} = \mathbf{0}$ and $\tau_1 \neq \tau_2$ for $\neg(\tau_1 = \tau_2)$. A formula will be called an *atom* if it involves no Boolean connectives, a *literal* if it is an atom or a negated atom, and *conjunctive* if it is a conjunction of literals. The *truth-relation* for $\mathcal{S}4_u$ is defined by setting:

$$\mathfrak{M} \models \tau_1 = \tau_2 \qquad \text{iff} \qquad \tau_1^{\mathfrak{M}} = \tau_2^{\mathfrak{M}},$$

and interpreting the Boolean connectives $\neg$ and $\wedge$ in the standard way. If $(T, \boldsymbol{S})$ is a topological frame, then φ is *satisfiable* over $(T, \boldsymbol{S})$ if $\mathfrak{M} \models \varphi$, for some model $\mathfrak{M} = (T, \boldsymbol{S}, \cdot^{\mathfrak{M}})$; if $\mathcal{F}$ is a class of topological frames, then φ is *satisfiable* over $\mathcal{F}$ if it is satisfiable over some $(T, \boldsymbol{S}) \in \mathcal{F}$. Similarly, *mutatis mutandis*, for the dual notion of *validity*. We denote by *Sat*$(\mathcal{S}4_u, \mathcal{F})$ the set of $\mathcal{S}4_u$-formulas that are satisfiable over $\mathcal{F}$.

In modal terms, $\mathcal{S}4_u$ in effect adds a 'universal modality' [26] to $\mathcal{S}4$, since an atom of the form $\tau = \mathbf{1}$ states that the modal formula corresponding to τ is true everywhere in the relevant space. It is well known (see, e.g., [1]) that the language of $\mathcal{S}4_u$ defined as above is as expressive as the 'standard' one that allows nested applications of universal modalities. In many cases, addition of the universal modality to a modal logic increases its computational complexity (e.g., the modal logic $\mathcal{K}$ with universal modality is EXPTIME-complete). For $\mathcal{S}4$ this turns out to be not the case. Denote by ALL the class of all topological frames and by ALEK the class of all Aleksandrov frames (that is, topological frames based on Aleksandrov spaces).

THEOREM 2 ([53, 2]). *$Sat(\mathcal{S}4_u, \text{ALL}) = Sat(\mathcal{S}4_u, \text{ALEK})$, and this set is* PSPACE-*complete.*

In contrast to Theorem 1, the equality in Theorem 2 cannot be extended to the set $Sat(\mathcal{S}4_u, \mathcal{F})$, where $\mathcal{F}$ is any class of topological frames over $\mathbb{R}^n$. Recall that a topological space T is *connected* just in case it is not the union of two non-empty, disjoint, open sets; a subset $X \subseteq T$ is *connected in T* just in case either it is empty, or the topological space X (with the subspace topology) is connected. If $X \subseteq T$, a maximal connected subset of X is called a *component* of X. Every set X has at least one component, and a set is connected just in case it has at most one component. Denote by CON the class of all frames over connected spaces. The $\mathcal{S}4_u$-formula

$$(v_1 \neq \mathbf{0}) \ \wedge \ (v_2 \neq \mathbf{0}) \ \wedge \ (v_1 \cup v_2 = \mathbf{1}) \ \wedge \ (v_1^- \cap v_2 = \mathbf{0}) \ \wedge \ (v_1 \cap v_2^- = \mathbf{0})$$

is satisfiable in a topological space T iff T is not connected. It follows that $Sat(\mathcal{S}4_u, \text{ALL}) \neq Sat(\mathcal{S}4_u, \text{CON})$. The formula above was used in [53] to axiomatize the logic (in the standard language of $\mathcal{S}4_u$) of connected spaces.

Observe that an Aleksandrov space T_F induced by $F = (W, R)$ is connected iff F is *connected* in the sense that between any two points $x, y \in W$ there is a path along the relation $R \cup R^{-1}$, where R^{-1} is the inverse of R. Denote by CONALEK the class of all connected Aleksandrov frames.

THEOREM 3. *$Sat(\mathcal{S}4_u, \text{CON}) = Sat(\mathcal{S}4_u, \text{CONALEK}) = Sat(\mathcal{S}4_u, T)$, for any connected dense-in-itself separable space T (in particular, for $T = \mathbb{R}^n$, $n \geq 1$). This set is* PSPACE-*complete.*

The equations in Theorem 3 were proved in [53], and the complexity result follows from Theorem 5 below. Although of the same complexity as $Sat(\mathcal{S}4_u, \text{ALL})$, $Sat(\mathcal{S}4_u, \text{CON})$ requires a subtler treatment. We illustrate this by the following example.

EXAMPLE 4. Denote by $\text{ALEK}^{\leq 1}$ the class of Aleksandrov frames induced by partial orders $F = (W, R)$ of depth 1, as in Fig. 1; i.e., R is the reflexive closure of a subset of $W_1 \times W_0$, where W_i is the set of points of depth i, $i = 0, 1$. Such partial orders will be called *quasi-saws*.

$Sat(\mathcal{S}4_u, \text{ALEK}^{\leq 1})$ is NP-complete because formulas in this set enjoy the *polysize model property*. More precisely, it is easy to see that every formula $\varphi \in Sat(\mathcal{S}4_u, \text{ALEK}^{\leq 1})$ is satisfied in a disjoint union of n many *m-brooms*, i.e., partial orders of the form $(\{x\} \cup W_0, R)$, where $|W_0| = m$ and R is the reflexive closure of $\{x\} \times W_0$, and both m and n are bounded by a linear function in $|\varphi|$. By contrast, formulas in $Sat(\mathcal{S}4_u, \text{CONALEK}^{\leq 1})$ may

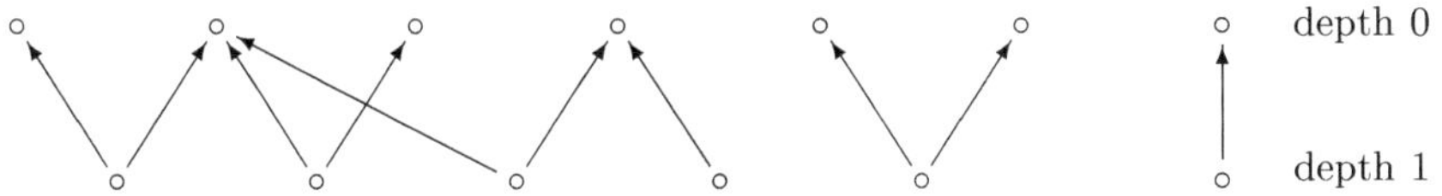

Figure 1. Quasi-saw.

require *exponential* satisfying models, and this set is PSPACE-complete [66]. We show how one can construct such formulas. Using n variables $v_1, \ldots, v_n$ one can represent (in binary) all natural numbers $< 2^n$. Now we can say that $\overline{v_n} \cap \cdots \cap \overline{v_1}$ (i.e., 0) and $v_n \cap \cdots \cap v_1$ (i.e., $2^n - 1$) are non-empty:

$$\overline{v_n} \cap \cdots \cap \overline{v_1} \neq \mathbf{0}, \qquad v_n \cap \cdots \cap v_1 \neq \mathbf{0},$$

and that the closure of the set representing a number m, $0 \leq m < 2^n - 1$, can only share points with the set representing $m + 1$:

$$(v_j \cap \overline{v_k})^- \subseteq v_j, \qquad (\overline{v_j} \cap \overline{v_k})^- \subseteq \overline{v_j}, \qquad \text{for all } n \geq j > k \geq 1,$$
$$(\overline{v_k} \cap v_{k-1} \cap \cdots \cap v_1)^- \subseteq (v_k \cap \overline{v_i}) \cup (\overline{v_k} \cap v_i), \qquad \text{for all } n \geq k > i \geq 1,$$

and that $2^n - 1$ is a closed set:

$$(v_n \cap \cdots \cap v_1)^- \subseteq v_n \cap \cdots \cap v_1.$$

As the space is connected, there is a path between 0 and $2^n - 1$, and this path must contain all the numbers $< 2^n$ (see Fig. 2). Using this idea, we can simulate any polynomial-space-bounded deterministic Turing machine.

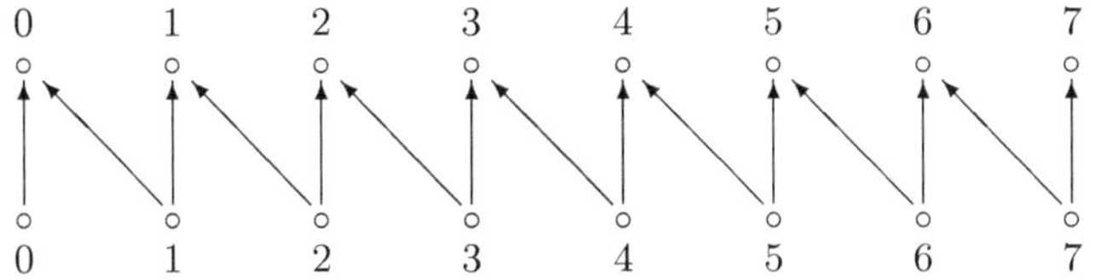

Figure 2. Satisfying the ‘counter formulas’ for $n = 3$.

Modal definability of separation properties and connectedness in $\mathcal{S}4_u$ and related hybrid logics, as well as their complexity, were studied in [57, 60].

3 Topological logics with connectedness

We now extend $\mathcal{S}4_u$ with an explicit *connectedness predicate* and denote the resulting language by $\mathcal{S}4_u c$. The *$\mathcal{S}4_u c$-formulas* are defined in the same way as the $\mathcal{S}4_u$-formulas, except that we have the additional clause

$$\varphi \quad ::= \quad \ldots \quad | \quad c(\tau) \quad | \quad \ldots,$$

where τ is an $\mathcal{S}4$-term. Given a topological model $\mathfrak{M} = (T, \boldsymbol{S}, \cdot^{\mathfrak{M}})$, the *truth-relation* for $\mathcal{S}4_u c$ is defined in the same way as for $\mathcal{S}4_u$, except that we have the additional clause

$$\mathfrak{M} \models c(\tau) \qquad \text{iff} \qquad \tau^{\mathfrak{M}} \text{ is connected in } T.$$

For example, most textbooks on general topology prove the following simple facts: (i) the union of two intersecting, connected sets is connected; (ii) any set sandwiched between a connected set and its closure is itself connected. These facts are expressible as $\mathcal{S}4_u c$-validities. That is, the formulas

$$c(v_1) \wedge c(v_2) \wedge (v_1 \cap v_2 \neq \mathbf{0}) \rightarrow c(v_1 \cup v_2), \tag{1}$$

$$c(v_1) \wedge (v_1 \subseteq v_2) \wedge (v_2 \subseteq v_1^-) \rightarrow c(v_2) \tag{2}$$

are valid in ALL.

Recalling that $\mathcal{S}4$ is a sub-language of $\mathcal{S}4_u c$, *Sat*($\mathcal{S}4_u c$, ALL) is certainly PSPACE-hard. But the matching upper bound holds only for the sublanguage $\mathcal{S}4_u c^1$ of $\mathcal{S}4_u c$ in which *at most one* subformula of the form $c(\tau)$ occurs with positive polarity.

THEOREM 5 ([31]). *Sat*($\mathcal{S}4_u c^1$, ALL) *is* PSPACE-*complete.*

Theorem 5 yields the promised result about $\mathcal{S}4_u$ interpreted over connected spaces: an $\mathcal{S}4_u$-formula φ is satisfiable in a connected space iff the $\mathcal{S}4_u c^1$-formula $\varphi \wedge c(\mathbf{1})$ is satisfiable in some topological space.

From a complexity-theoretic viewpoint, the main difference between the languages $\mathcal{S}4_u c$ and $\mathcal{S}4_u c^1$ is that when constructing a model for an $\mathcal{S}4_u c^1$-formula (using, say, a tableau-based technique) there is only one positive statement of the form $c(\tau)$ saying that points in τ have to be connected. We have seen above that connecting two points may require an exponentially long path. Nevertheless, 'connectivity' can be checked using a PSPACE-algorithm because it is not necessary to keep in memory all the points on the path. However, if two statements $c(\tau_1)$ and $c(\tau_2)$ have to be satisfied, then, while connecting two τ_1-points using a path, one has to check whether the τ_2-points on that path can be connected by a path, which, in turn, can contain another τ_1-point, and so on. And this situation can indeed happen if we have two positive occurrences of sub-formulas like $c(\tau_1)$ and $c(\tau_2)$.

THEOREM 6 ([31]). *Sat*($\mathcal{S}4_u c$, ALL) $=$ *Sat*($\mathcal{S}4_u c$, ALEK); *and this set is* EXPTIME-*complete.*

In fact, the lower bound holds already for ALEK$^{\leq 1}$. It can be proved by reduction of polynomial-space-bounded *alternating* Turing machines or satisfiability in logics like modal $\mathcal{K}$ with the universal modality. In either case, the crucial point in the proof is simulating large binary (*non-transitive*) trees. We have already seen how connectedness can help us generate quasi-saws representing an exponential counter. But now we also need *branching.* One idea of simulating both is as follows. We start by representing the root of the tree as a point v_0 (see Fig. 3), which is forced to be connected to an auxiliary point z by means of some $c(\tau_0)$. On the connecting path from v_0 to z we represent the two successors of the root by v_1 and v_2, which are forced to be connected in their turn to z by some other $c(\tau_1)$. On each of the two connecting paths, we again take two points representing the successors of v_1 and v_2, respectively. We treat these four points in the same way as v_0, reusing $c(\tau_0)$, and proceed in this way *ad infinitum* alternating between

τ_0 and τ_1 when forcing the paths which generate the required successors. Of course, in addition, certain information has to be passed from a node to its two successors (say, if $\Diamond\psi$ holds in the node, then ψ holds in one of its successors). Such information can be propagated along connected regions. Note now that all points are connected to z. Thus, to distinguish between the information we have to pass from distinct nodes of even (respectively, odd) level to their successors, we have to use *two* connectedness formulas of the form $c(f_i \cup a)$, $i = 0, 1$, in such a way that the f_i points form initial segments of the paths to z and a contains z. The f_i-segments are then used locally to pass information from a node to its successors without conflict. Note also that the points representing nodes of the tree belong to both f_0 and f_1 (except the root) and we have to separate them with auxiliary points in order to ensure proper tree structure (otherwise the 'tree' would collapse into a single node). For details the reader is referred to [31].

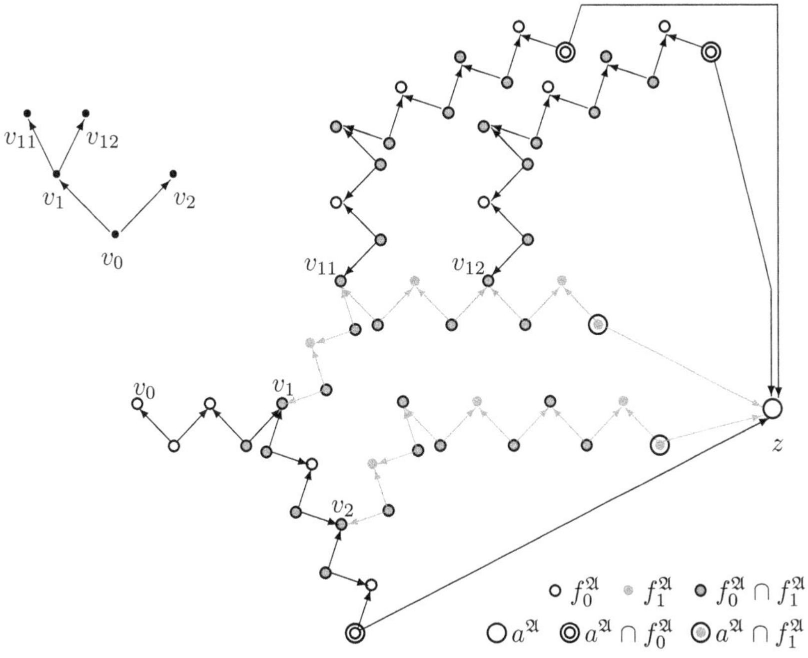

Figure 3. Encoding binary trees in $\mathcal{S}4_u c$.

To establish the upper bound for *Sat*($\mathcal{S}4_u c$, ALEK), we adapt the type-elimination technique first used to prove the EXPTIME upper bound for $\mathcal{PDL}$; see, e.g., [29]. Let a formula φ of $\mathcal{S}4_u c$ be given; but suppose for the moment that φ contains no occurrences of c. One can test the satisfiability of φ over ALEK by first computing the set of all φ-types (alias Hintikka sets), where a φ-type is a Boolean-saturated set of subterms of φ. Then one recursively eliminates all those φ-types t for which there is no witness type $t' \ni \tau$ for some $\tau^- \in t$. It can be shown that φ is satisfiable iff this elimination process terminates with a set of types corresponding to a model satisfying φ. Now suppose that φ involves some occurrences of c. Guess a set Ξ of subformulas of φ of the form $c(\tau)$ (those that one assumes to be true), and for each such $c(\tau)$, guess a φ-type t_τ containing τ. The elimination process described above can now be executed as before, except that one also eliminates those types t that contain a τ with $c(\tau) \in \Xi$ which 'cannot be connected' to t_τ (i.e., in the region corresponding to τ, one cannot find an $R \cup R^{-1}$-path from t to t_τ).

One can increase the expressive power of the connectedness predicate $c(\tau)$ by introducing the 'counting' predicates $c^{\leq k}(\tau)$ which state that τ has at most k connected components. We denote the language with such predicates by $\mathcal{S}4_u cc$. The $\mathcal{S}4_u cc$-*formulas* are defined in the same way as the $\mathcal{S}4_u$-formulas, except that we have the additional clause

$$\varphi \quad ::= \quad \dots \quad | \quad c^{\leq k}(\tau) \quad | \quad \dots,$$

where τ is an $\mathcal{S}4$-term and k a positive integer. Given a topological model $\mathfrak{M} = (T, \boldsymbol{S}, \cdot^{\mathfrak{M}})$, the *truth-relation* for $\mathcal{S}4_u cc$ is defined in the same way as for $\mathcal{S}4_u$, except that we have the additional clause:

$$\mathfrak{M} \models c^{\leq k}(\tau) \qquad \text{iff} \qquad \tau^{\mathfrak{M}} \text{ has at most } k \text{ components in } T.$$

We write $\neg c^{\leq k}(\tau)$ as $c^{\geq k+1}(\tau)$ and abbreviate $c^{\leq 1}(\tau)$ by $c(\tau)$. Thus, we may regard $\mathcal{S}4_u c$ as a sub-language of $\mathcal{S}4_u cc$. The numerical superscripts k in $c^{\leq k}$ are assumed to be coded in *binary* and so to have size $\lfloor \log k \rfloor + 1$.

The language $\mathcal{S}4_u cc$ is not essentially more expressive than $\mathcal{S}4_u c$. In particular, the $\mathcal{S}4_u cc$-literal $c^{\leq k}(\tau)$ is true in a model $\mathfrak{M}$ iff the $\mathcal{S}4_u c$-formula

$$(\tau = \bigcup_{1 \leq i \leq k} v_i) \ \wedge \ \bigwedge_{1 \leq i \leq k} c(v_i)$$

is true in some model $\mathfrak{M}'$ differing from $\mathfrak{M}$ at most in the assignments to the variables $v_1, \dots, v_k$. Thus, the $\mathcal{S}4_u cc$-literal $c^{\leq k}(\tau)$ may be 'encoded' by this $\mathcal{S}4_u c$-formula. Similarly, the $\mathcal{S}4_u cc$-literal $c^{\geq k}(\tau)$ may be likewise encoded by the $\mathcal{S}4_u c$-formula

$$(\tau = \bigcup_{1 \leq i \leq k} v_i) \ \wedge \ \bigwedge_{1 \leq i \leq k} (v_i \neq \mathbf{0}) \ \wedge \ \bigwedge_{1 \leq i < j \leq k} (\tau \cap v_i^- \cap v_j^- = \mathbf{0}).$$

Thus, any $\mathcal{S}4_u cc$-formula can be transformed into an equi-satisfiable $\mathcal{S}4_u c$-formula. However, this transformation involves a combinatorial explosion: the above formulas are exponentially larger than the literals they replace.

THEOREM 7 ([44]). *Sat*($\mathcal{S}4_u cc$, ALL) $=$ *Sat*($\mathcal{S}4_u cc$, ALEK)*; and this set is* NEXPTIME*-complete.*

The upper complexity bound follows by establishing an exponential model property of $\mathcal{S}4_u cc$. We remark that this exponential model property holds even though constraints of the form $c^{\geq k}(\tau)$ can be used to succinctly enforce regions with many components. The matching lower bound is proved by reduction of the $2^n \times 2^n$ tiling problem [64]. As we have seen in the proof sketch of Theorem 3, using n variables and a polynomial (in n) number of formulas, one can create a sequence of points in a model representing all natural numbers $< 2^n$, where only points representing m and $m+1$ may be neighbours (see Fig. 2). By using $2n$ variables $v_n, \ldots, v_1$ and $u_n, \ldots, u_1$, one can create all points of the $2^n \times 2^n$ grid (for additional formulas required see [44]) such that a point representing (i, j) has neighbours representing $(i-1, j)$, $(i+1, j)$, $(i, j-1)$ and $(i, j+1)$ and only these pairs. However, the constructed grid may contain 'defects' because the 'counter formulas' are unable to prevent numbers repeating as in the sequence $\ldots, m, m+1, m+2, m+1, m, \ldots$. A key point in the proof is the following. Using the terms

$$\tau_{black} = (v_0 \cap u_0) \cup (\overline{v_0} \cap \overline{u_0}) \quad \text{and} \quad \tau_{white} = (\overline{v_0} \cap u_0) \cup (v_0 \cap \overline{u_0})$$

we can 'colour' the grid in a chessboard manner. But then the constraints $c^{\leq 2^{n-1}}(\tau_{black})$ and $c^{\leq 2^{n-1}}(\tau_{white})$ will ensure that all points representing a pair (i, j) are in the same connected component of either τ_{black} or τ_{white}, and so we can 'cover' all points in this component with the same tile.

Returning to the language $\mathcal{S}4_u c$, it is natural to consider what happens when this language is interpreted over restricted classes of topological spaces. Perhaps the most salient such classes in this context are the singleton classes $\{\mathbb{R}^n\}$, for various n, as well as their union.

It is very easy to see that *Sat*$(\mathcal{S}4_u c, \mathbb{R})$ and *Sat*$(\mathcal{S}4_u c, \mathbb{R}^2)$ are both different from *Sat*$(\mathcal{S}4_u c, \textsc{Con})$ (and from each other). For instance, the formula

$$(3) \qquad \bigwedge_{1 \leq i \leq 3} c(v_i) \quad \wedge \bigwedge_{1 \leq i < j \leq 3} (v_i \cap v_j \neq \mathbf{0}) \quad \wedge \quad (v_1 \cap v_2 \cap v_3 = \mathbf{0})$$

is evidently satisfiable in $\mathbb{R}^n$ for all $n > 1$, but not satisfiable in $\mathbb{R}$, since connected, non-empty sets in $\mathbb{R}$ are simply intervals. Likewise, it is straightforward to write a formula satisfiable in $\mathbb{R}^n$ for all $n > 2$, but not satisfiable in $\mathbb{R}^2$. Let $v_{i,j}$, $1 \leq i < j \leq 5$, be distinct variables other than v_i, $1 \leq i \leq 5$; and let φ be the formula

$$(4) \qquad \bigwedge_{i \in \{j,k\}} (v_i \subseteq (v_{j,k})^\circ) \quad \wedge \quad \bigwedge_{1 \leq i \leq 5} (v_i \neq \mathbf{0}) \quad \wedge$$
$$\bigwedge_{\{i,j\} \cap \{k,l\} = \emptyset} (v_{i,j} \cap v_{k,l} = \mathbf{0}) \quad \wedge \quad \bigwedge_{1 \leq i < j \leq 5} c((v_{i,j})^\circ).$$

Then φ is not satisfiable in $\mathbb{R}^2$, since otherwise, one could easily embed the non-planar graph K_5 in the plane. On the other hand, it is straightforward to satisfy φ in, say, $\mathbb{R}^3$. Slightly less obviously, it turns out that *Sat*$(\mathcal{S}4_u c, \{\mathbb{R}^n \mid n > 0\})$ is different from *Sat*$(\mathcal{S}4_u c, \textsc{Con})$.

FACT 8 ([41], p. 137). If D_1 and D_2 are non-intersecting closed sets in $\mathbb{R}^n$, and points x and y are connected in $\overline{D_1}$ and also in $\overline{D_2}$, then x and y are connected in $\overline{D_1} \cap \overline{D_2}$.

Now consider the following formula:

(5) $$(v_1 \cap v_2 = \mathbf{0}) \;\wedge\; \bigwedge_{i=1,2} \big((v_i^- \subseteq v_i) \;\wedge\; c(\overline{v_i})\big) \;\wedge\; \neg c(\overline{v_1} \cap \overline{v_2}).$$

This formula is is not satisfied over any space $\mathbb{R}^n$, by Fact 8. However, it is satisfiable in many natural, connected topological spaces: e.g., let T be a torus, and let v_1 and v_2 be interpreted as rings in T, arranged as in Fig. 4.

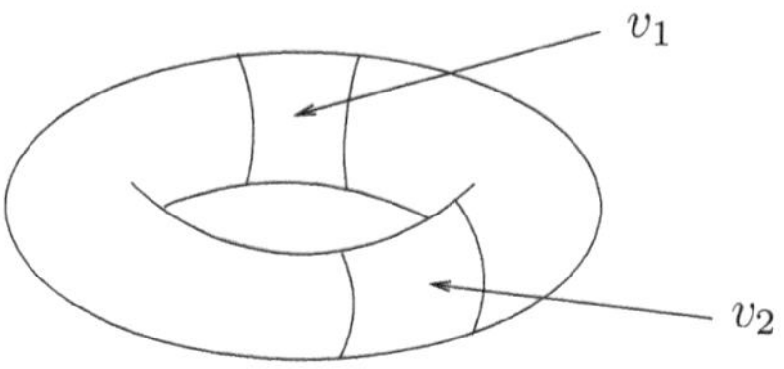

Figure 4. Two non-intersecting connected, closed sets v_1 and v_2 on a torus: note that $\overline{v_1}$ and $\overline{v_2}$ are connected, but $\overline{v_1} \cap \overline{v_2}$ is not.

Using an encoding of the topological interior and closure operators over $\mathbb{R}$ in standard temporal logic with 'since' and 'until' over $\mathbb{R}$ and Reynolds' [50] PSPACE-completeness result for this logic, one can prove the following:

THEOREM 9 ([31]). *$Sat(\mathcal{S}4_u c, \mathbb{R})$ and $Sat(\mathcal{S}4_u cc, \mathbb{R})$ are* PSPACE-*complete.*

Over higher-dimensional Euclidean spaces these languages turn out to be computationally more complex, because the proofs of the lower bounds in Theorems 6 and 7 can be restricted to such spaces:

THEOREM 10 ([31]). (i) *The sets $Sat(\mathcal{S}4_u c, \text{Con})$, $Sat(\mathcal{S}4_u c, \{\mathbb{R}^n \mid n > 2\})$ and $Sat(\mathcal{S}4_u c, \mathbb{R}^2)$ are all distinct; $Sat(\mathcal{S}4_u c, \text{Con})$ is* EXPTIME-*complete and the other two sets are* EXPTIME-*hard.*

(ii) *$Sat(\mathcal{S}4_u cc, \text{Con})$, $Sat(\mathcal{S}4_u cc, \{\mathbb{R}^n \mid n > 2\})$ and $Sat(\mathcal{S}4_u cc, \mathbb{R}^2)$ are all distinct; $Sat(\mathcal{S}4_u cc, \text{Con})$ is* NEXPTIME-*complete and the other two sets are* NEXPTIME-*hard.*

We conclude this section by mentioning two relevant research problems. First, it would be interesting to consider other modal logics with connectedness predicate (i.e. operator) $c(\tau)$, which is true in a Kripke model if any two distinct τ-points in the model are connected by a path of τ-points. For example, as in Theorem 6 one can show that basic modal logic $\mathcal{K}$ extended with the universal modality and connectedness predicate is EXPTIME-complete. The connectedness predicate can actually be expressed in the extension of $\mathcal{PDL}$ with converse programs and nominals, which is also EXPTIME-complete [14]. Another direction is to investigate the axiomatization problem for logics with connectedness predicate (see, e.g. [61]).

4 Regularized topological languages

So far, we have considered only frames $(T, \boldsymbol{S})$ in which $\boldsymbol{S}$ is the whole of 2^T—that is to say, frames in which every subset of the space counts as a region. When reasoning about spatial regions in practical situations, we may wish our variables to quantify only over 'sensible' subsets of space, corresponding to the regions potentially occupied by physical objects. In the same spirit, we may further wish to disregard differences between subsets of the space differing only with respect to boundary points. The following technical apparatus provides a convenient way to do this.

Let T be a topological space. A subset $X \subseteq T$ is called *regular closed* if $X = X^{\circ -}$. We denote the set of regular closed subsets of T by $\boldsymbol{RC}(T)$. It is easy to show that the regular closed subsets of T are in fact exactly those sets of the form $X^{\circ -}$, where X ranges over all subsets of T. The following fact is well-known (see, for example, [32], pp. 25–27).

FACT 11. Let T be a topological space. Then $\boldsymbol{RC}(T)$ is a Boolean algebra with top and bottom elements given by T and $\emptyset$, Boolean operations $\cdot, -$ given by $X \cdot Y = (X \cap Y)^{\circ -}$ and $-X = (\overline{X})^{-}$, and Boolean order $\leq$ given by the relation $\subseteq$.

In the context of $\mathbb{R}^2$, the regular closed sets are the closed sets with no 'filaments' or 'isolated points' (Fig. 5). Thus, we are led to consider logics interpreted over frames of the form $(T, \boldsymbol{RC}(T))$. We mention in passing

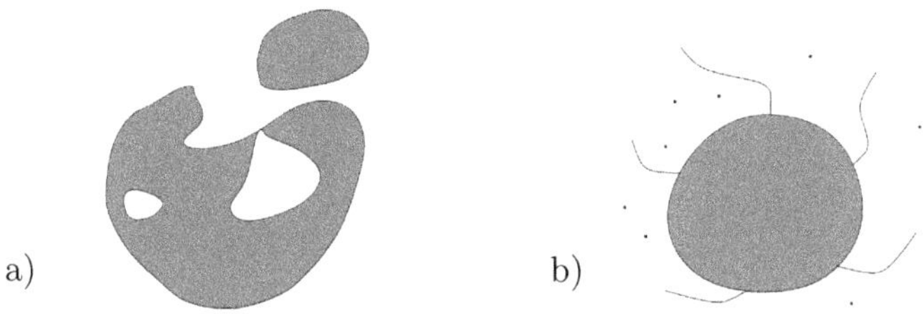

Figure 5. Shaded regions showing: a) a regular-closed subset of $\mathbb{R}^2$, and b) a (closed but) not regular-closed subset of $\mathbb{R}^2$.

that a set X is *regular open* if $X = X^{-\circ}$. The regular open subsets of T also form a Boolean algebra, $\boldsymbol{RO}(T)$, defined analogously to $\boldsymbol{RC}(T)$; in fact the map $X \mapsto X^{-}$ is a Boolean algebra isomorphism from $\boldsymbol{RO}(T)$ to $\boldsymbol{RC}(T)$. In this section, we speak only of regular closed sets; however, the same material can be presented (with minor changes) using regular open sets.

As $\boldsymbol{RC}(T)$ is not closed under complementation and intersection, it is not very natural to interpret $\mathcal{S}4$-terms over regular frames. This prompts us to define the term-language $\mathcal{B}$ as follows. Let $\mathcal{R} = \{r_i \mid i < \omega\}$ be a set of variables. The set of $\mathcal{B}$-*terms* is defined by:

$$\tau \quad ::= \quad r_i \quad | \quad -\tau \quad | \quad \tau_1 \cdot \tau_2.$$

We interpret $\mathcal{B}$-terms by taking variables to range over regular closed sets of topological spaces. More precisely, we confine attention to topological frames $(T, \boldsymbol{R})$, where $\boldsymbol{R}$ is a Boolean sub-algebra of $\boldsymbol{RC}(T)$. We call any such frame *regular*, and denote the class of regular frames by Reg. The class of regular frames based on connected topological spaces will be denoted by ConReg. We may then inductively define the *extension* $\tau^{\mathfrak{M}}$ of a term τ in a model $\mathfrak{M}$ over a regular frame by the equations:

$$(-\tau)^{\mathfrak{M}} = -\tau^{\mathfrak{M}}, \qquad\qquad (\tau_1 \cdot \tau_2)^{\mathfrak{M}} = \tau_1^{\mathfrak{M}} \cdot \tau_2^{\mathfrak{M}}.$$

Again, we have overloaded the symbols $\cdot$ and $-$: on the right-hand sides of these equations, they denote the Boolean algebra operations defined in Fact 11. We abbreviate $-((-\tau_1) \cdot (-\tau_2))$ by $\tau_1 + \tau_2$, $r_0 \cdot -r_0$ by $\mathbf{0}$, and $-\mathbf{0}$ by $\mathbf{1}$. The language of $\mathcal{B}$-terms can form the basis of topological logics just as well as $\mathcal{S}4$. In particular, we can introduce the languages $\mathcal{B}$, $\mathcal{B}c$ and $\mathcal{B}cc$ by defining their formulas as, respectively,

$$\begin{array}{lllllllll}
\varphi & ::= & \tau_1 = \tau_2 & \mid & \neg\varphi & \mid & \varphi_1 \wedge \varphi_2; & & \\
\varphi & ::= & \tau_1 = \tau_2 & \mid & \neg\varphi & \mid & \varphi_1 \wedge \varphi_2 & \mid & c(\tau_1); \\
\varphi & ::= & \tau_1 = \tau_2 & \mid & \neg\varphi & \mid & \varphi_1 \wedge \varphi_2 & \mid & c^{\leq k}(\tau_1),
\end{array}$$

where τ_1 and τ_2 range over $\mathcal{B}$-terms and k is a positive integer. The semantics of these predicates is exactly as for the languages $\mathcal{S}4_u$, $\mathcal{S}4_u c$ and $\mathcal{S}4_u cc$. We abbreviate $\tau_1 \cdot (-\tau_2) = \mathbf{0}$ by $\tau_1 \leq \tau_2$ (preferring this to $\tau_1 \subseteq \tau_2$).

Observe that the languages from the $\mathcal{B}$-family can be viewed as a *syntactic* restriction of the respective languages from the $\mathcal{S}4$-family, as follows. Let τ be a $\mathcal{B}$-term. Define the $\mathcal{S}4$-term $h(\tau)$ recursively by:

$$h(r_i) = v_i^{\circ -}, \quad h(\tau_1 \cdot \tau_2) = (h(\tau_1) \cap h(\tau_2))^{\circ -}, \quad h(-\tau_1) = \left(\overline{h(\tau_1)}\right)^{-};$$

and if φ is a $\mathcal{B}$-formula ($\mathcal{B}c$- or $\mathcal{B}cc$-formula), define $h(\varphi)$ to be the result of replacing each (maximal) $\mathcal{B}$-term τ occurring in φ by the corresponding $\mathcal{B}$-term $h(\tau)$. Thus, $h(\varphi)$ is an $\mathcal{S}4_u$-formula ($\mathcal{S}4_u c$- or $\mathcal{S}4_u cc$-formula, respectively). It is easy to check that a $\mathcal{B}cc$-formula φ is satisfiable over a frame $(T, \boldsymbol{RC}(T))$ iff $h(\varphi)$ is satisfiable over T.

The minimal logic $\mathcal{B}$ is as expressive as the modal logic $\mathcal{S}5$, with $\tau = \mathbf{1}$ playing the role of the $\mathcal{S}5$-box. Topologically, every satisfiable $\mathcal{B}$-formula φ is satisfied in a discrete topological space (= Aleksandrov frame of depth 0) with $\leq |\varphi|$ points. Hence, *Sat*($\mathcal{B}$, Reg) is NP-complete. It also follows that $\mathcal{B}$ does not distinguish between Reg, ConReg, $\boldsymbol{RC}(\mathbb{R}^n)$, $n \geq 1$.

The language $\mathcal{B}c$ is less trivial. For example, the smallest Aleksandrov model satisfying the formula $\neg c(r_1) \wedge c(\mathbf{1})$ is shown in Fig. 6. Another

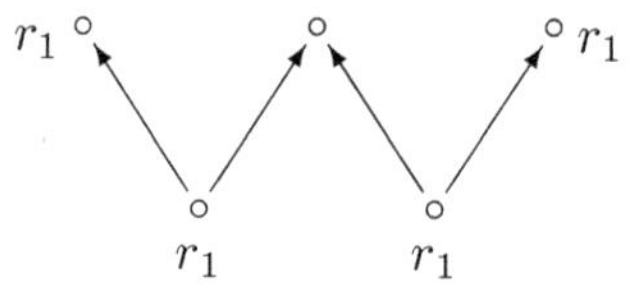

Figure 6. An Aleksandrov model for $\neg c(r_1) \wedge c(\mathbf{1})$.

important example is the formula $c(\tau_1) \wedge c(\tau_2) \wedge \neg c(\tau_1 + \tau_2)$, which says that both τ_1 and τ_2 are connected and do not intersect, i.e., $\tau_1^{\mathfrak{M}} \cap \tau_2^{\mathfrak{M}} = \emptyset$ in every model satisfying it. In fact, using formulas of this kind one can simulate binary trees (as in the proof of Theorem 6) and obtain the following rather surprising result:

THEOREM 12 ([31]). (i) *Sat*($\mathcal{B}c$, REG) $\neq$ *Sat*($\mathcal{B}c$, CONREG), *with both sets being* EXPTIME-*complete.*

(ii) *Sat*($\mathcal{B}cc$, REG) $\neq$ *Sat*($\mathcal{B}cc$, CONREG); *both are* NEXPTIME-*complete.*

It is also of interest to note that $\mathcal{B}c$ (and $\mathcal{B}cc$) can distinguish between $\boldsymbol{RC}(\mathbb{R})$, $\boldsymbol{RC}(\mathbb{R}^2)$ and $\boldsymbol{RC}(\mathbb{R}^n)$, for $n > 2$.

Consider now the language $\mathcal{B}c^\circ$ defined in the same way as $\mathcal{B}c$, except that the predicate c is replaced by the predicate c° with the interpretation:

$$\mathfrak{M} \models c^\circ(\tau) \quad \text{iff} \quad \text{the interior of } \tau^{\mathfrak{M}} \text{ is connected.}$$

It is a simple exercise in general topology to show that the analogues, in $\mathcal{B}c$ and $\mathcal{B}c^\circ$, of the formula (1), namely,

$$\begin{array}{rcl} c(r_1) \wedge c(r_2) \wedge (r_1 \cdot r_2 \neq \mathbf{0}) & \rightarrow & c(r_1 + r_2) \\ c^\circ(r_1) \wedge c^\circ(r_2) \wedge (r_1 \cdot r_2 \neq \mathbf{0}) & \rightarrow & c^\circ(r_1 + r_2) \end{array}$$

are both valid over REG. The language $\mathcal{B}c^\circ$ is a natural choice for describing arrangements in the Euclidean plane, particularly when variables are taken to range only over well-behaved regions. To understand the issues that arise in this context, consider the $\mathcal{B}c^\circ$-formula

$$(6) \quad \bigwedge_{1\le i\le 3} c^\circ(r_i) \;\wedge\; \bigwedge_{1\le i<j\le 3} (r_i \cdot r_j = \mathbf{0}) \;\wedge\; c^\circ\big(\sum_{1\le i\le 3} r_i\big) \;\wedge\; \bigwedge_{i=2,3} \neg c^\circ(r_1 + r_i).$$

Formula (6) 'says' that r_1, r_2 and r_3 are interior-connected, pairwise disjoint, regular closed sets having an interior-connected sum, such that the first forms an interior-connected sum with neither of the other two. This formula is satisfiable in $\boldsymbol{RC}(\mathbb{R}^2)$. For let $\mathfrak{M}$ be a model over $\boldsymbol{RC}(\mathbb{R}^2)$ in which

$$\begin{aligned} r_1^{\mathfrak{M}} &= \{(x,y) \in \mathbb{R}^2 \mid -1 \le x \le 0,\ -1-x \le y \le 1+x\}, \\ r_2^{\mathfrak{M}} &= \{(x,y) \in \mathbb{R}^2 \mid 0 < x \le 1,\ -1-x \le y \le \sin(1/x)\} \ \cup \\ &\qquad\qquad \{(0,y) \in \mathbb{R}^2 \mid -1 \le y \le 1\}, \\ r_3^{\mathfrak{M}} &= \{(x,y) \in \mathbb{R}^2 \mid 0 < x \le 1,\ \sin(1/x) \le y \le 1+x\} \ \cup \\ &\qquad\qquad \{(0,y) \in \mathbb{R}^2 \mid -1 \le y \le 1\}, \end{aligned}$$

as depicted in Fig. 7. It is easy to check that $(r_1+r_2+r_3)^\circ$ is the interior of the large triangle, and so is certainly connected, but that neither $(r_1+r_2)^\circ$ nor $(r_1+r_3)^\circ$ is connected. However, the regions r_2 and r_3 in Fig. 7 are rather 'wild': they cannot sensibly be used to represent regions of the plane occupied (or left unoccupied) by physical objects. The question therefore arises as to whether (6) is satisfiable if only 'tame' regions are allowed.

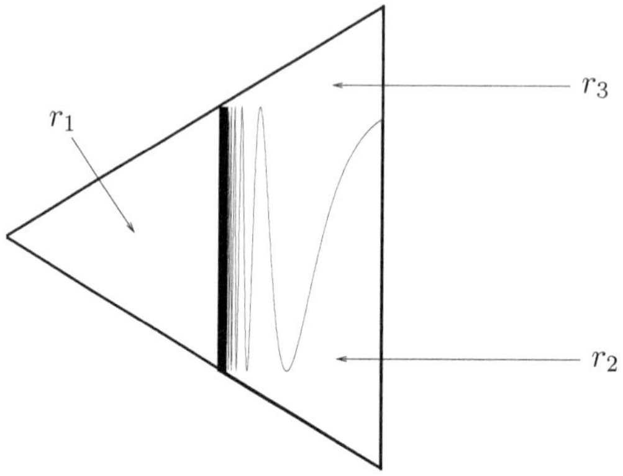

Figure 7. Three elements in $\boldsymbol{RC}(\mathbb{R}^2)$.

Any $(n-1)$-dimensional hyper-plane of $\mathbb{R}^n$ cuts $\mathbb{R}^n$ into two closed sets, in the obvious way, which we shall call *half-spaces*. It is easy to see that these half-spaces are regular closed, with each being the complement of the other in the Boolean algebra $\boldsymbol{RC}(\mathbb{R}^n)$. Hence, we can speak about the sums, products and complements of half-spaces in $\boldsymbol{RC}(\mathbb{R}^n)$.

A *basic polytope* in $\mathbb{R}^n$ is the product, in $\boldsymbol{RC}(\mathbb{R}^n)$, of finitely many half-spaces. A *polytope* in $\mathbb{R}^n$ is the sum, in $\boldsymbol{RC}(\mathbb{R}^n)$, of any finite set of basic polytopes. We denote the set of polytopes in $\mathbb{R}^n$ by $\boldsymbol{RCP}(\mathbb{R}^n)$; we call the polytopes in $\boldsymbol{RCP}(\mathbb{R}^2)$ *polygons*. Thus, polytopes (in our sense) may be unbounded, disconnected, and may have disconnected complements. (In alternative parlance, the elements of $\boldsymbol{RCP}(\mathbb{R}^n)$ are the regular closed *semi-linear* sets.) It is obvious that $\boldsymbol{RCP}(\mathbb{R}^n)$ is a Boolean sub-algebra of $\boldsymbol{RC}(\mathbb{R}^n)$. Polytopes are well-behaved in two crucial respects.

Let T be a topological space, and $\boldsymbol{M} \subseteq \boldsymbol{RC}(X)$. We call $\boldsymbol{M}$ *finitely decomposable* if, for all $R \in \boldsymbol{M}$, there exist $R_1, \ldots, R_n \in \boldsymbol{M}$ such that $R = R_1 + \cdots + R_n$. Let T be a topological space, $X \subseteq T$, and p a point on the frontier of X. An *end-cut to p in X* is a Jordan arc g in T such that $g(1) = p$ and $g([0,1[) \subseteq X$. We say that S *has curve-selection* if, for any point p in the frontier of X, there exists an end-cut in X to p. A set of subsets of T has *has curve-selection* if each of its members does.

LEMMA 13. *$\boldsymbol{RCP}(\mathbb{R}^n)$ is finitely decomposable, and has curve-selection.*

Indeed, basic polytopes are convex, and so trivially have curve selection. But if $R = R_1 + \cdots + R_n$, then $\partial R \subseteq \partial R_1 \cup \cdots \cup \partial R_n$ and $R^\circ \supseteq R_1^\circ \cup \cdots \cup R_n^\circ$, where ∂X denotes the frontier of X. The significance of these properties is that they affect the satisfiability of formula (6).

LEMMA 14 ([45], p. 40). *Let $\boldsymbol{M}$ be a finitely decomposable Boolean subalgebra of $\boldsymbol{RC}(\mathbb{R}^2)$ forming a closed basis for the usual topology on $\mathbb{R}^2$, and having curve-selection. Then (6) is not satisfiable over the frame $(\mathbb{R}^2, \boldsymbol{M})$.*

In particular, formula (6) is unsatisfiable over the frame $(\mathbb{R}^2, \boldsymbol{RCP}(\mathbb{R}^2))$; hence $Sat(\mathcal{B}c^\circ, \boldsymbol{RC}(\mathbb{R}^2)) \neq Sat(\mathcal{B}c^\circ, \boldsymbol{RCP}(\mathbb{R}^2))$. We remark that many other natural collections of 'tame' regions exhibit the property of finite decomposability and curve-selection—most notably, the regular closed *semi-algebraic* sets (see, e.g. [63]).

Unfortunately, little is known about these logics. In particular, it would be interesting to investigate the complexity of sets like $Sat(\mathcal{B}c, \boldsymbol{RC}(\mathbb{R}^2))$, $Sat(\mathcal{B}c, \boldsymbol{RCP}(\mathbb{R}^2))$, $Sat(\mathcal{B}c^\circ, \boldsymbol{RC}(\mathbb{R}^2))$, $Sat(\mathcal{B}c^\circ, \boldsymbol{RCP}(\mathbb{R}^2))$. The computational behaviour of $\mathcal{B}c^\circ$ over simpler classes of frames such as REG and CONREG remains also open for investigation.

5 Boolean contact algebras

By restricting interpretations of variables to regular closed (or open) sets and, correspondingly, the language of $\mathcal{S}4_u$ to its fragment $\mathcal{B}$, we considerably restrict the expressive capabilities of our spatial logics. In particular, Whitehead's 'extensive connection' [65] $C(\tau_1, \tau_2)$, which has historically played a prominent role in region-based theories of space, cannot be expressed by means of $\mathcal{B}$-formulas despite its very simple intended meaning:

$$\mathfrak{M} \models C(\tau_1, \tau_2) \quad \text{iff} \quad (\tau_2^{\mathfrak{M}})^- \cap (\tau_2^{\mathfrak{M}})^- \neq \emptyset.$$

In $\mathcal{S}4_u$, we clearly have $C(\tau_1, \tau_2) \equiv (\tau_1^- \cap \tau_2^- \neq \mathbf{0})$.

So we define the language $\mathcal{C}$ by extending $\mathcal{B}$ with the binary predicate C, interpreted as above. Thus the *$\mathcal{C}$-terms* are precisely the $\mathcal{B}$-terms and the *$\mathcal{C}$-formulas* are defined in the same way as the $\mathcal{B}$-formulas, except that we have the additional clause

$$\varphi \quad ::= \quad \dots \quad | \quad C(\tau_1, \tau_2) \quad | \quad \dots,$$

where τ_1 and τ_2 are $\mathcal{C}$-terms. Since Whitehead's term 'extensive connection' risks confusion with the standard topological notion of *connectedness*, we follow more recent usage and read $C(\tau_1, \tau_2)$ as 'τ_1 contacts τ_2.'

It turns out that $\mathcal{C}$ is adequate for the reconstruction of topology within a 'region-based' ontology, in the following sense. A *closed mereotopology* is a topological frame $(T, \boldsymbol{M})$, such that $\boldsymbol{M}$ is (i) a Boolean sub-algebra of $\boldsymbol{RC}(T)$; and (ii) $\boldsymbol{M}$ is a closed basis for the topology on T. It is easy to verify that, over any mereotopology, the following $\mathcal{C}$-formulas are valid:

(7) $\neg C(r, \mathbf{0})$,

(8) $(r \neq \mathbf{0}) \to C(r, r)$,

(9) $C(r, s) \to C(s, r)$,

(10) $C(r, s) \wedge (s \leq t) \to C(r, t)$,

(11) $C(r, s + t) \to C(r, s) \vee C(r, t)$.

Structures (in the first-order sense) satisfying the usual axioms of Boolean algebras together with the universal closures of (7)–(11) are known as

$\mathsf{DC}(r,s)$	$\neg C(r,s)$	r and s are disconnected
$\mathsf{EC}(r,s)$	$(r \cdot s = \mathbf{0}) \wedge C(r,s)$	r and s are externally connected
$\mathsf{EQ}(r,s)$	$r = s$	r and s are equal
$\mathsf{PO}(r,s)$	$(r \cdot s \neq \mathbf{0}) \wedge ((-r) \cdot s \neq \mathbf{0}) \wedge (r \cdot (-s) \neq \mathbf{0})$	r and s partially overlap
$\mathsf{TPP}(r,s)$	$(r \cdot (-s) = \mathbf{0}) \wedge C(r,-s)$	r is a tangential proper part of s
$\mathsf{NTPP}(r,s)$	$\neg C(r,-s)$	r is a non-tangential proper part of s
$\mathsf{TPP}^{-1}(r,s)$	$(s \cdot (-r) = \mathbf{0}) \wedge C(s,-r)$	s is a tangential proper part of r
$\mathsf{NTPP}^{-1}(r,s)$	$\neg C(s,-r)$	s is a non-tangential proper part of r

Table 1. The $\mathcal{RCC}$-8 relations in the language $\mathcal{C}$.

Boolean contact algebras. Thus, any mereotopology is a Boolean contact algebra. In fact we have a converse: every Boolean contact algebra is isomorphic to some mereotopology $(T, \boldsymbol{M})$, where T is a semiregular and compact T_0-space [16, 17]. Axiom sets corresponding to closed mereotopologies over spaces satisfying certain separation properties have also been obtained [16, 17, 19, 51]; see also [56].

The complexity of reasoning in $\mathcal{C}$ was studied in [66], where this logic was introduced under the name $\mathcal{BRCC}$-8 in recognition of the fact that it is able to express the eight relationships in Table 1, which have played an important role in the recent development of spatial logics.

THEOREM 15 ([66]). *Sat*$(\mathcal{C}, \textsc{Reg})$ *is* NP-*complete.*

This result follows from the fact that every satisfiable $\mathcal{C}$-formula φ can be satisfied in a frame belonging to $\textsc{Alek}^{<1}$ which is a disjoint union of linearly many (in the length of φ) forks; see Fig. 8.

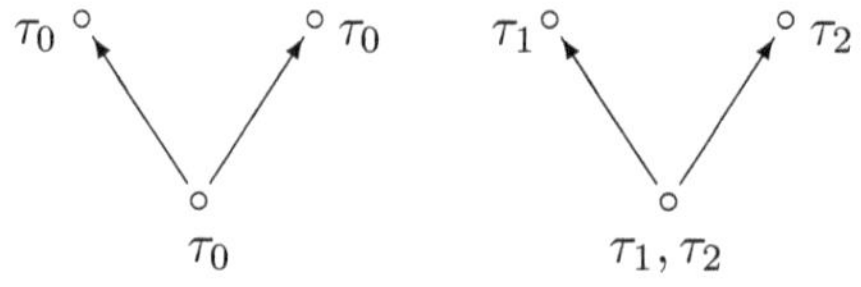

Figure 8. Satisfying $(\tau_0 \neq \mathbf{0}) \;\wedge\; C(\tau_1, \tau_2)$ in a disjoint union of forks.

THEOREM 16 ([66]). *Sat*$(\mathcal{C}, \textsc{ConReg}) =$ *Sat*$(\mathcal{C}, \boldsymbol{RC}(\mathbb{R}^n))$*; and this set is* $\textsc{PSpace}$-*complete.*

To show that the two sets coincide, one can use two observations: (i) every $\mathcal{C}$-formula satisfiable in a frame from $\textsc{ConReg}$ is satisfiable in a finite *saw* Aleksandrov model, i.e., a model induced by a partial order R on $\{x_0, z_1, x_1, z_2, x_2, \dots, z_n, x_n\}$ such that $z_i R x_{i-1}$ and $z_i R x_i$, for $1 \leq i \leq n$; and (ii) every saw Aleksandrov model can be embedded into $\mathbb{R}^n$. The first observation follows from the fact that every formula in *Sat*$(\mathcal{C}, \textsc{ConReg})$ is satisfiable in a connected quasi-saw Aleksandrov model, which can be transformed (by duplicating points) into a saw model. The proof of (ii) for $n = 1$ is illustrated in Fig. 9 (points of depth 0 correspond to closed intervals and points of depth 1 to the end-points of those intervals).

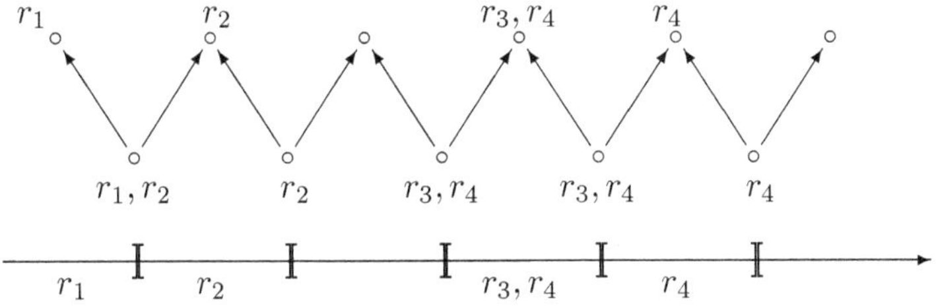

Figure 9. Embedding a saw model in $\mathbb{R}$.

The complexity result is proved similarly to Theorem 5; a full proof can be found in [31]. In fact, it turns out that $\mathcal{C}$ is powerful enough to model binary counters (as in Section 3).

We also mention here two results from [3, Section 9]: (i) axioms (7)–(11) together with the axioms for Boolean algebras axiomatize the validities of $\mathcal{C}$ over REG; and (ii) the extra axiom

$$(r \neq \mathbf{0}) \wedge (r \neq \mathbf{1}) \quad \rightarrow \quad C(r, -r)$$

is required to axiomatize $\mathcal{C}$-validities over CONREG.

As with the languages $\mathcal{S}4_u$ and $\mathcal{B}$, it is of interest to consider the extensions $\mathcal{C}c$ and $\mathcal{C}cc$ of $\mathcal{C}$ with the predicates c and $c^{\leq k}$. Surprisingly enough, these languages are of the same complexity as their $\mathcal{S}4_u$ counterparts:

THEOREM 17 ([31]). (i) *Sat*($\mathcal{C}c$, REG) *and* *Sat*($\mathcal{C}c$, CONREG) *are distinct and* EXPTIME-*complete*; *Sat*($\mathcal{C}c$, $\boldsymbol{RC}(\mathbb{R}^n)$), *for* $n \geq 2$, *is* EXPTIME-*hard.*

(ii) *Sat*($\mathcal{C}cc$, REG) *and* *Sat*($\mathcal{C}cc$, CONREG) *are distinct and* NEXPTIME-*complete*; *Sat*($\mathcal{C}cc$, $\boldsymbol{RC}(\mathbb{R}^n)$), *for* $n \geq 2$, *is* NEXPTIME-*hard.*

Another surprising result is that the satisfiability problem for $\mathcal{C}c$ (and $\mathcal{C}cc$) is reducible to the satisfiability problem for $\mathcal{B}c$ ($\mathcal{B}cc$, respectively). Clearly, two connected closed sets are in contact iff their union is connected; that is to say, the formula $c(\tau_1) \wedge c(\tau_2) \rightarrow \big(C(\tau_1, \tau_2) \;\leftrightarrow\; c(\tau_1 + \tau_2)\big)$ is a $\mathcal{C}cc$-validity. However, this 'reduction' of C to c assumes the arguments of $C(\tau_1, \tau_2)$ to be *connected*, which is not in general the case. Roughly, the idea behind the reduction is as follows. If $\mathfrak{M} \models C(\tau_1, \tau_2)$ then there are connected components X_i of $\tau_i^{\mathfrak{M}}$ such that $X_1 \cap X_2 \neq \emptyset$. So we can introduce fresh variables t_i for X_i, for which $\mathfrak{M} \models c(t_1 + t_2) \wedge c(t_1) \wedge c(t_2)$. On the other hand, if $\mathfrak{M} \models \neg C(\tau_1, \tau_2)$ then we can extend the Aleksandrov space T underlying $\mathfrak{M}$ with two extra points u_1 and u_2 that 'connect' all the points of τ_1 and τ_2, respectively, and consider the new connected sets $X_i = \tau_i^{\mathfrak{M}} \cup \{u_i\}$. By introducing fresh variables t_i for X_i, we then have $\mathfrak{M} \models \neg c(t_1 + t_2) \wedge c(t_1) \wedge c(t_2)$. For more details consult [31].

6 $\mathcal{RCC}$-8

In Sections 2 and 3, we considered topological logics based on the term-language $\mathcal{S}4$. In Section 4, we investigated the result of (in effect) restricting this term-language to those terms that denote regular closed sets. In this section, we consider topological languages in which terms have no structure at all: they are simply variables.

What topological primitives might we employ over such an impoverished term-language? The possibilities are almost endless; historically, however, one particular collection of primitives has held centre-stage. Define the language $\mathcal{RCC}$-8 as follows. The $\mathcal{RCC}$-8-*terms* are simply the variables. The $\mathcal{RCC}$-8-*formulas*, φ, are given by

$$\begin{aligned}\varphi \quad ::= \quad & r \,\mathsf{DC}\, s \;\mid\; r \,\mathsf{EC}\, s \;\mid\; r \,\mathsf{EQ}\, s \;\mid\; r \,\mathsf{PO}\, s \;\mid\; r \,\mathsf{TPP}\, s \;\mid\\ & r \,\mathsf{NTPP}\, s \;\mid\; r \,\mathsf{TPP}^{-1}\, s \;\mid\; r \,\mathsf{NTPP}^{-1}\, s \;\mid\; \neg\varphi \;\mid\; \varphi_1 \wedge \varphi_2,\end{aligned}$$

where r and s are variables (i.e. $\mathcal{RCC}$-8-terms). The semantics of $\mathcal{RCC}$-8 is defined according to Table 1, under the restriction that $\mathcal{RCC}$-8-terms are interpreted by non-empty, regular closed sets of topological spaces. $\mathcal{RCC}$-8 and similar formalisms were originally introduced in the area of knowledge representation and reasoning in AI, in particular, geographical information systems; see [20, 21, 55, 46]. The fact that it is actually a simple fragment of $\mathcal{S}4_u$ was first observed by Bennett [5]; see also [48, 42] (in fact, $\mathcal{RCC}$-8 can be embedded in $\mathcal{S}5$ [67]).

$\mathcal{RCC}$-8 is rather inexpressive. As was observed by Renz [47], we have:

THEOREM 18 ([47]). *Sat*$(\mathcal{RCC}\text{-8}, \textsc{Reg})$ = *Sat*$(\mathcal{RCC}\text{-8}, \textsc{ConReg})$ = *Sat*$(\mathcal{RCC}\text{-8}, \boldsymbol{RC}(\mathbb{R}^n))$, *for any* $n \geq 1$; *this set is* NP-*complete.*

NP-completeness follows from the fact that—similarly to $\mathcal{C}$—every satisfiable $\mathcal{RCC}$-8-formula φ is satisfied in a disjoint union of linearly many (in the length of φ) forks. Tractable fragments of $\mathcal{RCC}$-8 were analyzed in [48].

Actually, the NP-hardness result here arises entirely from the Boolean combinations available in formulas. Indeed, the following holds:

THEOREM 19 ([27]). *The problem of determining whether a conjunctive $\mathcal{RCC}$-8-formula is satisfiable in* Reg *is* NLogSpace-*complete.*

$\mathcal{RCC}$-8's lack of expressiveness at the level of the term-language opens up additional possibilities for restrictions on the topological frames considered, for it is perfectly natural to interpret $\mathcal{RCC}$-8 over topological frames $(T, \boldsymbol{M})$ in which $\boldsymbol{M}$ is a sub*set* (not necessarily a sub*algebra*) of $\boldsymbol{RC}(T)$. For instance, let $\boldsymbol{C}(T)$ be the set of non-empty, connected, regular closed subsets of the space T. It is easy to see that, for $n \geq 3$, *Sat*$(\mathcal{RCC}\text{-8}, \boldsymbol{C}(\mathbb{R}^n))$ = *Sat*$(\mathcal{RCC}\text{-8}, \boldsymbol{RC}(\mathbb{R}^n))$ [47]. However, this is not true for $n = 1$ or $n = 2$. For the latter case, we consider an example similar to (4). Let $r_{i,j}$, $1 \leq i < j \leq 5$, be distinct variables other than r_i, $1 \leq i \leq 5$; and let φ be the formula

$$\bigwedge_{i \in \{j,k\}} \mathsf{NTPP}(r_i, r_{j,k}) \quad \wedge \quad \bigwedge_{\{i,j\} \cap \{k,l\} = \emptyset} \mathsf{DC}(r_{i,j}, r_{k,l}).$$

Clearly, φ is not satisfiable over $\boldsymbol{C}(\mathbb{R}^2)$, again, because, if it were, one could construct a plane drawing of K_5.

Another salient frame over which to interpret $\mathcal{RCC}$-8 is the collection $\boldsymbol{D}$ of subsets of $\mathbb{R}^2$ homeomorphic to the closed unit disc. This interpretation is noteworthy for the following reason. Fix some topological frame $(T, \boldsymbol{S})$, and let $R_1, \ldots, R_8$ be the relations over $\boldsymbol{S}$ expressed by the $\mathcal{RCC}$-8-primitives, as specified in Table 1. It is routine to show that $R_1, \ldots, R_8$ are mutually exclusive and jointly exhaustive: any ordered pair of elements of $\boldsymbol{S}$ belongs to exactly one of these relations. But now consider the various *relative products* $R_i \circ R_j$. If each of these relative products is the *union* of some subset of $\{R_1, \ldots, R_n\}$, then these relations generate a finite relation algebra.

THEOREM 20 ([38]). *The relations expressed by the $\mathcal{RCC}$-8-predicates, interpreted over the set $\boldsymbol{D}$ of disc-homeomorphs in $\mathbb{R}^2$, are the atoms of a relation algebra.*

Schaefer *et al.* [52] analyse the relationship between *Sat*($\mathcal{RCC}$-8, $(\mathbb{R}^2, \boldsymbol{D})$) and the problem of determining the weak realizability of topological graphs. Let $G = (V, E)$ be a graph and R a set of (unordered) pairs of elements of E. We say that (G, R) is *weakly realizable* if there is a drawing of G such that only the pairs of edges allowed to cross are those occurring in R. Schaefer *et al.* show that the problem of determining weak realizability of graphs is in NP. This is a remarkable result, as it is known that some weakly realizable topological graphs require drawings in which the number of crossing points is bounded below by an exponential function of the size of the graph [33]. Using the close relationship between weak realizability of topological graphs and satisfiability of $\mathcal{RCC}$-8-formulas by closed discs, Schaefer *et al.* obtain:

THEOREM 21 ([52]). *The problem Sat($\mathcal{RCC}$-8, $(\mathbb{R}^2, \boldsymbol{D})$) is* NP*-complete.*

7 n-ary contact relation

As formulas in $\mathcal{RCC}$-8 and $\mathcal{C}$ are built from $\mathcal{B}$-terms using the *binary* predicates $\tau_1 = \tau_2$ and $C(\tau_1, \tau_2)$, they are not capable of expressing certain relations involving three or more regions. An obvious way of extending the expressive power of $\mathcal{C}$ in this direction is to generalize the contact predicate and consider the extension $\mathcal{C}^m$ of $\mathcal{B}$ with arbitrary n-ary contact relations $C(\tau_1, \ldots, \tau_n)$, for $n > 1$: the $\mathcal{C}^m$-formulas are defined in the same way as $\mathcal{B}$-formulas, except that we have the additional clause

$$\varphi \quad ::= \quad \ldots \quad | \quad C(\tau_1, \ldots, \tau_n) \quad | \quad \ldots .$$

The definition of the truth-relation is extended as follows:

$$\mathfrak{M} \models C(\tau_1, \ldots, \tau_n) \quad \text{iff} \quad (\tau_1^{\mathfrak{M}})^- \cap \cdots \cap (\tau_n^{\mathfrak{M}})^- \neq \emptyset.$$

Clearly, $\mathcal{C}^m$ can also be regarded as a fragment of $\mathcal{S}4_u$; indeed, the n-ary contact relation is definable in $\mathcal{S}4_u$ as $C(\tau_1, \ldots, \tau_n) \equiv (\tau_1^- \cap \cdots \cap \tau_n^- \neq \mathbf{0})$.

The extra expressive power of $\mathcal{C}^m$ as compared to $\mathcal{RCC}$-8 and $\mathcal{C}$ can be illustrated by the following formula

$$C(r_1, r_2, r_3) \quad \wedge \quad \bigwedge_{1 \le i < j \le 3} (r_i \cdot r_j = \mathbf{0}) \tag{12}$$

which says that boundaries of three regular closed sets r_1, r_2 and r_3 meet somewhere but the three sets have no common interior points. In particular, in order to satisfy it in Aleksandrov spaces, one requires a partial order of width 3 (see Fig. 10 a)) unlike for $\mathcal{C}$, where partial orders of width 2 (disjoint unions of forks) were enough. A model over $\boldsymbol{RC}(\mathbb{R}^2)$ satisfying (12) is depicted in Fig. 10 b): it interprets each r_i as a third of the disc; then the centre of the disc is the point in $r_1 \cap r_2 \cap r_3$ (i.e., a witness for $C(r_1, r_2, r_3)$).

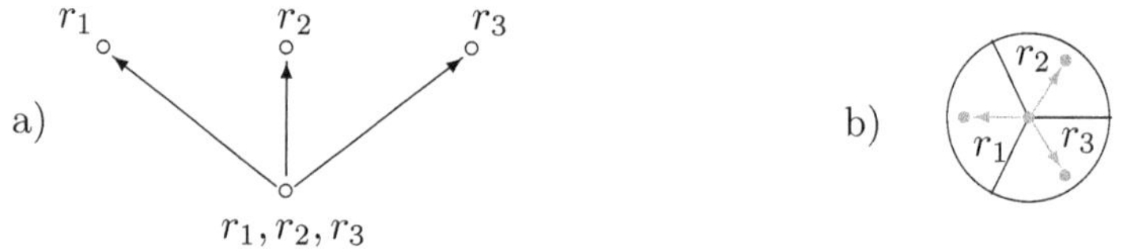

Figure 10. A 3-broom satisfying (12) and a satisfying model over $\boldsymbol{RC}(\mathbb{R}^2)$.

Despite the extra expressiveness, $\mathcal{C}^m$ is of the same complexity as $\mathcal{C}$:

THEOREM 22 ([23]). *Sat*($\mathcal{C}^m$, REG) *is* NP-*complete.*

This follows from the fact that every satisfiable $\mathcal{C}^m$-formula φ can be satisfied in an Aleksandrov model that is a disjoint union of linearly many (in $|\varphi|$) n-brooms, where n is the maximum arity of contact predicate in φ (each n-broom is a partial order $(\{z, x_1, \dots, x_n\}, R)$ with zRx_i, for all i).

It is also of interest to note that, unlike $\mathcal{C}$, $\mathcal{C}^m$ distinguishes between $\mathbb{R}$ and $\mathbb{R}^2$: formula (12) is clearly satisfiable in $\mathbb{R}^2$ but not in $\mathbb{R}$ (cf. Theorem 16).

THEOREM 23. *For* $n > 1$, *Sat*($\mathcal{C}^m, \boldsymbol{RC}(\mathbb{R})$) $\neq$ *Sat*($\mathcal{C}^m, \boldsymbol{RC}(\mathbb{R}^n)$) *and* *Sat*($\mathcal{C}^m, \boldsymbol{RC}(\mathbb{R}^n)$) $=$ *Sat*($\mathcal{C}^m$, CONREG). *All these sets are* PSPACE-*complete.*

The following complexity results for the extensions $\mathcal{C}^m c$ and $\mathcal{C}^m cc$ of $\mathcal{C}^m$ are immediate consequences of the results considered earlier in the paper:

COROLLARY 24. *Sat*($\mathcal{C}^m c$, REG), *Sat*($\mathcal{C}^m c$, CONREG) *are* EXPTIME-*complete; Sat*($\mathcal{C}^m cc$, REG), *Sat*($\mathcal{C}^m cc$, CONREG) *are* NEXPTIME-*complete.*

Finally, we note that every satisfiable $\mathcal{C}^m cc$-formula can be satisfied in an Aleksandrov model based on a frame of depth 1:

THEOREM 25. *For every finite Aleksandrov model* $\mathfrak{M}$ *induced by a quasi-order* (W, R), *there is an* ALEK$^{\le 1}$ *model* $\mathfrak{M}'$ *induced by* (W, R'), $R' \subseteq R$, *such that, for each* $\mathcal{B}$-*term* τ,

- $\tau^{\mathfrak{M}} = \tau^{\mathfrak{M}'}$;
- $\tau^{\mathfrak{M}}$ *and* $\tau^{\mathfrak{M}'}$ *have the same number of connected components.*

It follows that if $\mathfrak{M} \models \varphi$ *then* $\mathfrak{M}' \models \varphi$, *for every* $\mathcal{C}^m cc$-*formula* φ.

8 Conclusion

We conclude the paper with a table summarizing the complexity results considered above as well as the open problems. Merged cells in the table mean that the corresponding logics coincide, EXP stands for EXPTIME, and NEXP stands for NEXPTIME.

<table>
<tr><th></th><th>REG</th><th>CONREG</th><th>$\boldsymbol{RC}(\mathbb{R}^n)$
$n > 2$</th><th>$\boldsymbol{RC}(\mathbb{R}^2)$</th><th>$\boldsymbol{RC}(\mathbb{R})$</th></tr>
<tr><td>$\mathcal{RCC}$-8</td><td colspan="3" rowspan="3">NP
Thm. 18</td><td colspan="2"></td></tr>
<tr><td>$\mathcal{RCC}$-8c</td><td>?</td><td>≤PSPACE,≥NP</td></tr>
<tr><td>$\mathcal{RCC}$-8cc</td><td>?</td><td>≤PSPACE,≥NP</td></tr>
<tr><td>$\mathcal{B}$</td><td colspan="5">NP</td></tr>
<tr><td>$\mathcal{B}c$</td><td>EXP</td><td>EXP</td><td>?</td><td>?</td><td>≤PSPACE,≥NP</td></tr>
<tr><td>$\mathcal{B}cc$</td><td>NEXP</td><td>NEXP</td><td>?</td><td>?</td><td>≤PSPACE,≥NP</td></tr>
<tr><td>$\mathcal{C}$</td><td>NP
Thm. 15</td><td colspan="4">PSPACE
Thm. 16</td></tr>
<tr><td>$\mathcal{C}c$</td><td>EXP
Thm. 17</td><td>EXP
Thm. 17</td><td>≥EXP
Thm. 17</td><td>≥EXP
Thm. 17</td><td>PSPACE</td></tr>
<tr><td>$\mathcal{C}cc$</td><td>NEXP
Thm. 17</td><td>NEXP
Thm. 17</td><td>≥NEXP
Thm. 17</td><td>≥NEXP
Thm. 17</td><td>PSPACE</td></tr>
<tr><td>$\mathcal{C}^m$</td><td>NP
Thm. 22</td><td colspan="2">PSPACE
Thm. 23</td><td>PSPACE
Thm. 23</td><td>PSPACE</td></tr>
<tr><td>$\mathcal{C}^m c$</td><td>EXP</td><td>EXP</td><td>≥EXP</td><td>≥EXP</td><td>PSPACE</td></tr>
<tr><td>$\mathcal{C}^m cc$</td><td>NEXP</td><td>NEXP</td><td>≥NEXP</td><td>≥NEXP</td><td>PSPACE</td></tr>
<tr><th></th><th>ALL</th><th>CON</th><th>$\mathbb{R}^n$, $n > 2$</th><th>$\mathbb{R}^2$</th><th>$\mathbb{R}$</th></tr>
<tr><td>$\mathcal{S}4_u$</td><td>PSPACE
Thm. 2</td><td colspan="4">PSPACE
Thm. 3</td></tr>
<tr><td>$\mathcal{S}4_u c$</td><td>EXP
Thm. 6</td><td>EXP
Thm. 10</td><td>≥EXP</td><td>≥EXP</td><td>PSPACE</td></tr>
<tr><td>$\mathcal{S}4_u cc$</td><td>NEXP
Thm. 7</td><td>NEXP
Thm. 10</td><td>≥NEXP</td><td>≥NEXP</td><td>PSPACE
Thm. 9</td></tr>
</table>

Acknowledgements. The work on this paper was partially supported by the U.K. EPSRC research grants EP/E034942/1 and EP/E035248/1. We are grateful to Dimiter Vakarelov for comments and discussions.

BIBLIOGRAPHY

[1] M. Aiello and J. van Benthem. A modal walk through space. *J. of Applied Non-Classical Logics*, 12(3–4):319–364, 2002.

[2] C. Areces, P. Blackburn, and M. Marx. The computational complexity of hybrid temporal logics. *Logic J. of the IGPL*, 8:653–679, 2000.

[3] Ph. Balbiani, T. Tinchev, and D. Vakarelov. Modal logics for region-based theories of space. *Fundamenta Informaticae*, 81:29–82, 2007.

[4] M. Ben-Or, D. Kozen, and J. Reif. The complexity of elementary algebra and geometry. *J. of Computer and System Sciences*, 32:251–264, 1986.

[5] B. Bennett. Spatial reasoning with propositional logic. In *Proceedings of KR*, pages 51–62. Morgan Kaufmann, 1994.

[6] B. Bennett and I. Düntsch. Axioms, algebras and topology. In M. Aiello, I. Pratt-Hartmann, and J. van Benthem, editors, *Handbook of Spatial Logics*, pages 99–160. Springer, 2007.

[7] L. Biacino and G. Gerla. Connection structures. *Notre Dame J. of Formal Logic*, 32:242–247, 1991.

[8] N. Bourbaki. *General Topology, Part 1.* Hermann, Paris and Addison-Wesley, 1966.

[9] D. Cantone and V. Cutello. Decision algorithms for elementary topology I. Topological syllogistics with set and map constructs, connectedness and cardinailty composition. *Communications on Pure and Applied Mathematics*, XLVII:1197–1217, 1994.

[10] B. L. Clarke. A calculus of individuals based on 'connection'. *Notre Dame J. of Formal Logic*, 23:204–218, 1981.

[11] B. L. Clarke. Individuals and points. *Notre Dame J. of Formal Logic*, 26:61–75, 1985.

[12] A. G. Cohn and J. Renz. Qualitative spatial representation and reasoning. In F. van Hermelen, V. Lifschitz, and B. Porter, editors, *Handbook of Knowledge Representation*, pages 551–596. Elsevier, 2008.

[13] E. Davis. The expressivity of quantifying over regions. *J. of Logic and Computation*, 16:891–916, 2006.

[14] G. De Giacomo. *Decidability of Class-Based Knowledge Representation Formalisms*. PhD thesis, Università degli Studi di Roma 'La Sapienza', 1995.

[15] T. de Laguna. Point, line and surface as sets of solids. *The Journal of Philosophy*, 19:449–6, 1922.

[16] G. Dimov and D. Vakarelov. Contact algebras and region-based theory of space: A proximity approach, I. *Fundamenta Informaticae*, 74(2–3):209–249, 2006.

[17] G. Dimov and D. Vakarelov. Contact algebras and region-based theory of space: A proximity approach, II. *Fundamenta Informaticae*, 74(2–3):251–282, 2006.

[18] C. Dornheim. Undecidability of plane polygonal mereotopology. In A. Cohn, L. Schubert, and S. Shapiro, editors, *Proceedings of KR*, pages 342–353. Morgan Kaufmann, 1998.

[19] I. Düntsch and M. Winter. A representation theorem for Boolean contact algebras. *Theoretical Computer Science*, 347:498–512, 2005.

[20] M. Egenhofer and R. Franzosa. Point-set topological spatial relations. *International J. of Geographical Information Systems*, 5:161–174, 1991.

[21] M. Egenhofer and J. Herring. Categorizing topological relationships between regions, lines and point in geographic databases. Technical report, University of Maine, 1991.

[22] M. Fischer and M. Rabin. Super-exponential complexity of Presburger arithmetic. *Complexity of Computation, SIAM-AMS Proceedings*, 7:27–41, 1974.

[23] D. Gabelaia, R. Kontchakov, A. Kurucz, F. Wolter, and M. Zakharyaschev. Combining spatial and temporal logics: expressiveness vs. complexity. *J. of Artificial Intelligence Research (JAIR)*, 23:167–243, 2005.

[24] K. Gödel. Eine Interpretation des intuitionistischen Aussagenkalküls. *Ergebnisse eines mathematischen Kolloquiums*, 4:39–40, 1933.

[25] R. Goldblatt. Mathematical modal logic: A view of its evolution. In D. M. Gabbay and J. Woods, editors, *Handbook of the History of Logic, Volume 7: Logic and the Modalities in the Twentieth Century*, pages 1–98. Elsevier, 2006.

[26] V. Goranko and S. Passy. Using the universal modality: gains and questions. *J. of Logic and Computation*, 2:5–30, 1992.

[27] A. Griffiths. *Computational Properties Of Spatial Logics In The Real Plane*. PhD thesis, Depart. of Computer Science, University of Manchester, 2008. Forthcoming.

[28] A. Grzegorczyk. Undecidability of some topological theories. *Fundamenta Mathematicae*, 38:137–152, 1951.

[29] D. Harel, D. Kozen, and J. Tiuryn. *Dynamic Logic*. MIT Press, 2000.

[30] D. Hilbert. *Grundlagen der Geometrie*. B.G. Teubner, Leipzig and Berlin, 1909.

[31] R. Kontchakov, I. Pratt-Hartmann, F. Wolter, and M. Zakharyaschev. On the computational complexity of spatial logics with connectedness constraints. Submitted, 2008.

[32] S. Koppelberg. *Handbook of Boolean Algebras*, volume 1. North-Holland, 1989.

[33] J. Kratochvíl and J. Matoušek. String graphs requiring exponential representations. *J. of Combinatorial Theory, Series B*, 53:1–4, 1991.

[34] G. Kuper, L. Libkin, and J. Paredaens, editors. *Constraint Databases*. Springer, 2000.

[35] O. Kutz, H. Sturm, N.-Y. Suzuki, F. Wolter, and M. Zakharyaschev. Logics of metric spaces. *ACM Transactions on Computational Logic*, 4:260–294, 2003.

[36] R. Ladner. The computational complexity of provability in systems of modal logic. *SIAM J. on Computing*, 6:467–480, 1977.
[37] C. Lewis and C. Langford. *Symbolic Logic*. Appleton-Century-Crofts, New York, 1932.
[38] S. Li and M. Ying. Extensionality of the RCC8 composition table. *Fundamenta Informaticae*, 55(3–4):363–385, 2003.
[39] C. Lutz and F. Wolter. Modal logics of topological relations. *Logical Methods in Computer Science*, 2, 2006.
[40] J.C.C. McKinsey and A. Tarski. The algebra of topology. *Annals of Mathematics*, 45:141–191, 1944.
[41] M.H.A. Newman. *Elements of the Topology of Plane Sets of Points*. Cambridge, 1964.
[42] W. Nutt. On the translation of qualitative spatial reasoning problems into modal logics. In W. Burgard, T. Christaller, and A. Cremers, editors, *Proceedings of KI*, volume 1701 of *LNCS*, pages 113–124. Springer, 1999.
[43] I. Orlov. The calculus of compatibility of propositions. *Mathematics of the USSR, Sbornik*, 35:263–286, 1928. (In Russian).
[44] I. Pratt-Hartmann. A topological constraint language with component counting. *J. of Applied Non-Classical Logics*, 12:441–467, 2002.
[45] I. Pratt-Hartmann. First-order mereotopology. In M. Aiello, I. Pratt-Hartmann, and J. van Benthem, editors, *Handbook of Spatial Logics*, pages 13–97. Springer, 2007.
[46] D. Randell, Z. Cui, and A. Cohn. A spatial logic based on regions and connection. In B. Nebel, C. Rich, and W. Swartout, editors, *Proceedings of KR*, pages 165–176. Morgan Kaufmann, 1992.
[47] J. Renz. A canonical model of the region connection calculus. In A. Cohn, L. Schubert, and S. Shapiro, editors, *Proceedings of KR*, pages 330–341. Morgan Kaufmann, 1998.
[48] J. Renz and B. Nebel. On the complexity of qualitative spatial reasoning: A maximal tractable fragment of the region connection calculus. In *IJCAI*, pages 522–527, 1997.
[49] J. Renz and B. Nebel. Qualitative spatial reasoning using constraint calculi. In M. Aiello, I. Pratt-Hartmann, and J. van Benthem, editors, *Handbook of Spatial Logics*, pages 161–216. Springer, 2007.
[50] M. Reynolds. The complexity of the temporal logic over the reals. Manuscript; available at `http://www.csse.uwa.edu.au/ mark/research/Online/CORT.htm`, 2008.
[51] P. Roeper. Region-based topology. *J. of Philosophical Logic*, 26:251–309, 1997.
[52] M. Schaefer, E. Sedgwick, and D. Štefankovič. Recognizing string graphs in NP. *J. of Computer and System Sciences*, 67:365–380, 2003.
[53] V. Shehtman. "Everywhere" and "Here". *J. of Applied Non-Classical Logics*, 9, 1999.
[54] M. Sheremet, D. Tishkovsky, F. Wolter, and M. Zakharyaschev. From topology to metric: modal logic and quantification in metric spaces. In G. Governatori, I. Hodkinson, and Y. Venema, editors, *Advances in Modal Logic 6*, pages 429–448, 2006.
[55] T. Smith and K. Park. An algebraic approach to spatial reasoning. *International J. of Geographical Information Systems*, 6:177–192, 1992.
[56] J. Stell. Boolean connection algebras: A new approach to the Region Connection Calculus. *Artificial Intelligence*, 122:111–136, 2000.
[57] D. Sustretov. Topological semantics and decidability. arXiv:math/0703106v3 [math.LO], February 2008.
[58] A. Tarski. On the calculus of relations. *J. of Symbolic Logic*, 6(3):73–89, 1941.
[59] A. Tarski. What is Elementary Geometry? In *The Axiomatic Method, with Special Reference to Geometry and Physics*, pages 16–29. North-Holland, 1959.
[60] B. ten Cate, D. Gabelaia, and D. Sustretov. Modal languages for topology: expressivity and definability. arXiv:math/0610357v2 [math.LO], October 2006.
[61] D. Vakarelov. Region-based theory of space: algebras of regions, representation theory, and logics. In D. M. Gabbay, S. Goncharov, and M. Zakharyaschev, editors, *Mathematical Problems from Applied Logic II*, pages 267–348. Springer, 2007.
[62] J. van Benthem and G. Bezhanishvili. Modal logics of space. In M. Aiello, I. Pratt-Hartmann, and J. van Benthem, editors, *Handbook of Spatial Logics*, pages 217–298. Springer, 2007.

[63] L. van den Dries. *Tame Topology and O-Minimal Structures*, volume 248 of *London Mathematical Society Lecture Note Series*. Cambridge, 1998.

[64] P. van Emde Boas. The convenience of tilings. In A. Sorbi, editor, *Complexity, Logic and Recursion Theory*, volume 187 of *Lecture Notes in Pure and Applied Mathematics*, pages 331–363. Marcel Dekker Inc., 1997.

[65] A. N. Whitehead. *Process and Reality*. New York: The MacMillan Company, 1929.

[66] F. Wolter and M. Zakharyaschev. Spatial reasoning in RCC-8 with Boolean region terms. In W. Horn, editor, *Proceedings of ECAI*, pages 244–248. IOS Press, 2000.

[67] F. Wolter and M. Zakharyaschev. Qualitative spatio-temporal representation and reasoning: a computational perspective. In G. Lakemeyer and B. Nebel, editors, *Exploring AI in the New Millenium*, pages 175–216. Morgan Kaufmann, 2002.

[68] F. Wolter and M. Zakharyaschev. A logic for metric and topology. *J. of Symbolic Logic*, 70:795–828, 2005.

Roman Kontchakov
School of Computer Science and Information Systems
Birkbeck College
London WC1E 7HX
UK
roman@dcs.bbk.ac.uk

Ian Pratt-Hartmann
Department of Computer Science
Manchester University
Manchester M13 9PL
UK
ipratt@cs.man.ac.uk

Frank Wolter
Department of Computer Science
University of Liverpool
Liverpool L69 3BX
UK
frank@csc.liv.ac.uk

Michael Zakharyaschev
School of Computer Science and Information Systems
Birkbeck College
London WC1E 7HX
UK
michael@dcs.bbk.ac.uk

An interval logic for natural language semantics

Savas Konur

ABSTRACT. Most temporal logics, particularly interval temporal logics, are not expressive enough to capture meanings of natural language constructions, and they are not convenient to represent temporal expressions. In addition, these formal systems exhibit high computational complexity. In this paper we introduce a decidable *event-based interval logic*, called *EIL*. EIL can express the semantics of some natural language constructions.

Keywords: interval temporal logics, natural language semantics, temporal prepositions, decidability, complexity, tableau-methods

1 Introduction

In a sentence of natural language temporal information is stored in temporal constructions such as prepositions. In order to understand the semantics of a sentence in English or in any other language it is very important to capture temporal meanings. Sentences encoding temporal information usually speak of events and their temporal relations. A natural question that arises here is what is the computational complexity of determining logical relationships between sentences encoding temporal information?

This question is of theoretical interest, because events in sentences with temporal information are extended in time; and temporal logics which deal with extended events so-called interval temporal logics, typically exhibit high computational complexity. Generally speaking, these logics are not expressive enough to capture the meanings of natural language constructions, and therefore they are not convenient to represent temporal expressions.

The formal semantics of temporal constructions in English have been studied by various researchers [4, 7, 1, 6, 11, 10, 8]. In most of the cases the issues related to computational complexity and expressive power are rarely investigated. In fact, in many cases, the semantics of temporal constructions in a natural language are represented in a first-order language having variables which range over time-intervals and predicates which correspond to event-types and temporal order-relations. Such a logic can be easily shown to be undecidable.

The recent interest is using logical fragments of limited computational complexity, because there are evident practical and theoretical reasons for presenting the semantics of natural language constructions, using formal systems of limited expressive power. In particular, this is important to achieve the decidability of such systems.

In this paper, we introduce a decidable *event-based interval temporal logic*, called *EIL*. The logic EIL has a limited expressive power; yet it has affinity with the syntax of temporal constructions in English, and it is convenient for expressing the semantics of natural language constructions.

In the literature various methods have been proposed to achieve decidability for interval logics. However, most of the methods, such as translating interval logics into point-based ones, cause some syntactic and semantic restrictions. A major challenge in this area is thus to *genuinely* identify interval-based decidable logics, that is, logics which are not explicitly translated into point-based logics or other semantic restrictions.

Unlike many other interval logics, we do not translate the logic EIL into a point-based variant, and we therefore try to minimise semantic restrictions. Instead, we consider intervals as primitive objects of the model by allowing quantification over only interval objects. Another important feature of EIL is that it incorporates the notion of *duration*, which denotes the length of the time period at which an event occurs (that is, it starts and finishes).

In this paper we propose a tableau-based decision procedure for EIL, thus showing that its satisfiability problem is decidable. We, indeed, provide a complexity bound for satisfiability, showing that this problem can be solved in 2-NEXPTIME. The tableau method we introduce decides whether the given formula is satisfiable or not, and generates a model if the formula is satisfiable.

The plan of this paper is as follows. In Section 2 the syntax and semantics of EIL will be presented. In Section 3 we will give a depth bound for EIL models. In Section 4 a tableau system will be proposed for the logic. In Section 5 we will give some concluding remarks, and discuss some future work.

2 The logic EIL

In this section we present syntax and semantics of the logic EIL. In the rest of this paper we take an *interval* to be a closed, bounded and non-empty subset of the real line. More formally we say that an *interval* is a pair $[t_1, t_2]$ such that $t_1, t_2 \in \mathbb{R}$ and $t_1 \leq t_2$. We denote the set of all intervals $\{[t_1, t_2] : t_1 \leq t_2 \wedge t_1, t_2 \in \mathbb{R}\}$ by $\mathcal{I}$, and we use letters $I, J, ...,$ as intervals. It can be simply observed that intervals may be points. Note also the temporal domain is continuous. EIL formulas are evaluated relative to time-intervals. As will be seen later, having event-types in the syntax of the language will allow us to formalize event-based sentences of a natural language. EIL also incorporates the notion of duration (of an event).

Event types are denoted by the letters $e_1, e_2, e_3, ...$ We interpret an event e so that it is satisfied by all and only those time intervals over which e occurs. We will think of $\langle e \rangle$ as the occurrence of e over an interval J. Below we define some functions on $\mathcal{I}$.

DEFINITION 1. Let $J, I \in \mathcal{I}$ be the intervals $[a, b]$ and $[c, d]$, respectively, with $a \leq c \leq d \leq b$. The terms $init(J, I)$ and $fin(J, I)$ denotes the intervals $[a, c]$ and $[d, b]$, respectively, where $init$ and fin are partial functions.

In the sequel, let $\mathcal{E}$ be a finite set. We refer to elements of $\mathcal{E}$ as *event atoms.*

DEFINITION 2. Let $e \in \mathcal{E}$ be an event atom, ϕ, ψ be EIL formulas, $k \in \mathbb{R}$, and $\tau \in \{<, \leq, =, \geq, >\}$. The logic EIL is defined by induction as follows:

$$\phi ::= \top \mid \left\langle \textstyle\int e\tau k \right\rangle \phi \mid \left[\textstyle\int e\tau k \right] \phi \mid \left\langle \textstyle\int e\tau k \right\rangle_{<} \phi \mid \left[\textstyle\int e\tau k \right]_{<} \phi \mid$$
$$\left\langle \textstyle\int e\tau k \right\rangle_{>} \phi \mid \left[\textstyle\int e\tau k \right]_{>} \phi \mid \neg\phi \mid \phi \wedge \psi \mid \phi \vee \psi$$

The connectives $\rightarrow$ and $\leftrightarrow$ can be defined in usual way. For simplicity, we will denote $\left\langle \int e \geq 0 \right\rangle$ and $\left[\int e \geq 0 \right]$ as $\langle e \rangle$ and $[e]$, respectively.

Before giving the formal semantics of EIL formulas, we will define an EIL model.

DEFINITION 3. Let $\mathcal{I}$ be the set of all bounded, closed and non-empty intervals of real numbers, and $\mathcal{E}$ be a finite set of event atoms. An EIL **model** $\mathcal{M}$ is a finite subset of $\mathcal{I} \times \mathcal{E}$.

As can be seen from the construction an EIL model, intervals are *primitive* objects of the model.

DEFINITION 4. Let $\mathcal{M}$ be an EIL model, $|J|$ denote the length of the interval J, and $I \in \mathcal{I}$. The formal semantics of EIL formulas is then defined as follows:

- $\mathcal{M} \models_I \left\langle \int e\tau k \right\rangle \phi$ iff $\exists J \subseteq I$ such that $\langle J, e \rangle \in \mathcal{M}$ and $|J|\,\tau k$ and $\mathcal{M} \models_J \phi$;
- $\mathcal{M} \models_I \left[\int e\tau k \right] \phi$ iff $\forall J \subseteq I$ $\langle J, e \rangle \in \mathcal{M}$ and $|J|\,\tau k$ imply $\mathcal{M} \models_J \phi$;
- $\mathcal{M} \models_I \left\langle \int e\tau k \right\rangle_{<} \phi$ iff $\exists J \subseteq I$ such that $\langle J, e \rangle \in \mathcal{M}$ and $|J|\,\tau k$ and $\mathcal{M} \models_{fin(J,I)} \phi$;
- $\mathcal{M} \models_I \left[\int e\tau k \right]_{<} \phi$ iff $\forall J \subseteq I$ $\langle J, e \rangle \in \mathcal{M}$ and $|J|\,\tau k$ imply $\mathcal{M} \models_{fin(J,I)} \phi$;
- $\mathcal{M} \models_I \left\langle \int e\tau k \right\rangle_{>} \phi$ iff $\exists J \subseteq I$ such that $\langle J, e \rangle \in \mathcal{M}$ and $|J|\,\tau k$ and $\mathcal{M} \models_{init(J,I)} \phi$;
- $\mathcal{M} \models_I \left[\int e\tau k \right]_{>} \phi$ iff $\forall J \subseteq I$ $\langle J, e \rangle \in \mathcal{M}$ and $|J|\,\tau k$ imply $\mathcal{M} \models_{init(J,I)} \phi$;
- $\mathcal{M} \models_I \neg\phi$ iff not $\mathcal{M} \models_I \phi$;
- $\mathcal{M} \models_I \phi \wedge \psi$ iff $\mathcal{M} \models_I \phi$ and $\mathcal{M} \models_I \psi$;
- $\mathcal{M} \models_I \phi \vee \psi$ iff $\mathcal{M} \models_I \phi$ or $\mathcal{M} \models_I \psi$.

One important characteristic of EIL formulas is the 'quasi-guarded' nature of the quantification they feature. Thus, for example, the formula $\langle e \rangle \phi$ existentially quantifies over intervals satisfying the event e (similarly for universal formulas). So it does not quantify over all subintervals of the current interval of evaluation without restriction. However, many modal logics, such as HS [5] and CDT [12], lack the 'quasi-guarded' character of the quantification that EIL formulas feature. This feature is very important to guarantee the decidability.

Before ending this section we will show how EIL represents English sentences including the temporal constructions. Consider the following sentences (in a fragment of English):

(2.1) `A warning is received during every control period until the water level becomes normal.`

(2.2) `After a drop in the water level, a warning is received during every control period until the water level becomes normal.`

The meaning of (2.1) is that, within the given temporal context I, there is an interval J over which the water level is normal; over every interval J', which is subsumed by the initial segment of I up to the beginning of I, a control period occurs; and J' subsumes some interval over which a warning is received. The sentence (2.1) is translated into EIL as follows:

(2.3) $$\langle normal \rangle_{<} [control] \langle warning \rangle \top.$$

The sentence (2.2) can be represented by the following EIL formula:

(2.4) $$\langle drop \rangle_{>} \langle normal \rangle_{<} [control] \langle warning \rangle \top.$$

Let's look at how EIL represents the event-based English sentences including duration. Consider the following sentence:

(2.5) `John solved a problem in less than ten minutes during every lunch break.`

This sentence can be translated into EIL as follows:

(2.6) $$[break] \langle \textstyle\int solve \leq 10 \rangle$$

3 Finding a depth limit for EIL models

In this section we show that the depth of an EIL model is exponentially bounded by the length of a given formula φ whose satisfiability is checked.

We remark that the condition in Definition 3 that models are finite subsets of $\mathcal{I} \times \mathcal{E}$ is significant. Because there might be some EIL formulas which cannot be satisfied in a finite model. Consider, for example, the $\langle e \rangle \top \wedge [e] \langle e \rangle \top$. This formula is not satisfiable in a finite model; because it implies that every occurrence of e over an interval J requires another e to occur over a subinterval of J. Therefore, the formula is unsatisfiable in a finite model.

Below we will show how to normalize an EIL formula to the desired form.

LEMMA 5. *Every EIL formula is logically equivalent to one in which* $\neg$ *appears only in subformula of the form* $\bot$ $(= \neg\top)$.

Proof. The proof is trivial for $\bot$. In an EIL formula $\neg$ can be moved inwards as follows:

$$\neg\left\langle \int e\tau k\right\rangle \phi \equiv \left[\int e\tau k\right] \neg\phi;\ \neg\left[\int e\tau k\right] \phi \equiv \left\langle \int e\tau k\right\rangle \neg\phi;$$
$$\neg\left\langle \int e\tau k\right\rangle_{<} \phi \equiv \left[\int e\tau k\right]_{<} \neg\phi;\ \neg\left[\int e\tau k\right]_{<} \phi \equiv \left\langle \int e\tau k\right\rangle_{<} \neg\phi;$$
$$\neg\left\langle \int e\tau k\right\rangle_{>} \phi \equiv \left[\int e\tau k\right]_{>} \neg\phi;\ \neg\left[\int e\tau k\right]_{>} \phi \equiv \left\langle \int e\tau k\right\rangle_{>} \neg\phi.$$ ■

By means of Lemma we can normalize the forms of EIL formulas.

DEFINITION 6. Given an EIL formula φ and a non-empty model $\mathcal{M}$, the **depth** of $\mathcal{M}$ is the greatest m for which there exist $J_1 \subseteq ... \subseteq J_m$ such that for all i, $1 \leq i \leq m$ and for some $e \in \mathcal{E}$, $\langle J_i, e\rangle \in \mathcal{M}$. The depth of an empty model is defined to be 0.

LEMMA 7. *Let* φ *be an EIL formula.* φ *can be satisfied in a model* $\mathcal{M}$ *which is exponentially bounded by the length of* φ.

Proof. Assume φ has the form guaranteed by Lemma 5. Let m be the number of existential subformulas of φ ($\left\langle \int e\tau k\right\rangle \phi$, $\left\langle \int e\tau k\right\rangle_{<} \phi$ and $\left\langle \int e\tau k\right\rangle_{>} \phi$), and n be the number of universal subformulas of φ ($\left[\int e\tau k\right] \psi$, $\left[\int e\tau k\right]_{<} \psi$ and $\left[\int e\tau k\right]_{>} \psi$). An existential subformula $\left\langle \int e\tau k\right\rangle \phi$, similarly $\left\langle \int e\tau k\right\rangle_{<} \phi$ and $\left\langle \int e\tau k\right\rangle_{>} \phi$, implies that $\mathcal{M}$ contains an entry $\langle J, e\rangle$ for some interval J. Since φ is satisfiable, by semantics ϕ must be true at J. From this we can conclude that J subsumes a chain of intervals which satisfy event atoms occuring in ϕ. The length of such a chain is thus bounded by $|\varphi|$. So, every existential subformula of φ implies a chain of intervals, whose length is bounded by $|\varphi|$. In the worst case, these chains are aligned one under the other, and construct a longer chain, which is bounded by $m\,|\varphi|$.

Moreover, a universal subformula of the type $\left[\int e\tau k\right] \psi$ implies that ψ is true at each interval J satisfying e. We can therefore conclude that each J subsumes a chain of intervals which satisfy event atoms occuring in ψ ($\left[\int e\tau k\right]_{<} \psi$ and $\left[\int e\tau k\right]_{>} \psi$ can be considered similarly). In the worst case, the bound on the length of the whole chain increases to $m\,|\varphi|^2$. If we repeat the same step for the remaining $n-1$ universal subformulas we will see that the bound becomes k^n, where $k = m\,|\varphi|$. Since $m < |\varphi|$ and $n < |\varphi|$, we can conclude that $k^n < |\varphi|^{2|\varphi|}$. It easily follows that the depth bound of the model $\mathcal{M}$ is $2^{p(|\varphi|)}$, where p is a fixed polynomial. ■

We remark that the exponential depth bound that we have found is not optimal. Here we have based our calculations into the worst case to find an upper limit, even if this case may be never encountered.

In the next section we will show that the size of this model is doubly exponential by the length of φ. We will actually derive the model from the tableau generated by a tableau procedure.

4 A tableau system procedure for EIL

In this section we propose a terminating tableau system for the logic EIL, thus showing that its satisfiability problem is decidable. Indeed, the satisfiability problem for EIL is in 2-NEXPTIME. This is proved by building models whose sizes are exponentially bounded.

In the following, we define a tableau-based decision procedure for EIL, and analyze its computational complexity. Then, we prove its soundness and completeness. The procedure is based on an *expansion strategy.* The expansion strategy involves three rules: the *interval relation rule,* which nondeterministically guesses the interval relation among nodes in the graph, the *existential node expansion rule*, which expands existential subformulas in a node and the *universal node expansion rule*, which expands universal subformulas in a node. A *blocking* condition guarantees the termination of the method.

4.1 Preliminary notions

In the following we introduce some preliminary notions which will be used throughout the rest of the paper.

DEFINITION 8.

A **successor** of a node v is a node w such that there is an edge from v to w. A **path** is a sequence of nodes $v_1, ..., v_k$ such that for all $1 \leq i < k$, v_{i+1} is a successor of v_i. The **depth** of a node v is the maximum number of edges of a path from the root node to v.

DEFINITION 9.

A **decorated graph** $\mathcal{G}$ is a graph in which every node has a decoration. For a node $v \in \mathcal{G}$, a **decoration** $\lambda(v)$ is a 5-tuple $([b_v, e_v], \rho(v), \mathcal{K}(v), \mathcal{L}(v), \mathcal{L}'(v))$, where b_v (e_v) is a *constraint variable* denoting the beginning (ending) of the interval represented by the node v, $\rho(v)$ denotes the label of the node v (where $\rho(v) \in \mathcal{E}$), $\mathcal{K}(v)$ denotes a formula associated with the node v, and $\mathcal{L}(v)$ and $\mathcal{L}'(v)$ denote a set of subformulas associated with the node v.

DEFINITION 10.

A **temporal constraint** is a relation involving constraint variables which denote interval endpoints.

For example, the temporal constraint $b_v \geq b_u, e_v \leq e_u$ shows an interval relation between $[b_v, e_v]$ and $[b_u, e_u]$.

DEFINITION 11.

A **tableau** for a given formula φ is a tuple $\langle \mathcal{G}, \mathcal{C} \rangle$, where $\mathcal{G}$ denotes a decorated graph, and $\mathcal{C}$ denotes the set of temporal constraints in the graph $\mathcal{G}$.

4.2 Tableau method

Let φ be a formula to be checked for satisfiability over an interval I_0. The *initial tableau* for φ is the tuple $\langle v_0, \mathcal{C}_0 \rangle$, where v_0 is the initial graph with the decoration $\lambda(v_0) = ([b_{v_0}, e_{v_0}], \rho(v_0), \mathcal{K}(v_0), \mathcal{L}(v_0), \mathcal{L}'(v_0))$ such that $\rho(v_0) = root$, $\mathcal{K}(v_0) = \varphi$, $\mathcal{L}(v_0) = \emptyset$, $\mathcal{L}'(v_0) = \emptyset$, and $\mathcal{C}_0$ is the initial set of temporal constraints such that $\mathcal{C}_0 = \{b_{v_0} = start(I_0), e_{v_0} = end(I_0)\}$. Assume Q denotes the queue of nodes in $\mathcal{G}$ awaiting processing. Then, the inital value of Q is $\{v_0\}$.

A tableau for φ is a tuple $\langle \mathcal{G}, \mathcal{C} \rangle$, where $\mathcal{C}$ is obtained by expanding the initial constraint set $\mathcal{C}_0$ with temporal constraints in the existing nodes, and the decorated graph $\mathcal{G}$ is obtained by expanding the initial node v_0 through successive applications of the *expansion strategy* to existing nodes until no node remains to process. In other words, the expansion strategy is applied to every node in Q until $Q = \emptyset$. When a node is selected, it is removed from Q.

During the application of the expansion strategy to a node, we need to solve the temporal constraints in $\mathcal{C}$. Remember that each node of the graph represents an interval. For our purposes, we model intervals as pairs of endpoints, which are distinct numbers on the real line. Let $T = \{b_{v_1}, ..., b_{v_n}, e_{v_1}, ..., e_{v_n}\}$ be a set of constraint variables. The constraints of a tableau can be represented as a Simple Temporal Problem [3]. Given that n is the number of variables the complexity of a solution to a STP (if there is any) can be found in $\mathcal{O}(n^3)$ time and $\mathcal{O}(n^2)$ space. If the set of temporal constraints in $\mathcal{C}$ is inconsistent, then a solution will not be found, and we say $\mathcal{C}$ is not satisfiable.

In order to avoid infinite paths, and therefore to have a finite satisfying model we need to guarantee the termination of the proposed tableau method below. In the following we give a suitable *stopping condition* for the tableau procedure:

DEFINITION 12. A tableau $\langle \mathcal{G}, \mathcal{C} \rangle$ is **closed** if one of the following conditions hold:

- $\perp \in \mathcal{L}(v)$ for some node v in $\mathcal{G}$,
- $\mathcal{C}$ is not satisfiable,
- The depth of the shortest path $v_0 \rightarrow ... \rightarrow v$ is more than $|\varphi|^2$ for some node v in $\mathcal{G}$ (where v_0 is the root node.)

DEFINITION 13. A tableau is **open** if it is not closed.

Once the tableau procedure terminates, we check whether the tableau generated is open. For a given formula φ if there is an open tableau, then φ is satisfiable, and the satisfying model $\mathcal{M}$ is derived from the tableau. We do this by picking some solution σ, which assigns real values to constraint variables in $\mathcal{C}$. Let $J_v = [\sigma(b_v), \sigma(e_v)]$ be the interval represented by a node v of $\mathcal{G}$. We construct a model $\mathcal{M}$ as follows: $\mathcal{M} = \{\langle J_v, \rho(v) \rangle \mid \text{for any } v \in \mathcal{G}$ s.t. $\rho(v) \notin \{root, -\}\}$. If the tableau is closed, then φ is unsatisfiable.

Expansion strategy.

Let $\langle \mathcal{G}, \mathcal{C} \rangle$ be a tableau, v be a node in $\mathcal{G}$ with $\lambda(v) = ([b_v, e_v], \rho(v), \mathcal{K}(v), \mathcal{L}(v), \mathcal{L}'(v))$, and Q be the queue of nodes awaiting processing. We say the **expansion strategy** for a node v is defined as follows:

If the tableau is open, apply the following rules:

Rule 1. Set $Q := Q \setminus \{v\}$, and apply the **interval relation rule** to the node v.

Rule 2. Let the Disjunctive Normal Form (DNF) of $\mathcal{K}(v)$ be $\psi_1 \vee ... \vee \psi_n$ where $\psi_i = \psi_{i1} \wedge ... \wedge \psi_{in_i}$ ($n \geq 1, 1 \leq i \leq n$ and $n_i \geq 1$). Select some i, and set $\mathcal{L}'(v) := \{\psi_{i1}, ..., \psi_{in_i}\}$, $\mathcal{L}(v) := \mathcal{L}(v) \cup \mathcal{L}'(v)$ and $\mathcal{K}(v) := \top$.

Rule 3. Apply the **universal node expansion rule** to the node v.

Rule 4. Apply the **existential node expansion rule** to the node v.

Interval relation rule.

The *interval relation rule* guesses the interval relation between the given node and all other nodes in the graph. In [1] Allen introduced well-known thirteen different binary relations between intervals on a linear ordering, which are *before, meets, overlaps, starts, during, finishes, equals, finished by, during by, started by, overlapped by, met by* and *after*.

Let $\langle \mathcal{G}, \mathcal{C} \rangle$ be a tableau, and v be a node in $\mathcal{G}$ with $\lambda(v) = ([b_v, e_v], \rho(v), \mathcal{K}(v), \mathcal{L}(v), \mathcal{L}'(v))$. Assume τ' is the corresponding inverted operator of τ (where $\tau \in \{<, \leq, =, \geq, >\}$). The *interval relation rule* for a node v is defined as follows:

For any node u (except v) in $\mathcal{G}$

If there is no edge from u to v, or from v to u, then nondeterministically guess the interval relation between u and v:

v *before* u : Set $\mathcal{C} := \mathcal{C} \cup \{e_v < b_u\}$.

v *meets* u : Set $\mathcal{C} := \mathcal{C} \cup \{e_v = b_u\}$.

v *non-strict-during* u : Set $\mathcal{C} := \mathcal{C} \cup \{b_v \geq b_u, e_v \leq e_u\}$, and add an edge from u to v ($u \rightarrow v$).

- **if** $\rho(v) = e$ and $\left[\int e\tau k\right] \psi \in \mathcal{L}(u)$, **then** set either *i)* $\mathcal{C} := \mathcal{C} \cup \{(e_v - b_v)\tau' k\}$; or *ii)*$\mathcal{C} := \mathcal{C} \cup \{(e_v - b_v)\tau k\}$ and $\mathcal{K}(v) := \mathcal{K}(v) \wedge \psi$.

- **if** $\rho(v) = e$ and $\left[\int e\tau k\right]_< \psi \in \mathcal{L}(u)$, **then** for every $\left[\int e\tau k\right]_< \psi \in \mathcal{L}(u)$ do either *i)* set $\mathcal{C} := \mathcal{C} \cup \{(e_v - b_v)\tau' k\}$; or *ii)* set $\mathcal{C} := \mathcal{C} \cup \{(e_v - b_v)\tau k\}$, add an immediate successor w with $\rho(w) = -$, $\mathcal{K}(w) = \psi$, $\mathcal{L}(w) = \emptyset$, $\mathcal{L}'(w) = \emptyset$, set $\mathcal{C} := \mathcal{C} \cup \{b_w = e_v, e_w = e_u\}$, add an edge from u to w and v to w ($u \rightarrow w, v \rightarrow w$), and set $Q := Q \cup \{w\}$.

- **if** $\rho(v) = e$ and $\left[\int e\tau k\right]_> \psi \in \mathcal{L}(u)$, **then** follow the step above; except that rather than setting $\mathcal{C} := \mathcal{C} \cup \{b_w = e_v, e_w = e_u\}$, set $\mathcal{C} := \mathcal{C} \cup \{b_w = b_v, e_w = b_u\}$.

v *overlaps* u : Set $\mathcal{C} := \mathcal{C} \cup \{b_v < b_u < e_v < e_u\}$, and add an edge from u to v ($u \rightarrow v$).

Given two intervals J_1 and J_2, we say J_1 *non-strict-during* J_2 if J_1 is a non-strict subinterval of J_2. Once we guess the interval relation as "non-strict-during", we do not need to consider the relations "equals", "during", "starts", "started-by", "finishes" and "finished-by". The cases where v "after" u, v "met-by" u, v "includes" u and v "overlapped-by" u can be dealt with similarly. Note that in the interval relation rule, we consider the possibility that $\mathcal{L}(u)$ of an existing node u includes a universal subformula which might update the decoration of the node v.

Please note that when we denote an interval relation between two nodes, such as v "during" u, we mean this interval relation holds between the intervals represented by these nodes. For simplicty, we will use this adaption.

Universal node expansion rule.

The *universal node expansion rule* expands all universal subformulas in $\mathcal{L}'(v)$. Let $\langle \mathcal{G}, \mathcal{C} \rangle$ be a tableau, and v be a node in $\mathcal{G}$ with $\lambda(v) = ([b_v, e_v], \rho(v), \mathcal{K}(v), \mathcal{L}(v), \mathcal{L}'(v))$. Assume τ' is the corresponding inverted operator of τ (where $\tau \in \{<, \leq, =, \geq, >\}$). The *universal node expansion rule* for a node v is defined as follows:

For every $\xi \in \mathcal{L}'(v)$

if $\xi = \left[\int e\tau k\right] \psi$, **then for every** node u (except v) in $\mathcal{G}$ with $\rho(u) = e$ and u *non-strict-during* v, set either *i)* $\mathcal{C} := \mathcal{C} \cup \{(e_u - b_u)\tau' k\}$; or *ii)* $\mathcal{C} := \mathcal{C} \cup \{(e_u - b_u)\tau k\}$, $\mathcal{K}(u) := \mathcal{K}(u) \wedge \psi$ and $Q := Q \cup \{u\}$.

if $\xi = \left[\int e\tau k\right]_< \psi$, **then for every** node u (except v) in $\mathcal{G}$ with $\rho(u) = e$ and u *non-strict-during* v, do either *i)* set $\mathcal{C} := \mathcal{C} \cup \{(e_u - b_u)\tau' k\}$; or *ii)* set $\mathcal{C} := \mathcal{C} \cup \{(e_u - b_u)\tau k\}$, add an immediate successor w with $\rho(w) = -$, $\mathcal{K}(w) = \psi$, $\mathcal{L}(w) = \emptyset$, $\mathcal{L}'(w) = \emptyset$, set $\mathcal{C} := \mathcal{C} \cup \{b_w = e_u, e_w = e_v\}$ and set $Q := Q \cup \{w\}$.

where u "non-strict-during" v is true if $b_w \geq b_v, e_w \leq e_v \in \mathcal{C}$. The case where $\xi = \left[\int e\tau k\right]_> \psi$ can be dealt with similarly. As a result of applying the universal node expansion rule, some of the existing nodes might be revisited, which means we re-execute the expansion strategy for these nodes. In this case, interval relations will not be guessed again; but their decoration might get updated.

Existential node expansion rule.

The *existential node expansion rule* expands all existential subformulas in $\mathcal{L}'(v)$. Let $\langle \mathcal{G}, \mathcal{C} \rangle$ be a tableau, and v be a node in $\mathcal{G}$ with $\lambda(v) = ([b_v, e_v], \rho(v), \mathcal{K}(v), \mathcal{L}(v), \mathcal{L}'(v))$. Assume τ' is the corresponding inverted operator of τ (where $\tau \in \{<, \leq, =, \geq, >\}$). The *existential node expansion rule* for a node v is defined as follows:

For every $\xi \in \mathcal{L}'(v)$

if $\xi = \left\langle \int e\tau k \right\rangle \psi$, **then** add an immediate successor w with $\rho(w) = e$, $\mathcal{K}(w) = \psi$, $\mathcal{L}(w) = \emptyset$, $\mathcal{L}'(w) = \emptyset$, set $\mathcal{C} := \mathcal{C} \cup \{b_w \geq b_v, e_w \leq e_v, (e_w - b_w)\tau k\}$, and set $Q := Q \cup \{w\}$.

if $\xi = \left\langle \int e\tau k \right\rangle_< \psi$, **then** add two immediate successors w, w' with $\rho(w) = e$, $\mathcal{K}(w) = \emptyset$, $\mathcal{L}(w) = \emptyset$, $\mathcal{L}'(w) = \emptyset$ and $\rho(w') = -$, $\mathcal{K}(w') = \psi$, $\mathcal{L}(w') = \emptyset$,

$\mathcal{L}'(w') = \emptyset$, set $\mathcal{C} := \mathcal{C} \cup \{b_w \geq b_v, e_w \leq e_v, b_{w'} = e_w, e_{w'} = e_v\}$, and set $Q := Q \cup \{w, w'\}$.

The case where $\xi = \langle \int e\tau k \rangle_> \psi$ can be dealt with similarly. The existential node expansion rule creates a new node (or nodes). In the next run, we apply the expansion strategy to this node, and the decoration of this node gets updated according to Rule 2.

4.3 Soundness and completeness

The soundness and completeness of the proposed tableau method is proved below. But we first prove the termination of the method.

THEOREM 14. *The tableau method for EIL terminates.*

Proof. Let $\langle \mathcal{G}, \mathcal{C} \rangle$ be a tableau constructed by the tableau procedure for a given a formula φ. By the stopping condition in the tableau procedure every node of $\mathcal{G}$ has a finite outgoing degree and every branch of it is of finite length. Therefore, the tableau method terminates. ■

THEOREM 15. *Let φ be an EIL formula which has the form guaranteed by Lemma 5. φ is satisfiable iff there is an open tableau for φ.*

Proof. Soundness $(\Leftarrow)$ **:**

Suppose $\langle \mathcal{G}, \mathcal{C} \rangle$ is an open tableau for φ. We pick some solution $\sigma : \mathcal{V} \to \mathbb{R}$, which assigns real values to constraint variables in $\mathcal{C}$. Let $J_v = [\sigma(b_v), \sigma(e_v)]$ be the interval represented by the node v of $\mathcal{G}$. We construct a model $\mathcal{M}$ as follows: $\mathcal{M} = \{\langle J_v, \rho(v) \rangle \mid \text{for any } v \in \mathcal{G} \text{ s.t. } \rho(v) \notin \{root, -\}\}$.

Now we show that $\mathcal{M} \models_{I_0} \varphi$ (where I_0 is the initial interval). We claim that for every v in $\mathcal{G}$, $\mathcal{M} \models_{J_v} \mathcal{L}(v)$. We show, by structural induction, that $\phi \in \mathcal{L}(v)$ implies $\mathcal{M} \models_{J_v} \phi$. Note that, by construction of the tableau, $\mathcal{L}(v)$ comprises the formulas are of the forms $\top$, $\bot$, $\langle \int e\tau k \rangle \psi$, $\langle \int e\tau k \rangle_< \psi$, $\langle \int e\tau k \rangle_> \psi$, $[\int e\tau k]\, \psi$, $[\int e\tau k]_< \psi$ and $[\int e\tau k]_> \psi$.

Base Case:

$\phi = \top$: Trivial

$\phi = \bot$: Since $\langle \mathcal{G}, \mathcal{C} \rangle$ is an open tableau, by definition 12 and 13, $\bot \notin \mathcal{L}(v)$.

Inductive Case:

$\phi = \langle \int e\tau k \rangle \psi$: By the existential node expansion rule, there exists a node w with $\rho(w) = e$ and $\mathcal{K}(w) = \psi$. In addition, $\mathcal{C}$ contains $b_w \geq b_v, e_w \leq e_v$ and $(e_w - b_w)\tau k$. Let ψ be $\psi_1 \vee ... \vee \psi_n$ where $\psi_i = \psi_{i1} \wedge ... \wedge \psi_{in_i}$ ($n \geq 1, 1 \leq i \leq n$ and $n_i \geq 1$). By Rule 2, $\psi_{i1}, .., \psi_{in_i} \in \mathcal{L}(w)$ for some i ($1 \leq i \leq n$). By the inductive hypothesis, $\mathcal{M} \models_{J_w} \psi_{i1} \wedge ... \wedge \mathcal{M} \models_{J_w} \psi_{in_i}$. Therefore, $\mathcal{M} \models_{J_w} \psi$. By construction, we have $\langle J_w, e \rangle \in \mathcal{M}$ with $|J_w|\, \tau k$ and $J_w \subseteq J_v$. Thus, $\mathcal{M} \models_{J_v} \phi$.

$\phi = \langle \int e\tau k \rangle_< \psi$ and $\phi = \langle \int e\tau k \rangle_> \psi$: Similar to the case $\phi = \langle \int e\tau k \rangle \psi$.

$\phi = [\int e\tau k]\, \psi$: By the construction of $\mathcal{M}$, for any $J \in \mathcal{I}$ if $\langle J, e \rangle \in \mathcal{M}$, then there exists a node u in $\mathcal{G}$ such that $J_u = J$. According to the universal node expansion rule (or the interval relation rule) if $J_u \subseteq J_v$, then we do either: *i)* set $\mathcal{C} := \mathcal{C} \cup \{(e_u - b_u)\tau' k\}$ (τ' is the corresponding inverted operator of τ); or *ii)* set $\mathcal{C} := \mathcal{C} \cup \{(e_u - b_u)\tau k\}$ and $\mathcal{K}(u) := \mathcal{K}(u) \wedge \psi$.

Assume $|J_u|\tau k$ is false. Whatever the choice is, it is trivial to see that $\langle J_u, e\rangle \in \mathcal{M}$, $J_u \subseteq J_v$ and $|J_u|\tau k$ imply $\mathcal{M} \models_{J_u} \psi$. Assume $|J_u|\tau k$ is true. In this case, option i mentioned above cannot have been selected. Otherwise, $\mathcal{C}$ would contain $\{(e_u - b_u)\tau' k\}$, and it would result in an inconsistency. So option ii has been taken. In this case, we set $\mathcal{C} := \mathcal{C} \cup \{(e_u - b_u)\tau k\}$ and $\mathcal{K}(u) := \mathcal{K}(u) \wedge \psi$. Let ψ be $\psi_1 \vee ... \vee \psi_n$ where $\psi_i = \psi_{i1} \wedge ... \wedge \psi_{in_i}$ ($n \geq 1, 1 \leq i \leq n$ and $n_i \geq 1$). By Rule 2, $\psi_{i1}, .., \psi_{in_i} \in \mathcal{L}(u)$ for some i ($1 \leq i \leq n$). By the inductive hypothesis, $\mathcal{M} \models_{J_u} \psi$. By construction, we have $\langle J_u, e\rangle \in \mathcal{M}$. We also know that $J_u \subseteq J_v$ and $|J_u|\tau k$. Therefore, for any witness J_u, $\langle J_u, e\rangle \in \mathcal{M}$, $J_u \subseteq J_v$ and $|J_u|\tau k$ imply $\mathcal{M} \models_{J_u} \psi$. Thus, $\mathcal{M} \models_{J_u} \phi$.

$\phi = \left[\int e\tau k\right]_< \psi$ and $\phi = \left[\int e\tau k\right]_> \psi$: Similar to the case $\phi = \left[\int e\tau k\right] \psi$.

We have proved that for every v in $\mathcal{G}$, $\mathcal{M} \models_{J_v} \mathcal{L}(v)$. In particular, $\mathcal{M} \models_{I_0} \mathcal{L}(v_0)$. We know that $\mathcal{K}(v_0) = \varphi$. Now assume $\varphi = \varphi_1 \vee ... \vee \varphi_n$, where $\varphi_i = \varphi_{i1} \wedge ... \wedge \varphi_{in_i}$ ($n \geq 1$, $1 \leq i \leq n$ and $n_i \geq 1$). According to Rule 2, $\mathcal{L}(v_0) = \{\varphi_{i1}, ..., \varphi_{in_i}\}$ for some value of i. Therefore, we can easily conclude that $\mathcal{M} \models_{I_0} \varphi$.

Completeness ($\Rightarrow$) **:**

Let φ be a satisfiable formula, and I_0 be an interval. By Lemma 7 φ can be satisfied by a model $\mathcal{M}$, which has a depth bound of $2^{p(|\varphi|)}$ for a fixed polynomial p, such that $\mathcal{M} \models_{I_0} \varphi$. We will show that there is an open tableau $\langle \mathcal{G}, \mathcal{C}\rangle$ for φ.

The *initial tableau* for φ is the tuple $\langle v_0, \mathcal{C}_0\rangle$, where v_0 is the initial graph such that $\mathcal{K}(v_0) = \varphi$ and $\mathcal{L}(v_0) = \emptyset$, and $\mathcal{C}_0$ is the initial set of temporal constraints such that $\mathcal{C}_0 = \{b_{v_0} = start(I_0), e_{v_0} = end(I_0)\}$. According to the expansion strategy we apply the interval relation rule to the node v_0 as $\mathcal{L}(v_0)$ is empty. But since there is only one node, $\mathcal{K}(v_0)$ does not get updated. Let the disjunctive normal form of $\mathcal{K}(v_0) = \varphi$ be $\varphi_1 \vee ... \vee \varphi_n$, where $\varphi_i = \varphi_{i1} \wedge ... \wedge \varphi_{in_i}$ ($n \geq 1$, $1 \leq i \leq n$ and $n_i \geq 1$). Since $\mathcal{M} \models_{I_0} \varphi$, $\mathcal{M} \models_{I_0} \varphi_i$ for at least one value of i. So in Rule 2 we pick this value of i, so that $\mathcal{L}(v_0) = \{\varphi_{i1}, ..., \varphi_{in_i}\}$.

Now, we claim that for each node v in $\mathcal{G}$, there exists an interval J_v such that $\mathcal{M} \models_{J_v} \mathcal{L}(v)$ (Once we pick a witness J_v, it remains assigned to the node v until the tableau procedure terminates.) We prove the claim by induction on the stage in tableau construction at which the node v was created.

Base case:

Above we have shown that $\mathcal{M} \models_{I_0} \varphi_i$ for some value of i, and $\mathcal{L}(v_0) = \{\varphi_{i1}, ..., \varphi_{in_i}\}$. So, it is trivial to see $\mathcal{M} \models_{I_0} \mathcal{L}(v_0)$.

Inductive case:

Case 1: Let w be a node in $\mathcal{G}$ such that $\rho(w) = e$. Then w must have been created by the existential node expansion rule applied to a node v of which w is a successor node. After the node w has been created, we apply the expansion strategy to the node w. So we first apply the interval relation rule. Let us consider two cases:

i) Application of the interval relation rule adds no material to $\mathcal{L}(w)$: Assume $\mathcal{L}(w) = \{\psi_0\}$ where $\psi_0 = \psi_{01} \wedge ... \wedge \psi_{0n_0}$ $(n_0 \geq 1)$. In this case, $\mathcal{L}(v)$ must contain $\xi = \langle \int e\tau k \rangle \, \psi$ where ψ has the form $\psi_0 \vee ... \vee \psi_l$ $(l \geq 0)$ (If $\mathcal{L}(v)$ contained $\langle \int e\tau k \rangle_< \psi$ or $\langle \int e\tau k \rangle_> \psi$, then the existential rule would set $\rho(w) = -$. But we already know that $\rho(w) = -$. So, $\mathcal{L}(v)$ can contain neither $\langle \int e\tau k \rangle_< \psi$ nor $\langle \int e\tau k \rangle_> \psi$). By the inductive hypothesis a witness J_v is defined such that $\mathcal{M} \models_{J_v} \mathcal{L}(v)$. Let J_w be an interval for the node w. Thus, $\mathcal{M} \models_{J_w} \psi$.

When the existential rule was applied to v, we set $\mathcal{K}(w) := \psi$ and $\mathcal{C} := \mathcal{C} \cup \{b_w \geq b_v, e_w \leq e_v, (e_w - b_w)\tau k\}$. According to Rule 2 we select some of the disjunct of ψ, and extend $\mathcal{L}(w)$ with this disjunct. It is clear that ψ_0 is the subformula which was selected. So, $\mathcal{M} \models_{J_w} \psi_0$. Hence, $\mathcal{M} \models_{J_w} \mathcal{L}(w)$.

ii) Application of the interval relation rule adds some material to $\mathcal{L}(w)$: Assume $\mathcal{L}(w) = \{\psi_0, \psi_1, ..., \psi_m\}$ where $\psi_i = \psi_{i1} \wedge ... \wedge \psi_{im_i}$ $(0 \leq i \leq m$ and $m_i \geq 1)$, ψ_0 has been added to $\mathcal{L}(w)$ by applying the existential rule in v, and $\psi_1, ..., \psi_m$ have been added to $\mathcal{L}(w)$ by applying the interval relation rule to the node w. Above we have shown that $\mathcal{M} \models_{J_w} \psi_0$.

According to the interval relation rule we guess the interval relation between w and any node in $\mathcal{G}$. Assume for any $1 \leq j \leq m$ ψ_j has been added to $\mathcal{L}(w)$ as a result of guessing the interval relation between w and a node u_j. Since $\mathcal{K}(w)$, and therefore $\mathcal{L}(w)$, has been updated, this relation must have been "non-strict-during". In this case, $\mathcal{L}(u_j)$ must contain $\xi = [\int e\tau k] \, \psi$, where ψ has the form $\psi_j \vee ... \vee \psi_{j+l}$ $(l \geq 0)$. By the inductive hypothesis we have picked a witness J_{u_j} such that $\mathcal{M} \models_{J_{u_j}} \mathcal{L}(u_j)$; thus $\mathcal{M} \models_{J_{u_j}} \xi$. We know that $J_w \subseteq J_{u_j}$ because in the interval rule we have guessed the relation between J_w and J_{u_j}as "non-strict-during" (As we can see in the interval rule, $\mathcal{C}$ has been updated according to the corresponding non-deterministic choice of the relation.) We also know that $|J_w| \, \tau k$ because we have selected the option *ii* in the interval relation rule, and set $\mathcal{C} := \mathcal{C} \cup \{(e_w - b_w)\tau k\}$ (Otherwise, $\mathcal{K}(w)$ could not have been updated). Therefore, $\mathcal{M} \models_{J_w} \psi$.

When the interval rule was applied to w, we set $\mathcal{K}(w) := \mathcal{K}(w) \wedge \psi$. It is clear that ψ_j was selected when the Rule 2 the expansion strategy was applied. Thus, for any $1 \leq j \leq m$ $\mathcal{M} \models_{J_w} \psi_j$. Hence, $\mathcal{M} \models_{J_w} \mathcal{L}(w)$.

So, we have shown that once a node w is created, and the expansion strategy is applied, it is true that $\mathcal{M} \models_{J_w} \mathcal{L}(w)$. However, when new nodes are added to $\mathcal{G}$, $\mathcal{L}(w)$ might get updated through the application of the universal node expansion rule in these nodes. So, we must show that whenever new material is added to $\mathcal{L}(w)$, $\mathcal{M} \models_{J_w} \mathcal{L}(w)$ remains true.

Now, assume $\mathcal{L}(w) = \{\psi_0, ..., \psi_m, \psi_{m+1}, ..., \psi_{m+n}\}$ where $\psi_i = \psi_{i1} \wedge ... \wedge \psi_{in_i}$ $(0 \leq i \leq m+n$ and $n_i \geq 1)$, and $\psi_{m+1}, ..., \psi_{m+n}$ have been added to $\mathcal{L}(w)$ by applying the universal node expansion rule to some nodes in $\mathcal{G}$. Above we have shown that $\mathcal{M} \models_{J_w} \{\psi_0, ..., \psi_m\}$. Assume for any $m+1 \leq$

$k \leq m+n$, ψ_k has been added to $\mathcal{L}(w)$ by applying the universal node expansion rule to a node u_k in $\mathcal{G}$. In this case, $\mathcal{L}(u_k)$ must contain $\xi = \langle e \rangle \psi$, where ψ has the form $\psi_k \vee ... \vee \psi_{k+l}$ $(l \geq 0)$. By the inductive hypothesis we have picked a witness J_{u_k} such that $\mathcal{M} \models_{J_{u_k}} \mathcal{L}(u_k)$; thus $\mathcal{M} \models_{J_{u_k}} \xi$. We know that $J_w \subseteq J_{u_k}$. We also know that $|J_w| \tau k$ because we have selected the option *ii* of the universal rule, and set $\mathcal{C} := \mathcal{C} \cup \{(e_w - b_w)\tau k\}$ (Otherwise, $\mathcal{K}(w)$ could not have been updated.) Therefore, $\mathcal{M} \models_{J_w} \psi$.

When the universal rule was applied to u_k, we set $\mathcal{K}(w) := \mathcal{K}(w) \wedge \psi$. It is clear that ψ_k was selected when Rule 2 of the expansion strategy was applied. So, for any $m+1 \leq k \leq m+n$ $\mathcal{M} \models_{J_w} \psi_k$. Hence, $\mathcal{M} \models_{J_w} \mathcal{L}(w)$.

Case 2: Let w be a node in $\mathcal{G}$ such that $\rho(w) = -$. Assume $\mathcal{L}(w) = \{\psi_0\}$ where $\psi_0 = \psi_{01} \wedge ... \wedge \psi_{0n_0}$ $(n_0 \geq 1)$. Then, the dummy node w must have been created by either the existential node expansion rule, the interval relation rule, or the universal node expansion rule. If it has been created by the existential rule, then $\mathcal{L}(v)$ of a node v of which w is a successor node must contain either $\xi = \langle \int e\tau k \rangle_< \psi$ or $\xi = \langle \int e\tau k \rangle_> \psi$. Otherwise, $\mathcal{L}(u)$ of a node u at which the interval relation rule or the universal rule has been applied contains either $\xi = [\int e\tau k]_< \psi$ or $\xi = [\int e\tau k]_> \psi$. In each case, by the inductive hypothesis a witness J_v (J_u) is defined such that $\mathcal{M} \models_{J_v} \mathcal{L}(v)$ $(\mathcal{M} \models_{J_u} \mathcal{L}(u))$. By construction, there exists a node w' with $\rho(w') = e$. Let $J_{w'}$ be an interval for the w'. It is trivial to see that $J_{w'} \subseteq J_v$ $(J_{w'} \subseteq J_u)$ and $|J_{w'}| \tau k$. Since $\mathcal{M} \models_{J_v} \xi$, $\mathcal{M} \models_{J_w} \psi$, where $J_w = \hbar(J_{w'}, J_v)$ or $(J_w = \hbar(J_{w'}, J_u))$. Here the partial function $\hbar$ is fin if $\xi = \langle \int e\tau k \rangle_< \psi$, and it is $init$, otherwise.

When any of these rules (existential rule, universal rule and interval rule) was applied, we set $\mathcal{K}(w) := \mathcal{K}(w) \wedge \psi$. Suppose ψ have the form $\psi_0 \vee ... \vee \psi_l$ $(l \geq 0)$. Since ψ_0 is the selected disjunct of ψ in Rule 2, $\mathcal{M} \models_{J_w} \psi_0$. Hence, $\mathcal{M} \models_{J_w} \mathcal{L}(w)$.

Therefore, we have proved that for each node v in $\mathcal{G}$, there exists an interval J_v such that $\mathcal{M} \models_{J_v} \mathcal{L}(v)$.

Meanwhile, we know the depth of the model $\mathcal{M}$ is at most of order $2^{p(|\varphi|)}$ by the assumption. Since for any node v in $\mathcal{G}$, $\mathcal{M} \models_{J_v} \mathcal{L}(v)$, $\perp$ cannot be contained in $\mathcal{L}(v)$. As we have a witness J_v for each node v, we must have a solution for $\mathcal{C}$. Therefore, $\mathcal{C}$ must be satisfiable. Because none of the conditions in Definition 12 holds, it follows that $\langle \mathcal{G}, \mathcal{C} \rangle$ is an open tableau. ■

4.4 Computational complexity

THEOREM 16. *The satisfiability problem for EIL is in 2-NEXPTIME.*

Proof. In Theorem 14 we show that the proposed method terminates. Now, we analyse its computational complexity. We now give a bound on the size of any tableau for φ.

The out degree of any node is bounded by $|\varphi|$. The depth of the longest path in the tableau is bounded by $2^{p(|\varphi|)}$ for a fixed polynomial p by Lemma

7. Therefore, the size of the tableau is bounded by $|\varphi|^{2^{p(|\varphi|)}} = 2^{2^{p(|\varphi|)} log_2 |\varphi|}$. So, the tableau procedure builds a tableau of size $2^{2^{p'(|\varphi|)}}$ for some fixed polynomial p'. We can say that if an EIL formula φ is satisfiable, then the tableau procedure construct a graph, from which a satisfying model $\mathcal{M}$ is extracted, which has doubly exponential size by the length of φ. ■

5 Conclusion

In this paper we introduced an interval temporal logic EIL to represent meanings of sentences in English. EIL has affinity with the syntax of temporal constructions in English, and which is convenient for expressing the semantics of natural language constructions. EIL is interpreted over a linear time flow with only finitely many events able to occur over a bounded-time interval. EIL also employs the notion of duration.

In order to bound models we showed that the depth of an EIL model is exponentially bounded by the length of a given formula. We also proposed a terminating tableau system for the logic EIL, thus showing that its satisfiability problem is decidable. Indeed, it was proved that the satisfiability problem for EIL is in 2-NEXPTIME. This was proved by building models, which have doubly exponential size by the length of the given formula.

The future research directions include extending EIL with states and state models to specify real-time system requirements, finding a lower bound for the complexity of the satisfiability problem, introducing an axiomatization system to complement the semantic view, and comparing expressive powers of EIL with the related interval temporal logics.

BIBLIOGRAPHY

[1] Allen J. F. Maintaining Knowledge about Temporal Intervals. *Communications of the ACM*, vol. 26, pp. 832-843, 1983.

[2] Crouch R., Pulman S. G. Time and Modality in a Natural Language Interface to a Planning System. *Artificial Intelligence*, vol. 63, num. 1-2, pp. 265-304, 1993.

[3] Dechter R., Meiri I. and Pearl J. Temporal Constraint Networks. *Artificial Intelligence*, vol. 49, pp. 61-95, 1991.

[4] Dowty D. *Word Meaning and Montague Grammar*. Dordrecht: D. Reidel, 1979.

[5] Halpern J. Y. and Shoham Y. A Propositional Modal Logic of Time Intervals. *Journal of the ACM*, vol. 38, num. 4, pp. 935-962, 1991.

[6] Hwang C. H. and Schubert L. K. Interpreting Tense, Aspect and Time Adverbials. In *Proceedings of the First International Conference on Temporal Logic*, LNCS, Springer, vol. 827, pp. 238-274, 1994.

[7] Kamp H. Events, Instants and Temporal Reference. *Semantics from Different Points of View*, Springer, pp. 376-417, 1979.

[8] Konur S. A Decidable Temporal Logic for Events and States. *Thirteenth International Symposium on Temporal Representation and Reasoning (TIME 2006)*, vol.15-17, pp. 36-41, 2006.

[9] Pratt-Hartmann I. and Bree D. *The Expressive Power of the English Temporal Preposition System.* University of Manchester, Department of Computer Science Technical Report, UMCS-93-1-7, 1993.

[10] Pratt-Hartman I. Temporal Prepositions and their logic. Artificial Intelligence, vol. 166(1-2), pp. 1-36, 2005.

[11] Ter Meulen A. G. *Representing Time in Natural Language.* MIT Press, Cambridge, 1996.
[12] Venema Y. A Modal Logic for Choppping Intervals. *Journal of Logic and Computation*, vol. 1, pp. 453-476, 1991.

Savas Konur
Department of Computer Science
University of Manchester
Manchester, M13 9PL, UK
konurs@cs.man.ac.uk

Completeness of the finitary Moss logic

CLEMENS KUPKE, ALEXANDER KURZ AND YDE VENEMA

ABSTRACT. We give a sound and complete derivation system for the valid formulas in the finitary version of Moss' coalgebraic logic, for coalgebras of arbitrary type.

Keywords: coalgebra, modal logic, coalgebraic logic, completeness

1 Introduction

Generalizing Kripke models and frames, coalgebras provide a general, category theoretic account of state-based evolving systems. This point of view was emphasized by Rutten [22], who developed, in analogy with Universal Algebra, the basics of Universal Coalgebra as a general theory of systems. One of the strengths of the coalgebraic approach is that a substantial part of the theory of systems can be developed uniformly in a functor T (on the category Set of sets and functions), which intuitively represents the type of the transition system. For example, as discovered by Aczel [2], any functor T induces a canonical notion of bisimilarity on T-coalgebras.

The research programme of Coalgebraic Logic is to extend this uniform approach to logics for specifying and reasoning about the behavior of coalgebras. This research direction was initiated by Moss [18], who described a logic for T-coalgebras uniformly for all set functors T (satisfying a mild condition). Moss' fascinating idea was, roughly, to take T itself as a modality. In the case of the power set functor $\mathcal{P}$, this modality, denoted as ∇, has surfaced in modal logic from time to time, for instance in Fine's work [9] on normal forms. It can be defined using the standard box and diamond: With $\alpha \in \mathcal{PL}$ a set of formulas, the formula $\nabla\alpha$ can be seen as an abbreviation: $\nabla\alpha = \Box \bigvee \alpha \wedge \bigwedge \Diamond\alpha$, where $\Diamond\alpha$ denotes the set $\{\Diamond a \mid a \in \alpha\}$. The *semantics* of ∇ can be expressed in terms of the so-called Egli-Milner *lifting* of the satisfaction relation $\Vdash \subseteq S \times \mathcal{L}$ between states and formulas to a relation $\overline{\Vdash}$ between $\mathcal{P}S$ (sets of states) and $\mathcal{PL}$ (sets of formulas):

(1) $\mathbb{S}, s \Vdash \nabla\alpha$ iff $\sigma(s) \overline{\Vdash} \alpha$,

where $\sigma : S \to \mathcal{P}S$ denotes the successor function. Since one may associate a reasonable notion of relation lifting with other set functors as well, the observation (1) paves the way for generalization to an arbitrary functor T. Moss shows that his coalgebraic logic, based on a modality ∇_T, is invariant under bisimilarity, and, in the presence of infinitary conjunctions, characterizes bisimilarity.

The operator ∇_T associated with an arbitrary functor T looks strikingly different from the usual $\Box$ and $\Diamond$ modalities. Following on from [18], attention turned to the question how to obtain modal languages for T-coalgebras which use more standard modalities [15, 21, 11], and how to find derivation systems for these formalisms. This approach is now usually described in terms of predicate liftings [20, 24] or, equivalently, Stone duality [6, 16]. For a while, this approach displaced the interest in Moss' logic and the relationship between the two was not completely clear.

Interest in Moss' logic revived when it became clear that even in standard modal logic, a ∇-based approach has some advantages. In fact, independently of Moss' work, Janin & Walukiewicz [12] already observed that the connectives ∇ and $\vee$ may in some sense replace the set $\{\Box, \Diamond, \wedge, \vee\}$. This observation, which is closely linked to fundamental automata-theoretic constructions, lies at the heart of the theory of the modal μ-calculus, and has many applications, see for instance [8, 23]. Generalizing the link between fixpoint logics and automata theory to the coalgebraic level of generality, Kupke & Venema [14] generalized some of these observations to show that many fundamental results in automata theory are really theorems of universal coalgebra.

This paper addresses the main problem left open by Moss [18]. Moss' approach focuses on semantics, and he provides only some sound logical principles which do not constitute a complete syntactic calculus. As a first result in the direction of a derivation system for ∇ modalities, Palmigiano & Venema [19] gave a complete axiomatization for the cover modality (i.e., $\nabla_{\mathcal{P}}$ for the power set functor $\mathcal{P}$). This calculus was streamlined by Bílková, Palmigiano & Venema [5] into a formulation that admits a straightforward generalization to an arbitrary set functor T.

Our main contribution here is a uniform completeness proof. That is, in this paper we provide, uniformly in the functor T, a derivation system $\mathbf{M}$ which is sound and complete with respect to the semantics of the coalgebraic language based on the modality ∇_T. The main idea of the completeness proof is based on the Stone duality approach to coalgebraic logic and, as a byproduct, we also see how Moss' language fits into this approach.

In the Stone duality approach to coalgebraic logic, the relationship between logic and semantic is based on the following situation

(2)

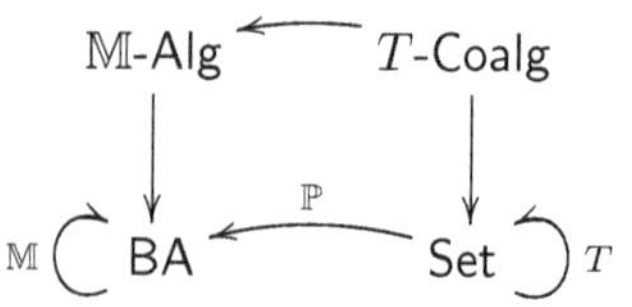

where $\mathbb{M}$ is the functor on Boolean algebras given by the proof system of the logic under consideration and $\mathbb{P}$ is the contravariant powerset functor. The semantics of the logic appears in this setting as a natural transformation $\delta : \mathbb{M}\mathbb{P} \to \mathbb{P}T$ (using δ, $\mathbb{P}$ lifts to a functor in the upper row, which maps a T-coalgebra to its 'complex' $\mathbb{M}$-algebra). The proof system is complete if δ is injective (Proposition 48). One advantage of this approach is its flexibility. For example, descriptive-general-frame semantics corresponds to replacing Set by Stone spaces. On the algebra side, one can treat positive logic by replacing BA by distributive lattices or infinitary logic (like in Moss' original work) by replacing BA by complete atomic Boolean algebras. This paper treats the case of BA and Set which is of particular interest to us and leaves the others for future work. This means, in particular, that we will concentrate on the finitary version of Moss' logic first introduced in [27].

2 Preliminaries

In this section we settle on notation and terminology, and we introduce the finitary version of Moss' logic. For background on coalgebra the reader is referred to [26].

General Two categories play a major role in our paper: the category Set with sets as objects and functions as arrows, and the category BA of Boolean algebras and homomorphisms. The categories Set and BA are related by the contravariant functor $\mathbb{P} : \mathsf{Set} \to \mathsf{BA}$, by the forgetful functor $U : \mathsf{BA} \to \mathsf{Set}$, and by the left adjoint $\mathbb{F}$ of U mapping a set X to the free Boolean algebra over X. We write P for $U\mathbb{P}$, $\mathbf{2}$ for the two-element Boolean algebra and 1 for a one-element set.

Coalgebra A coalgebra (over Set) for a functor $T : \mathsf{Set} \to \mathsf{Set}$, also called T-coalgebra, is a pair (S, σ) where S is a set (of "states") and $\sigma : S \to TS$ is a function (the "transition structure"). A T-coalgebra morphism from a T-coalgebra (S_1, σ_1) to a T-coalgebra (S_2, σ_2) is a function $f : S_1 \to S_2$ such that $Tf \circ \sigma_1 = \sigma_2 \circ f$.

For a modal logician, the prime examples of coalgebras are Kripke frames and Kripke models. Bisimulations between Kripke structures also have their natural coalgebraic generalization: a relation Z between the carrier sets of two coalgebras is a bisimulation if for all $(s_1, s_2) \in Z$, the pair $(\sigma_1(s_1), \sigma_2(s_2))$ belongs to the *relation lifting* $\overline{Z}$ of Z.

DEFINITION 1. Let T be a set functor. Given a binary relation Z between two sets S_1 and S_2, we define the relation $\overline{Z} \subseteq TS_1 \times TS_2$ as follows:

$$\overline{Z} := \{((T\pi_1)\phi, (T\pi_2)\phi) \mid \phi \in TZ\},$$

where $\pi_i : Z \to S_i$ for $i = 1, 2$ are the projection functions.

In this paper we will confine attention to set functors that are *standard* (that is, inclusions are mapped to inclusions), and that *preserve weak pullbacks*. We will not define the latter property, but simply note that it is equivalent to requiring that the composition of two bisimulations is again a bisimulation, or, equivalently, that for all relations Z_1, Z_2 we have $\overline{Z_1 \circ Z_2} = \overline{Z_1} \circ \overline{Z_2}$ (and it will be apparent from the development below that this property is essential to work with the Moss modality). The requirement of standardness is not essential and only serves to keep the notation a bit smoother. The class of standard and weak pullback preserving functors includes the ones that are used to model infinite words, infinite binary trees,

Kripke frames and probabilistic transition systems as coalgebras. A more detailed discussion of these examples can be found in [14]. For reasons of space limitations we cannot go into further detail here.

CONVENTION 2. Throughout this paper we fix a standard and weak pullback preserving set functor T.

The following fact lists the properties of relation lifting that we use in our paper. (Here $\mathrm{Gr}(f) \subseteq S \times S'$ denotes the graph of a function $f : S \to S'$.) For proofs we refer to [18] and references therein.

FACT 3. Let T be a set functor that is standard and weak pullback preserving. Then relation lifting
(1) extends T: $\overline{\mathrm{Gr}(f)} = \mathrm{Gr}(Tf)$, and preserves the diagonal: $\overline{Id_S} = Id_{TS}$;
(2) is monotone: $R \subseteq Q$ implies $\overline{R} \subseteq \overline{Q}$;
(3) commutes with taking restrictions: $\overline{R\restriction_{U\times U'}} = \overline{R}\restriction_{TU\times TU'}$;
(4) preserves composition: $\overline{R \circ Q} = \overline{R} \circ \overline{Q}$, and converse: $\overline{(R^{\smile})} = (\overline{R})^{\smile}$;

We let T_ω denote the *finitary*, or, *ω-accessible*, version of T, that is, the set functor T_ω which agrees with T on finite sets, while for an infinite set X,

$$T_\omega(X) := \bigcup\{TY \mid Y \in \mathcal{P}_\omega(X)\}.$$

On maps, T_ω simply agrees with T. It is not hard to see that T_ω is a well-defined subfunctor of T (cf. [4, p.314]) and that $T_\omega X \subseteq TX$ for all sets X. Furthermore, as any standard set functor preserves finite intersections ([4, III, Prop. 4.6]), for any set X, and any element $\alpha \in T_\omega X$, there is a *smallest, finite* subset $X_0 \subseteq X$ such that $\alpha \in T_\omega X_0$. This set X_0 is called the *base* of α, notation: $Base(\alpha)$.

Moss' language

DEFINITION 4. Given a set X of proposition letters, we define the following. $\mathcal{L}_0(X)$ is the smallest superset of X which is closed under taking negations and finitary conjunctions and disjunctions. $\mathcal{L}_{n+1}(X) := \mathcal{L}_0(\{\nabla\alpha \mid \alpha \in T_\omega\mathcal{L}_n(X)\})$ is the smallest set containing the formula $\nabla\alpha$ for each $\alpha \in T_\omega\mathcal{L}_n(X)$, which is closed under taking negations and finitary conjunctions and disjunctions. $\mathcal{L}(X) := \bigcup_{n\in\omega}\mathcal{L}_n(X)$ is the set of *formulas in* X; in case $X = \emptyset$ we write $\mathcal{L}_n$ and $\mathcal{L}$ instead of $\mathcal{L}_n(\emptyset)$ or $\mathcal{L}(\emptyset)$. The *depth* of a formula a is the smallest n such that $\phi \in \mathcal{L}_n$.

We write $\top := \bigwedge\emptyset$ and $\bot := \bigvee\emptyset$. Then by definition, $\top$ and $\bot$ belong to every layer of the language. While it is not hard to prove that $\mathcal{L}_n \subseteq \mathcal{L}_{n+1}$, for all $n \in \omega$, it is in general not the case that $X \subseteq \mathcal{L}_n$ for $n > 0$.

It will occasionally be useful to think of $\mathcal{L}_0(X)$ as the (carrier of the) absolutely free algebra of Boolean type, or the *Boolean term algebra*, generated by X, and of $\mathcal{L}_{n+1}$ as the Boolean term algebra generated by the set $\{\nabla\alpha \mid \alpha \in T_\omega\mathcal{L}_n\}$.

The language can be seen as an initial algebra for a functor.

PROPOSITION 5. *Let M be the set functor $Id + Id \times Id + Id \times Id + T_\omega$. Then $(\mathcal{L}, \neg, \wedge, \vee, \nabla)$ is the initial M-algebra.*

REMARK 6. For the category theoretic minded reader we note that, identifying formulas up to Boolean equivalence, Moss' language $\mathcal{L}$ is the initial algebra for the functor $\mathbb{L} = \mathbb{F}T_\omega U : \mathsf{BA} \to \mathsf{BA}$.

While we will refer to the above language as *Moss'* coalgebraic language, there are actually some differences. The most important of these is that by defining $\nabla\alpha$ to be a formula only for elements $\alpha \in T_\omega\mathcal{L}$ (rather than for all $\alpha \in T\mathcal{L}$), we construct a language that is *finitary* in the sense that every formula has a finite number of *subformulas*. This notion can be defined inductively, the key clause being that the subformulas of $\nabla\alpha$ are given as the closure of the set $Base(\alpha)$ under subformulas.

Concerning the semantics of $\mathcal{L}$, we only give the clause for the ∇ modality.

DEFINITION 7. Given a coalgebra $\mathbb{S} = (S, \sigma)$, we define $s \Vdash \nabla\alpha$ if $\sigma(s)\overline{\Vdash}\alpha$.

EXAMPLE 8. Let Prop be a set of propositional variables and recall that coalgebras for the functor $K = \mathcal{P}\text{Prop} \times \mathcal{P}$ correspond to Kripke models. Then any formula $\nabla_K\alpha$ is of the form $\nabla_K\alpha = \nabla_K(P, A)$ where $P \subseteq$ Prop is a set of proposition letters and $A \subseteq \mathcal{L}$ is a finite set of formulas. If the set Prop is finite it is easy to see that one can define a translation t of formulas in $\mathcal{L}$ into the basic modal language by putting

$$t(\nabla_K(P, A)) := \bigwedge_{p\in P} p \wedge \bigwedge_{p\notin P} \neg p \wedge \bigwedge_{a\in A} \Diamond t(a) \wedge \Box(\bigvee_{a\in A} t(a))$$

such that $(S, \sigma), s \Vdash a$ iff $(S, \sigma), s \Vdash t(a)$ for all $a \in \mathcal{L}$.

The semantics of a ∇-formula can be also expressed using the following natural transformation which plays a central role in our paper.

DEFINITION 9. We define a natural transformation $\rho : TP \to PT$ by putting $\rho_X(\Phi) := \{\alpha \in TX \mid \alpha\overline{\in}\Phi\}$.

REMARK 10. ρ is natural if T preserves weak-pullbacks. This is also true if one replaces the contravariant P with the covariant $\mathcal{P}$.

In order to gain some intuitions about the ∇-operator and the transformation ρ, the reader is invited to prove the following easy lemma.

LEMMA 11. *For any $\nabla\alpha \in \mathcal{L}$ we have $s \Vdash \nabla\alpha$ iff $s \in \sigma^{-1} \circ \rho_S(T\mu(\alpha))$, where $\mu : \mathcal{L} \to \mathcal{P}S$ is the function that maps a formula to its semantics.*

REMARK 12. Following on from Remark 6, freely extending ρ to Boolean algebras yields a natural transformation $\gamma : \mathbb{LP} \to \mathbb{P}T$. γ allows us to associate with any coalgebra (S, σ) a 'complex $\mathbb{L}$-algebra' $\mathbb{LP}S \stackrel{\gamma_S}{\to} \mathbb{P}TS \stackrel{\mathbb{P}\sigma}{\to} \mathbb{P}S$. Denote by $\mathcal{L}'$ the language $\mathcal{L}$ quotiented by Boolean equivalence. Then $\mathcal{L}'$ is the initial $\mathbb{L}$-algebra. For each coalgebra (S, σ), initiality of $\mathcal{L}'$ gives us a map $[\![\cdot]\!] : \mathcal{L}' \to \mathbb{P}S$ interpreting elements of $\mathcal{L}'$ as propositions on S. This definition agrees with Definition 7 (because γ is the free extension of ρ).

3 The derivation system

In this section we will define and discuss the derivation system **M**. Before we can provide the actual definition of **M**, we need a few preparatory remarks and definitions.

First of all, it will be convenient for us to have the derivation system operating on *inequalities*, that is, expressions of the form $a \preceq b$, with $a, b \in \mathcal{L}$. The main reason for this is that we like our system to stay close to equational reasoning. Indeed, in any logic with an underlying algebraic semi-lattice structure, inequalities can be seen as (special) *equations*: we may for instance identify the inequality $a \preceq b$ with the equation $a \wedge b \approx a$. Conversely, we may think of an equation $a \approx b$ as a *pair* of inequalities $a \preceq b$, $b \preceq a$.

DEFINITION 13. An inequality $a \preceq b$ is *valid in a coalgebra* $\mathbb{S} = (S, \sigma)$, notation: $\mathbb{S} \Vdash a \preceq b$, if $\mathbb{S}, s \Vdash a$ implies $\mathbb{S}, s \Vdash b$ for all $s \in S$, and *valid* simpliciter if it is valid in every coalgebra, notation: $a \models b$.

Note that the set of valid formulas can be obtained from the set of valid inequalities: a formula a is true in every state in every coalgebra iff the inequality $\top \preceq a$ is valid.

In the sequel we will need symbols to refer to formulas ($\mathcal{L}$), and to elements of the sets $\mathcal{P}_\omega\mathcal{L}$, $T_\omega\mathcal{L}$, $T_\omega\mathcal{P}_\omega\mathcal{L}$ and $\mathcal{P}_\omega T_\omega\mathcal{L}$. For convenience we fix our notation for such objects as follows:

$\mathcal{L}$	$a, b, c, \ldots$	$T_\omega\mathcal{L}$	$\alpha, \beta, \gamma \ldots$
$\mathcal{P}_\omega\mathcal{L}$	$\phi, \psi, \ldots$	$T_\omega\mathcal{P}_\omega\mathcal{L}$	$\Phi, \Psi, \ldots$
$\mathcal{P}_\omega T_\omega\mathcal{L}$	$A, B, C \ldots$		

The same notation will be used for variants where $\mathcal{L}$ is replaced by an arbitrary set or $\mathcal{P}_\omega, T_\omega$ are replaced by $\mathcal{P}, T$.

An important role in the definition of **M** is played by the notion of a *slim redistribution.*

DEFINITION 14. A set $\Phi \in T\mathcal{P}(X)$ is a *redistribution* of a set $A \in \mathcal{P}T(X)$ if $A \subseteq \rho_X(\Phi)$. In case $A \in \mathcal{P}_\omega T_\omega(X)$, we call a redistribution Φ *slim* if $\Phi \in T_\omega\mathcal{P}_\omega\big(\bigcup_{\alpha \in A} \mathit{Base}(\alpha)\big)$. The set of slim redistributions of A is denoted as $\mathit{SRD}(A)$.

A special case

Our derivation system is given in Definition 15. It turns out, however, that we can give a somewhat simpler version in case the functor T restricts to finite sets (that is, if TX is finite whenever X is finite). This simpler system is the direct generalization of the system for $T = \mathcal{P}$ (that is, where the coalgebras are Kripke structures) given by Bílková, Palmigiano and the third author in [5]. [1]

M is given as follows. On top of a complete set of axioms and rules for classical propositional logic, and the cut rule (from $a \preceq b$ and $b \preceq c$ derive $a \preceq c$), it has the axioms and derivation rules given in Table 1.

Let us hasten to give some explanation of the system. To start with, the reader may be slightly puzzled by our formulation of the derivation rule

[1] In [5] it was shown that for $T = \mathcal{P}$ axiom (∇4) is derivable from (∇1)-(∇3). We recently discovered that this is also true for the case of an arbitrary functor T.

(∇1)	From $\alpha \overline{\preceq} \beta$ infer $\vdash \nabla\alpha \preceq \nabla\beta$
(∇2)	$\bigwedge\{\nabla\alpha \mid \alpha \in A\} \preceq \bigvee\{\nabla(T\bigwedge)(\Phi) \mid \Phi \in SRD(A)\}$
(∇3)	$\nabla(T\bigvee)(\Phi) \preceq \bigvee\{\nabla\alpha \mid \alpha \overline{\in} \Phi\}$
(∇4)	From $\vdash \top \preceq \bigvee\phi$ infer $\vdash \top \preceq \bigvee\{\nabla\alpha \mid \alpha \in T\phi\}$

Table 1. Axioms and rules of the system **M**, if T restricts to finite sets

(∇1), since its premiss '$\alpha \overline{\preceq} \beta$' uses syntax that has not been defined as part of the object language. The proper way to read this premiss is as follows: 'the relation $Z := \{(a, b) \in Base(\alpha) \times Base(\beta) \mid\vdash a \preceq b\}$ is such that $(\alpha, \beta) \in \overline{Z}$'. In order to see this, note that using Fact 3(3) one can show that for all $\alpha, \beta \in T_\omega\mathcal{L}$ and all $Z \subseteq \mathcal{L} \times \mathcal{L}$

$$(\alpha, \beta) \in \overline{Z} \quad \text{iff} \quad (\alpha, \beta) \in \overline{Z'},$$

where $Z' := Z \restriction_{Base(\alpha) \times Base(\beta)}$ is the restriction of Z to the finite sets $Base(\alpha)$ and $Base(\beta)$. An alternative formulation of this rule would therefore say that 'if there is a relation $Z \subseteq Base(\alpha) \times Base(\beta)$ such that $(\alpha, \beta) \in \overline{Z}$, and $\vdash a \preceq b$ for all $(a, b) \in Z$, then infer $\vdash \nabla\alpha \preceq \nabla\beta$'. But the presentation in Table 1 is shorter and reveals more clearly that the rule is in fact the inequality version of a *congruence* rule. Our discussion shows that (∇1) is a *finitary* rule, because its set of premisses can be assumed to be contained in the finite set $Base(\alpha) \times Base(\beta)$ if we want to derive $\nabla\alpha \preceq \nabla\beta$.

The axioms (∇2) and (∇3) could in fact both be replaced with identities, since in both cases, the reverse inequality of the axiom can be derived as a theorem. In order to be able to *read* the axioms ∇2 and ∇3, recall that $\bigwedge$ and $\bigvee$ are maps from $\mathcal{P}_\omega\mathcal{L}$ to $\mathcal{L}$, so that $T\bigwedge : T\mathcal{P}_\omega\mathcal{L} \to T\mathcal{L}$, and likewise for $T\bigvee$. Hence for $\Phi \in T_\omega\mathcal{P}_\omega\mathcal{L}$, $(T\bigwedge)(\Phi)$ and $(T\bigvee)(\Phi)$ belong to $T_\omega\mathcal{L}$, and thus $\nabla(T\bigwedge)(\Phi)$ and $\nabla(T\bigvee)(\Phi)$ are well-formed formulas. In addition, if T restricts to finite sets, every $A \in \mathcal{P}_\omega T_\omega\mathcal{L}$ can have at most finitely many slim redistributions, and every $\Phi \in T_\omega\mathcal{P}_\omega\Phi$ can have at most finitely many lifted members. So the two axioms (∇2) and (∇3) are at least well-defined. What these axioms have in common further is that they can be seen as *distributive principles.* This is the clearest in the case of (∇3), which states that ∇ distributes over certain disjunctions. In the case of (∇2) the distributivity is a bit more involved, but basically, the axiom states that any conjunction of ∇s can be replaced with a disjunction of ∇s of conjunctions.

Finally, although the formulation of (∇4) does not use the actual symbol, it is here that the interaction of the coalgebraic modality with negation is dealt with. To see why this is so, observe that the conclusion of (∇4) implies that $\neg\nabla\beta \preceq \bigvee\{\nabla\alpha \mid \beta \neq \alpha \in T\phi\}$.

The general case

In the case of a general functor, that is, one that does not necessarily restricts to finite sets, some of the axioms and rules in Table 1 above may

$$(\nabla 1) \quad \frac{\{b_1 \preceq b_2 \mid (b_1, b_2) \in Z\}}{\nabla\alpha \preceq \nabla\beta} \; (\alpha, \beta) \in \overline{Z}$$

$$(\nabla 2) \quad \frac{\{\nabla(T\bigwedge)(\Phi) \preceq a \mid \Phi \in SRD(A)\}}{\bigwedge\{\nabla\alpha \mid \alpha \in A\} \preceq a}$$

$$(\nabla 3) \quad \frac{\{\nabla\alpha \preceq a \mid \alpha \overline{\in} \Phi\}}{\nabla(T\bigvee)(\Phi) \preceq a}$$

$$(\nabla 4) \quad \frac{\{a \wedge \nabla\alpha' \preceq \bot \mid \alpha' \in T_\omega(\phi), \alpha' \neq \alpha\} \qquad \top \preceq \bigvee\phi}{a \preceq \nabla\alpha}$$

Table 2. Axioms and rules of the system **M**

involve ill-defined syntax. In particular, none of the disjunctions on the right-hand side of the axioms ($\nabla 2$) and ($\nabla 3$) will be taken over a *finite* set. (Algebraically, however, it will often be convenient to think of e.g. ($\nabla 2$) as stating that *in case* that the least upper bound given on the right hand side *exists*, it is greater than the object denoted by the left hand side.) The solution is to replace the axioms ($\nabla 2$) and ($\nabla 3$) with *infinitary derivation rules* (and to do something similar for the conclusion of ($\nabla 4$)), according to the following principle. An axiom of the form $s \preceq \bigvee_{i \in I} t_i$ is replaced with the derivation rule: 'from $\{\vdash t_i \preceq a \mid i \in I\}$, infer $\vdash s \preceq a$'. Applying this principle to the above axiom system, we obtain the following derivation system.

DEFINITION 15. The derivation system **M** is given by the axioms and derivation rules of Table 2, on top of a complete set of axioms and rules for classical propositional logic, and the cut rule.

A *derivation* is a well-founded tree, labelled with inequalities, such that the leaves of the tree are labelled with axioms of **M**, whereas with each parent node we may associate a derivation rule of which the conclusion labels the parent node itself, and the premisses label its children. If $\mathcal{D}$ is a derivation of the inequality $a \preceq b$, we write $\mathcal{D} \vdash_{\mathbf{M}} a \preceq b$. We write $\vdash_{\mathbf{M}} a \preceq b$ if we want to suppress the actual derivation and we write $a \equiv b$ if $\vdash_{\mathbf{M}} a \preceq b$ and $\vdash_{\mathbf{M}} b \preceq a$.

The main result

THEOREM 16. *Let T be a standard functor that preserves weak pullbacks. Then for any pair a and b of formulas in $\mathcal{L}$:*

$$\vdash_{\mathbf{M}} a \preceq b \iff a \models b.$$

4 Soundness

Soundness is the direction from left to right of Theorem 16. It is proved by induction on the complexity of derivations. The key steps are to show that the rules (∇1)–(∇4) preserve validity.

First we consider the rule (∇1). Suppose that $\mathbb{S} \Vdash a \preceq b$ for all pairs (a, b) belonging to some relation $Z \subseteq \mathcal{L} \times \mathcal{L}$ such that $(\alpha, \beta) \in \overline{Z}$. From the first assumption it follows that $\Vdash \circ Z \subseteq \Vdash$, and so, by the properties of relation lifting, we see that $\overline{\Vdash} \circ \overline{Z} \subseteq \overline{\Vdash}$. In order to show that $\mathbb{S} \Vdash \nabla\alpha \preceq \nabla\beta$, take an arbitrary state s such that $\mathbb{S}, s \Vdash \nabla\alpha$. Hence, by the truth definition of ∇, we see that $\sigma(s)\ \overline{\Vdash}\ \alpha$, and so from $(\alpha, \beta) \in \overline{Z}$ we may infer that $(\sigma(s), \beta) \in \overline{\Vdash} \circ \overline{Z} \subseteq \overline{\Vdash}$. But then, again by the truth definition of ∇, we see that, indeed, $\mathbb{S}, s \Vdash \nabla\beta$.

For the rule (∇2), fix a set $A \in \mathcal{P}_\omega T_\omega \mathcal{L}$, and some formula $a \in \mathcal{L}$. Suppose that $\mathbb{S}$ validates all the premisses of the rule, that is, $\mathbb{S} \Vdash \nabla(T_\omega \bigwedge)(\Phi) \preceq a$, for all slim redistributions Φ of A. In order to prove that $\mathbb{S}$ validates the conclusion of (∇2), assume that $\mathbb{S}, s \Vdash \bigwedge\{\nabla\alpha \mid \alpha \in A\}$. Clearly it suffices to come up with a slim redistribution Φ_s of A such that $\mathbb{S}, s \Vdash \nabla(T \bigwedge)(\Phi)$.

For the definition of Φ_s, first associate, with any state t in $\mathbb{S}$, the finite set

$$\phi(t) := \{b \in \bigcup_{\alpha \in A} Base(\alpha) \mid \mathbb{S}, t \Vdash b\},$$

and define $\Phi_s := (T\phi)(\sigma(s))$.

First we show that $\mathbb{S}, s \Vdash \nabla(T \bigwedge)(\Phi_s)$. For that purpose, observe that by definition of ϕ, the map $\bigwedge \circ \phi : S \to \mathcal{L}$ is such that $\mathrm{Gr}(\bigwedge \circ \phi) \subset \Vdash$. From this it follows by the properties of relation lifting that $\mathrm{Gr}\big((T \bigwedge) \circ (T\phi)\big) \subseteq \overline{\Vdash}$. In other words, for every element $\tau \in TS$ we have that $\tau\ \overline{\Vdash}\ \big((T \bigwedge) \circ (T\phi)\big)(\tau)$. Taking $\tau = \sigma(s)$, we obtain immediately by the definitions that $\mathbb{S}, s \Vdash \nabla(T \bigwedge)(\Phi_s)$.

In order to see that Φ_s is a slim redistribution of A, observe that by definition of ϕ, $\mathrm{Gr}(\phi) \circ \in^{\smile} = \Vdash$ when restricted to elements of $\bigcup_{\alpha \in A} Base(\alpha)$. Then by the properties of relation lifting, it follows that $\mathrm{Gr}(T\phi \circ \overline{\in}^{\smile}) = \overline{\Vdash}$. But then for every $\alpha \in A$ it follows from $\sigma(s)\ \overline{\Vdash}\ \alpha$ that there is some object Ψ such that the pair $(\sigma(s), \Psi)$ belongs to the relation $\mathrm{Gr}(T\phi)$, and $\alpha \overline{\in} \Psi$. From the first fact it follows that $\Psi = \Phi_s$, and so we find that each $\alpha \in A$ is a lifted member of Φ_s. In other words, Φ_s is a redistribution of A; but then by its definition it is slim.

In order to understand the soundness of (∇3), first consider the statement $\mathbb{S}, s \Vdash \bigvee \phi$. This statement can be reformulated equivalently by saying that the pair (s, ϕ) belongs to the relation $\Vdash \circ \in$, since there is some element $a \in \phi$ such that $s \Vdash a$. Alternatively, $s \Vdash \bigvee \phi$ iff $(s, \phi) \in \Vdash \circ \mathrm{Gr}(\bigvee)^{\smile}$. In other words, we find that the relations $\Vdash \circ \in$ and $\Vdash \circ \mathrm{Gr}(\bigvee)^{\smile}$ coincide. From this it follows that

(3) $\overline{\Vdash \circ \in} = \overline{\Vdash \circ \mathrm{Gr}(\bigvee)^{\smile}}$.

Fix some object $\Phi \in T_\omega \mathcal{P}_\omega \mathcal{L}$ and some formula a, and suppose that the coalgebra $\mathbb{S}$ validates all the premisses of (∇3), i.e., $\mathbb{S} \Vdash \nabla\alpha \preceq a$, for all $\alpha \overline{\in} \Phi$. In order to prove that $\mathbb{S}$ also validates the conclusion of the rule, take an arbitrary state s such that $\mathbb{S}, s \Vdash \nabla(T \bigvee)(\Phi)$. From this it follows that $(\sigma(s), (T\bigvee)(\Phi))$ belongs to the relation $\overline{\Vdash}$, and so $(\sigma(s), \Phi)$ belongs to $\overline{\Vdash} \circ \mathrm{Gr}(T\bigvee)^{\smile} = \overline{\Vdash \circ \mathrm{Gr}(\bigvee)^{\smile}}$. But then by (3), $(\sigma(s), \Phi)$ belongs to the relation $\overline{\Vdash \circ \in} = \overline{\Vdash} \circ \overline{\in}$. In other words, there is some object α such that $\sigma(s) \overline{\Vdash} \alpha$ and $\alpha \overline{\in} \Phi$. Clearly then $\mathbb{S}, s \Vdash \nabla\alpha$, and so by the assumption we have $\mathbb{S}, s \Vdash a$.

Finally, for the rule (∇4), fix some finite set ϕ of formulas. It suffices to prove that, for an arbitrary T-coalgebra $\mathbb{S} = (S, \sigma)$, if $\mathbb{S} \Vdash \top \preceq \bigvee \phi$, then for every point $s \in \mathbb{S}$ we can find an $\alpha \in T(\phi)$ such that $\mathbb{S}, s \Vdash \nabla\alpha$. From the assumption it follows that every state in $\mathbb{S}$ satisfies some formula in ϕ. We may formulate this using a function $f : S \to \phi$ such that $\mathbb{S}, s \Vdash f(s)$, for all $s \in S$, or, equivalently, $\mathrm{Gr}(f) \subseteq \Vdash$. But then by the properties of relation lifting, we find that $\mathrm{Gr}(Tf) \subseteq \overline{\Vdash}$. Now consider an arbitrary state s in $\mathbb{S}$, and let $\alpha \in T(\phi)$ be the element $(Tf)(\sigma(s))$. Then $(\sigma(s), \alpha) \in \mathrm{Gr}(Tf) \subseteq \overline{\Vdash}$, and so by the truth definition of ∇, we find that $\mathbb{S}, s \Vdash \nabla\alpha$. That is, we have found our α.

5 Completeness

The completeness proof will use a standard coalgebraic technique, namely to prove completeness via one-step-completeness. This is well-known in domain theory (see e.g. Abramsky [1]) and was introduced to coalgebra by Pattinson [20]. Subsequently, it was used in for instance [7, 13, 17].

The main idea is the following. First, we show that Moss' logic $(\mathcal{L}, \equiv)$ can be stratified into layers $(\mathcal{L}_n, \equiv_n)$, with all layers at $n+1$ arising in a uniform way from layers at n (Proposition 37). This uniform construction can be described by means of a 'one-step version' of the derivation system **M**. Technically, it is described by a functor $\mathbb{M} : \mathsf{BA} \to \mathsf{BA}$, which constructs $(\mathcal{L}_{n+1}/{\equiv_{n+1}})$ as $\mathbb{M}(\mathcal{L}_n/{\equiv_n})$. Our main technical result consists of showing that this one-step proof system is complete in a suitable sense (Proposition 48). Then, using a standard argument, completeness follows from one-step completeness (Proposition 50).

REMARK 17. Continuing from Remark 12, the proof system **M** defines a quotient $\mathbb{L} \to \mathbb{M}$. Then $\delta_X : \mathbb{M}\mathbb{P}X \to \mathbb{P}TX$ is given by factoring $\gamma : \mathbb{L}\mathbb{P} \to \mathbb{P}T$. $\mathbb{M}$-algebras are the Boolean algebras with operator for the Moss modality. The initial $\mathbb{M}$-algebra $\mathcal{M}$ is Lindenbaum algebra of Moss' logic (Proposition 39) and δ_X is injective (Proposition 48), which then implies completeness.

5.1 A one-step proof system

Recall the definition of $\mathcal{L}_n(X)$ from Definition 4.

DEFINITION 18. Let $\mathfrak{L}(X) = \mathcal{L}_0\{\nabla\alpha \mid \alpha \in T_\omega X\}$. In the following we consider $\mathfrak{L}$ to be a functor $\mathsf{Set} \to \mathsf{Set}$, which maps $f : X \to Y$ to the function

$\mathfrak{L}(f) : \mathfrak{L}(X) \to \mathfrak{L}(Y)$ that extends the map $\nabla\alpha \mapsto \nabla(Tf)(\alpha)$ via Boolean operations.

$\mathfrak{L}$ constructs formulas step-wise: $\mathcal{L}_n = \mathfrak{L}^n(\mathcal{L}_0)$. Next we show how to construct $\equiv_n$ step-wise. In order to smooth our presentation, it is convenient in the following definition to assume that the generators are already closed under Boolean operations.

DEFINITION 19. Let A be the carrier of an algebra for the Boolean signature and let $R \subseteq \mathcal{L}_0(A) \times \mathcal{L}_0(A)$ be a set of pairs called *relations*. Using the laws of Boolean algebra, with pairs $(a, a') \in R$ as additional axioms $a \preceq a'$, one may generate a congruence relation $\equiv_R$ on the set $A \times A$. We say that the pair (A, R) is a *presentation of* the Boolean algebra $A/{\equiv_R}$ and denote this algebra as $\mathrm{BA}\langle A, R\rangle$. A homomorphism $f : A \to B$ of algebras for the Boolean signature is a presentation morphism from $(A, \equiv_R)$ to $(B, \equiv_S)$ if $a_1 \equiv_R a_2$ implies $f(a_1) \equiv_S f(a_2)$ for all $a_1, a_2 \in A$. The category of presentations and presentation morphisms is denoted by PRS.

The notion of a presentation morphism is motivated by the following lemma which is not difficult to prove.

LEMMA 20. *Let $f : (A, R) \to (B, S)$ be a presentation morphism. Then the function $[f] : A/{\equiv_R} \to B/{\equiv_S}$ that maps the equivalence class of an element $a \in A$ to the equivalence class of $f(a)$ is well-defined. Moreover $[f]$ is a Boolean homomorphism.*

EXAMPLE 21. The *standard presentation* of a Boolean algebra $\mathbb{B}$ is the pair $(U\mathbb{B}, \leq)$ where $\leq$ is the relation on terms over $U\mathbb{B}$ induced by the partial order of $\mathbb{B}$.

The derivation system $\mathbf{M}$ is essentially a 'one-step' derivation system since in every rule involving the modality, every occurrence of α is under the scope of exactly one ∇. The following definition makes this precise.

DEFINITION 22. Let (X, R) be a presentation. The *one-step proof system* $\mathbf{M}(X, R)$ is the version of $\mathbf{M}$ in which all inequalities $b_1 \preceq b_2$ from R (that is, with $(b_1, b_2) \in R$) are additional axioms, and in which *only* elements from X and $\mathfrak{L}(X)$ may be used.[2] We denote the associated notion of derivability by $\vdash_{\mathbf{M}(X,R)}$. Furthermore, for a_1, a_2 in $\mathfrak{L}(X)$, we write $a_1 \preceq_{\mathbf{M}(X,R)} a_2$ if $\vdash_{\mathbf{M}(X,R)} a_1 \preceq a_2$; and $a_1 \equiv_{\mathbf{M}(X,R)} a_2$ iff $a_1 \preceq_{\mathbf{M}(X,R)} a_2$ and $a_2 \preceq_{\mathbf{M}(X,R)} a_1$. We let $\mathbb{M}(X, R)$ denote the Boolean algebra presented by $(\mathfrak{L}(X), \preceq_{\mathbf{M}(X,R)})$.

In case (X, R) is the standard representation of a Boolean algebra $\mathbb{A}$, we write $\mathbf{M}(\mathbb{A})$ for the one-step proof system based on the standard presentation of $\mathbb{A}$, and $\vdash_{\mathbf{M}(\mathbb{A})}$, $\preceq_{\mathbf{M}(\mathbb{A})}$, $\equiv_{\mathbf{M}(A)}$ and $\mathbb{M}(\mathbb{A})$ for the associated notions.

The next subsection shows that $\mathbb{M}(\mathbb{A})$ is not only a Boolean algebra, but that $\mathbb{M}$ is a functor on the category of Boolean algebras. This will allow us, in Section 5.3, to recover the Lindenbaum algebra of $\mathbf{M}$ as the union of algebras $(\mathbb{M}^n\mathbf{2})_{n<\omega}$, where $\mathbb{M}^{n+1}\mathbf{2} = \mathbb{M}(\mathbb{M}^n\mathbf{2})$.

[2] In $(\nabla 1)$ – $(\nabla 4)$, the b_i range over X, the a, a' over $\mathfrak{L}(X)$, $\alpha, \alpha' \in T_\omega X$, $\phi \subseteq X$.

Sections 5.2 and 5.3 are rather technical and are needed mainly[3] to deduce completeness from one-step completeness. Our main result, the one-step completeness (Proposition 48), does not depend on Sections 5.2 and 5.3 and the reader might wish to go directly to Section 5.4.

5.2 Technical interlude: $\mathbb{M}$ as a functor of Boolean algebras

We want to define a functor $\mathbb{M} : \mathsf{BA} \to \mathsf{BA}$ associated with the one-step proof system. As a first step in this direction we note that $\mathfrak{L}$ can be seen as a functor on the category PRS. The proof-theoretic content of this is that a morphism $f : (A, R) \to (B, S)$ between presentations extends to a map of derivations between the one-step proof systems $\mathbf{M}(A, R)$ and $\mathbf{M}(B, S)$.

PROPOSITION 23. *Let (A, R) and (B, S) be presentations in* PRS *and let $f : (A, R) \to (B, S)$ be a presentation morphism. The function $\mathfrak{L}(f)$ is a presentation morphism from $(\mathfrak{L}(A), \preceq_{\mathbf{M}(A,R)})$ to $(\mathfrak{L}(B), \preceq_{\mathbf{M}(B,S)})$, i.e., for all $a', a'' \in \mathfrak{L}(A)$,*

(4) $\vdash_{\mathbf{M}(A,R)} a' \preceq a''$ *implies* $\vdash_{\mathbf{M}(B,S)} \mathfrak{L}(f)(a') \preceq \mathfrak{L}(f)(a'')$.

Proof. Let $c', c'' \in \mathfrak{L}(A)$ or $c', c'' \in A$. One shows by structural induction on the derivation of $c' \preceq_{\mathbf{M}(A,R)} c''$ that substituting, in the derivation, each occurrence of $a \in A$ with $f(a) \in B$ and each occurrence of $a' \in \mathfrak{L}(A)$ with $\mathfrak{L}f(a') \in \mathfrak{L}(B)$ yields a proof of $\mathfrak{L}(f)(c') \preceq_{\mathbf{M}(B,S)} \mathfrak{L}(f)(c'')$ or $f(c') \preceq_{\mathbf{M}(B,S)} f(c'')$, respectively. Let us give some details of the induction argument (the case in which the derivation ends by an application of $(\nabla 2)$ is similar to the cases $(\nabla 1)$, $(\nabla 3)$ and $(\nabla 4)$ and has been omitted due to space limitations).

Case $c' \preceq_{\mathbf{M}(A,R)} c''$ is a (Boolean) axiom. If $c', c'' \in \mathfrak{L}(A)$ it is clear from the definition of $\mathfrak{L}$ that $\mathfrak{L}(f)(c') \preceq_{\mathbf{M}(B,S)} \mathfrak{L}(f)(c'')$ is an axiom as well. If $c, c'' \in A$, $f(c') \preceq_{\mathbf{M}(B,S)} f(c'')$ follows from the fact that f is a morphism in PRS.

Case $c' \preceq_{\mathbf{M}(A,R)} c''$ is obtained by a derivation that ends by the application of some rule R for classical propositional logic. Then $\mathfrak{L}(f)(c') \preceq_{\mathbf{M}(B,S)} \mathfrak{L}(f)(c'')$ or $f(c') \preceq_{\mathbf{M}(B,S)} f(c'')$ can be easily obtained by applying the inductive hypothesis to the premises of R.

Case $(\nabla 1)$ Suppose $c' = \nabla\alpha$ and $c'' = \nabla\beta$ and suppose that there is a derivation $\mathcal{D}$ of $c' \preceq_{\mathbf{M}(A,R)} c''$ such that $(\nabla 1)$ is the last ruled applied in $\mathcal{D}$, i.e., $\mathcal{D}$ ends with

$$\frac{\{b_1 \preceq b_2 \mid (b_1, b_2) \in Z\}}{\nabla\alpha \preceq \nabla\beta}\ (\alpha, \beta) \in \overline{Z}$$

Because f is a Boolean homomorphism we get $f(b_1) \leq f(b_2)$ for all $(b_1, b_2) \in Z$. Moreover one can easily calculate that $(\nabla(Tf)(\alpha), \nabla(Tf)(\beta)) \in \overline{Z'}$

[3] But Propositions 26 (via 23), 28 and 37 have a proof-theoretic interpretation and are of independent interest.

with $Z' := \{(f(b_1), f(b_2)) \mid (b_1, b_2) \in Z\}$. Therefore we can apply rule (∇1) again: from the premisses $\{a_1 \preceq a_2 \mid (a_1, a_2) \in Z'\}$ we obtain $\mathfrak{L}(f)(\nabla\alpha) = \nabla(Tf)(\alpha) \preceq_{\mathbf{M}(B,S)} \nabla(Tf)(\beta) = \mathfrak{L}(f)(\nabla\beta)$ as required.

Case (∇3) Suppose $b' = \nabla(T\bigvee)(\Phi)$ and suppose that there is a derivation $\mathcal{D}$ of $\nabla(T\bigvee)(\Phi) \preceq_{\mathbf{M}(A,R)} b''$ that ends with the following rule:

$$\frac{\{\nabla\alpha \preceq b'' \mid \alpha\overline{\in}\Phi\}}{\nabla(T\bigvee)(\Phi) \preceq b''}$$

By the inductive hypothesis we have

(5) $\nabla(Tf)(\alpha) \preceq_{\mathbf{M}(B,S)} \mathfrak{L}(f)(b'')$ for all $\alpha\overline{\in}\Phi$.

Furthermore we get

$$(6) \qquad \mathfrak{L}(f)\left(\nabla(T\bigvee)(\Phi)\right) \overset{\text{by Def.}}{=} \nabla(Tf)((T\bigvee)(\Phi))$$

$$(7) \qquad \overset{f \text{ is hom.}}{=} \nabla(T\bigvee)((T\mathcal{P}f)(\Phi)).$$

Moreover the following chain of equivalences holds:

$$\begin{aligned} \alpha'\overline{\in}(T\mathcal{P}f)(\Phi) \quad &\text{iff} \quad (\alpha', \Phi) \in \overline{(\in \circ \mathrm{Gr}(\mathcal{P}f)^{\smile})} = \overline{\mathrm{Gr}(f)^{\smile} \circ \in} \\ &\text{iff} \quad \alpha' = (Tf)(\alpha) \text{ for some } \alpha\overline{\in}\Phi \end{aligned}$$

The latter equivalence together with (5) yields $\nabla\alpha' \preceq_{\mathbf{M}(B,S)} \mathfrak{L}(f)(b'')$ for all $\alpha' \overline{\in} (T\mathcal{P}f)(\Phi)$. By applying rule ($\nabla$3) we obtain

$$\mathfrak{L}(f)\left(\nabla(T\bigvee)(\Phi)\right) \overset{(7)}{=} \nabla(T\bigvee)((T\mathcal{P}f)(\Phi)) \preceq_{\mathbf{M}(B,S)} \mathfrak{L}(f)(b'')$$

as required.

Case(∇4) Consider now the case that $b'' = \nabla\alpha$ and that there is a derivation $\mathcal{D}$ of $b' \leq_{\mathbf{M}(A,R)} \nabla\alpha$ that ends as follows:

$$\frac{\{b' \wedge \nabla\alpha' \preceq \bot \mid \alpha' \in T_\omega(\phi), \alpha' \neq \alpha\} \qquad \top \preceq \bigvee\phi}{b' \preceq \nabla\alpha}$$

Again we can inductively assume that

(8) $\mathfrak{L}(f)(b') \wedge \nabla(Tf)(\alpha') \preceq \bot$ for all $\alpha' \in T_\omega(\phi)$ s.t. $\alpha' \neq \alpha$.

Let us put $\psi := f[\phi]$. Then by the fact that f is a homomorphism of Boolean algebras we get $\top \preceq \bigvee\psi$. It is clear that for all $\beta' \in T_\omega(\psi)$ such that $\beta' \neq Tf(\alpha)$ there is an $\alpha' \neq \alpha$ such that $Tf(\alpha') = \beta'$. Together with (8) this implies that $\mathfrak{L}(f)(b') \wedge \nabla\beta' \preceq \bot$ for all $\beta' \in T_\omega(\psi)$ such that $(Tf)(\alpha) \neq \beta'$. Now we can apply rule (∇4) which yields $\mathfrak{L}(f)(b') \preceq \nabla(Tf)(\alpha)$. This is what we had to show, because $\mathfrak{L}(f)(\nabla\alpha) = \nabla(Tf)(\alpha)$. ∎

As a consequence, one obtains that the value of $\mathbb{M}$ does not depend on a choice of presentation.

PROPOSITION 24. *If* (X, R) *and* (X', R') *generate isomorphic Boolean algebras, then* $\mathbb{M}(X, R) \cong \mathbb{M}(X', R')$.

Proof. Clearly it suffices to prove that $\mathbb{M}(X, R) \cong \mathbb{M}(B, \leq)$, where $(B, \leq)$ is the standard presentation of $\mathbb{B} := \mathrm{BA}\langle X, R\rangle$. Recall that B consists of equivalence classes of the set $\mathcal{L}_0(X)$, let $f : \mathcal{L}_0(X) \to B$ be the quotient map, and let $m : B \to \mathcal{L}_0(X)$ be a function such that $f \circ m = \mathrm{id}_B$. Then it is clear that f and m are presentation morphisms between (X, R) and $(B, \leq)$. Then by Proposition 23 also $\mathfrak{L}(f)$ and $\mathfrak{L}(m)$ are presentation morphisms and $[\mathfrak{L}(f)]$ is an surjective BA-homomorphism from $\mathbb{M}(X, R)$ to $\mathbb{M}(B \leq)$. In order to prove the claim we show that $[\mathfrak{L}(f)]$ is also injective. Note first that for all $x \in X$ we have $m(f(x)) \equiv_R x$. Using axiom $(\nabla 1)$ it can be shown that this implies $\mathfrak{L}(m)(\mathfrak{L}(f)(\nabla\alpha)) \equiv \nabla\alpha$ for all $\nabla\alpha \in \mathfrak{L}(X)$ which can be inductively extended to $\mathfrak{L}(m)(\mathfrak{L}(f)(a)) \equiv a$ for all $a \in \mathfrak{L}(X)$. But then for all $a, a' \in \mathfrak{L}(A)$ such that $\mathfrak{L}(f)(a) \equiv \mathfrak{L}(f)(a')$ we have

$$a \equiv \mathfrak{L}(m)(\mathfrak{L}(f)(a)) \equiv \mathfrak{L}(m)(\mathfrak{L}(f)(a')) \equiv a',$$

which implies that $[\mathfrak{L}(f)]$ has to be injective as required. ■

Recall that, given a Boolean algebra $\mathbb{A}$ with standard presentation $(A, \leq)$, we let $\mathbb{M}\mathbb{A}$ denote the Boolean algebra $\mathbb{M}(A, \leq)$ (Definition 22). $\mathbb{M}$ is thus an operation on the class of Boolean algebras. We will now see that in fact it is (or can be extended to) a *functor* on the category of Boolean algebras.

DEFINITION 25. The quotient map from $\mathfrak{L}(U\mathbb{A})$ to $\mathbb{M}\mathbb{A} = (\mathfrak{L}(U\mathbb{A})/{\equiv_{\mathbf{M}(\mathbb{A})}})$ will be denoted by $\mathfrak{q}_{\mathbb{A}}$.

Another consequence of Proposition 23 is:

PROPOSITION 26. $\mathbb{M}$ *is a functor on the category of Boolean algebras.*

Proof. Let $f : \mathbb{A} \to \mathbb{B}$ be a Boolean homomorphism and let A and B denote the underlying sets of $\mathbb{A}$ and $\mathbb{B}$, respectively. Obviously any homomorphism $f : \mathbb{A} \to \mathbb{B}$ is a presentation morphism from $(A, \leq)$ to $(B, \leq)$ where $(A, \leq)$ and $(B, \leq)$ denote the standard presentations of $\mathbb{A}$ and $\mathbb{B}$. So by Proposition 23, $\mathfrak{L}(f) : (\mathfrak{L}(A), \preceq_{\mathbf{M}(A,\leq)}) \to (\mathfrak{L}(B), \preceq_{\mathbf{M}(B,\leq)})$ is also a presentation morphism and we can define a Boolean homomorphism $\mathbb{M}f : \mathbb{M}\mathbb{A} \to \mathbb{M}\mathbb{B}$ by putting $\mathbb{M}f := [\mathfrak{L}(f)]$. It is easy to see that this $\mathbb{M}$ satisfies the usual functor conditions, i.e., that $\mathbb{M}(\mathrm{id}) = \mathrm{id}$ and $\mathbb{M}(f \circ g) = \mathbb{M}f \circ \mathbb{M}g$. ■

REMARK 27. Wrt Remark 12, we note that $\mathbb{M}$ is a quotient $\mathbb{L} \to \mathbb{M}$.

It turns out that $\mathbb{M}$ has some nice properties, which will be of use later on. In particular, we will show that $\mathbb{M}$ is *finitary* (or ω-accessible) which means, proof-theoretically, that for any Boolean algebra $\mathbb{A}$, a derivation of $\vdash_{\mathbf{M}(\mathbb{A})} a_1 \preceq a_2$ can be carried out in a *finite* subalgebra of $\mathbb{A}$. A fairly easy consequence of this is the second useful property given below, namely, that $\mathbb{M}$ *preserves embeddings*.

PROPOSITION 28. $\mathbb{M}$ *is a finitary functor that preserves embeddings.*

Proof. Fix a Boolean algebra $\mathbb{A}$ with carrier set $A := U\mathbb{A}$. Given two elements $a_1, a_2 \in \mathfrak{L}(A)$, consider the collection of elements of A that occur as *subformulas* of a_1 and a_2. It follows from our earlier remarks on subformulas that this is a *finite* set, which then generates a finite subalgebra $\mathbb{A}'$ of $\mathbb{A}$. By definition we have $a_1, a_2 \in \mathfrak{L}(A')$.

We claim

(9) $\vdash_{\mathbf{M}(\mathbb{A})} a_1 \preceq a_2$ iff $\vdash_{\mathbf{M}(\mathbb{A}')} a_1 \preceq a_2$.

The interesting direction of (9) is from left to right. The key observation here is that from the fact that $\mathbb{A}'$ is a finite subalgebra of $\mathbb{A}$, we may infer the existence of a *surjective* homomorphism $f : \mathbb{A} \to \mathbb{A}'$ such that $f(a') = a'$ for all $a' \in A'$. (In other words, $\mathbb{A}'$ is a *retract* of $\mathbb{A}$.) There are various ways to prove this statement; here we refer to Sikorski's theorem that complete Boolean algebras are injective [25]. But if f is a homomorphism, by Proposition 23 it follows from $\vdash_{\mathbf{M}(\mathbb{A})} a_1 \preceq a_2$ that $\vdash_{\mathbf{M}(\mathbb{A}')} \mathfrak{L}(f)(a_1) \preceq \mathfrak{L}(f)(a_2)$. Since $a_1, a_2 \in \mathfrak{L}(A')$ and f restricts to the identity on A', we may conclude that $\mathfrak{L}(f)(a_i) = a_i$, for both $i = 1, 2$. Thus, indeed, $\vdash_{\mathbf{M}(\mathbb{A}')} a_1 \preceq a_2$. Using the fact that every BA is the directed colimit (union) of finite Boolean algebras, the finitariness of $\mathbb{M}$ follows by a standard argument.

For the second part of the proof, let $e : \mathbb{A} \to \mathbb{B}$ be an embedding. Without loss of generality we will assume that e is actually the inclusion (that is, $\mathbb{A}$ is a subalgebra of $\mathbb{B}$). In order to prove that $\mathbb{M}e : \mathbb{MA} \to \mathbb{MB}$ is also injective, it suffices to prove the following, for all $a_1, a_2 \in A$:

(10) $\vdash_{\mathbf{M}(\mathbb{B})} a_1 \preceq a_2$ implies $\vdash_{\mathbf{M}(\mathbb{A})} a_1 \preceq a_2$.

But the proof of (10) simply follows from two applications of (9). ■

As a straightforward corollary of Proposition 28, we obtain the existence of an *initial* $\mathbb{M}$-algebra. Furthermore, this initial algebra is obtained as the union of the initial $\mathbb{M}$-sequence to be defined now.

DEFINITION 29. We define $j_0 : \mathbf{2} \to \mathbb{M}\mathbf{2}$ to be the unique embedding of the two-element Boolean algebra $\mathbf{2}$ into $\mathbb{M}\mathbf{2}$, and inductively we define $j_{n+1} : \mathbb{M}^n\mathbf{2} \to \mathbb{M}(\mathbb{M}^n\mathbf{2})$ to be $\mathbb{M}j_n$. Let $\mathcal{M}$ be the colimit of the sequence $(\mathbb{M}^n\mathbf{2})_{n<\omega}$.

We take the liberty to consider $\mathcal{M}$ as an $\mathbb{M}$-algebra, a Boolean algebra, or a set, depending on the context. Since $\mathbb{M}$ is finitary (Proposition 28), we have the following.

COROLLARY 30. *$\mathcal{M}$ is the initial $\mathbb{M}$-algebra.*

As $\mathbb{M}$ preserves embeddings, all maps in the initial sequence $(\mathbb{M}^n\mathbf{2})_{n<\omega}$ are injective. This means that we can consider the initial $\mathbb{M}$-algebra as a *union* of its approximants $\mathbb{M}^n\mathbf{2}$.

5.3 Technical interlude: stratification of the Moss logic

As we will see now, the one-step version of $\mathbf{M}$ allows for a layer-wise construction of the (inter)derivability relation between formulas.

DEFINITION 31. For each n, we define relations $\leq_n$ and $\equiv_n$ on $\mathcal{L}_n$. For $n = 0$, we simply let $\leq_0$ on the set $\mathcal{L}_0$ of all closed Boolean formulas denote derivability (in Boolean logic). Inductively, we define $\leq_{n+1}$ as derivability in the one-step proof system $\mathbf{M}(\mathcal{L}_n, \leq_n)$. Finally, $a \equiv_n b \Leftrightarrow (a \leq_n b$ and $a \leq_n b)$.

The following proposition reveals the crucial role of $\mathbb{M}$ in this stratification. Its proof proceeds via a straightforward inductive argument, of which the inductive step is an immediate consequence of Proposition 24.

PROPOSITION 32. *For all n, $\mathcal{L}_n/{\equiv_n} \cong \mathbb{M}^n\mathbf{2}$.*

DEFINITION 33. For every $n \in \omega$ we let $\mathfrak{q}_n$ be the quotient map from $\mathcal{L}_n$ onto $U\mathbb{M}^n\mathbf{2} \cong \mathcal{L}_n/{\equiv_n}$. Furthermore we let $i_0 : \mathcal{L}_0 \to \mathcal{L}_1$ be the obvious embedding, and inductively we define $i_{n+1} = \mathfrak{L}(i_n)$.

PROPOSITION 34. *We have $\mathcal{L}_n = \mathfrak{L}^n(\mathcal{L}_0)$ and for all $n \in \omega$ the map i_n is the inclusion of $\mathcal{L}_n$ into $\mathcal{L}_{n+1}$.*

Due to lack of space we omit the simple induction argument. The next proposition establishes a connection between the embeddings $j_n : \mathbb{M}^n\mathbf{2} \to \mathbb{M}^{n+1}\mathbf{2}$ and the inclusions $i_n : \mathcal{L}_n \to \mathcal{L}_{n+1}$. The proof is based on the following lemma. (i) is proved using (∇1) and (ii) follows from Proposition 26.

LEMMA 35. *For all $n < \omega$, we have (i) $\mathfrak{q}_{n+1} = \mathfrak{q}_{\mathbb{M}^n\mathbf{2}} \circ \mathfrak{L}(\mathfrak{q}_n)$ and (ii) $Uj_n \circ \mathfrak{q}_{\mathbb{M}^{n-1}\mathbf{2}} = \mathfrak{q}_{\mathbb{M}^n\mathbf{2}} \circ \mathfrak{L}(Uj_{n-1})$:*

$$
\begin{array}{ccc}
\mathfrak{L}(\mathcal{L}_n) & \xrightarrow{\ \mathfrak{q}_{n+1}\ } & U\mathbb{M}^{n+1}\mathbf{2} \\
& {\scriptstyle \mathfrak{L}(\mathfrak{q}_n)} \searrow \quad \nearrow {\scriptstyle \mathfrak{q}_{\mathbb{M}^n\mathbf{2}}} & \\
& \mathfrak{L}(U\mathbb{M}^n\mathbf{2}) &
\end{array}
\qquad
\begin{array}{ccc}
\mathfrak{L}(U\mathbb{M}^{n-1}\mathbf{2}) & \xrightarrow{\ \mathfrak{L}(Uj_{n-1})\ } & \mathfrak{L}(U\mathbb{M}^n\mathbf{2}) \\
{\scriptstyle \mathfrak{q}_{\mathbb{M}^{n-1}\mathbf{2}}} \downarrow & & \downarrow {\scriptstyle \mathfrak{q}_{\mathbb{M}^n\mathbf{2}}} \\
U\mathbb{M}^n\mathbf{2} & \xrightarrow[\ Uj_n\ ]{} & U\mathbb{M}(\mathbb{M}^n\mathbf{2})
\end{array}
$$

PROPOSITION 36. *For all $n \in \omega$ we have $Uj_n \circ \mathfrak{q}_n = \mathfrak{q}_{n+1} \circ i_n$:*

$$
\begin{array}{ccc}
\mathcal{L}_n & \xrightarrow{\ i_n\ } & \mathcal{L}_{n+1} \\
{\scriptstyle \mathfrak{q}_n} \downarrow & & \downarrow {\scriptstyle \mathfrak{q}_{n+1}} \\
U\mathbb{M}^n\mathbf{2} & \xrightarrow[\ Uj_n\ ]{} & U\mathbb{M}^{n+1}\mathbf{2}
\end{array}
$$

Proof. The case $n = 0$ can be easily checked. Consider now $n = m+1$ and some $\nabla\alpha \in \mathcal{L}_{m+1} = \mathfrak{L}^m(\mathcal{L}_0)$. Then

$$
\begin{aligned}
j_{m+1}(\mathfrak{q}_{m+1}(\nabla\alpha)) &\overset{\text{(i)}}{=} j_{m+1}(\mathfrak{q}_{\mathbb{M}^m\mathbf{2}}(\mathfrak{L}(\mathfrak{q}_m)(\nabla\alpha)) \\
&\overset{\text{(ii)}}{=} \mathfrak{q}_{\mathbb{M}^{m+1}\mathbf{2}}(\mathfrak{L}(j_m)(\mathfrak{L}(\mathfrak{q}_m)(\nabla\alpha)) \\
&= \mathfrak{q}_{\mathbb{M}^{m+1}\mathbf{2}}(\mathfrak{L}(j_m \circ \mathfrak{q}_m)(\nabla\alpha)) \\
&\overset{\text{I.H.}}{=} \mathfrak{q}_{\mathbb{M}^{m+1}\mathbf{2}}(\mathfrak{L}(\mathfrak{q}_{m+1} \circ i_m)(\nabla\alpha)) \\
&\overset{\text{(i)}}{=} (\mathfrak{q}_{m+2} \circ i_{m+1})(\nabla\alpha)
\end{aligned}
$$

■

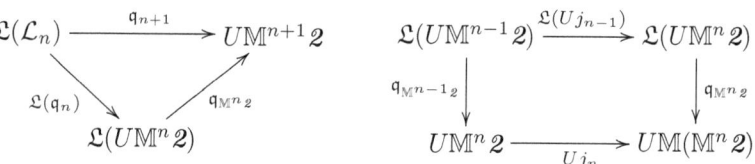

The following proposition is crucial. It shows that if we have $\vdash_{\mathbf{M}} a \preceq b$ for formulas a, b of depth n, there always is a derivation that does not employ formulas of depth greater than n. This is typical for axiomatisations where each variable is under the scope of precisely one modal operator. The situation here is slightly more complicated than usual since our rules allow infinite sets of premises.

PROPOSITION 37. *Let a and b be formulas. Then*

1. *If $a, b \in \mathcal{L}_n$, and $a \leq_m b$ for some $m > n$, then $a \leq_n b$.*

2. *If $a, b \in \mathcal{L}_n$, then $\vdash_{\mathbf{M}} a \preceq b$ iff $a \leq_n b$.*

Proof. For Part 1 of the proposition, it suffices to confine attention to the case where $m = n + 1$, which is a consequence of Proposition 36 and the fact that the injective BA-morphism j_n reflects the order.

Part 2 of the proposition is proved by induction on the complexity of derivations in $\mathbf{M}$. Here we discuss a sample case of the inductive step, namely, where the last applied rule was $(\nabla 4)$:

$$\frac{\{a \wedge \nabla\alpha' \preceq \bot \mid \alpha' \in T_\omega(\phi), \alpha' \neq \alpha\} \qquad \top \preceq \bigvee \phi}{a \preceq \nabla\alpha}$$

Inductively, there is some natural number k such that $\top \leq_k \phi$. Let $m := \max\{d(b) \mid b \in \phi \cup \{a\}\}$, where $d(b)$ denotes the depth of the formula b. Then clearly m is (well-defined as) a finite natural number since ϕ is a finite set by assumption. Then $\phi \subseteq \mathcal{L}_m$, and so $\nabla\beta \in \mathcal{L}_{m+1}$ for all $\beta \in T_\omega\phi$. Since also $a \in \mathcal{L}_m$, by the first part of the proposition we obtain $a \wedge \nabla\alpha' \leq_{m+1} \bot$ for all $\alpha' \in T_\omega(\phi) \setminus \{\alpha\}$. Thus, with $p = \max\{m+1, k\}$, we see that all premisses of the final rule are p-derivable. But then the conclusion is $p+1$-derivable, and then, by part 1, n-derivable (where we assumed that $a, \nabla\alpha \in \mathcal{L}_n$). ■

The next proposition shows that $\mathcal{M}$ is the Lindenbaum algebra of $\mathcal{L}$ modulo the proof system $\mathbf{M}$. To see this, recall from Proposition 5 that $(\mathcal{L}, \neg, \wedge, \vee, \nabla)$ is the initial algebra for the functor $M = Id + Id \times Id + Id \times Id + T_\omega$. Since $\mathcal{M}$ can also be seen as an algebra of this kind, it follows from initiality of $\mathcal{L}$ that there is a unique quotient $\mathfrak{q} : \mathcal{L} \to \mathcal{M}$. The next proposition states that the kernel of $\mathfrak{q}$ is the interderivability relation $\equiv$ according to the proof system $\mathbf{M}$.

LEMMA 38. *The quotient maps $\mathfrak{q}_n : \mathcal{L}_n \to U\mathbb{M}^n\mathbf{2}$ are the restrictions of $\mathfrak{q} : \mathcal{L} \to \mathcal{M}$. More precisely denote by $k_n : M^n\mathbf{2} \to \mathcal{M}$ the embeddings of the initial sequence of $\mathcal{M}$ and by $l_n : \mathcal{L}_n \to \mathcal{L}$ the inclusions. Then the claim is that $q \circ l_n = k_n \circ \mathfrak{q}_n$.*

Proof. To prove this we need to observe that, by definition, $\mathfrak{q} : \mathcal{L} \to \mathcal{M}$ is the unique map for which $\mathfrak{q} \circ l_n = f_n$, where the $f_n : \mathcal{L}_n \to \mathcal{M}$ are given inductively by $f_{n+1} = \mu \circ \mathfrak{q}_{\mathcal{M}} \circ \mathfrak{L}(f_n)$ and $\mu : \mathbb{M}\mathcal{M} \to \mathcal{M}$ is the structure map of the $\mathbb{M}$-algebra $\mathcal{M}$. We then proceed to show by induction that $k_n \circ \mathfrak{q}_n = f_n$. Indeed, $f_{n+1} = \mu \circ \mathfrak{q}_{\mathcal{M}} \circ \mathfrak{L}(f_n) \stackrel{\text{indhyp}}{=} \mu \circ \mathfrak{q}_{\mathcal{M}} \circ \mathfrak{L}(k_n) \circ \mathfrak{L}\mathfrak{q}_n = \mu \circ \mathbb{M}k_n \circ \mathfrak{q}_{\mathbb{M}^n\mathbf{2}} \circ \mathfrak{L}\mathfrak{q}_n = k_{n+1} \circ \mathfrak{q}_{\mathbb{M}^n\mathbf{2}} \circ \mathfrak{L}\mathfrak{q}_n = k_{n+1} \circ \mathfrak{q}_{n+1}$. where the third equation holds by definition of $\mathbb{M}k_n = [\mathfrak{L}(k_n)]$ and the last equation is an instance of Lemma 35(i). ■

PROPOSITION 39. *The kernel of the quotient map* $\mathfrak{q} : \mathcal{L} \to \mathcal{M}$ *is* $\equiv$.

Proof. Recall that $\mathcal{M}$ can be seen as the union of the initial sequence $(\mathbb{M}^n\mathbf{2})_{n<\omega}$, that $\mathcal{L} = \bigcup \mathcal{L}_n$, and that the maps $\mathfrak{q}_n : \mathcal{L}_n \to \mathbb{M}^n\mathbf{2}$ map formulas to their equivalence classes. By Lemma 38, each $\mathfrak{q}_n$ is the restriction of $\mathfrak{q} : \mathcal{L} \to \mathcal{M}$ to $\mathcal{L}_n \to \mathbb{M}^n\mathbf{2}$. Then, $\mathfrak{q}(a) = \mathfrak{q}(b)$ iff there is an n such that $\mathfrak{q}_n(a) = \mathfrak{q}_n(b)$. This, in turn, is equivalent to $a \equiv_n b$ (by Definition 33) and then to $a \equiv b$ by Proposition 37. ■

5.4 Semantics of $\mathbb{M}$ and Moss algebras

Let us first summarise the two preceding sections in the following diagram:

$$\begin{array}{ccccccc}
\mathcal{L}_0 & \xrightarrow{i_0} & \mathcal{L}_1 & \xrightarrow{i_1} & \mathcal{L}_2 & \cdots & \mathcal{L} \\
\downarrow{\scriptstyle \mathfrak{q}_0} & & \downarrow{\scriptstyle \mathfrak{q}_1} & & \downarrow{\scriptstyle \mathfrak{q}_2} & & \downarrow{\scriptstyle \mathfrak{q}} \\
\mathbf{2} & \xrightarrow{j_0} & \mathbb{M}\mathbf{2} & \xrightarrow{j_1} & \mathbb{M}^2\mathbf{2} & \cdots & \mathcal{M}
\end{array}$$

$\mathcal{L}$ is the union of the $\mathcal{L}_n$. $\mathbb{M}$ is a functor on BA, $\mathbf{2}$ is the initial BA and $\mathcal{M}$ is the colimit of the sequence $\mathbb{M}^n\mathbf{2}$. Since $\mathbb{M}$ preserves injections and is finitary, $\mathcal{M}$ is the initial $\mathbb{M}$-algebra and can be considered to be the union of the $\mathbb{M}^n\mathbf{2}$. $\mathfrak{q}$ is the quotient of $\mathcal{L}$ wrt interderivability $\equiv$ and the $\mathfrak{q}_n$ are the restrictions of $\mathfrak{q}$.

Thus, up to interderivability, we can work with $\mathbb{M}$ and $\mathcal{M}$ instead of $\mathcal{L}$. In this section, we define the semantics of the logic directly in terms of $\mathbb{M}$ and show that it agrees with the previously given one.

The relationship between $\mathbb{M}$ and T is provided by a natural transformation $\delta : \mathbb{M}\mathbb{P} \to \mathbb{P}T$. For the definition of δ, recall the natural transformation ρ from Definition 9.

DEFINITION 40. Given a set S, define the map $\tilde{\rho} : \mathfrak{L}(PS) \to PTS$ as follows. For $\alpha \in T_\omega PS$, we let $\tilde{\rho}(\nabla\alpha) = \rho(\alpha)$ and then extend it freely to Boolean terms.

The soundness of the one-step proof system is enshrined in the next proposition. The proof is essentially the same as that of the soundness direction in Theorem 16.

PROPOSITION 41. $a_1 \equiv_{\mathbb{M}(\mathbb{P}X)} a_2$ *implies* $\tilde{\rho}(a_1) = \tilde{\rho}(a_2)$, *for* $a_i \in \mathfrak{L}(PX)$.

By Proposition 41, the following is well-defined.

DEFINITION 42. Given a set X, let $\delta_X : \mathfrak{L}(PX)/{\equiv_{\mathbb{M}(\mathbb{P}X)}} \to PTX$ be the map given by $\tilde{\rho}_X = \delta_X \circ \mathfrak{q}_{\mathbb{P}(X)}$.

PROPOSITION 43. *The collection of maps given by Definition 42 form a natural transformation* $\delta : \mathbb{M}\mathbb{P} \to \mathbb{P}T$.

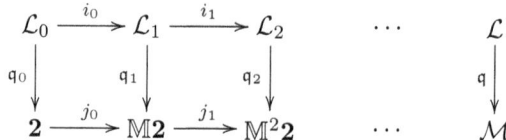

Proof. We need to show that for each X, δ_X is a (Boolean) homomorphism, and that δ is natural. Both proofs are straightforward. ■

REMARK 44. Continuing from Remark 27, δ is given by factoring $\gamma : \mathbb{LP} \to \mathbb{P}T$ (Remark 6) through $\mathbb{LP} \to \mathbb{MP}$.

The natural transformation δ allows us to associate with a coalgebra (S, σ) its 'complex $\mathbb{M}$-algebra' $\mathbb{P}\sigma \circ \delta_S : \mathbb{MP}S \to \mathbb{P}S$. Recall that $\mathcal{M}$ denotes the initial $\mathbb{M}$-algebra. For each coalgebra (S, σ), initiality of $\mathcal{M}$ gives us a map

(11) $[\![\cdot]\!] : \mathcal{M} \to \mathbb{P}S$

interpreting elements of $\mathcal{M}$ as propositions on S. Note that this map is an arrow in the category of Boolean algebras.

The next proposition ensures that the coalgebraic semantics of $\mathcal{M}$ (see (11)) and of $\mathcal{L}$ (Definition 7) agree.

PROPOSITION 45. *Denote by* $\mathfrak{q} : \mathcal{L} \to \mathcal{M}$ *the quotient map. Given a coalgebra* (S, σ) *and* $a \in \mathcal{L}$*, we have* $s \Vdash a$ *iff* $s \in [\![\mathfrak{q}(a)]\!]$.

Proof. The semantic map $\mu : \mathcal{L} \to \mathcal{P}S$ can be written as $\mu = f \circ \mathfrak{q}$ for some $f : \mathcal{M} \to \mathcal{P}S$ by putting $f(\mathfrak{q}(a)) := \mu(a)$. The function f is well-defined because of soundness of our logic: if $\mathfrak{q}(a) = \mathfrak{q}(a')$ then $a \equiv a'$ by Proposition 39 and therefore by soundness we get $\mu(a) = \mu(a')$. Using Lemma 11 it is not difficult to see that f is in fact an $\mathbb{M}$-algebra morphism from the initial $\mathbb{M}$-algebra $\mathcal{M}$ to the $\mathbb{M}$-algebra $(\mathbb{P}S, \mathbb{P}\sigma \circ \delta_S)$. Therefore by initiality we get $f = [\![\cdot]\!]$ and thus $\mu(a) = [\![\mathfrak{q}(a)]\!]$. ■

REMARK 46. We have now finished the functorial presentation of Moss' logic. A central role play the *Moss algebras*, that is, the algebras for the functor $\mathbb{M}$. In the case of $T = \mathcal{P}$, the category of Moss algebras is isomorphic to the category of Boolean algebras with operators. (11) corresponds to the fact that formulas are evaluated on a Kripke frame S by the morphism from the Lindenbaum BAO $\mathcal{M}$ to the complex algebra $\mathbb{P}S$ of S. The completeness proof in the next section generalises the well-known fact that we have an injection (iso for finite X) $d_X : \mathbb{KP}X \to \mathbb{P}\mathcal{P}X$ where $\mathbb{K}$ is the functor $\mathsf{BA} \to \mathsf{BA}$ mapping $\mathbb{A}$ to the algebra freely generated by $\Box a, a \in \mathbb{A}$, modulo the equations expressing that $\Box$ preserves finite meets (d_X is given by $\Box a \mapsto \{b \subseteq X \mid b \subseteq a\}$).

5.5 One-step completeness

Completeness of $\mathbf{M}$ is enshrined in the injectivity of δ_X. To show this we use the following basic fact about Boolean algebras.

LEMMA 47. *Let* $\mathbb{A}$ *and* $\mathbb{B}$ *be Boolean algebras and* $f : \mathbb{A} \to \mathbb{B}$ *be a homomorphism. Furthermore assume that* $\mathbb{A}$ *is join-generated by* $\mathcal{G} \subseteq A$*, i.e., assume that for every* $a \in A$ *we have* $a = \bigvee\{b \in \mathcal{G} \mid b \leq a\}$*. Then* $f(b) \neq \perp_{\mathbb{B}}$ *for all* $b \in \mathcal{G}$ *implies that* f *is injective.*

Proof. In order to prove the claim note first that for all $a \in A$ we clearly have $\bot_{\mathbb{A}} < a$ implies $\bot_{\mathbb{B}} < f(a)$. Let now a, a' be elements of A such that $a \neq a'$. By our assumption we have w.l.o.g. that there is some $b \in \mathcal{G}$ with $b \leq a$ and $b \nleq a'$. Therefore $\bot_{\mathbb{A}} < \neg a' \wedge b$ which implies by our first observation and the fact that f is a homomorphism that $\bot_{\mathbb{B}} < \neg f(a') \wedge f(b)$ and thus $f(b) \nleq f(a')$. On the other hand we clearly have $f(b) \leq f(a)$ which yields $f(a) \neq f(a')$. As a, a' where assumed to be arbitrary we showed that f is injective. ∎

PROPOSITION 48. *For every set X, the map $\delta_X : \mathbb{MP}(X) \to \mathbb{P}T(X)$ is an embedding.*

Proof. The basic idea of the proof is to work with the map $T\eta : TX \to TPX$, where we write $\eta_X : X \to PX$ for the singleton map $x \mapsto \{x\}$. The crucial property is that

(12) $\rho_X \circ T(\eta_X) = \eta_{TX}$.

The proof of (12) is based on the observation that $\mathrm{Gr}(\eta_X) \circ \in^{\smile} = Id_X$. From this it follows by the properties of relation lifting, that $\mathrm{Gr}(T\eta_X) \circ \overline{\in}^{\smile} = Id_{TX}$ and thus $\rho_X(T\eta_X(\alpha)) = \{\beta \mid \beta \mathrel{\overline{\in}} T\eta_X(\alpha)\} = \{\alpha\}$.

We define the set of "T-singletons" by putting $\mathcal{G} := \{\nabla T(\eta_X)(\alpha) \mid \alpha \in T_\omega X\}$. In order to prove the proposition it now suffices to show that the Boolean algebra $\mathbb{MP}(X)$ is *join-generated* by the T-singletons:

(13) $\forall a \in \mathbb{MP}(X).\ a = \bigvee \{\nabla\beta \in \mathcal{G} \mid \nabla\beta \leq a\}$.

(Note that the algebra $\mathbb{MP}(X)$ need not be complete. The intended reading of (13) is that every element of $\mathbb{MP}(X)$ is the join of the T-singletons below it, not that every set of T-singletons has a join.) To see why the injectivity of δ_X follows from this, note that by (12) we have $\delta_X(\nabla T\eta(\alpha)) = \rho_X(T\eta(\alpha)) = \{\alpha\} \neq \emptyset = \bot_{\mathbb{P}(TX)}$ for all $\nabla T\eta(\alpha) \in \mathcal{G}$. Therefore an application of Lemma 47 yields that f is injective.
Turning to the proof of (13), we distinguish cases as to the nature of the element a.

Case 1: Consider first an element of $\mathbb{MP}(X)$ of the form $\nabla\beta$ with $\beta \in T_\omega U\mathbb{P}X$. It can be easily shown that

$$\nabla\beta = \nabla(T_\omega \bigvee)(T_\omega \mathcal{P}_\omega \eta_X(\beta)) = \bigvee_{\gamma \in \rho_{\mathcal{P}X}(T_\omega \mathcal{P}_\omega \eta(\beta))} \nabla\gamma$$

where the first equality follows from the fact that $\bigvee \circ \mathcal{P}_\omega \eta_X = \mathrm{id}_{\mathcal{P}_\omega X}$ and the second equality is an instance of axiom (∇3). Furthermore one calculates (a detailed proof can be found in [10]) that

$$\rho_{\mathcal{P}X}(T_\omega \mathcal{P}_\omega \eta(\beta)) = T_\omega \eta[\rho_X(\beta)] = \{T_\omega \eta(\alpha) \mid \alpha \in \rho_X(\beta)\}.$$

By combining the latter equality with the preceding ones we obtain

$$\nabla\beta = \bigvee_{\alpha \in \rho_X(\beta)} \nabla(T\eta)(\alpha),$$

which shows that $\nabla\beta$ is the join of elements of $\mathcal{G}$.

Case 2: Consider now an element of $\mathbb{MP}(X)$ of the form $\neg\nabla\beta$. Let $\mathbb{B}$ be the subalgebra of $\mathbb{P}(X)$ generated by $Base(\beta) \subseteq_\omega U\mathbb{P}(X)$. We write B for the carrier of $\mathbb{B}$. As $Base(\beta)$ is finite, the Boolean algebra $\mathbb{B}$ is finite as well. Let $\phi \in \mathcal{P}_\omega(X)$ be the (finite) set of atoms of $\mathbb{B}$. Then clearly $\bigvee\phi = \top$, while $a \wedge a' = \bot$ for any two distinct $a, a' \in \phi$. Furthermore $\bigvee$ induces an isomorphism from $\mathcal{P}_\omega\phi$ to B that lifts to an isomorphism $T\bigvee$ between $T_\omega\mathcal{P}_\omega\phi$ and $T_\omega B$. As $Base(\beta) \subseteq B$ we have $\beta \in T_\omega B$ and thus there exists some $\Phi_\beta \in T_\omega\mathcal{P}_\omega\phi$ such that $(T\bigvee)(\Phi_\beta) = \beta$. Now axiom ($\nabla 3$) entails that

$$(14)\quad \nabla\beta = \bigvee\{\nabla\gamma \mid \gamma \in T_\omega\phi, \gamma\overline{\in}\Phi_\beta\}.$$

Our claim is now that

$$(15)\qquad \top \preceq \bigvee\{\nabla\gamma \mid \gamma \in T_\omega\phi\} \qquad \text{and}$$

$$(16)\qquad \nabla\gamma \wedge \nabla\gamma' \preceq \bot \qquad \text{for all } \gamma, \gamma' \in T_\omega\phi \quad \text{s.t. } \gamma \neq \gamma'.$$

Items (14), (15) and (16) together entail that

$$(17)\quad \neg\nabla\beta = \bigvee\{\nabla\gamma \mid \gamma \in T_\omega\phi \text{ and not } \gamma\overline{\in}\Phi_\beta\}.$$

From Case 1 we know that all elements $\nabla\gamma$ that occur on the righthand side of equation (17) can be written as joins of elements of $\mathcal{G}$ and therefore the same applies to $\neg\nabla\beta$.

We now turn to the proof of (15) and (16). First note that because $\bigvee\phi = \top$ an application of ($\nabla 4$) shows that $\top \preceq \bigvee\{\nabla\gamma \in T_\omega\phi \mid \gamma \in T_\omega\phi\}$ which proves (15). For the proof of (16) consider $\gamma, \gamma' \in T_\omega\phi$ with $\gamma \neq \gamma'$. Let $\Phi \in SRD(C)$ be a slim redistribution of $C := \{\gamma, \gamma'\}$. We want to show that $\nabla(T\bigwedge)(\Phi) \preceq \bot$. As Φ is an arbitrary redistribution of C this will imply by ($\nabla 2$) that $\nabla\gamma \wedge \nabla\gamma' \preceq \bot$ as required.

By assumption we have $(\gamma, \Phi), (\gamma', \Phi) \in \overline{\in}$. This shows that $\emptyset \notin Base(\Phi)$. Suppose now for a contradiction that $B_\Phi := Base(\Phi) \subseteq \mathcal{P}\phi$ contains only singleton sets and put $\in' := \in\restriction_{\phi \times B_\Phi}$. Then $\in' \circ \in'^{\smile} \subseteq Id_\phi$. As a consequence we get $(\gamma, \gamma') \in \overline{\in' \circ (\in')^{\smile}} \subseteq Id_{T\phi}$ which means $\gamma = \gamma'$ — a contradiction. Hence we can assume that B_Φ contains at least one set $\psi^* \subseteq \phi$ such that $|\psi^*| > 1$.

In order to prove that $\nabla(T\bigwedge)(\Phi) \preceq \bot$, define a function $d : \mathcal{P}\phi \to \mathcal{P}\phi$ by letting

$$d(\psi) := \begin{cases} \emptyset & \text{if } \ |\psi| \geq 1 \\ \psi & \text{if } \ |\psi| = 1 \\ \phi & \text{if } \ \psi = \emptyset \end{cases}$$

It follows from our assumptions on the set ϕ that $\vdash \bigwedge\psi \preceq \bigvee d(\psi)$, for all $\psi \in \mathcal{P}\phi$. Then an application of axiom ($\nabla 1$) shows that $\nabla(T\bigwedge)(\Phi) \preceq \nabla(T\bigvee)(Td(\Phi))$. Because $Base(Td(\Phi)) = d[B_\Phi]$ and $d(\psi^*) = \emptyset$ we obtain $\emptyset \in Base(Td(\Phi))$. It is now a matter of routine checking that $A := \{\alpha \in T_\omega U\mathbb{P}(X) \mid \alpha\overline{\in}Td(\Phi)\} = \emptyset$. By axiom ($\nabla 3$) we have $\nabla(T\bigvee)(Td(\Phi)) \preceq$

$\bigvee_{\alpha \in A} \nabla\alpha = \bot$, and thus $\nabla(T \bigwedge)(\Phi) \preceq \bot$ which finishes the argument for proving (16) and hence also of (17).

Case 3: Consider an element of $\mathbb{MP}(X)$ of the form $\bigwedge_{i \in I} \nabla\beta_i$ for some finite set $A = \{\nabla\beta_i \in T_\omega U\mathbb{P}(X) \mid i \in I\}$. Then by axiom $(\nabla 2)$ we have

$$\begin{array}{rcl} \bigwedge_{i \in I} \nabla\beta_i & = & \bigvee_{\Phi \in SRD(A)} \nabla(T_\omega \bigwedge)(\Phi) \\ & \stackrel{\text{Case 1}}{=} & \bigvee_{\Phi \in SRD(A)} \{\nabla(T\eta)(\alpha) \in \mathcal{G} \mid \nabla(T\eta)(\alpha) \leq \nabla(T_\omega \bigwedge)(\Phi)\}. \end{array}$$

Finally, the general case (that is, for an arbitrary element of $\mathbb{MP}(X)$) can be obtained from the cases above using standard Boolean reasoning. ■

As a corollary of the proof of Proposition 48 we obtain the following one-step normal form theorem.

COROLLARY 49. *For $a \in \mathfrak{L}PX$ we have*

$$a = \bigvee\{\nabla(T\eta)(\alpha) \mid \alpha \in \delta_X(a) \cap T_\omega X\}.$$

In case that T preserves finite sets, the join is finite for finite sets X and can be expressed in the language. Induction along the sequence of the $\mathcal{L}_n$ then yields a normal form theorem for $\mathcal{L}$.

5.6 Completeness

The following proposition, going back to [13], is a standard result in coalgebra based on δ being injective and $\mathbb{M}$ being finitary and preserving injective maps.

PROPOSITION 50. *Suppose $a \not\leq b$ in the initial $\mathbb{M}$-algebra. Then there is a T-coalgebra (S, σ) and $s \in S$ such that $s \Vdash a$ and not $s \Vdash b$.*

Proof. (Sketch) To explain the idea of the proof assume first that a final T-coalgebra $\zeta : Z \to TZ$ exists. Then we would prove that the unique $\mathbb{M}$-algebra morphism $[\![\cdot]\!]$ from the initial $\mathbb{M}$-algebra to $\mathbb{MP}Z \xrightarrow{\delta} T\mathbb{P} \xrightarrow{\zeta^{-1}} \mathbb{P}Z$ is injective. Indeed, $a \not\leq b$ then implies $[\![a]\!] \not\subseteq [\![b]\!]$, ie there is $z \in Z$ such that $(Z, \zeta), z \Vdash a$ and $(Z, \zeta), z \not\Vdash b$.

Since the assumption of the existence of a final coalgebra excludes important examples such as Kripke frames or models, we replace the final coalgebra by the corresponding final sequence $(T^n 1)_{n<\omega}$, which is defined as follows. We denote by $1 = T^0 1$ the final object in Set. The map $p_0 : T1 \to 1$ is given by finality and $p_{n+1} : T(T^n 1) \to T^n 1$ is defined to be Tp_n. It is easy to see that each p_n is surjective. We think of the $T^n 1$ as approximating the final coalgebra. (Indeed, if we let run the final sequence through all ordinals, we obtain the final coalgebra as a limit if it exists, see [3].) In the same way as any coalgebra $\xi : X \to TX$ has a unique arrow into the final coalgebra, there are canonical 'n-step behavior maps', that is, arrows

$\xi_n : X \to T^n 1$ to the approximants of the final coalgebra: $\xi_0 : X \to 1$ is given by finality and $\xi_{n+1} = T(\xi_n) \circ \xi$.

Recall that we may consider $\mathcal{M}$, the initial $\mathbb{M}$-algebra, as a union of its approximants $\mathbb{M}^n\mathbf{2}$. Elements of $\mathbb{M}^n\mathbf{2}$ correspond to formulas of depth n and we define their semantics wrt the final sequence of T as a BA-morphism $[\![-]\!]_n : \mathbb{M}^n\mathbf{2} \to \mathbb{P}T^n 1$ as follows.

$$(18)\qquad \begin{array}{ccccccc} \mathbb{P}1 & \xrightarrow{\mathbb{P}p_0} & \cdots & \mathbb{P}T^n 1 & \xrightarrow{\mathbb{P}p_n} & \mathbb{P}T^{n+1}1 & \cdots \\ \uparrow{\scriptstyle [\![-]\!]_0} & & & \uparrow{\scriptstyle [\![-]\!]_n} & & \uparrow{\scriptstyle [\![-]\!]_{n+1}} & \\ \mathbf{2} & \xrightarrow[j_0]{} & \cdots & \mathbb{M}^n\mathbf{2} & \xrightarrow[j_n]{} & \mathbb{M}^{n+1}\mathbf{2} & \cdots \end{array}$$

$[\![-]\!]_0$ is given by initiality (and is actually the identity). For the definition of $[\![-]\!]_{n+1}$, recall that $\delta_{T^n 1} : \mathbb{M}\mathbb{P}T^n 1 \to \mathbb{P}T^{n+1}1$, and assume inductively that $[\![-]\!]_n : \mathbb{M}^n\mathbf{2} \to \mathbb{P}T^n 1$ has been defined, so that $\mathbb{M}([\![-]\!]_n) : \mathbb{M}^{n+1}\mathbf{2} \to \mathbb{M}\mathbb{P}T^n 1$. Composing these two maps, we obtain $[\![-]\!]_{n+1} := \delta_{T^n 1} \circ \mathbb{M}([\![-]\!]_n)$.

Observe that the semantics of a formula is independent of the particular approximant we choose (all squares in the diagram commute). Moreover, given a coalgebra $\xi : X \to TX$ and $a \in \mathbb{M}^n\mathbf{2}$, the semantics via the initial $\mathbb{M}$-algebra and the semantics via the final sequence coincide: $[\![a]\!]_{(X,\xi)} = \xi_n^{-1}([\![a]\!]_n)$. Since δ is injective (Proposition 48) and $\mathbb{M}$ preserves embeddings (Proposition 28), a straightforward inductive proof shows that all $[\![-]\!]_n$, $n \in \omega$, are injective.

To show the claim now, suppose $a \not\leq b$ in the initial $\mathbb{M}$-algebra. We find an approximant $\mathbb{M}^n\mathbf{2}$, in which $a \not\leq_n b$. Choosing a half-inverse h of p_0, we let $\xi : T^n 1 \to TT^n 1$ be $T^n(h)$. ξ provides $T^n 1$ with T-coalgebra structure. Now injectivity of $[\![-]\!]_n$ shows that $(T^n 1, \xi)$ provides a counter-example for $a \leq b$. ■

The proof of Theorem 16 is now a corollary. Reasoning by contraposition, take formulas $a, b \in \mathcal{L}$ such that $\not\models_{\mathbf{M}} a \leq b$. By Proposition 39, $a \not\leq b$ in $\mathcal{M}$. Now, completeness follows from Propositions 50 and 45.

Acknowledgements. Clemens Kupke is supported by NWO under FOCUS/BRICKS grant 642.000.502; Alexander Kurz is partially supported by EPSRC grant EP/C014014/1; and the research of Yde Venema has been made possible by VICI grant 639.073.501 of the Netherlands Organization for Scientific Research (NWO).

BIBLIOGRAPHY

[1] S. Abramsky. Domain theory in logical form. *Ann. Pure Appl. Logic*, 51, 1991.

[2] P. Aczel. *Non-Well-Founded Sets*. CSLI, Stanford, 1988.

[3] J. Adámek and V. Koubek. On the greatest fixed point of a set functor. *Theoret. Comput. Sci.*, 150, 1995.

[4] J. Adámek and V. Trnková. *Automata and Algebras in Categories*. Kluwer, 1990.

[5] M. Bílková, A. Palmigiano, and Y. Venema. Proof systems for the coalgebraic cover modality, 2008. Same volume.

[6] M. Bonsangue and A. Kurz. Duality for logics of transition systems. In *FoSSaCS'05.*
[7] C. Cîrstea and D. Pattinson. Modular proof systems for coalgebraic logics. *Theoret. Comp. Sci.*, 338, 2007.
[8] G. D'Agostino and G. Lenzi. On modal μ-calculus with explicit interpolants. *Journal of applied logic*, 338, 2006.
[9] K. Fine. Normal forms in modal logic. *Notre Dame Journal of Formal Logic*, 16, 1975.
[10] B. Jacobs. Trace Semantics for Coalgebras. In *CMCS'04.*
[11] B. Jacobs. Many-sorted coalgebraic modal logic: a model-theoretic study. *Theor. Inform. Appl.*, 35, 2001.
[12] D. Janin and I. Walukiewicz. Automata for the modal μ-calculus and related results. In *Proc. MFCS'95.*
[13] C. Kupke, A. Kurz, and D. Pattinson. Algebraic semantics for coalgebraic logics. In *CMCS'04.*
[14] C. Kupke and Y. Venema. Coalgebraic automata theory: basic results. *Logical Methods in Computer Science*, to appear.
[15] A. Kurz. Specifying coalgebras with modal logic. *Theoret. Comput. Sci.*, 260, 2001.
[16] A. Kurz. Coalgebras and their logics. *SIGACT News*, 37, 2006.
[17] A. Kurz and D. Petrişan. Functorial coalgebraic logic: The case of many-sorted varieties. In *CMCS'08.*
[18] L. Moss. Coalgebraic logic. *Ann. Pure Appl. Logic*, 96, 1999.
[19] A. Palmigiano and Y. Venema. Nabla algebras and Chu spaces. In *CALCO'07.*
[20] D. Pattinson. Coalgebraic modal logic: Soundness, completeness and decidability of local consequence. *Theoret. Comput. Sci.*, 309, 2003.
[21] M. Rößiger. From modal logic to terminal coalgebras. *Theoret. Comput. Sci.*, 260, 2001.
[22] J. Rutten. Universal coalgebra: A theory of systems. *Theoret. Comput. Sci.*, 249, 2000.
[23] L. Santocanale and Y. Venema. Completeness for flat modal fixpoint logics (extended abstract). In *LPAR'07.*
[24] L. Schröder. Expressivity of Coalgebraic Modal Logic: The Limits and Beyond. In *FoSSaCS'05.*
[25] R. Sikorski. A theorem on extensions of homomorphisms. *Annals of Polish Mathematical Society*, 21, 1948.
[26] Y. Venema. Algebras and coalgebras. In P. Blackburn, J. van Benthem, and F. Wolter, editors, *Handbook of Modal Logic*, pages 331–426. Elsevier, 2006.
[27] Y. Venema. Automata and fixed point logics: a coalgebraic perspective. *Inf. Comput.*, 204, 2006.

Clemens Kupke
Department of Computing
Imperial College London
180 Queen's Gate, London SW7 2AZ
UK
`ckupke@doc.ic.ac.uk`

Alexander Kurz
Department of Computer Science
University of Leicester
UK
`kurz@mcs.le.ac.uk`

Yde Venema
Institute for Logic, Language and Computation,
Universiteit van Amsterdam,
Plantage Muidergracht 24, 1018 TV Amsterdam
The Netherlands
Y.Venema@uva.nl

On axiomatising products of Kripke frames, part II

AGI KURUCZ

ABSTRACT. We generalise some results of [7, 5] and show that if L is an α-modal logic (for some ordinal $\alpha \geq 3$) such that (i) L contains the product logic $\mathbf{K}^\alpha$ and (ii) the product of α-many trees of depth one and with arbitrary large finite branching is a frame for L, then any axiomatisation of L must contain infinitely many propositional variables. As a consequence we obtain that product logics like $\mathbf{K}^\alpha$, $\mathbf{K4}^\alpha$, $\mathbf{S4}^\alpha$, $\mathbf{GL}^\alpha$, and $\mathbf{Grz}^\alpha$ cannot be axiomatised using finitely many propositional variables, whenever $\alpha \geq 3$.

Keywords: many-dimensional modal logic, axiomatisation

1 Introduction and results

We consider the problem whether certain propositional α-modal logics can be axiomatised by a (possibly infinite) set of α-modal formulas containing only finitely many propositional variables altogether. By an *α-modal formula*, for any non-zero ordinal α, we mean any formula built up from propositional variables using the Booleans and the modal operators $\Diamond_\beta$ and $\Box_\beta$ for $\beta < \alpha$. A set L of α-modal formulas is called a (normal) *α-modal logic* if it contains all propositional tautologies and the formulas $\Box_\beta(p \to q) \to (\Box_\beta p \to \Box_\beta q)$, for $\beta < \alpha$, and is closed under the rules of Substitution, Modus Ponens and Necessitation $\varphi/\Box_\beta\varphi$, for $\beta < \alpha$. Given an α-modal logic L and a set Σ of α-modal formulas, we say that Σ *axiomatises* L if L is the smallest α-modal logic containing Σ.

In what follows we assume that the reader is familiar with the basics of possible world semantics for multimodal logics (see e.g. [2]). The α-modal logics we deal with are 'α-dimensional' in the sense that they have among their frames *α-dimensional product frames*: The *product* of 1-frames $\mathfrak{F}_\beta = (W_\beta, R_\beta)$, $\beta < \alpha$ is the α-frame $\mathfrak{F} = (W, \bar{R}_\beta)_{\beta<\alpha}$, where W is the Cartesian product of the W_β, $\beta < \alpha$, and for each $\beta < \alpha$, $\bar{R}_\beta$ is the following binary relation on W:

$$(u_\gamma)_{\gamma<\alpha}\bar{R}_\beta(v_\gamma)_{\gamma<\alpha} \quad \text{iff} \quad u_\beta R_\beta v_\beta \quad \text{and} \quad u_\gamma = v_\gamma, \quad \text{for } \gamma \neq \beta < \alpha.$$

For finite α, we will use the notation $\mathfrak{F} = \mathfrak{F}_0 \times \cdots \times \mathfrak{F}_{\alpha-1}$. The *product* of Kripke complete unimodal logics L_β, $\beta < \alpha$ is the α-modal logic determined by the class of all those α-dimensional product frames whose βth component is a frame for L_β, for each $\beta < \alpha$. For example, $\mathbf{K}^\alpha$ is the α-modal logic

© *2008, the author(s).*

of all α-dimensional product frames, and $\mathbf{S5}^\alpha$ is that of all α-dimensional products of equivalence relation frames. Products of modal logics were introduced by Segerberg [10] and Shehtman [11] and have been extensively studied, see [2] for further references and [8] for a more recent survey.

It is not hard to show (see e.g. [8]) that product logics of any dimension are recursively enumerable whenever for each component logic the class of its frames is definable by a recursive set of first-order sentences. Gabbay and Shehtman [3] showed that many 2-dimensional product logics ($\mathbf{K}^2$ and $\mathbf{S5}^2$ among them) and products of **Alt** (the logic of functional frames) in any finite dimension are finitely axiomatisable. For higher dimensions, that is, for $\alpha \geq 3$, no other 'positive' axiomatisation result is known. On the 'negative' side, the non-finite axiomatisability of $\mathbf{S5}^\alpha$ was proved by Johnson [6] and that of $\mathbf{K}^\alpha$ by Kurucz [7]. These results were generalised by Hirsch *et al.* [5] who showed that no α-modal logic between $\mathbf{K}^\alpha$ and $\mathbf{S5}^\alpha$ can be axiomatised finitely. Here we show that, for many of these logics, any axiomatisation actually must contain infinitely many propositional variables.

Throughout, an *n-fan* (for $n < \omega$) is a unimodal (reflexive or irreflexive) tree of depth 1 having n leaves (see Figure 1).

Figure 1. Reflexive and irreflexive 4-fans

THEOREM 1. *Let $\alpha \geq 3$ and L be any α-modal logic containing $\mathbf{K}^\alpha$. If the product of α arbitrarily large finite fans is a frame for L, then L is not axiomatisable using finitely many propositional variables.*

Well-known examples of modal logics having fans among their frames are **K**, **K4** (the logic of transitive frames), **S4** (the logic of reflexive and transitive frames), **GL** (the logic of irreflexive and transitive frames without infinite ascending chains), **Grz** (the logic of reflexive and transitive frames without infinite ascending chains of distinct points), so we have the following:

COROLLARY 2. *None of the logics $\mathbf{K}^\alpha$, $\mathbf{K4}^\alpha$, $\mathbf{S4}^\alpha$, $\mathbf{GL}^\alpha$, $\mathbf{Grz}^\alpha$ is axiomatisable using finitely many propositional variables, whenever $\alpha \geq 3$.*

Theorem 1 does not apply to product logics where some components have transitive frames with some restriction on their width. An important example of this kind is $\mathbf{S5}^\alpha$. Johnson's [6] non-finite axiomatisability result was obtained in an algebraic setting: he proved that the modal algebras corresponding to $\mathbf{S5}^\alpha$ (*representable diagonal-free cylindric algebras of dimension* α) have a non-finitely axiomatisable equational theory, whenever $\alpha \geq 3$. Representable cylindric algebras of dimension α (modal algebras of $\mathbf{S5}^\alpha$ plus diagonal constants) have been extensively studied in algebraic logic. Strengthening earlier results of Monk [9], Andréka [1] proved (among other strong non-finitisability properties) that any possible axiomatisation of their equational theory must contain infinitely many variables, if $\alpha \geq 3$. She also left open, however, whether one needed infinitely many variables in the diagonal-free case.

The rest of the paper is devoted to the proof of Theorem 1. To begin with, given $m < \omega$, we call a modal model $\mathfrak{M} = (\mathfrak{F}, \vartheta)$ *m-generated* if there are at most m different propositional variables p such that $\vartheta(p) \neq \emptyset$. We plan to proceed as follows. For every $0 < k < \omega$, we will define an α-frame $\mathfrak{F}_k$ such that:

(a) If $2k > 2^m$ then $\mathfrak{M} \models L$ for every m-generated model $\mathfrak{M}$ based on $\mathfrak{F}_k$.

(b) $\mathfrak{F}_k \not\models \mathbf{K}^\alpha$.

This will prove Theorem 1 because of the following. Suppose that Σ axiomatises L and Σ contains m propositional variables, for some $m < \omega$. Let $2k > 2^m$ and take an α-frame $\mathfrak{F}_k$ satisfying (a). Let $\mathfrak{M}$ be an arbitrary model based on $\mathfrak{F}_k$. Let $\mathfrak{M}_m$ be another model over $\mathfrak{F}_k$ that is the same as $\mathfrak{M}$ on propositional variables occurring in Σ, and $\emptyset$ otherwise. Then $\mathfrak{M}_m$ is clearly m-generated and $\mathfrak{M}_m \models \Sigma$ iff $\mathfrak{M} \models \Sigma$. So by (a), we have $\mathfrak{M}_m \models L$. As $\Sigma \subseteq L$, we obtain $\mathfrak{M}_m \models \Sigma$, and so $\mathfrak{M} \models \Sigma$. This holds for any model $\mathfrak{M}$ over $\mathfrak{F}_k$, so $\mathfrak{F}_k \models \Sigma$ follows. Therefore, $\{\varphi : \mathfrak{F}_k \models \varphi\}$ is an α-modal logic containing Σ, and so we have $\mathfrak{F}_k \models L$. As $\mathbf{K}^\alpha \subseteq L$, this implies $\mathfrak{F}_k \models \mathbf{K}^\alpha$, contradicting (b).

2 Frames

In this section we construct the α-frames $\mathfrak{F}_k$ via some steps, and show that they have property (a). To make things clearer, we make two simplifications. First, we work with $\alpha = 3$ and then explain how to extend everything to any $\alpha \geq 3$ (see Remark 7). And second, we deal with products of irreflexive fans only and then, also in Remark 7, we explain how to extend the proof to the reflexive cases. In drawing 'three-dimensional' pictures of 3-frames, we adopt the following convention in drawing three accessibility relations:

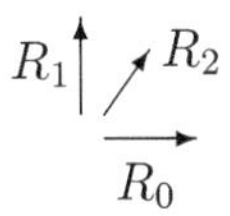

Fix some $0 < k < \omega$. We define a (rooted) 3-frame $\mathfrak{G}_k = (G, R_0^{\mathfrak{G}}, R_1^{\mathfrak{G}}, R_2^{\mathfrak{G}})$ as follows (see also Figure 2):

$$
\begin{aligned}
G &= \{r_{000}, d_{010}, d_{001}, d_{110}, d_{101}, i_{011}, i_{100}, i_{111}^1, i_{111}^2\}_{i<k} \\
R_0^{\mathfrak{G}} &= \{(r_{000}, i_{100}), (d_{010}, d_{110}), (d_{001}, d_{101}), (i_{011}, i_{111}^1), (i_{011}, i_{111}^2)\}_{i<k} \\
R_1^{\mathfrak{G}} &= \{(r_{000}, d_{010}), (i_{100}, d_{110}), (d_{001}, i_{011}), (d_{101}, i_{111}^1), (d_{101}, i_{111}^2)\}_{i<k} \\
R_2^{\mathfrak{G}} &= \{(r_{000}, d_{001}), (i_{100}, d_{101}), (d_{010}, i_{011}), (d_{110}, i_{111}^1), (d_{110}, i_{111}^2)\}_{i<k}
\end{aligned}
$$

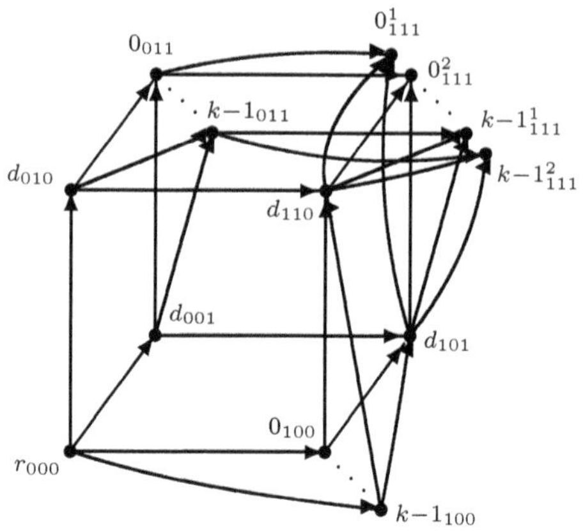

Figure 2. The 3-frame $\mathfrak{G}_k$.

Recall that, given two 3-frames $\mathfrak{H} = (W, S_i^{\mathfrak{H}})_{i<3}$ and $\mathfrak{G} = (V, S_i^{\mathfrak{G}})_{i<3}$, a function $f : W \to V$ is called a *p-morphism* from $\mathfrak{H}$ to $\mathfrak{G}$, if for all $u, v \in W$, $i < 3$, $uS_i^{\mathfrak{H}}v$ implies $f(u)S_i^{\mathfrak{G}}f(v)$ (*forward condition*), and for all $x \in V$, $i < 3$, the following

$$BC_i(x): \qquad \forall u \in W,\, y \in V \,\big(f(u) = x \text{ and } xS_i^{\mathfrak{G}}y \implies \exists v \in W,\ uS_i^{\mathfrak{H}}v \text{ and } f(v) = y\big)$$

hold (*backward condition*). If f is onto then we say that $\mathfrak{G}$ is a *p-morphic image* of $\mathfrak{H}$. It is a well-known property that the validity of modal formulas in frames is preserved under taking p-morphic images.

Below we show that $\mathfrak{G}_k$ is a p-morphic image of a product of finite fans. To this end, for any $n < \omega$, consider the irreflexive n-fan $\mathfrak{H}_n = (H_n, R^{\mathfrak{H}_n})$, where $H_n = \{u, z_0, \ldots, z_{n-1}\}$ and $R^{\mathfrak{H}_n} = \{(u, z_0), \ldots, (u, z_{n-1})\}$.

CLAIM 3. $\mathfrak{G}_k$ is a p-morphic image of $\mathfrak{H}_k \times \mathfrak{H}_{2k} \times \mathfrak{H}_{2k}$.

Proof. For all $j, \ell < 2k$, let $(j +_k \ell)$ denote the sum of j and ℓ modulo k. We define a function g on $H_k \times H_{2k} \times H_{2k}$ as follows, for all $i < k$, $j, \ell < 2k$ (see also Figure 3 where each point in $H_k \times H_{2k} \times H_{2k}$ is labelled with its g-image):

$$g(z_i, z_j, z_\ell) = \begin{cases} (j +_k \ell)^1_{111}, & \text{if } i \text{ is odd and either } j, \ell < k \\ & \qquad\qquad \text{or } k \le j, \ell < 2k, \\ (j +_k \ell)^1_{111}, & \text{if } i \text{ is even and either } j < k \le \ell < 2k, \\ & \qquad\qquad \text{or } \ell < k \le j < 2k, \\ (j +_k \ell)^2_{111}, & \text{if } i \text{ is even and either } j, \ell < k \\ & \qquad\qquad \text{or } k \le j, \ell < 2k, \\ (j +_k \ell)^2_{111}, & \text{if } i \text{ is odd and either } j < k \le \ell < 2k, \\ & \qquad\qquad \text{or } \ell < k \le j < 2k, \end{cases}$$

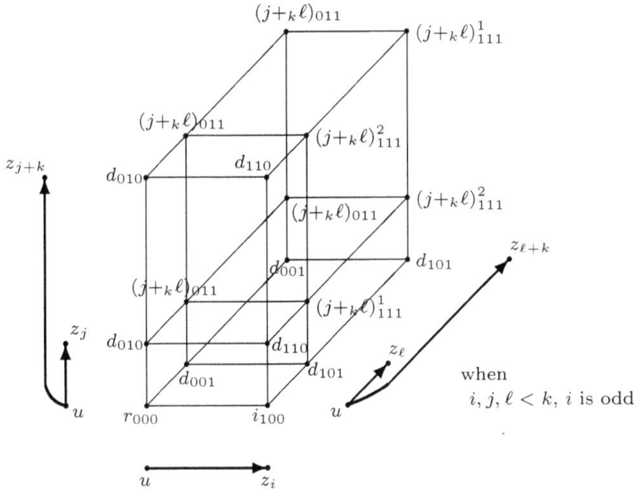

Figure 3. The p-morphism $g : \mathfrak{H}_k \times \mathfrak{H}_{2k} \times \mathfrak{H}_{2k} \to \mathfrak{G}_k$.

$$g(u, u, u) = r_{000}, \qquad g(z_i, u, u) = i_{100},$$
$$g(u, u, z_\ell) = d_{001}, \qquad g(u, z_j, u) = d_{010},$$
$$g(z_i, u, z_\ell) = d_{101}, \qquad g(z_i, z_j, u) = d_{110},$$
$$g(u, z_j, z_\ell) = (j +_k \ell)_{011}.$$

A tedious but straightforward computation shows that g is a p-morphism from $\mathfrak{H}_k \times \mathfrak{H}_{2k} \times \mathfrak{H}_{2k}$ onto $\mathfrak{G}_k$. Here we go through two of the trickiest cases. For $BC_2(d_{110})$, we need to show that for all $i < k$, $j < 2k$, $n < k$, there exist $\ell_1, \ell_2 < 2k$ such that $g(z_i, z_j, z_{\ell_1}) = n^1_{111}$ and $g(z_i, z_j, z_{\ell_2}) = n^2_{111}$. Given such i, j, n, we always have an $s < k$ such that $s +_k j = n$. Now if either i is odd and $j < k$, or i is even and $k \le j < 2k$, then $\ell_1 = s$ and $\ell_2 = s + k$ will do. In any other case, take $\ell_1 = s + k$ and $\ell_2 = s$.

For $BC_0(n_{011})$, $n < k$, we need to show that for all $j, \ell < 2k$ such that $j +_k \ell = n$ there exist $i_1, i_2 < k$ such that $g(z_{i_1}, z_j, z_\ell) = n^1_{111}$ and $g(z_{i_2}, z_j, z_\ell) = n^2_{111}$. Now if either $j, \ell < k$ or $k \le j, \ell < 2k$ then choose $i_1 < k$ to be odd and $i_2 < k$ to be even. In any other case, choose $i_1 < k$ to be even and $i_2 < k$ to be odd. ■

Next, we 'ruin' $\mathfrak{G}_k$ a bit by adding some more points and arrows to it. We define the 3-frame $\mathfrak{F}_k = (F, R_0^{\mathfrak{F}}, R_1^{\mathfrak{F}}, R_2^{\mathfrak{F}})$ as follows (see also Figure 4):

$$F = G \cup \{a_{010}, a_{001}, i_{110}, i_{101}\}_{i<k}$$
$$R_0^{\mathfrak{F}} = R_0^{\mathfrak{G}} \cup \{(a_{010}, i_{110}), (a_{001}, i_{101})\}_{i<k}$$
$$R_1^{\mathfrak{F}} = R_1^{\mathfrak{G}} \cup \{(r_{000}, a_{010}), (i_{100}, i_{110})(a_{001}, i_{011}), \\ (i_{101}, j^2_{111}), (i_{101}, \ell^1_{111})\}_{i,j,\ell<k,\ \ell\neq i}$$
$$R_2^{\mathfrak{F}} = R_2^{\mathfrak{G}} \cup \{(r_{000}, a_{001}), (i_{100}, i_{101})(a_{010}, i_{011}), \\ (i_{110}, j^1_{111}), (i_{110}, \ell^2_{111})\}_{i,j,\ell<k,\ \ell\neq i}$$

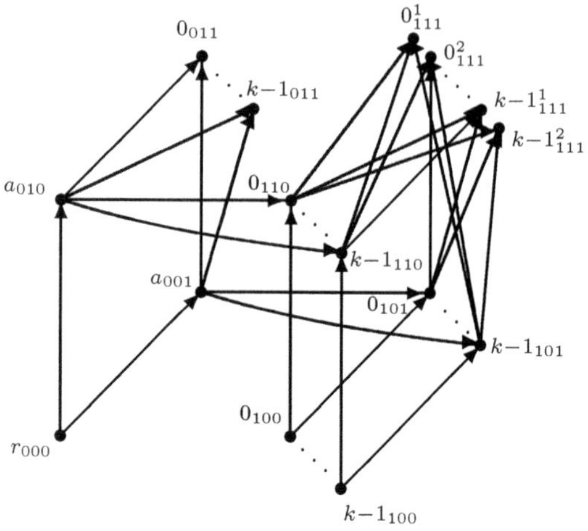

Figure 4. The arrows in $\mathfrak{F}_k$ that are not present in $\mathfrak{G}_k$.

Though, as we shall see in Section 3, $\mathfrak{F}_k \not\models \mathbf{K}^3$ and so it cannot be a p-morphic image of any product frame, it is 'almost' such:

CLAIM 4. $\mathfrak{F}_k$ is a p-morphic image of a subframe $\mathfrak{H}$ of $\mathfrak{H}_k \times \mathfrak{H}_{2k+1} \times \mathfrak{H}_{2k+1}$.

Proof. We give a proof for $k \geq 4$ only (for $k = 3$ a slightly different function would work). Let

$$H = (H_k \times H_{2k+1} \times H_{2k+1}) - (H_k \times \{z_{2k}\} \times \{z_{2k}\}),$$

and let $\mathfrak{H}$ be the subframe of $\mathfrak{H}_k \times \mathfrak{H}_{2k+1} \times \mathfrak{H}_{2k+1}$ having H as its domain. Take the p-morphism $g : \mathfrak{H}_k \times \mathfrak{H}_{2k} \times \mathfrak{H}_{2k} \to \mathfrak{G}_k$ defined in the proof of Claim 3. We define a function g^+ on H such that g^+ is an extension of g, that is, for every $x \in H_k \times H_{2k} \times H_{2k}$, $g^+(x) = g(x)$.

For the 'new' points we define g^+ as follows, for all $i < k$, $j < 2k$ (see also Figure 5):

$$g^+(u, z_{2k}, u) = a_{010}, \qquad g^+(u, u, z_{2k}) = a_{001},$$

$$g^+(u, z_{2k}, z_j) = g^+(u, z_j, z_{2k}) = \begin{cases} j_{011}, & \text{if } j < k, \\ (j-k)_{011}, & \text{if } k \leq j < 2k, \end{cases}$$

$$g^+(z_i, z_{2k}, u) = i_{110}, \qquad g^+(z_i, u, z_{2k}) = i_{101},$$

$$g^+(z_i, z_{2k}, z_j) = \begin{cases} j^1_{111}, & \text{if } i = j \text{ or } i = j-k, \\ j^1_{111}, & \text{if } i \neq j,\ i \text{ is odd and } j < k, \\ (j-k)^1_{111}, & \text{if } i \neq j-k,\ i \text{ is even and } k \leq j < 2k, \\ j^2_{111}, & \text{if } i \neq j,\ i \text{ is even and } j < k, \\ (j-k)^2_{111}, & \text{if } i \neq j-k,\ i \text{ is odd and } k \leq j < 2k, \end{cases}$$

$$g^+(z_i, z_j, z_{2k}) = \begin{cases} j^2_{111}, & \text{if } i = j \text{ or } i = j-k, \\ j^2_{111}, & \text{if } i \neq j,\ i \text{ is odd and } j < k, \\ (j-k)^2_{111}, & \text{if } i \neq j-k,\ i \text{ is even and } k \leq j < 2k, \\ j^1_{111}, & \text{if } i \neq j,\ i \text{ is even and } j < k, \\ (j-k)^1_{111}, & \text{if } i \neq j-k,\ i \text{ is odd and } k \leq j < 2k. \end{cases}$$

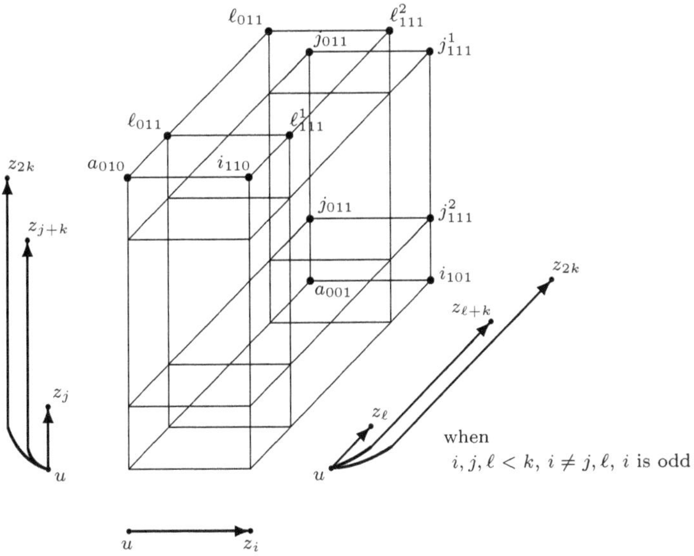

Figure 5. The p-morphism $g^+ : \mathfrak{H} \to \mathfrak{F}_k$ on the 'new' points.

Then it is straightforward to show that g^+ is a p-morphism from $\mathfrak{H}$ onto $\mathfrak{F}_k$. Here we give two sample cases only. As concerns $BC_0(i_{011})$, for each $i < k$, we have four new pre-images of i_{011}: (u, z_i, z_{2k}), (u, z_{i+k}, z_{2k}), (u, z_{2k}, z_i), and (u, z_{2k}, z_{i+k}). Take first (u, z_i, z_{2k}). We need to show that there exist $j_1, j_2 < k$ such that $g^+(z_{j_1}, z_i, z_{2k}) = i^1_{111}$ and $g^+(z_{j_2}, z_i, z_{2k}) = i^2_{111}$. Now if $k \geq 4$ then we can choose $j_1 < k$ to be even and different from i and $j_2 < k$ to be odd. For (u, z_{i+k}, z_{2k}), we need to show that there exist $j_1, j_2 < k$ such that $g^+(z_{j_1}, z_{i+k}, z_{2k}) = i^1_{111}$ and $g^+(z_{j_2}, z_{i+k}, z_{2k}) = i^2_{111}$. To this end, choose $j_1 < k$ to be odd and different from i and $j_2 < k$ to be even. The other two cases are similar.

For $BC_2(i_{110})$, $i < k$ we need to show that for all $n < k$ there exists $j_1 < 2k$ such that $g(z_i, z_{2k}, z_{j_1}) = n^1_{111}$, for all $n < k$, $n \neq i$, there exists $j_1 < 2k$ such that $g(z_i, z_{2k}, z_{j_2}) = n^2_{111}$. Now if $i = n$ then take $j_1 = n$. If $i \neq n$ and i is odd then take $j_1 = n$ and $j_2 = n + k$, and if $i \neq n$ and i is even then take $j_1 = n + k$ and $j_2 = n$. ∎

CLAIM 5. Let $\mathfrak{F}$ be a 3-frame, and suppose that $f : \mathfrak{F}_k \to \mathfrak{F}$ is an onto p-morphism such that $f(i^m_{111}) = f(j^n_{111})$ for some $i \neq j$ or $n \neq m$. Then $\mathfrak{F}$ is a p-morphic image of $\mathfrak{H}_k \times \mathfrak{H}_{2k+1} \times \mathfrak{H}_{2k+1}$.

Proof. We again give a proof for $k \geq 4$ only (for $k = 3$ a slightly different function would work). Take the p-morphism $g^+ : \mathfrak{H} \to \mathfrak{F}_k$ defined in the proof of Claim 4. We define a function h on $H_k \times H_{2k+1} \times H_{2k+1}$ such that h is an extension of $g^+ \circ f$, that is, for every $x \in H$, $h(x) = fg^+(x)$.

Let i, j, m, n be as in the assumption of the claim. For the 'new' points we define h as follows:

$$h(u, z_{2k}, z_{2k}) = f(i_{011}),$$

$$h(z_\ell, z_{2k}, z_{2k}) = \begin{cases} f(i^m_{111}) = f(j^n_{111}), & \text{if } \ell = i, \\ f(i^1_{111}), & \text{if } \ell < k,\ \ell \neq i \text{ and } \ell \text{ is odd}, \\ f(i^2_{111}), & \text{if } \ell < k,\ \ell \neq i \text{ and } \ell \text{ is even}. \end{cases}$$

It is straightforward to check that h is a p-morphism from $\mathfrak{H}_k \times \mathfrak{H}_{2k+1} \times \mathfrak{H}_{2k+1}$ onto $\mathfrak{F}$. Here is the trickiest case only. For $BC_0\big(f(i_{011})\big)$, there is no problem with the 'old' h-pre-images of $f(i_{011})$ (those that are in H), as the composition of p-morphisms is a p-morphism. As concerns the only new one, (u, z_{2k}, z_{2k}), we need to show that there exist $j_1, j_2 < k$ such that $h(z_{j_1}, z_{2k}, z_{2k}) = f(i^1_{111})$ and $h(z_{j_2}, z_{2k}, z_{2k}) = f(i^2_{111})$. Now if $k \geq 4$ then we can choose both j_1 and j_2 to be different from i and such that $j_1 < k$ is odd and $j_2 < k$ is even. ∎

Now we can show that $\mathfrak{F}_k$ satisfies property (a):

LEMMA 6. *Let L be a 3-modal logic such that $\mathfrak{H}_k \times \mathfrak{H}_{2k+1} \times \mathfrak{H}_{2k+1} \models L$ for some $k < \omega$. If $2k > 2^m$ then $\mathfrak{M} \models L$ for every m-generated model $\mathfrak{M}$ over $\mathfrak{F}_k$.*

Proof. Fix some k, m with $2k > 2^m$. Let $\mathfrak{M} = (\mathfrak{F}_k, \vartheta)$ be such that $\vartheta(p_j) = \emptyset$ for every propositional variable p_j with $j \geq m$.

We call two points in $\mathfrak{F}_k$ $\equiv$-*equivalent* iff no 3-modal formula can distinguish them in $\mathfrak{M}$, that is, for all $a, b \in F$, we let

(1) $\quad a \equiv b \quad \iff \quad \big(\forall \text{ formula } \varphi,\ \ a \in \vartheta(\varphi) \Leftrightarrow b \in \vartheta(\varphi)\big).$

For every $a \in F$, let $[a]$ denote the $\equiv$-class of a, and let $A = \{[a] : a \in F\}$. We define a 3-frame $\mathfrak{A}_{\mathfrak{M}} = (A, S_0, S_1, S_2)$ by taking, for $i < 3$,

$$[a] S_i [b] \quad \iff \quad \exists a' \in [a],\ b' \in [b],\ a' R_i^{\mathfrak{F}} b'.$$

We claim that the function

$$f(a) = [a], \qquad a \in F$$

is a p-morphism from $\mathfrak{F}_k$ onto $\mathfrak{A}_{\mathfrak{M}}$. This is a straightforward consequence of duality theory and the finiteness of $\mathfrak{F}_k$, but we give a short direct proof here. The forward condition holds by the definition of S_i. For the backward condition, observe that since $\mathfrak{F}_k$ is finite, there are finitely many formulas $\varphi_0, \ldots, \varphi_{n-1}$ such that

(2) $\quad a \equiv b \quad \iff \quad \big(\forall j < n,\ \ a \in \vartheta(\varphi_j) \Leftrightarrow b \in \vartheta(\varphi_j)\big)$

(these $\vartheta(\varphi_j)$ are the atoms of the algebra of $\mathfrak{M}$-definable subsets of F). Now take some $i < 3$, $a, b \in F$ such that $[a] S_i [b]$, and let $a' \in [a]$. Then there are $a'' \in [a]$, $b'' \in [b]$ with $a'' R_i^{\mathfrak{F}} b''$. Let φ be the 'atomic type' of b'', that is,

$$\varphi = \bigwedge_{j<n,\, b'' \in \vartheta(\varphi_j)} \varphi_j \quad \wedge \bigwedge_{j<n,\, b'' \notin \vartheta(\varphi_j)} \neg \varphi_j.$$

Then $b'' \in \vartheta(\varphi)$. Therefore $a'' \in \vartheta(\Diamond_i\varphi)$, and so $a' \in \vartheta(\Diamond_i\varphi)$. So there is some b' such that $a' R_i^{\mathfrak{F}} b'$ and $b' \in \vartheta(\varphi)$. Now $b' \equiv b''$ follows by (2).

Next, define F_{111} as the subset of F containing all 'dead ends':

$$F_{111} = \{i_{111}^1, i_{111}^2\}_{i<k}.$$

We define an equivalence relation $\equiv_m$ on F_{111} by taking, for all $a, b \in F_{111}$,

$$a \equiv_m b \qquad \Longleftrightarrow \qquad \big(\forall j < m,\ \ a \in \vartheta(p_j) \ \Leftrightarrow\ b \in \vartheta(p_j)\big).$$

Now recall the definition of $\equiv$ from (1). An easy induction on formulas (using that $a \notin \vartheta(\Diamond_i\psi)$, for any formula ψ, $a \in F_{111}$, $i < 3$) shows that

$$(3) \quad \forall a, b \in F_{111}, \quad \big(a \equiv_m b \quad \Longrightarrow \quad a \equiv b\big).$$

As the cardinality of F_{111} is $2k$ and there are 2^m many $\equiv_m$-classes, by the pigeonhole principle and (3), there exist $a \neq b \in F_{111}$ such that $a \equiv b$, and so $f(a) = f(b)$. Therefore, the 3-frame $\mathfrak{A}_{\mathfrak{M}}$ and the p-morphism f satisfy the conditions of Claim 5, and so $\mathfrak{A}_{\mathfrak{M}}$ is a p-morphic image of $\mathfrak{H}_k \times \mathfrak{H}_{2k+1} \times \mathfrak{H}_{2k+1}$. As by assumption $\mathfrak{H}_k \times \mathfrak{H}_{2k+1} \times \mathfrak{H}_{2k+1} \models L$, we obtain that $\mathfrak{A}_{\mathfrak{M}} \models L$ as well. In particular, $\mathfrak{M}' \models L$ for the model $\mathfrak{M}' = (\mathfrak{A}_{\mathfrak{M}}, \vartheta')$ defined by taking, for each propositional variable p, $\vartheta'(p) = \{f(a) : a \in \vartheta(p)\}$. As f is a p-morphism between models $\mathfrak{M}$ and $\mathfrak{M}'$, $\mathfrak{M} \models L$ follows, as required. ■

REMARK 7. If $\alpha \geq 3$ then we can extend the 3-frames $\mathfrak{G}_k$ and $\mathfrak{F}_k$ above to α-frames by taking $R_\beta^{\mathfrak{F}} = R_\beta^{\mathfrak{G}} = \emptyset$, for $3 \leq \beta < \alpha$. Then in Claims 3–5 and Lemma 6 we should use α-dimensional product frames, where the βth component is $\mathfrak{H}_0 = (\{u\}, \emptyset)$ whenever $3 \leq \beta < \alpha$.

If the logic L in Theorem 1 is such that it has products of arbitrarily large finite fans among its frames, but some (or all) of these fans are reflexive, then in Claims 3–5 and Lemma 6 we have to define the corresponding relations in $\mathfrak{G}_k$ and $\mathfrak{F}_k$ and the corresponding 'fan-components' in the product frames to be reflexive as well. Then everything goes through with not much change in the arguments. In particular, when proving (3) above by induction on formulas, we need to use that if i is one of the 'reflexive coordinates' then for any formula ψ and any $a \in F_{111}$, $a \in \vartheta(\Diamond_i\psi)$ iff $a \in \vartheta(\psi)$.

3 Formulas

In this section we prove property (b) of our frames, that is, that $\mathfrak{F}_k \not\models \mathbf{K}^\alpha$, for any $0 < k < \omega$. We do this by showing, for each k, a 3-modal formula that is valid in all α-dimensional product frames but fails in $\mathfrak{F}_k$.

To this end, for each $0 < k < \omega$, we define Φ_k to be the following first-order sentence of the language having binary predicates R_0, R_1 and R_2 (see also Figure 6):

$$\Phi_k: \quad \forall\, vyzx_0 \ldots x_{k-1} \Big[vR_1y \wedge vR_2z \wedge \bigwedge_{i<k} vR_0x_i \;\longrightarrow$$

$$\exists\, uy_0 \ldots y_{k-1}z_0 \ldots z_{k-1}u_0 \ldots u_{k-1} \Big(yR_2u \wedge zR_1u \wedge$$

$$\bigwedge_{i<k} (yR_0y_i \wedge zR_0z_i \wedge uR_0u_i \wedge x_iR_1y_i \wedge z_iR_1u_i \wedge x_iR_2z_i \wedge y_iR_2u_i)\Big)\Big].$$

Note that Φ_1 is the well-known 'cubifying' property of $\geq$ 3-dimensional product frames (see [4, 3.2.68] and [7, 8]).

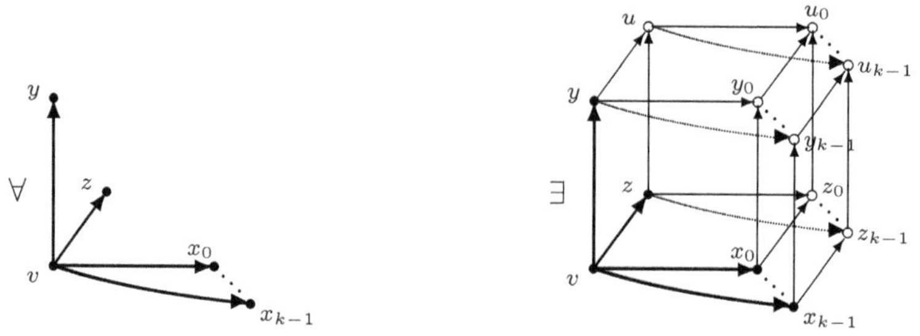

Figure 6. The first-order sentence Φ_k.

It is easy to check the following claim.

CLAIM 8. For any $0 < k < \omega$, Φ_k is true in every α-dimensional product frame.

These first-order properties are modally definable. Namely, for every $0 < k < \omega$, consider the following 3-modal formula φ_k:

$$\Big[\Diamond_1(\Box_0p_{10} \wedge \Box_2p_{12}) \wedge \Diamond_2(\Box_0p_{20} \wedge \Box_1p_{21}) \wedge \bigwedge_{i<k} \big(\Diamond_0(\Box_1p^i_{01} \wedge \Box_2p^i_{02})$$

$$\wedge\, \Box_0\Box_1(p^i_{01} \wedge p_{10} \rightarrow \Box_2q_i) \wedge \Box_0\Box_2(p^i_{02} \wedge p_{20} \rightarrow \Box_1r_i)\big)\Big]$$

$$\longrightarrow\; \Diamond_1\Diamond_2\big(p_{12} \wedge p_{21} \wedge \bigwedge_{i<k} \Diamond_0(q_i \wedge r_i)\big).$$

CLAIM 9. For every $0 < k < \omega$ and every α-frame $\mathfrak{F}$, Φ_k is true in $\mathfrak{F}$ iff $\mathfrak{F} \models \varphi_k$.

Proof. We prove the harder right-to-left direction only. Fix some k and suppose that $\mathfrak{F} = (W, S^{\mathfrak{F}}_\beta)_{\beta<\alpha}$ is an α-frame such that $\mathfrak{F} \models \varphi_k$. Let $v, y, z, x_0, \ldots, x_{k-1}$ in W be given as in Φ_k. We define a model $\mathfrak{M} = (\mathfrak{F}, \vartheta)$ over $\mathfrak{F}$ as follows.

$$\vartheta(p^i_{01}) = \{w \in W : x_iS^{\mathfrak{F}}_1w\}, \qquad \vartheta(p^i_{02}) = \{w \in W : x_iS^{\mathfrak{F}}_2w\}, \quad \text{for } i < k,$$

$$\vartheta(p_{10}) = \{w \in W : yS^{\mathfrak{F}}_0w\}, \qquad \vartheta(p_{12}) = \{w \in W : yS^{\mathfrak{F}}_2w\},$$

$$\vartheta(p_{20}) = \{w \in W : zS^{\mathfrak{F}}_0w\}, \qquad \vartheta(p_{21}) = \{w \in W : zS^{\mathfrak{F}}_1w\},$$

$$\vartheta(q_i) = \{w \in W : \exists s \in \vartheta(p^i_{01}) \cap \vartheta(p_{10})\ sS_2^{\mathfrak{F}}w\}, \quad \text{for } i < k,$$
$$\vartheta(r_i) = \{w \in W : \exists s \in \vartheta(p^i_{02}) \cap \vartheta(p_{20})\ sS_1^{\mathfrak{F}}w\}, \quad \text{for } i < k.$$

It is routine to check that the antecedent of φ_k holds in $\mathfrak{M}$ at point v. Thus, by assumption, $\Diamond_1\Diamond_2\big(p_{12} \wedge p_{21} \wedge \bigwedge_{i<k}\Diamond_0(q_i \wedge r_i)\big)$ also holds in $\mathfrak{M}$ at v. This implies that there are points $u, u_0, \ldots, u_{k-1}$ such that $yS_2^{\mathfrak{F}}u$, $zS_1^{\mathfrak{F}}u$, $uS_0^{\mathfrak{F}}u_i$, and $q_i \wedge r_i$ holds in $\mathfrak{M}$ at point u_i, for each $i < k$. By unfolding the definitions of $\vartheta(q_i)$ and $\vartheta(r_i)$, we obtain worlds $y_0, \ldots, y_{k-1}, z_0, \ldots, z_{k-1}$ as required. ■

LEMMA 10. *For any* $0 < k < \omega$, $\mathfrak{F}_k \not\models \mathbf{K}^\alpha$.

Proof. By Claims 8 and 9, it is enough to show that Φ_k fails in $\mathfrak{F}_k$. To this end, take the following 'fork' in $\mathfrak{F}_k$:

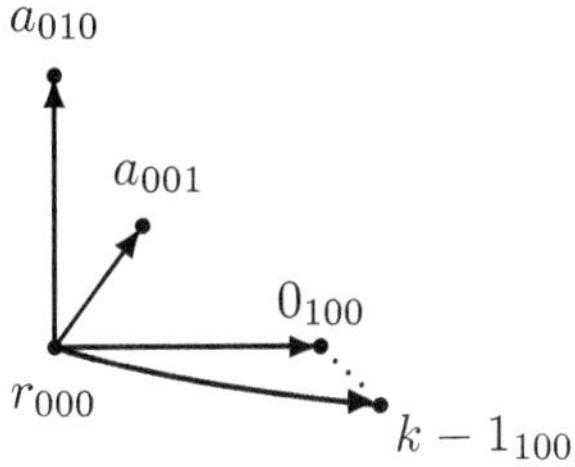

that is, let $v = r_{000}$, $y = a_{010}$, $z = a_{001}$, and $x_i = i_{100}$ for $i < k$. Now the only points in $\mathfrak{F}_k$ suitable for u are i_{011}, for all $i < k$. We will show that none of them can be 'extended' with other points as required. To this end, fix some $i < k$ and let $u = i_{011}$. On the one hand, the only points in $\mathfrak{F}_k$ suitable for u_i are i^1_{111} and i^2_{111}. On the other, the only points in $\mathfrak{F}_k$ suitable for y_i and z_i are i_{110} and i_{101}, respectively. But $(i_{101}, i^1_{111}) \notin R_1^{\mathfrak{F}}$ and $(i_{110}, i^2_{111}) \notin R_2^{\mathfrak{F}}$. ■

Acknowledgements. I am grateful to Rob Goldblatt for his careful reading of the preliminary version and for his suggestions.

BIBLIOGRAPHY

[1] H. Andréka. Complexity of equations valid in algebras of relations. Part I: strong non-finitizability. *Annals of Pure and Applied Logic*, 89:149–209, 1997.

[2] D.M. Gabbay, A. Kurucz, F. Wolter, and M. Zakharyaschev. *Many-Dimensional Modal Logics: Theory and Applications*, volume 148 of *Studies in Logic and the Foundations of Mathematics*. Elsevier, 2003.

[3] D.M. Gabbay and V. Shehtman. Products of modal logics. Part I. *Journal of the IGPL*, 6:73–146, 1998.

[4] L. Henkin, J.D. Monk, and A. Tarski. *Cylindric Algebras, Part II*, volume 115 of *Studies in Logic and the Foundations of Mathematics*. North Holland, 1985.

[5] R. Hirsch, I. Hodkinson, and A. Kurucz. On modal logics between $\mathbf{K} \times \mathbf{K} \times \mathbf{K}$ and $\mathbf{S5} \times \mathbf{S5} \times \mathbf{S5}$. *Journal of Symbolic Logic*, 67:221–234, 2002.

[6] J.S. Johnson. Nonfinitizability of classes of representable polyadic algebras. *Journal of Symbolic Logic*, 34:344–352, 1969.

[7] A. Kurucz. On axiomatising products of Kripke frames. *Journal of Symbolic Logic*, 65:923–945, 2000.
[8] A. Kurucz. Combining modal logics. In P. Blackburn, J. van Benthem, and F. Wolter, editors, *Handbook of Modal Logic*, volume 3 of *Studies in Logic and Practical Reasoning*, pages 869–924. Elsevier, 2007.
[9] J.D. Monk. Nonfinitizability of classes of representable cylindric algebras. *Journal of Symbolic Logic*, 34:331–343, 1969.
[10] K. Segerberg. Two-dimensional modal logic. *Journal of Philosophical Logic*, 2:77–96, 1973.
[11] V. Shehtman. Two-dimensional modal logics. *Mathematical Notices of the USSR Academy of Sciences*, 23:417–424, 1978. (Translated from Russian).

Agi Kurucz
Department of Computer Science
King's College London
Strand, London WC2R 2LS,
U.K
`agi.kurucz@kcl.ac.uk`

A modal perspective on monadic second-order alternation hierarchies

ANTTI KUUSISTO

ABSTRACT. We establish that the quantifier alternation hierarchy of formulae of Second-Order Propositional Modal Logic (*SOPML*) induces an infinite corresponding semantic hierarchy over the class of finite directed graphs. This is a response to an open problem posed in [4] and [8]. We also provide modal characterizations of the expressive power of Monadic Second-Order Logic (*MSO*) and address a number of points that should promote the potential advantages of viewing *MSO* and its fragments from the modal perspective.

Keywords: monadic second-order logic, second-order propositional modal logic, finite model theory, alternation hierarchies

1 Introduction

In this paper we investigate the expressive power of Second-Order Propositional Modal Logic (*SOPML*), which is a modal logic extended with propositional quantifiers ranging over sets of possible worlds. Modal logics with propositional quantifiers have been investigated by a variety of researchers, see [7, 8, 9, 10, 11, 12, 13, 14, 22, 23] for example.

Johan van Benthem [4] and Balder ten Cate [8] raise the question whether the quantifier alternation hierarchy of *SOPML*-formulae induces an ascending corresponding hierarchy of definable classes of Kripke frames. This is an interesting question, especially as ten Cate shows in [8] that formulae of *SOPML* admit a prenex normal form representation. In this paper we prove that the semantic counterpart of the quantifier alternation hierarchy of *SOPML*-formulae is infinite over the class of finite directed graphs. This automatically implies that the semantic hierarchy is infinite over arbitrary Kripke frames.

Alternation hierarchies have received a lot of attention in finite model theory, see [16, 18, 19, 20, 21, 24] for example. As *SOPML* is a semantically natural fragment of *MSO* (see Theorem 6 in [8]), we feel that our result is also relatively interesting from the point of view of finite model theory.

Our main tool in answering the the question of van Benthem and ten Cate is a theorem of Schweikardt [21] which states that the alternation hierarchy of Monadic Second-Order Logic is strict over the class of grids. Inspired by the approach of Matz and Thomas in [19], we employ an approach based on *strong first-order reductions* in order to transfer the result of Schweikardt to a special class of finite directed graphs we define. Over this class the expressive power of *SOPML* coincides with that of *MSO*, whence we easily obtain the desired result that the alternation hierarchy of *SOPML* is infinite over finite directed graphs. The precise definition of strong first-order reductions (found in [18]) is of no particular importance for the present paper, as we give a virtually self-contained exposition of all our results.

As a by-product of our investigations we obtain a simple, effective procedure (inspired by the approach of ten Cate [8]) that translates *MSO*-sentences to equivalent formulae of Second-Order Propositional Modal Logic with Universal Modality (*SOPML*(E)). This implies that the expressive power of *SOPML*(E) on finite/arbitrary relational structures coincides with that of *MSO*, and a trivial adaptation of our argument shows that replacing universal modality E with difference modality D does not change the picture. Such modal perspectives on *MSO* could turn out interesting from the point of view of finite model theory.

The paper is structured as follows: In Section 2 we fix the notation and discuss a number of preliminary issues. In Section 3 we show that *MSO* = *SOPML*(E) with regard to expressive power. Using an approach analogous to that in Section 3, we then define in Section 4 a special class of directed graphs over which *MSO* and *SOPML* coincide in expressive power. In Section 5 we first work with *MSO*, transferring the result of Schweikardt to our special class of directed graphs. Then, using the connection created in Section 4, it is easy to establish that the *SOPML* alternation hierarchy is infinite over directed graphs.

2 Preliminary considerations

In this section we introduce technical notions that occupy a central role in the rest of the discourse.

2.1 Syntax and semantics

With a model we mean a model of predicate logic. We only consider models associated with a relational vocabulary. With a relational vocabulary we mean a vocabulary with relation symbols and constant symbols only.

We fix countable sets VAR_{FO} and VAR_{SO} of first-order and second-order variables, respectively. Naturally we assume that the sets are disjoint. We let $VAR = VAR_{FO} \cup VAR_{SO}$. We let lower-case symbols x, y, z denote first-order variables. Upper-case symbols X, Y, Z denote second-order variables. A union f of two functions $f_{FO} : VAR_{FO} \longrightarrow Dom(M)$ and $f_{SO} : VAR_{SO} \longrightarrow \mathcal{P}(Dom(M))$, where M is a model and $Dom(M)$ its domain, is called an *assignment*. Monadic Second-Order Logic is interpreted in terms of models and assignments in the usual way: We write $M, f \models \varphi$ when model M satisfies *MSO*-formula φ under assignment f.

Let *PROP* denote a countable set of *proposition variables*. We let symbols $p_x, p_y, p_z, p_X, p_Y, p_Z$ denote proposition variables. Let $\mathcal{S} = \mathcal{S}_0 \cup \mathcal{S}_1 \cup \mathcal{S}_2 \cup \mathcal{S}_3$ be a relational vocabulary with set $\mathcal{S}_0$ of constant symbols and sets $\mathcal{S}_1$ and $\mathcal{S}_2$ of unary and binary relation symbols respectively; set $\mathcal{S}_3$ contains the relation symbols of higher arities. The language $L(\mathcal{S})$ of Second-Order Propositional Modal Logic associated with vocabulary $\mathcal{S}$ is determined by the following recursive definition:

$$\varphi ::= c_i \mid p_x \mid p_j \mid \neg\varphi \mid \varphi_1 \wedge \varphi_2 \mid \Diamond_k \varphi \mid \triangle_l(\varphi_1, ..., \varphi_{n-1}) \mid \exists p_x\, \varphi$$

such that $c_i \in \mathcal{S}_0$, $p_x \in PROP$, $P_j \in \mathcal{S}_1$, $R_k \in \mathcal{S}_2$, and $R_l \in \mathcal{S}_3$ is an n-ary relation symbol.

In order to interpret formulae of Second-Order Propositional Modal Logic, we need the notion of a *pointed model*:

DEFINITION 1. A pointed model is a pair (M, w), where M is a model and $w \in Dom(M)$.

We also need objects that interpret free occurrences of proposition variables $p_x \in PROP$: Any mapping $V : PROP \longrightarrow \mathcal{P}(Dom(M))$, where M is a model, is called a *valuation*.

Let $\mathcal{S}$ be a vocabulary and M an $\mathcal{S}$-model with $w \in Dom(M) = W$. Let V be a related valuation. We let $\Vdash$ denote the modal truth relation, which we define in the following way:

$$\begin{array}{lcl}
(M,w),V \Vdash c_i & \Leftrightarrow & w = c_i^M \\
(M,w),V \Vdash p_j & \Leftrightarrow & w \in P_j^M \\
(M,w),V \Vdash p_x & \Leftrightarrow & w \in V(p_x) \\
(M,w),V \Vdash \neg\varphi & \Leftrightarrow & (M,w),V \nVdash \varphi \\
(M,w),V \Vdash \varphi \wedge \psi & \Leftrightarrow & (M,w),V \Vdash \varphi \text{ and } (M,w),V \Vdash \psi \\
(M,w),V \Vdash \exists p_x \varphi & \Leftrightarrow & \exists U \subseteq W((M,w),V[p_x \mapsto U] \Vdash \varphi) \\
(M,w),V \Vdash \Diamond_k\, \varphi & \Leftrightarrow & \exists u \in W(wR_k u \text{ and } (M,u) \Vdash \varphi) \\
(M,w),V \Vdash \triangle_l\, (\varphi_1, ..., \varphi_{n-1}) & \Leftrightarrow & \exists u_1, ..., u_{n-1} \in W \text{ such that} \\
 & & R_l(w, u_1, ..., u_{n-1}) \text{ and} \\
 & & \forall i < n((M,u_i),V \Vdash \varphi_i)
\end{array}$$

If a formula φ does not contain free occurrences of proposition variables, we may drop valuation V and write $(M, w) \Vdash \varphi$. An *SOPML*-formula without free proposition variables is an *SOPML-sentence*. We extend the definition of relation $\Vdash$ to models in the following way:

$$M \Vdash \varphi \;\Leftrightarrow\; \text{for all } w \in W,\; (M,w) \Vdash \varphi$$

We also extend the truth relation of predicate logic to cover pointed models. We define

$$(M,w) \models \varphi(x) \;\Leftrightarrow\; M, [x \mapsto w] \models \varphi(x),$$

where $\varphi(x)$ is a formula with exactly one free variable, x.

Let H_p be a class of *pointed* models. We say that *SOPML*-sentence φ *defines class* C of pointed models *with respect to* H_p if $C = \{(M,w) \in H_p \mid (M,w) \Vdash \varphi\}$. We write $MOD_{H_p}(\varphi) = C$. Similarly, we say that *MSO*-formula $\psi(x)$ *defines class* C of pointed models *with respect to* H_p if $C = \{(M,w) \in H_p \mid (M,w) \models \psi(x)\}$. Formula $\psi(x)$ is required to contain exactly one free first-order variable and no free second-order variables. We write $MOD_{H_p}(\psi(x)) = C$.

Let H be a class of models. We say that $SOPML$-sentence φ *defines class* C of models *with respect to* H if $C = \{M \in H \mid M \Vdash \varphi\}$. This corresponds to the notion of global definability. We write $MOD_H(\varphi) = C$. Similarly, we say that MSO-sentence ψ *defines class* C of models *with respect to* H if $C = \{M \in H \mid M \models \psi\}$. We write $MOD_H(\psi) = C$.

When we informally leave out parentheses when writing formulae, the order of preference of logical connectives is such that unary connectives have the highest priority and then come $\wedge, \vee, \rightarrow, \leftrightarrow$ in the given order.

When a subindex of a symbol $(c_i, p_i, R_i, \Diamond_i$ etc.) is irrelevant or understood from the context, we may leave it unwritten.

2.2 Grids and graphs

Two classes of structures have a central role in the considerations that follow:

DEFINITION 2. Let $m, n \in \mathbb{N}_{\geq 1}$ and let $D_m^n = \{1, 2, ..., m\} \times \{1, 2, ..., n\}$. Define binary relations S_1^{Gr} and S_2^{Gr} such that S_1^{Gr} contains exactly the pairs of type $((i,j),(i+1,j)) \in D_m^n \times D_m^n$ and S_2^{Gr} exactly the pairs of type $((i,j),(i,j+1)) \in D_m^n \times D_m^n$. A structure $Gr = (D_m^n, S_1^{Gr}, S_2^{Gr})$, where $m, n \in \mathbb{N}_{\geq 1}$, is called a *grid*. Grid $Gr = (D_m^n, S_1^{Gr}, S_2^{Gr})$ is said to *correspond to* an $m \times n$-matrix. Element $(1,1)$ is referred to as the *top left element*. We let $GRID$ denote the class of grids. Note that this class is not closed under isomorphism.

The other class of structures we shall consider is that of (non-empty) *directed graphs*. We define a directed graph to be a structure of type (W, R), where $W \neq \emptyset$ is a finite set and $R \subseteq W \times W$ a binary relation. When we refer to a graph we always mean a finite directed graph. We let $GRAPH$ denote the class of finite directed graphs.

2.3 Alternation hierarchies

Intuitively, the levels of the *monadic second-order quantifier alternation hierarchy* measure the number of alternations of existential and universal second-order quantifiers of MSO-formulae in prenex normal form. (An MSO-formula in prenex normal form consists of a vector of second-order quantifiers, followed by a first order part.) It is natural to classify $SOPML$-formulae in an analogous way.

Below, we give formal definitions of alternation hierarchies. We only define levels containing formulae that begin with an existential quantifier, as this suffices for the purposes of this article.

Let $L_{FO}(\mathcal{S} \cup VAR_{SO})$ denote the first-order language associated with relational vocabulary $\mathcal{S} \cup VAR_{SO}$. We define $\Sigma_0(\mathcal{S}) = L_{FO}(\mathcal{S} \cup VAR_{SO})$ and let

$$\Sigma_{n+1}(\mathcal{S}) = \{\exists X_1, ..., \exists X_k \neg\varphi \mid k \in \mathbb{N} \text{ and } \varphi \in \Sigma_n(\mathcal{S})\}.$$

Sets $\Sigma_n(\mathcal{S})$ are levels of the syntactic alternation hierarchy of MSO.

We write Σ_n instead of $\Sigma_n(\mathcal{S})$ when the vocabulary is clear from the context. With $[\Sigma_n]$ we refer to the equivalence closure of Σ_n. In other words, $\varphi \in [\Sigma_n]$ iff φ is equivalent to some formula $\varphi' \in \Sigma_n$.

Levels of the syntactic alternation hierarchy are associated with natural semantic counterparts: Let H be a subclass of $\mathcal{S}$-structures. We define

$$\underline{\Sigma_n}(H) = \{C \in \mathcal{P}(H) \mid MOD_H(\varphi) = C \text{ for some sentence } \varphi \in \Sigma_n(\mathcal{S})\}.$$

Similarly, we let

$$\underline{\Sigma_n}(H_p) \\ = \{C \in \mathcal{P}(H_p) \mid MOD_{H_p}(\varphi(x)) = C \text{ for some formula } \varphi(x) \in \Sigma_n(\mathcal{S})\},$$

where H_p is a class of pointed $\mathcal{S}$-models.

We then deal with the quantifier alternation hierarchies of *SOPML*. The zeroeth level of the syntactic hierarchy of *SOPML* contains all quantifier free *SOPML*-formulae, and any formula $\exists p_{x_1}, ..., \exists p_{x_k} \neg\varphi$ belongs to level $n+1$ iff φ belongs to the n-th level. We let $\Sigma_n^{ML}(\mathcal{S})$ denote the n-th level of this hierarchy. On the semantic side we define

$$\underline{\Sigma_n^{ML}}(H) \\ = \{C \in \mathcal{P}(H) \mid MOD_H(\varphi) = C \text{ for some sentence } \varphi \in \Sigma_n^{ML}(\mathcal{S})\},$$

where H is a subclass of the class of $\mathcal{S}$-models. Similarly, we define

$$\underline{\Sigma_n^{ML}}(H_p) \\ = \{C \in \mathcal{P}(H_p) \mid MOD_{H_p}(\varphi) = C \text{ for some sentence } \varphi \in \Sigma_n^{ML}(\mathcal{S})\},$$

where H_p is a class of pointed $\mathcal{S}$-models.

If for all $n \in \mathbb{N}$ there exists a $k \in \mathbb{N}$ such that $\underline{\Sigma_n}(K) \neq \underline{\Sigma_k}(K)$, we say that *the alternation hierarchy of MSO is infinite on* K. We define infinity of *SOPML* alternation hierarchies analogously.

3 SOPML(E) = MSO

In this section we show that Second-Order Propositional Modal Logic with Universal Modality (*SOPML*(E)) has the same expressive power as *MSO*. This result is closely related to the fact that hybrid logic $\mathcal{H}(\downarrow, E)$ is expressively complete for first-order logic (see [3] and the references therein). In fact, in the light of the results in [1, 2, 8], the result is not surprising.

In order to establish that *SOPML*(E) is expressively complete for *MSO*, we define a simple translation from the set of *MSO*-formulae to the set of *SOPML*(E)-formulae. The translation was inspired by a very similar translation defined in [8].

We begin with a formal definition of logic *SOPML*(E) (cf. *SOEPDL* in [22]): Let $\mathcal{S}$ be a relational vocabulary and let $L(\mathcal{S})$ denote the related second-order propositional modal language. We extend language $L(\mathcal{S})$ to a new language $L^E(S)$ in the following way:

$$\begin{array}{lcl} \varphi \in L(S) & \Rightarrow & \varphi \in L^E(S) \\ \varphi \in L^E(S) & \Rightarrow & \langle E \rangle \varphi \in L^E(S) \end{array}$$

The truth definition of $SOPML$ is extended by setting

$$(M,w),V \Vdash \langle E\rangle\varphi \;\Leftrightarrow\; \exists u \in Dom(M)((M,u),V \Vdash \varphi).$$

Next we prepare ourselves for an important auxiliary result (Lemma 3), which we then prove.

Let M be a model and $f : VAR \longrightarrow Dom(M) \cup \mathcal{P}(Dom(M))$ a related assignment. Define $PROP = \{p_x \mid x \in VAR_{FO}\} \cup \{p_X \mid X \in VAR_{SO}\}$. We let V_f denote the valuation mapping from $PROP$ to $\mathcal{P}(Dom(M))$ such that $V_f(p_x) = \{f(x)\}$ and $V_f(p_X) = f(X)$ for all $p_x, p_X \in PROP$.

Consider the following formula:

$$uniq(p_x) = \langle E\rangle p_x \wedge \forall p_y(\langle E\rangle(p_y \wedge p_x) \rightarrow [E](p_x \rightarrow p_y)),$$

where $[E]$ stands for $\neg\langle E\rangle\neg$. The formula states that proposition p_x is satisfied by exactly one point of the model.

We define the following translation TR from the set of MSO-formulae to the set of $SOPML(E)$-formulae:

$$\begin{array}{lcl}
TR(P(y)) & = & \langle E\rangle(p \wedge p_y) \\
TR(Y(z)) & = & \langle E\rangle(p_Y \wedge p_z) \\
TR(R_i(y,z)) & = & \langle E\rangle(p_y \wedge \Diamond_i p_z) \\
TR(R_j(x_1,...,x_n)) & = & \langle E\rangle(p_{x_1} \wedge \triangle_j(p_{x_2},...,p_{x_n})) \\
TR(y = z) & = & \langle E\rangle(p_y \wedge p_z) \\
TR(c = y) & = & \langle E\rangle(c \wedge p_y) \\
TR(y = c) & = & \langle E\rangle(p_y \wedge c) \\
TR(c_{i_1} = c_{i_2}) & = & \langle E\rangle(c_{i_1} \wedge c_{i_2}) \\
TR(\neg\psi) & = & \neg TR(\psi) \\
TR(\psi \wedge \varphi) & = & TR(\psi) \wedge TR(\varphi) \\
TR(\exists z(\psi)) & = & \exists p_z(uniq(p_z) \wedge TR(\psi)) \\
TR(\exists Z(\psi)) & = & \exists p_Z(TR(\psi))
\end{array}$$

LEMMA 3. *For all MSO-formulae φ,*

$$M, f[x \mapsto w] \models \varphi \quad\Leftrightarrow\quad (M,w), V_f[p_x \mapsto \{w\}] \Vdash TR(\varphi)$$

for all models $M = (W,R)$, all points $w \in W$ and all assignments $f : VAR \longrightarrow W \cup \mathcal{P}(W)$.

Proof. We prove the claim by induction on the structure of formula φ. The basis of the induction is established by a straightforward argument. The case where $\varphi = \neg\psi$ for some formula ψ is trivial, as is the case where φ has a conjunction as its main connective. Therefore we may proceed directly to the case where $\varphi = \exists z(\psi)$.

Assume first that $M, f[x \mapsto w] \models \exists z(\psi)$ (we assume w.l.o.g. that $z \neq x$). Thus $M, f[z \mapsto u, x \mapsto w] \models \psi$ for some $u \in W$. Therefore $(M,w), V_f[p_z \mapsto \{u\}, p_x \mapsto \{w\}] \Vdash TR(\psi)$ by the induction hypothesis. Thus $(M,w), V_f[p_x \mapsto \{w\}] \Vdash \exists p_z(uniq(p_z) \wedge TR(\psi))$, as required.

Assume then that $(M, w), V_f[p_x \mapsto \{w\}] \Vdash \exists p_z(uniq(p_z) \wedge TR(\psi))$. Therefore $(M, w), V_f[p_z \mapsto U, p_x \mapsto \{w\}] \Vdash uniq(p_z) \wedge TR(\psi)$ for some set $U \subseteq W$. As $(M, w), V_f[p_z \mapsto U, p_x \mapsto \{w\}] \Vdash uniq(p_z)$, we have $U = \{u\}$ for some $u \in W$. Therefore $(M, w), V_f[p_z \mapsto \{u\}, p_x \mapsto \{w\}] \Vdash TR(\psi)$. Thus $M, f[z \mapsto u, x \mapsto w] \models \psi$ by the induction hypothesis, and therefore $M, f[x \mapsto w] \models \exists z(\psi)$, as required.

Finally, the argument for the case where formula φ is of type $\exists Z(\psi)$, for some formula ψ, is straightforward. ■

We are now ready for the main results of this section:

THEOREM 4. *A subclass K of a class C of pointed models is definable w.r.t. C by an MSO-formula if and only if K is definable w.r.t. C by an SOPML(E)-sentence.*

Proof. Let $\varphi(x)$ be an arbitrary *MSO*-formula with exactly one free variable. Let $M = (W, R, ...)$ be an arbitrary model and $f : VAR \longrightarrow W \cup \mathcal{P}(W)$ an arbitrary assignment. We have the following equivalence by Lemma 3:

$$M, f[x \mapsto w] \models \varphi \quad \Leftrightarrow \quad (M, w), V_f[p_x \mapsto \{w\}] \Vdash TR(\varphi)$$

We also have the following equivalence:

$$\begin{array}{l} (M, w), V_f[p_x \mapsto \{w\}] \Vdash TR(\varphi) \\ \Leftrightarrow \\ (M, w) \Vdash \exists p_x(p_x \wedge uniq(p_x) \wedge TR(\varphi)) \end{array}$$

By the two equivalences, it is clear that $\exists p_x(p_x \wedge uniq(p_x) \wedge TR(\varphi))$ is the desired *SOPML*(E)-sentence equivalent to φ.

For the converse, if φ is an *SOPML*(E)-sentence, the desired *MSO*-formula is $St_x(\varphi)$, where St_x denotes the required trivial generalization of the standard translation operator (see [5] for the definition of standard translation). ■

THEOREM 5. *A subclass K of a class C of models is definable w.r.t. C by an MSO-sentence if and only if K is definable w.r.t. C by an SOPML(E)-sentence.*

Proof. Let φ be an arbitrary *MSO*-sentence. Notice that $TR(\varphi)$ does not contain any free proposition variables. We have the following equivalences:

$$\begin{array}{rcl} M \models \varphi & \Leftrightarrow & M, f[x \mapsto w] \models \varphi \text{ for all } w \in W \\ & \Leftrightarrow & (M, w), V_f[p_x \mapsto \{w\}] \Vdash TR(\varphi) \text{ for all } w \in W \\ & \Leftrightarrow & (M, w) \Vdash TR(\varphi) \text{ for all } w \in W \\ & \Leftrightarrow & M \Vdash TR(\varphi) \end{array}$$

where the second equivalence follows from Lemma 3.

For the converse, $\forall x St_x(\psi)$ is the desired *MSO*-sentence equivalent to *SOPML*(E)-sentence ψ. ■

A trivial adaptation of the approach in this section leads to the realization that with regard to expressive power, *SOPML*$(D) =$ *MSO*, where D is the difference modality.

4 Simulating globality

The local nature of *SOPML* (cf. Proposition 4 of [8]) limits its expressive power. In this section we define a class of structures over which this is not the case. The key point is to insist that each structure contains a point which connects to every point of the structure:

DEFINITION 6. Let $S = (W, R, ...)$ be a structure with a binary relation R. Assume there is a point $w \in W$ such that wRu for all $u \in W$. We call such a point u a *localizer*. Structures with a localizer are called *localized.* If (M, w) is a pointed model where w is a localizer, we say that (M, w) is *l-pointed.*

The notions of a localizer and a localized model resemble the notions of a *spypoint* and a *spypoint model* applied in the hybrid logic literature (see [2, 6]).

We then prepare ourselves for the next result (Lemma 7) by defining local analogues of formula $uniq(p_x)$ and translation TR defined in Section 3.

Let $uniq'(p_x)$ be the following formula:

$$\Diamond p_x \wedge \forall p_y(\Diamond(p_y \wedge p_x) \rightarrow \Box(p_x \rightarrow p_y)),$$

where $\Box$ stands for $\neg\Diamond\neg$. It is easy to see that if (W, R) is a directed graph with a localizer $w \in W$, then $((W, R), w), [p_x \mapsto U] \Vdash uniq'(p_x)$ if and only if $U = \{u\}$ for some $u \in W$.

Consider translation TR defined in Section 3. Replace the occurrences of the universal diamond $\langle E \rangle$ by $\Diamond$, and also replace $uniq(p_z)$ by $uniq'(p_z)$. Denote this new translation by LTR.

The following lemma is a local analogue of Lemma 3:

LEMMA 7. *For all MSO-formulae* φ*,*

$$M, f[x \mapsto w] \models \varphi \quad \Leftrightarrow \quad (M, w), V_f[p_x \mapsto \{w\}] \Vdash LTR(\varphi)$$

for all localized models $M = (W, R)$*, all localizers* $w \in W$ *and all assignments* $f : VAR \longrightarrow W \cup \mathcal{P}(W)$.

Proof. The proof is essentially the same as that of Lemma 3. ∎

The following lemma is a local analogue of Theorem 4:

LEMMA 8. *Let* C *be a class of l-pointed models. A class* $K \subseteq C$ *of l-pointed models is definable w.r.t.* C *by an MSO-formula if and only if* K *is definable w.r.t.* C *by an SOPML-sentence.*

Proof. Let *MSO*-formula φ define K w.r.t. C. Formula $\exists p_x(p_x \wedge uniq'(p_x) \wedge LTR(\varphi))$ is the desired *SOPML*-sentence equivalent to φ. The proof is essentially the same as that of Theorem 4. Instead of using Lemma 3, however, we apply the analogous lemma that fits the framework without universal modality, i.e., Lemma 7. ∎

Let C be a class of localized models. Let φ be an *SOPML*-sentence such that for each model $M \in C$ there exists at least one point $w \in Dom(M)$ that satisfies φ, and moreover, every point w that satisfies φ is a localizer. We say that φ *fixes localizers on* C.

The following lemma is a local analogue of Theorem 5:

LEMMA 9. *Let C be a class of localized models and assume there exists some SOPML-sentence φ that fixes localizers on C. A class $K \subseteq C$ of localized models is definable w.r.t C by an MSO-sentence if and only if K is definable w.r.t. C by an SOPML-sentence.*

Proof. Let ψ be an *MSO*-sentence that defines K w.r.t. C. Let $M \in C$ and let $U \subseteq Dom(M)$ be the set of points $w \in Dom(M)$ such that $(M, w) \Vdash \varphi$. We have the following equivalences:

$$\begin{aligned} M \models \psi &\Leftrightarrow \forall w \in U(M, f[x \mapsto w] \models \psi) \\ &\Leftrightarrow \forall w \in U((M, w), V_f[p_x \mapsto \{w\}] \Vdash LTR(\psi)) \\ &\Leftrightarrow \forall w \in U((M, w) \Vdash LTR(\psi)) \\ &\Leftrightarrow \forall w \in U((M, w) \Vdash \varphi \rightarrow LTR(\psi)) \\ &\Leftrightarrow M \Vdash \varphi \rightarrow LTR(\psi), \end{aligned}$$

where the second equivalence follows from Lemma 7.

For the converse, $\forall x St_x(\pi)$ is the desired *MSO*-sentence equivalent to *SOPML*-sentence π. ■

5 Alternation hierarchy of SOPML is infinite

In this section we prove that the *SOPML* alternation hierarchy is infinite over the class of finite directed graphs. The following theorem from [21] is the most important tool we shall use in the elaborations below:

THEOREM 10. *For all $n \in \mathbb{N}_{\geq 1}$ we have $\underline{\Sigma_n}(GRID) \neq \underline{\Sigma_{n+1}}(GRID)$.*

While a similar result holds for directed graphs[1], on *words* and *labelled trees*, for example, the alternation hierarchy of *MSO* is known to collapse to level Σ_1 (see [17] for a recent survey of related results). This explains why we use grids in the elaborations below.

In Subsection 5.1 we show how to encode grids as *localized grid graphs* (see Definition 11). In Subsection 5.2 we then transfer the result of Theorem 10 to localized grid graphs (Proposition 16) and l-pointed localized grid graphs (Proposition 17). The transferred results will be needed in Subsection 5.3, where we show that the alternation hierarchy of *SOPML* is infinite over pointed directed graphs (Theorem 18) and ordinary directed graphs (Theorem 19).

---

[1]See [18, 19]. In [18], the result for directed graphs is established via a reduction from the class of grids to a certain subclass of directed graphs. Let us call this class C. While we could prove Proposition 16 via a reduction from class C, we instead prove it via a direct reduction from the class of grids. The two alternative approaches are very similar, but the approach via a direct reduction from the class of grids has presentational advantages over the approach via a reduction from class C.

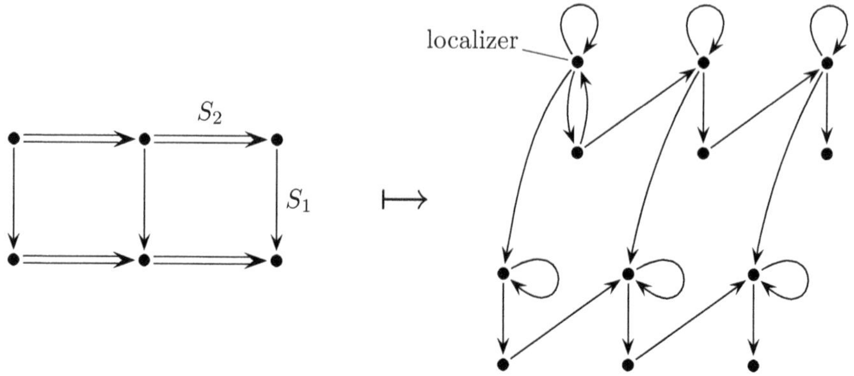

Figure 1. The figure shows a grid and its encoding. The localizer connects to each point of the graph; for the sake of clarity, most arrows originating from the localizer have not been drawn.

5.1 Encoding grids as localized grid graphs

In this subsection we define a map that encodes grids as localized directed graphs.

DEFINITION 11. Mapping $\alpha : GRID \longrightarrow GRAPH$ transforms grid Gr to a directed graph (W, R) such that $W = (Dom(Gr)\times\{0\}) \cup (Dom(Gr)\times\{1\})$ and

$$\begin{array}{rcl} R & = & \{\ ((a,0),(a,0)) \mid a \in Dom(Gr)\} \\ & \cup & \{\ ((a,0),(a,1)) \mid a \in Dom(Gr)\} \\ & \cup & \{\ ((a,0),(b,0)) \mid (a,b) \in S_1^{Gr}\} \\ & \cup & \{\ ((a,1),(b,0)) \mid (a,b) \in S_2^{Gr}\} \\ & \cup & \{\ ((t,0),(a,i)) \mid a \in Dom(Gr),\ \ i \in \{0,1\}\ \} \\ & \cup & \{\ ((t,1),(t,0))\ \}, \end{array}$$

where $t = (1,1)$ is the top left element of grid Gr. We call structures in the isomorphism closure of $\alpha(GRID)$ *localized grid graphs.* We let LGG denote this class of structures. We let LGG_p denote the corresponding class of l-pointed grid graphs. See Figure 1 for an example of a grid and the corresponding localized grid graph.

Point $(t,0)$ sees every point in graph $\alpha(Gr)$, i.e., it is a localizer. This property enables us to overcome difficulties resulting from the local nature of *SOPML*. We define the following formula:

$$\psi_{t_0}(x) = xRx \wedge \exists y(xRy \wedge yRx \wedge x \neq y)$$

The formula asserts that $x = (t,0)$. Insisting that $(t,1)R(t,0)$ will help us with a number of technical issues, such as defining formula

$$\psi_{t_1}(x) = \neg xRx \wedge \exists y(xRy \wedge yRx),$$

which asserts that $x = (t,1)$.

We then show that encoding $\alpha : GRID \longrightarrow GRAPH$ is injective:

LEMMA 12. *Encoding $\alpha : GRID \longrightarrow GRAPH$ is injective in the following sense: If $\alpha(Gr)$ and $\alpha(Gr')$ are isomorphic, then $Gr = Gr'$.*

Proof. Let $\alpha(Gr) = (W, R) = G$ and $\alpha(Gr') = (W', R') = G'$ for some grids Gr and Gr'. Assume $f : W \longrightarrow W'$ is an isomorphism between the graphs. Let k be the number of elements $w \in W$ with a reflexive loop. It is clear that Gr corresponds to an $m \times n$-matrix such that $m \cdot n = k$ (cf. Definition 2). The number of points $w' \in W'$ with a reflexive loop must also be k, as the two graphs are isomorphic. Thus grid Gr' corresponds to an $m' \times n'$-matrix such that $m' \cdot n' = k$. To conclude the proof it suffices to show that $n = n'$.

We shall show that for each $i \in \mathbb{N}_{\geq 1}$ there is a first-order formula φ_i such that for all $M \in GRID$ we have $\alpha(M) \models \varphi_i$ iff M corresponds to a $j \times i$-matrix for some j. The claim of the lemma follows from this: As $G \cong G'$, they satisfy the same first-order sentences. Thus there is some i such that both graphs G and G' satisfy sentence φ_i. Thus $n = i = n'$.

We then show how to define formulae φ_i. We deal with the case where $i = 1$ separately: We let $\varphi_1 = \exists x(\psi_{t_1}(x) \wedge \exists^{=1} y(xRy))$, where $\exists^{=1} y$ stands for "there exists exactly one y". We then consider the cases where $i \geq 2$. We first define the following formulae:

$$\begin{array}{lcl} \pi_2(x) & = & \exists y \exists z(\psi_{t_1}(y) \wedge yRz \wedge \neg zRy \wedge zRx \wedge \neg xRx) \\ succ(x,y) & = & \exists z(xRz \wedge zRy \wedge \neg yRy) \end{array}$$

We then define φ_i (where $i \geq 2$) in the following way:

$$\varphi_i = \exists x_2, ..., x_i \Big(\pi_2(x_2) \wedge \big(\bigwedge_{2 \leq r < i} succ(x_r, x_{r+1}) \big) \wedge \neg \exists y(x_i R y) \Big)$$

It is relatively easy to see that formulae φ_i have the desired meaning. ■

5.2 MSO alternation hierarchy over localized grid graphs

In this subsection we show that results analogous to Theorem 10 hold for localized grid graphs (Proposition 16) and l-pointed grid graphs (Proposition 17).

We begin by showing how to transform any grid-formula $\varphi_1 \in \Sigma_n$ into a graph-formula $\varphi_2 \in \Sigma_n$ that says the same about localized grid graphs as φ_1 says about grids:

LEMMA 13. *For every grid-formula φ_1 there exists a graph-formula φ_2 such that for all grids Gr and all assignments $f : VAR \rightarrow Dom(Gr) \cup \mathcal{P}(Dom(Gr))$,*

$$Gr, f \models \varphi_1 \;\Leftrightarrow\; \alpha(Gr), f' \models \varphi_2,$$

where valuation f' is defined such that for all $x, X \in VAR$, all $a \in Dom(Gr)$ and all $A \subseteq Dom(Gr)$ we have $f'(x) = (a, 0) \;\Leftrightarrow f(x) = a$ and $f'(X) = A \times \{0\} \;\Leftrightarrow f(X) = A$. Furthermore, for all $n \in \mathbb{N}$, if $\varphi_1 \in \Sigma_n$, then $\varphi_2 \in \Sigma_n$.

Proof. We begin by showing how to define φ_2 in the case where φ_1 is atomic. If φ_1 is of type $x = y$ or type $X(y)$, we let $\varphi_2 = \varphi_1$. If φ_1 is of type xS_1y, we let φ_2 be the following formula:

$$\begin{array}{ll} & \psi_{t_0}(x) \wedge \psi_{t_0}(y) \rightarrow \bot \\ \wedge & \psi_{t_0}(x) \wedge \neg\psi_{t_0}(y) \rightarrow \forall z(zRy \rightarrow (\psi_{t_0}(z) \vee z = y)) \\ \wedge & \neg\psi_{t_0}(x) \wedge \psi_{t_0}(y) \rightarrow \bot \\ \wedge & \neg\psi_{t_0}(x) \wedge \neg\psi_{t_0}(y) \rightarrow xRy \wedge x \neq y \end{array}$$

If φ_1 is of type xS_2y, we define φ_2 to be the following formula:

$$\begin{array}{ll} & \psi_{t_0}(x) \wedge \psi_{t_0}(y) \rightarrow \bot \\ \wedge & \psi_{t_0}(x) \wedge \neg\psi_{t_0}(y) \rightarrow \exists u(\psi_{t_1}(u) \wedge uRy) \\ \wedge & \neg\psi_{t_0}(x) \wedge \psi_{t_0}(y) \rightarrow \bot \\ \wedge & \neg\psi_{t_0}(x) \wedge \neg\psi_{t_0}(y) \rightarrow \exists z(xRz \wedge \neg zRz \wedge zRy) \end{array}$$

For the sake of induction, assume $\varphi_1 = \neg\pi_1$. By the induction hypothesis there exists a graph-formula π_2 such that $Gr, f \models \pi_1 \;\Leftrightarrow\; \alpha(Gr), f' \models \pi_2$ for all grids Gr and related assignments f. Let $\varphi_2 = \neg\pi_2$. Similarly, in the case where $\varphi_1 = \pi_1 \wedge \pi_1'$, let $\varphi_2 = \pi_2 \wedge \pi_2'$, where graph-formulae π_2, π_2' are again chosen by the induction hypothesis. In the case where $\varphi_1 = \exists x(\pi_1)$, let $\varphi_2 = \exists x(xRx \wedge \pi_2)$. Finally, in the case $\varphi_1 = \exists X(\pi_1)$, let $\varphi_2 = \exists X(\forall x(X(x) \rightarrow xRx) \wedge \pi_2)$. ■

Our next aim is to show that for each graph-sentence $\varphi_2 \in \Sigma_n$, there exists a grid-sentence $\varphi_1 \in \Sigma_n$ that says the same about grids as φ_2 says about localized grid graphs. In order to establish this, we first need to address a number of technical issues.

We define a new set of symbols $VAR' = VAR_{FO} \cup (VAR_{SO} \times \{0\}) \cup (VAR_{SO} \times \{1\}) \cup (VAR_{SO} \times \{t_0\}) \cup (VAR_{SO} \times \{t_1\})$. Naturally we choose our symbols such that the above five sets making up VAR' are disjoint. We associate each first-order variable with an index such that $VAR_{FO} = \{x_1, x_2, ...\}$. We denote the new second-order variables of type $(X, 0)$, $(X, 1)$, (X, t_0) and (X, t_1) by X^0, X^1, X^{t_0} and X^{t_1} respectively.

Let Gr be a grid. We partition the domain of grid graph $\alpha(Gr)$ into four sets:

$$\begin{array}{lcl} V_{t_0} & = & \{\ ((1,1),0)\ \} \\ V_{t_1} & = & \{\ ((1,1),1)\ \} \\ V_0 & = & \{\ ((x,y),0) \in Dom(\alpha(Gr)) \mid (x,y) \neq (1,1)\} \\ V_1 & = & \{\ ((x,y),1) \in Dom(\alpha(Gr)) \mid (x,y) \neq (1,1)\} \end{array}$$

Now let $\kappa : \mathbb{N}_{\geq 1} \longrightarrow \{0, 1, t_0, t_1\}$ be a function. We say that assignment $f : VAR \longrightarrow Dom(\alpha(Gr)) \cup \mathcal{P}(Dom(\alpha(Gr)))$ is *of type* κ if $f(x_i) \in V_{\kappa(i)}$ for all $i \in \mathbb{N}_{\geq 1}$. We call function κ an *assignment type*.

Each assignment $f : VAR \longrightarrow Dom(\alpha(Gr)) \cup \mathcal{P}(Dom(\alpha(Gr)))$ is associated with a related assignment $f_{Gr} : VAR' \longrightarrow Dom(Gr) \cup \mathcal{P}(Dom(Gr))$ defined in the following way:

$$\forall a \in Dom(Gr)\Big(f_{Gr}(x) = a \;\Leftrightarrow\; \big(f(x) = (a,0) \text{ or } f(x) = (a,1)\big)\Big)$$

for first-order variables $x \in VAR'$. For second-order variables $X^0, X^1 \in VAR'$ we let

$$\begin{array}{rcl} f_{Gr}(X^0) & = & \{a \in Dom(Gr) \mid (a,0) \in f(X)\} \setminus \{t\} \\ f_{Gr}(X^1) & = & \{a \in Dom(Gr) \mid (a,1) \in f(X)\} \setminus \{t\}, \end{array}$$

where $t = (1,1)$ is the top left element of grid Gr. For second-order variables X^{t_i}, where $i \in \{0,1\}$, we let

$$f_{Gr}(X^{t_i}) = \begin{cases} \{t\} & \text{if } (t,i) \in f(X), \text{ where } t = (1,1) \in Dom(Gr) \\ \emptyset & \text{otherwise} \end{cases}$$

We are now ready for the following lemma:

LEMMA 14. *For every graph-formula φ_2 and every assignment type κ there exists a grid-formula φ_1^κ such that for all grid graphs $\alpha(Gr)$ and assignments $f : VAR \to Dom(\alpha(Gr)) \cup \mathcal{P}(Dom(\alpha(Gr)))$ of type κ,*

$$Gr, f_{Gr} \models \varphi_1^\kappa \;\Leftrightarrow\; \alpha(Gr), f \models \varphi_2.$$

Furthermore, for all $n \in \mathbb{N}$, if $\varphi_2 \in \Sigma_n$, then also $\varphi_1^\kappa \in \Sigma_n$.

Proof. First assume that φ_2 is atomic. If φ_2 is $x_i = x_j$, then we let

$$\varphi_1^\kappa = \begin{cases} x_i = x_j & \text{when } \kappa(i) = \kappa(j) \\ \bot & \text{when } \kappa(i) \neq \kappa(j) \end{cases}$$

If $\varphi_2 = x_i R x_j$, we define φ_1^κ according to the following table:

$(\kappa(i), \kappa(j))$	φ_1^κ	$(\kappa(i), \kappa(j))$	φ_1^κ
$(0,0)$	$x_i = x_j \vee x_i S_1 x_j$	$(0,t_0)$	$\bot$
$(0,1)$	$x_i = x_j$	$(t_0,0)$	$\top$
$(1,0)$	$x_i S_2 x_j$	$(0,t_1)$	$\bot$
$(1,1)$	$\bot$	$(t_1,0)$	$\exists z(topleft(z) \wedge z S_2 x_j)$

$(\kappa(i), \kappa(j))$	φ_1^κ	$(\kappa(i), \kappa(j))$	φ_1^κ
$(1,t_0)$	$\bot$	(t_0,t_0)	$\top$
$(t_0,1)$	$\top$	(t_0,t_1)	$\top$
$(1,t_1)$	$\bot$	(t_1,t_0)	$\top$
$(t_1,1)$	$\bot$	(t_1,t_1)	$\bot$

where $topleft(z)$ denotes formula $\neg\exists x(x S_1 z \vee x S_2 z)$. Finally, if $\varphi_2 = X(x_i)$, we let $\varphi_1^\kappa = X^{\kappa(i)}(x_i)$. We now have a basis for an argument by induction.

If $\varphi_2 = \neg\pi_2$, we use π_2 and the induction hypothesis to find π_1^κ. We then let $\varphi_1^\kappa = \neg\pi_1^\kappa$. Similarly, if $\varphi_2 = \pi_2 \wedge \chi_2$, we use the induction hypothesis to find π_1^κ and χ_1^κ and let $\varphi_1^\kappa = \pi_1^\kappa \wedge \chi_1^\kappa$.

In the case where $\varphi_2 = \exists x(\pi_2)$ we apply the induction hypothesis to π_2 in order to find formulae $\pi_1^{\kappa[x \mapsto i]}$, where $i \in \{0, 1, t_0, t_1\}$, such that

$$Gr, f_{Gr} \models \pi_1^{\kappa[x \mapsto i]} \;\Leftrightarrow\; \alpha(Gr), f \models \pi_2$$

holds for all grid graphs $\alpha(Gr)$ and valuations f of type $\kappa[x \mapsto i]$. We then use these four formulae and define φ_1^κ to be the following formula:

$$\begin{array}{ll} \exists x & \Big(topleft(x) \wedge \pi_1^{\kappa[x \mapsto t_0]} \\ \vee & topleft(x) \wedge \pi_1^{\kappa[x \mapsto t_1]} \\ \vee & \neg topleft(x) \wedge \pi_1^{\kappa[x \mapsto 0]} \\ \vee & \neg topleft(x) \wedge \pi_1^{\kappa[x \mapsto 1]} \Big) \end{array}$$

Finally, if $\varphi_2 = \exists X(\pi_2)$, we find a grid formula π_1^κ corresponding to π_2 by the induction hypothesis and set $\varphi_1^\kappa = \exists X^0 \exists X^1 \exists X^{t_0} \exists X^{t_1}(\chi \wedge \pi_1^\kappa)$, where χ is the conjunction of formulae $\forall x(X^0(x) \vee X^1(x) \rightarrow \neg topleft(x))$ and $\forall x(X^{t_0}(x) \vee X^{t_1}(x) \rightarrow topleft(x))$. ■

COROLLARY 15. *For every graph-sentence φ_2 there exists a grid-sentence φ_1 such that for all grid graphs $\alpha(Gr)$,*

$$Gr \models \varphi_1 \Leftrightarrow \alpha(Gr) \models \varphi_2.$$

Sentence φ_1 can be chosen such that it is on the same level of the second-order quantifier alternation hierarchy as φ_2.

Proof. Choose an arbitrary κ and apply Lemma 14. ■

The next two propositions will be needed later on, but they are also interesting in their own right as they characterize the *MSO* alternation hierarchy with respect to *localized* graphs.

PROPOSITION 16. *For all $n \in \mathbb{N}_{\geq 1}$ we have $\underline{\Sigma_n}(LGG) \neq \underline{\Sigma_{n+1}}(LGG)$.*

Proof. Fix an arbitrary positive integer n. By Theorem 10 there is a class of grids $C \in \underline{\Sigma_{n+1}}(GRID) \setminus \underline{\Sigma_n}(GRID)$. Let $\varphi_1 \in \Sigma_{n+1}$ define C w.r.t. class $GRID$. We apply Lemma 13 to find a graph-sentence $\varphi_2 \in \Sigma_{n+1}$ such that $Gr \models \varphi_1 \Leftrightarrow \alpha(Gr) \models \varphi_2$ for all grids Gr. It is clear that φ_2 defines, with respect to the class of localized grid graphs, the isomorphism closure of class $\alpha(C)$.

We then show that there exists no graph-sentence $\psi_2 \in \Sigma_n$ that defines the isomorphism closure of class $\alpha(C)$ w.r.t. class LGG. For assume ψ_2 exists. Use Corollary 15 to choose the related grid-sentence ψ_1. Now, since α is injective, grid-sentence $\psi_1 \in \Sigma_n$ defines class C w.r.t. the class of grids. This is a contradiction. ■

PROPOSITION 17. *For all $n \in \mathbb{N}_{\geq 1}$ we have $\underline{\Sigma_n}(LGG_p) \neq \underline{\Sigma_{n+1}}(LGG_p)$.*

Proof. Fix an arbitrary $n \in \mathbb{N}_{\geq 1}$. By Proposition 16 there exists some sentence $\pi \in \Sigma_{n+1}$ that defines some class $C \in \underline{\Sigma_{n+1}}(LGG) \setminus \underline{\Sigma_n}(LGG)$ w.r.t LGG. Thus the l-pointed version C_p of C is definable w.r.t. LGG_p by formula $(x = x) \wedge \pi$, which is obviously in $[\Sigma_{n+1}]$.

Assume that C_p is definable w.r.t. LGG_p by some formula $\varphi(x) \in \Sigma_n$. Let $\varphi(x) = \overline{Q}\psi(x)$, where $\overline{Q}$ is a vector of second-order quantifiers and $\psi(x)$ is a first-order formula. Sentence $\overline{Q}(\exists x(\psi_{t_0}(x) \wedge \psi(x))) \in \Sigma_n$ defines class C w.r.t. LGG. This contradicts our assumption. ■

5.3 Alternation hierarchy of SOPML over directed graphs

We now prove that the alternation hierarchy of *SOPML* is infinite. We first show this for pointed graphs and then for graphs.

THEOREM 18. *The alternation hierarchy of SOPML over pointed directed graphs is infinite.*

Proof. Fix an arbitrary $n \in \mathbb{N}_{\geq 1}$. Then apply Proposition 17 in order to find some class $H_p \in \underline{\Sigma_{n+1}}(LGG_p) \setminus \underline{\Sigma_n}(LGG_p)$ of l-pointed grid graphs. By Lemma 8 there exists an *SOPML*-sentence that defines class H_p w.r.t. class LGG_p.

Now, class H_p cannot be definable w.r.t. class LGG_p by any *SOPML*-sentence on the n-th level of the alternation hierarchy of *SOPML*. For assume that $\varphi \in \Sigma_n^{ML}$ defines H_p w.r.t LGG_p. Now $St_x(\varphi)$ is an *MSO*-formula in Σ_n that defines H_p w.r.t. LGG_p. ■

THEOREM 19. *The alternation hierarchy of SOPML over directed graphs is infinite.*

Proof. Fix an arbitrary $n \in \mathbb{N}$. By Proposition 16 there exists a class $H \in \underline{\Sigma_{n+3}}(LGG) \setminus \underline{\Sigma_{n+2}}(LGG)$ of localized grid graphs. We shall first establish that class H is *SOPML*-definable w.r.t. *LGG*.

Consider the following *SOPML*-sentence:

$$\psi = \forall p_x(p_x \rightarrow \Diamond p_x) \wedge \forall p_x(p_x \rightarrow \exists p_y(\neg p_y \wedge \Diamond(p_y \wedge \Diamond p_x)))$$

To see that ψ fixes localizers on *LGG*, notice that the only point u of a localized grid graph that satisfies conditions uRu and $\exists v(v \neq u \wedge uRv \wedge vRu)$ is the localizer. As sentence ψ fixes localizers on *LGG*, Lemma 9 implies that class H is definable w.r.t *LGG* by some *SOPML*-sentence.

Assume then, for contradiction, that $H \in \underline{\Sigma_n^{ML}}(LGG)$. Thus there exists an *SOPML*-sentence $\pi \in \Sigma_n^{ML}$ that defines class H w.r.t *LGG*. It is easy to see that therefore *MSO*-sentence $\varphi = \forall x(St_x(\pi))$ defines H w.r.t *LGG*. To conclude the proof, it now suffices to show that there is an *MSO*-sentence in Σ_{n+2} that is equivalent to φ.

Let $\pi = \overline{\exists p_y ... p_v}(\pi')$, where $\overline{\exists p_y ... p_v}$ is a vector of proposition quantifiers and π' a quantifier-free formula. Consider the following sentence:

$$\forall X \overline{\exists P_y ... P_v}\Big(\forall x\Big(X(x) \wedge \forall z(X(z) \rightarrow x = z) \rightarrow St_x(\pi')\Big)\Big)$$

It is easy to see that this sentence is equivalent to φ and in Σ_{n+2}. ■

As the class of Kripke frames is a superclass of the class of finite directed graphs, we immediately obtain the following corollary:

COROLLARY 20. *The alternation hierarchy of SOPML over Kripke frames is infinite.*

6 Concluding remarks

We have shown that the quantifier alternation hierarchy of *SOPML* induces an infinite corresponding semantic hierarchy over the class of finite directed graphs (Theorem 19). While establishing the result, we have defined the notion of a localized structure and characterized the *MSO* alternation hierarchy over localized (finite directed) graphs. Theorem 19 answers a longstanding open problem from [4] (also addressed in [8]). The result is also relatively interesting from the point of view of finite model theory, as *SOPML* is a semantically natural fragment of *MSO* (cf. Theorem 6 in [8]).

In addition to obtaining the results related to alternation hierarchies, we have observed that with regard to expressive power, $MSO = SOPML(E) = SOPML(D)$. Connections of this kind offer an interesting modal perspective on *MSO*. For example, they suggest alternative approaches to *MSO*-games (see [15] for the definition).

Finally, our techniques do not directly yield *strictness* of the hierarchy of *SOPML*. The reason for this is that an *MSO*-formula $\varphi \in \Sigma_n$ cannot necessarily be translated to an *SOPML*-formula in Σ_n^{ML}, as in the general case the first-order quantifiers of φ translate to second-order quantifiers. Therefore, it remains to be investigated whether the *SOPML* alternation hierarchy is strict over finite directed graphs.

Acknowledgements. The author was supported by the Graduate School in Mathematical Logic and Algebra, MALJA, grant no. 118815 from the Academy of Finland. The author wishes to thank Professor Lauri Hella and the referees for their feedback.

BIBLIOGRAPHY

[1] C. Areces, P. Blackburn, and M. Marx. Hybrid logic is the bounded fragment of first order logic. In *Proc. of WOLLIC*, pages 33-50, 1999.

[2] C. Areces, P. Blackburn, and M. Marx. Hybrid Logics: Characterization, Interpolation and Complexity. *Journal of Symbolic Logic*, 66(3):977-1010, 2001.

[3] C. Areces and B. ten Cate. Hybrid Logics. In P. Blackburn, F. Wolter, and J. van Benthem, editors, *Handbook of Modal Logic*. Elsevier, 2006.

[4] J. van Benthem. *Modal Logic and Classical Logic*. Bibliopolis, 1983.

[5] P. Blackburn, M. de Rijke, and Y. Venema. *Modal Logic*. Cambridge University Press, 2001.

[6] P. Blackburn and J. Seligman. Hybrid Languages. *Journal of Logic, Language and Information*, 4(3):251-272, 1995.

[7] R.A. Bull. On Modal Logic with Propositional Quantifiers. *Journal of Symbolic Logic*, 34(2):257-263, 1969.

[8] B. ten Cate. Expressivity of Second Order Propositional Modal Logic. *Journal of Philosophical Logic*, 35(2):209-223, 2006.

[9] K. Fine. Propositional Quantifiers in Modal Logic. *Theoria*, 36:336-346, 1970.

[10] M. Fitting. Interpolation for First Order S5. *Journal of Symbolic Logic*, 67(2):621-634, 2002.

[11] D. Kaplan. S5 with Quantifiable Propositional Variables. *Journal of Symbolic Logic*, 35(2):352-363, 1970.

[12] P. Kremer. Defining Relevant Implication in a Propositionally Quantified S4. *Journal of Symbolic Logic*, 62(4):1057-1069, 1997.

[13] P. Kremer. On the Complexity of Propositional Quantification in Intuitionistic Logic. *Journal of Symbolic Logic*, 62(2):529-544, 1997.

[14] S. Kripke. A Completeness Theorem in Modal Logic. *Journal of Symbolic Logic*, 21:1-14, 1959.

[15] L. Libkin. *Elements of Finite Model Theory*. Springer, 2004.

[16] J.A. Makowsky and Y.B. Pnueli. Arity and Alternation in Second-Order Logic. *Annals of Pure and Applied Logic*, 78(1-3):189-202, 1996.

[17] O. Matz and N. Schweikardt. Expressive power of monadic logics on words, trees, pictures, and graphs. In J. Flum, E. Grädel and T. Wilke, editors, *Logic and Automata: History and Perspectives*, Texts in Logic and Games, pages 531-552. Amsterdam University Press, 2007.

[18] O. Matz, N. Schweikardt, and W. Thomas. The Monadic Quantifier Alternation Hierarchy over Grids and Graphs. *Information and Computation*, 179(2):356-383, 2002.

[19] O. Matz and W. Thomas. The Monadic Quantifier Alternation Hierarchy over Graphs is Infinite. In *Proc. of LICS*, pages 236-244, 1997.

[20] N. Schweikardt. The Monadic Quantifier Alternation Hierarchy over Grids and Pictures. In *Proc. of CSL*, pages 441-460, 1997.

[21] N. Schweikardt. The Monadic Second-Order Quantifier Alternation Hierarchy over Grids and Pictures. Diplomarbeit, Johannes Gutenberg-Universität Mainz, 1997.

[22] N.V. Shilov and K. Yi. On Expressive and Model Checking Power of Propositional Program Logics. In *Ershov Memorial Conference*, pages 39-46, 2001.

[23] C. Stirling. Games and Modal Mu-Calculus. In *Proc. of TACAS*, pages 298-312, 1996.

[24] L.J. Stockmeyer. The Polynomial-Time Hierarchy. *Theoretical Computer Science*, 3(1):1-22, 1976.

Antti Kuusisto
Department of Mathematics and Statistics
University of Tampere
Kanslerinrinne 1 A, 33014 Tampere,
Finland
antti.j.kuusisto@uta.fi

Modal logics for mereotopological relations

YAVOR NENOV AND DIMITER VAKARELOV

ABSTRACT. We present a complete axiomatization of a logic denoted by MTML (Mereo-Topological Modal Logic) based on the following set of mereotopological relations: *part-of, overlap, underlap, contact, dual contact* and *interior part-of.* We prove completeness theorems for MTML with respect to several classes of models including the standard topological models over the set of regular-closed subsets of arbitrary topological spaces. We show that MTML possesses fmp with respect to a class of non-standard models, which implies its decidability. In this way we propose also a solution of the main open problem, formulated in [17] to find a decidable modal logic for topological relations.

Keywords: mereotopology, spatial relations, spatial reasoning, modal logic, completeness theorems, decidability.

Introduction

This paper can be considered as an application of modal logic to mereotopology. Mereotopology is an extension of mereology with some relations of topological nature. Mereology is an ontological discipline which can be characterized shortly as a theory of "Parts and Wholes" (see [21] for a general reference to mereology). Typical in mereology are the relations "part-of", "overlap" and "underlap". One of the basic mereological systems is Lesnewski's mereology, but as Tarski showed, the mathematical equivalent of mereology are complete Boolean algebras (see [21] for this fact). In Boolean formulation the part-off relation coincides with the Boolean ordering $x \leq y$, the overlap relation xOy can be defined by $x.y \neq 0$ (where "." is the Boolean multiplication) and the underlap (dual overlap) $x\widehat{O}y$ is defined by $x\widehat{O}y$ iff $x^*.y^* \neq 0$ iff $x + y \neq 1$ (x^* is the Boolean complement of x). Mereology, however, is not capable for describing some relations between individuals as, for instance, one individual to be in a contact with another one. Adding to mereology contact-like relations goes back to de Laguna [8] and Whitehead [26]. The intention of de Laguna and Whitehead was to use mereology for building of a new, point-free theory of space as an extension of mereology with the relation of contact (or "connection" in Whitehead terminology). The primitive objects of the new theory of space are called regions and it is called "point-free", because points are not taken as primitives but are definable by means of regions, contact and some mereological relations. As Tarski showed (see [21]) standard point models of the new theory of space are regular closed (or open) sets of some topological spaces with a topological definition of the contact relation. This motivates some authors to call the extension of mereology with the contact relation (or some of its derivatives) "mereotopology". Mereotopology is often called also a "region-based theory of space", because it is on the base of the Whitehedian approach to the theory of space. Since mereology can be identified in some sense with the theory of Boolean algebras, mereotopology can be identified with the theory of *contact algebras*, which are Boolean algebras with an additional relation C called contact (see [10]). The reader can find more about mereotopology, region-based theory of space and the related logics in the papers [2, 3, 17, 19, 25, 27]. This field of research is closely related to some applied areas as Qualitative Spatial Reasoning (QSR), Knowledge Representation (KR) and Geographical Information Systems (GIS). A survey on the research in QSR and related subareas in KR and GIS can be find in [6, 7].

The main aim of the present paper is to build a multimodal logic interpreted in frames related to mereotopology. The standard frames for such a logic will be in the form $(W, R_1, \ldots, R_n)$, where W is a nonempty set of regular closed sets of a given topological space and the relations R_i are certain mereotopological relations between regions. Logics of such kind have been considered for the first time by Lutz and Wolter in [17]. The relations which Lutz and Wolter considered are the well-known 8 Egenhofer-Franzosa RCC-8 topological relations between regions [13]. However, all considered logics in [17] are undecidable and one of the main open problems formulated in [17] was to find decidable modal logics based on a reasonable set of mereotopological relations. We present such a logic, based on the following mereotopological relations:

(I) the mereological relations: overlap O, underlap (the dual overalap) $\widehat{O}$, part-of $\leq$ and converse part-of $\geq$,

(II) the mereotopological relations: contact C, dual contact $\widehat{C}$, interior part-of $\ll$ and its converse $\gg$.

Frames, based on such kind of relations, are called in this paper *mereotopological structures*. The modal logic corresponding to the class of all mereotopological structures is called mereotopological modal logic and is denoted by MTML. We denote box and diamond modalities of MTML by $[R]$ and $\langle R\rangle$, where $R \in \{O, \widehat{O}, \leq, \geq, C, \widehat{C}, \ll, \gg\}$. Additionally we include the universal modality $[U]$.

Motivations to chose mereotopological structures as a semantical basis of MTML are, among others, the following. The relations from group (I) are the most typical mereological relations. Moreover the corresponding modal logic (introduced in [23] under the name *modal logic of set relations*) was decidable and our aim was to extend it with some mereotopological relations, preserving the completeness theorem and decidability. The first attempt was by adding the contact relation and this was done in Nenov's master thesis [18]. Still the obtained logic was complete with respect to its intended topological semantics and decidable. Then, we decided to extend further the language by modalities, corresponding to dual contact $\widehat{C}$, interior part-of $\ll$ and its converse $\gg$. Note that these relations are definable in contact algebras: $x\widehat{C}y \leftrightarrow x^*Cy^*$ and $x \ll y \leftrightarrow x\overline{C}y^*$. This fact shows that all relations of mereotopological structures are definable in contact algebras, which makes possible to use the corresponding representation theory developed in [10, 12]. Let us note that RCC-8 relations - the semantic base of Lutz-Wolter modal logic ([17]) (LW-logic for short and for later references), are definable in our mereotopological structures, while the converse is not true: for instance, dual overlap is not definable in RCC-8 (this will be discussed with more details in the main text). This does not imply, however, that all modalities from LW-logic are definable in MTML. In fact MTML and LW-logic are incomparable in the sense that neither of the two can be considered as a part of the other. But LW-logic is much more expressive: it possesses difference modality, and hence definable nominals. Moreover, since the relations in RCC-8 are jointly exhaustive and pairwise disjoint (JEPD), all Boolean combinations of them are expressible by sums of the basic 8 relations, and hence their corresponding modalities are definable. So in LZ-logic one can work with quite enough different modalities. In LMTM we have 9 basic modalities and also we may define new modalities by the sums and compositions but not by complements and intersections of the base relations. Maybe just the closure with complements and intersections of the basic modalities of the LZ-logic is one of the reasons of its undecidability. We can see later that if we can allow modalities of MTML corresponding to Boolean combinations of the base modalities, we can interpret LW-logic in MTML and obtain in this way that the resulting extension is undecidable.

Let us now discuss what kind of reasoning can be expressed in MTML. Note that MTML and LW-logic are similar as logical formalisms: both are modal logics over frames which elements are spatial regions. Since the propositional variables in modal logics are interpreted by subsets of a given frame, in general, these two logics propose reasoning for sets of regions. For instance, the formula $[U](p \Rightarrow \langle C\rangle q)$ expresses the fact that each region from the set p is in a contact with some region from the set q. Another example: the formula $[U](p \Rightarrow [\geq]p)$ expresses the fact that the set of regions p is closed with respect to part-of relation: if $x \in p$ and y is a part of x, then y is in p. The frame condition (Con) $xOx \wedge yOy \rightarrow x\widehat{O}y \vee xCy$ is true in the frame of all closed regions in a topological space iff the space is connected. This condition is modally definable in MTML by the modal formula $\langle O\rangle([\widehat{O}]p \wedge [C]q) \Rightarrow [U]([O](p \vee q)$, which distinguished connected from non-connected topological spaces. This is an example of a property expressible in MTML but not in the LW-logic.

As it was mentioned in [17], LW-logic is similar to the Halpern and Shoham's temporal logic [16], based semantically on the Allen's 13 relations between time intervals. The same can be said also for MTML. Allen's relations are relations not between time points but between time intervals, which over the real line are closed regions. The interpretation of RCC-8 and our mereotopological structures over the real line gives a temporal meaning of LW-logic and MTML. Let us mention yet another logic with a similar nature: the hyperboolean modal logics introduced in [15]. The frames of this logics are arbitrary Boolean algebras and Boolean algebras of sets, which relates these logics to LW-logic and MTML.

The main aim of this paper is to give a finite normal and complete axiomatization of MTML and to prove its decidability. The axiomatization goes through several steps. We first give an abstract characterization of the relations $\leq, O, \widehat{O}, C, \widehat{C}, \ll$ by means of a finite set of first-order sentences, introducing in this way an abstract, point-free semantics for MTML. We prove that each abstract mereotopological structure $\underline{W}$ is representable in a contact algebra, and then, applying the topological representation theory of contact algebras developed in [10], we show that $\underline{W}$ can be isomorphically embedded into the contact algebra of regular closed subsets of some topological space. The method of the proof of this characterization is based on a considerable generalization of the Stone representation theory of distributive lattices (see [1]). The obtained results for mereotopological structures have also some independent interest for mereotopology: they can be considered as a kind of first-order logic for mereotopological relations disregarding the Boolean structure of regions. The obtained abstract semantics of MTML cannot give, however, a direct axiomatization of the logic, because one of its axioms is not modally definable. That is why we introduce a nonstandard semantics of MTML which leads to an easy and complete axiomatization. Then, by using the Segerberg's bulldozer techniques [20], we prove the equivalence of the standard and nonstandard semantics for MTML. Finally, applying the method of filtration to the non-standard models of MTML we prove its decidability. We show, however, that MTML does not possess fmp with respect to its standard semantics.

We propose as standard reference books: [4, 5, 20] for modal logic, [14] for topology, and [1] for Stone representation theory.

1 The first-order logic of mereotopological structures

1.1 Contact algebras, topological and relational representation

DEFINITION 1. [10] By a *Contact Algebra* (CA) we will mean any system $\underline{B} = (B, C) = (B, 0, 1, ., +, *, C)$, where $(B, 0, 1, ., +, *)$ is a non-degenerate Boolean algebra with a complement denoted by "$*$" and C – a binary relation in B, called *contact* and satisfying the following axioms:

$(C1)$ $xCy \rightarrow x, y \neq 0$, $\quad$ $(C2)$ $xCy \rightarrow yCx$,

$(C3)$ $xC(y+z) \leftrightarrow xCy$ or xCz, $\quad$ $(C4)$ $x.y \neq 0 \rightarrow xCy$.

The algebra $\underline{B}$ is *connected* if it satisfies the axiom of connectedness

(Con) $x \neq 0$, $y \neq 0$ and $x + y = 1 \rightarrow xCy$.

The complement of C is denoted by $\overline{C}$.

EXAMPLES 2. **Examples of contact algebras.**

• **(1). Topological example: the CA of regular closed sets.** Let X be an arbitrary topological space. A subset a of X is *regular closed* if

$a = Cl(Int(a))$, where Cl and Int are the standard topological closure and interior operations in X. The set of all regular closed subsets of X will be denoted by $RC(X)$. It is a well-known fact that regular closed sets with the operations

$$a + b = a \cup b,\ a.b = Cl(Int(a \cap b)),\ a^* = Cl(X \setminus a),\ 0 = \varnothing \text{ and } 1 = X$$

form a Boolean algebra. If we define the contact by $a\, C_X\, b$ iff $a \cap b \neq \varnothing$, then $RC(X)$ with the above contact is a contact algebra. If X is a connected space then $RC(X)$ is a connected contact algebra. The following representation theorem is a special case of Theorem 5.1 from [10].

THEOREM 3. *Every (connected) contact algebra* $\underline{B}$ *can be isomorphically embedded into the contact algebra* $RC(X)$ *over some (connected) topological space* X.

• **(2). Non-topological example, related to Kripke semantics of modal logic.** Let (X, R) be a reflexive and symmetric modal frame and let $B(X)$ be the Boolean algebra of all subsets of X. Define a contact C_R between two subsets $a, b \in B(X)$ by aC_Rb iff $(\exists x \in a)(\exists y \in b)(xRy)$. Then we have that $B(X)$ equipped with the contact C_R is a contact algebra, called the contact algebra over the frame (W, R) [12, 9]. If (W, R) is connected in a graph sense (every two points are connected by an R-sequence), then $B(X)$ is a connected contact algebra [12]. Moreover the following representation theorem is true:

THEOREM 4. *[12] Every contact algebra can be isomorphically embedded into the contact algebra of some reflexive and symmetric frame* (W, R).

1.2 Mereotopological structures

DEFINITION 5. Let (B, C) be a contact algebra. We define in B the following relations:

• *part-of* $a \leq b$ iff $a.b^* = 0$ ($\leq$ is the standard Boolean ordering) we denote the converse of $\leq$ by $\geq$.

• *overlap* aOb iff $a.b \neq 0$,

• *underlap* (dual overlap) $a\widehat{O}b$ iff a^*Ob^* iff $a + b \neq 1$,

• *dual contact* $a\widehat{C}b$ iff a^*Cb^*,

• *interior part-of* $a \ll b$ iff $a\overline{C}b^*$. The converse of $\ll$ is denoted by $\gg$.

The complements of the above relations are denoted by $\not\leq, \not\geq, \overline{O}, \overline{\widehat{O}}, \overline{\widehat{C}}\ \not\ll, \not\gg$.

The proof of the following lemma is straightforward.

LEMMA 6. *The relations* $\leq, O, \widehat{O}, C, \widehat{C}, \ll$ *satisfy the following first-order conditions:*

(≤ 0)	$a \leq b$ *and* $b \leq a \to a = b,$	(≤ 1)	$a \leq a,$
(≤ 2)	$a \leq b$ *and* $b \leq c \to a \leq c,$		
$(O1)$	$aOb \to bOa,$	$(\widehat{O}1)$	$a\widehat{O}b \to b\widehat{O}a,$
$(O2)$	$aOb \to aOa,$	$(\widehat{O}2)$	$a\widehat{O}b \to a\widehat{O}a,$
$(\overline{O} \leq)$	$a\overline{O}a \to a \leq b,$	$(\overline{\widehat{O}} \leq),$	$b\overline{\widehat{O}}b \to a \leq b,$
$(O \leq)$	aOb *and* $b \leq c \to aOc,$	$(\widehat{O} \leq)$	$c \leq a$ *and* $a\widehat{O}b \to c\widehat{O}b,$
$(O\widehat{O})$	aOa *or* $a\widehat{O}a,$	$(\leq O\widehat{O})$	$c\overline{O}a$ *and* $c\overline{\widehat{O}}b \to a \leq b,$
(C)	$aCb \to bCa,$	$(\widehat{C})$	$a\widehat{C}b \to b\widehat{C}a,$
$(CO1)$	$aOb \to aCb,$	$(\widehat{C}\widehat{O}1)$	$a\widehat{O}b \to a\widehat{C}b,$
$(CO2)$	$aCb \to aOa,$	$(\widehat{C}\widehat{O}2)$	$a\widehat{C}b \to a\widehat{O}a,$
$(C \leq)$	aCb *and* $b \leq c \to aCc,$	$(\widehat{C} \leq)$	$a\widehat{C}b$ *and* $c \leq b \to a\widehat{C}c,$
$(\ll\leq 1)$	$a \ll b \to a \leq b,$		
$(\ll\leq 2)$	$a \leq b$ *and* $b \ll c \to a \ll c,$	$(\ll\leq 3)$	$a \ll b$ *and* $b \leq c \to a \ll c,$
$(\ll O)$	$a\overline{O}a \to a \ll b,$	$(\ll \widehat{O})$	$b\overline{\widehat{O}}b \to a \ll b,$
$(\ll CO)$	aCb *and* $b \ll c \to aOc,$	$(\ll \widehat{C}\widehat{O})$	$c \ll a$ *and* $a\widehat{C}b \to c\widehat{O}b,$
$(\ll C\widehat{O})$	$c\overline{C}a$ *and* $c\overline{\widehat{O}}b \to a \ll b,$	$(\ll \widehat{C}O)$	$c\overline{O}a$ *and* $c\overline{\widehat{C}}b \to a \ll b.$

DEFINITION 7. Let $\underline{W} = (W, \leq, O, \widehat{O}, C, \widehat{C}, \ll)$, $W \neq \varnothing$, be a relational system. Then $\underline{W}$ is called a *mereotopological structure* if it satisfies the first-order conditions of lemma 6; $\underline{W}$ is called a *standard mereotopological structure* if there exists a contact algebra (B, C) such that $W \subseteq B$ and the relations $\leq, O, \widehat{O}, C, \widehat{C}, \ll$ coincide with those that are defined in Definition 5; $\underline{W}$ is called *completely standard* if the algebra (B, C) is the contact algebra of regular closed subsets of some topological space; if in addition $W = B$, then the (standard, completely standard) mereotopological structure is called *full.*

The following lemma is an easy consequence of Theorem 3.

LEMMA 8. *A mereotopological structure is standard iff it is completely standard.*

In the next section we will show that each mereotopological structure is a standard one, and in view of Lemma 8 that it is completely standard.

REMARKS 9. (1) Let us note that the axioms $(\overline{O} \leq)$, $(\overline{\widehat{O}} \leq)$, $(O2)$ and $(\widehat{O}2)$ follow from the remaining and can be skipped. We preserve them in the definition, because they are part of an important subset of the axioms characterizing mereological relations.

(2) We can establish some duality between the relations in a mereotopological structure and their axioms. We divide the relations in a dual pairs as follows: $(\leq - \geq)$, $(O - \widehat{O})$, $(C - \widehat{C})$ and $(\ll - \gg)$. Note also that the set of axioms is closed with respect to this duality and very often we may skip some proofs which are "dual" to given ones.

(3) We adopt the standard definitions of isomorphism and embedding between mereotopological structures and two isomorphic structures are treated as identical. Thus, for instance, a structure which is isomorphic to a standard structure will be called also a standard structure. We say that a mereotopological structure $\underline{W}$ is embeddable into a contact algebra (B, C) if there exists an isomorphic embedding of $\underline{W}$ into the full mereotopological structure over (B, C).

(4) It can be seen that the axiom of connectedness for contact algebras can be expressed by the following axiom in the language of mereotopological relations

(Con) $aOa \wedge bOb \rightarrow a\widehat{O}b \vee aCb$.

It is natural to call a mereotopological structure *connected* if it satisfies the axiom (Con). Note that mereotopological structures over connected topological spaces are connected. Another non-topological example can be obtained from connected contact algebras over frames (W, R) with a reflexive, symmetric and connected relation R (see Examples 2(2)).

The following lemma lists some easy consequences of the axioms of mereotopological structures which sometimes we will use later on without explicit reference.

LEMMA 10. *(COO)* $aCb \rightarrow aOa$ *and* bOb, $(\widehat{C}\widehat{O}\widehat{O})$ $a\widehat{C}b \rightarrow a\widehat{O}a$ *and* $b\widehat{O}b$, $(\leq\leq O)$ $a \leq a'$, $b \leq b'$, $aOb \rightarrow a'Ob'$, $(\leq\leq C)$ $a \leq a'$, $b \leq b'$, $aCb \rightarrow a'Cb'$, $(\geq\geq \widehat{O})$ $a \geq a'$, $b \geq b'$, $a\widehat{O}b \rightarrow a'\widehat{O}b'$, $(\geq\geq \widehat{C})$ $a \geq a'$, $b \geq b'$, $a\widehat{C}b \rightarrow a'\widehat{C}b'$, $(\leq\ll\leq)$ $a \leq a'$, $a' \ll b'$, $b' \leq b \rightarrow a \ll b$.

1.3 Representation theory for mereotopological structures

In this section we will develop a representation theory for mereotopological structures by a generalization of the representation theory for distributive lattices. First we will do this for a subsystem of mereotopological structures which we call *mereological structures*.

Mereological structures and a characterization of mereological relations $\leq, O, \widehat{O}$.

DEFINITION 11. A system $\underline{W} = (W, \leq, O, \widehat{O})$ is called a *mereological structure* if it satisfies the axioms of mereotopological structure containing only the relations $\leq, O$ and $\widehat{O}$.

Obviously every mereotopological structure is a mereological structure. Mereological structures was introduced and studied in another context, name and notations in [22] from which we will use some results.

DEFINITION 12. [22] Let $\underline{W}$ be a mereological structure and A be a subset of W.

- A is called a *$\leq$-set* if $(\forall x, y \in W)(x \in A \text{ and } x \leq y \rightarrow y \in A)$,
- A is called a *$\geq$-set* if $(\forall x, y \in W)(x \in A \text{ and } x \geq y \rightarrow y \in A)$,
- A is a *filter* if A is a $\leq$-set and $(\forall x, y \in A)(xOy)$,
- A is an *ideal* if A is a $\geq$-set and $(\forall x, y \in A)(x\widehat{O}y)$.
- A is a *good filter* if A is a filter and $(\forall x, y \notin A)(x\widehat{O}y)$,
- A is a *good ideal* if A is an ideal and $(\forall x, y \notin A)(xOy)$.

We denote by $GF(\underline{W})$ the set of good filters of $\underline{W}$. Similarly $GI(\underline{W})$ will denote the set of good ideals of $\underline{W}$.

The given definitions of a filter, good filter, ideal and a good ideal are generalizations of the standard notions of a filter, prime filter, ideal and a prime ideal from the theory of distributive lattices (see [1]).

Note that $\varnothing$ and W are both $\leq$- and $\geq$-sets. Define for $x \in W$:
$[x) = \{y \in W : x \leq y\}$, $(x] = \{y \in W : x \geq y\}$.

LEMMA 13. *[22] (i) The set $[x)$ (the set $(x]$) is the smallest $\leq$-set ($\geq$-set) containing x.*

(ii) If A, B are $\leq$-sets ($\geq$-sets) then $A \cup B$ and $A \cap B$ are $\leq$-sets ($\geq$-sets). If A is a $\leq$-set ($\geq$-set) then $-A = W \setminus A$ is a $\geq$-set ($\leq$-set).

(iii) Let A be a $\leq$-set ($\geq$-set). Then $A \cup [x)$ ($A \cup (x]$) is the smallest $\leq$-set ($\geq$-set) containing A and x. In particular the set $[x) \cup [y)$ (the set $(x] \cup (y]$) is the smallest $\leq$-set ($\geq$-set) containing x and y.

(iv) The set $[x)$ is a filter iff xOx (The set $(x]$ is an ideal iff $x\widehat{O}x$).

(v) Let A be a filter (ideal). Then $A \cup [x)$ is a filter iff xOx and $(\forall y \in A)(xOy)$, ($A \cup (x]$ is an ideal iff $x\widehat{O}x$ and $(\forall y \in A)(x\widehat{O}y)$).

Let $A \neq \varnothing$ be a filter (ideal). Then $A \cup [x)$ is a filter iff $(\forall y \in A)(xOy)$, ($A \cup (x]$ is an ideal iff $(\forall y \in A)(x\widehat{O}y)$).

(vi) The set $[x) \cup [y)$ is a filter iff xOy. The set $(x] \cup (y]$ is an ideal iff $x\widehat{O}y$.

(vii) Let $\{A_i : i \in I\}$ be a non-empty family of filters (ideals), linearly ordered by set-inclusion. Then the set $A = \bigcup_{i \in I} A_i$ is a filter (ideal).

(viii) Let A be a filter (ideal). Then A is a good filter (good ideal) iff $-A = W \setminus A$ is an ideal (filter).

DEFINITION 14. Let $\underline{W}$ be a mereological structure. A pair $\Gamma = (A, B)$ of subsets of W is called a filter-ideal pair, if A is a filter, B is an ideal and $A \cap B = \varnothing$). Γ is called a good filter-ideal pair if A is a good filter and B is a good ideal. Γ is called a complete filter-ideal pair if $A \cup B = W$. Obviously every complete pair is a good pair. If Γ denotes a filter-ideal pair, then Γ_1 will denote its filter part and Γ_2 will denote its ideal part. If Γ, Δ are filter-ideal pairs, we will define the ordering relation $\Gamma \subseteq \Delta$ iff $\Gamma_i \subseteq \Delta_i$, $i = 1, 2$.

LEMMA 15. *Let (F, I) be a filter-ideal pair. Then for every $x \in W$ either (1) $F \cup [x)$ is a filter and $(F \cup [x)) \cap I = \varnothing$ or (2) $I \cup (x]$ is an ideal and $F \cap (I \cup (x]) = \varnothing$.*

Proof. Suppose that the assumptions of the lemma are fulfilled and that neither (1) nor (2) are true, so we have $\neg(1)$ and $\neg(2)$. Note that $\neg(1)$ is equivalent to: (a) $F \cup [x)$ is not a filter or (a') $(F \cup [x)) \cap I \neq \varnothing$. By Lemma 13(v) (a) is equivalent to the disjunction: (a1) $x\overline{O}x$ or (a2) $(\exists y \in F)(x\overline{O}y)$. It is easy to see that (a') is equivalent to (a3) $x \in I$.

In a similar way $\neg(2)$ is equivalent to: (b) $I \cup (x]$ is not an ideal or (b') $F \cap (I \cup (x]) \neq \varnothing$. By Lemma 13(v) (b) is equivalent to the disjunction: (b1) $x\overline{\widehat{O}}x$ or (b2) $(\exists z \in I)(x\overline{\widehat{O}}z)$. Also, it is easy to see that (b') is equivalent to: (b3) $x \in F$.

So $\neg(1) \leftrightarrow (a1)$ or $(a2)$ or $(a3)$ and $\neg(2) \leftrightarrow (b1)$ or $(b2)$ or $(b3)$ and we have to consider all combinations $(ai)(bj)$ for $i \neq j, i, j = 1, 2, 3$ and in each case to obtain a contradiction.

Case (a1)(b1): $x\overline{O}x$ and $x\overline{\widehat{O}}x$. This contradicts axiom $(O\widehat{O})$.

Case (a1)(b2): $x\overline{O}x$ and $x\overline{\widehat{O}}z$, $z \in I$. From $x\overline{O}x$ by $(\overline{O} \leq)$ we get $x \leq z$ and consequently $x \in I$. From $x, z \in I$ we obtain $x\widehat{O}z$ which contradicts $x\overline{\widehat{O}}z$.

Case (a1)(b3): $x\overline{O}x$ and $x \in F$, which implies xOx - a contradiction.

The cases **(a2)(b1)** and **(a3)(b1)** can be considered in a dual way.

Case (a2)(b2): $y \in F$, $x\overline{O}y$, $z \in I$ and $x\overline{\widehat{O}}z$. From $x\overline{O}y$ and $x\overline{\widehat{O}}z$ we get by axiom $(\leq O\widehat{O})$ that $y \leq z$. Conditions $z \in I$ and $y \leq z$ imply $y \in I$. But $y \in F$ and $y \in I$ imply that $F \cap I \neq \varnothing$ - a contradiction.

Case (a2)(b3): $y \in F$, $x\overline{O}y$ and $x \in F$. Conditions $x \in F$ and $y \in F$ imply xOy, which contradicts $x\overline{O}y$.

In a similar (dual) way one can consider the **case (a3)(b2)**.

Case (a3)(b3): $x \in I$, $x \in F$. This case contradicts the condition $F \cap I = \varnothing$. ■

The following lemma generalizes the separation lemma for filters and ideals from the theory of distributive lattices (see [1]).

LEMMA 16. *[22]* **Separation Lemma.** *Let (F_0, I_0) be a filter-ideal pair. Then there exists a complete (and consequently a good) filter-ideal pair (F, I) extending the pair (F_0, I_0).*

Proof. Let $M = \{(F, I) : (F_0, I_0) \subseteq (F, I)\}$. It follows by Lemma 13(vii) that the conditions of the Zorn Lemma for the set M ordered by the relation $\subseteq$ are fulfilled, and hence M has a maximal element (F, I). Applying Lemma 15 to (F, I) we obtain that (F, I) is a complete filter-ideal pair. This implies that $F = -I$ and $I = -F$, which by Lemma 13 (viii) implies that F is a good filter and that I is a good ideal. ■

LEMMA 17. *[22]* **Characterization Lemma for the relations $\leq, O, \widehat{O}$.** *Let $\underline{W}$ be a mereological structure. Then for all $x, y \in W$ we have:*

(i) $x \leq y \leftrightarrow (\forall A \in GF(W))(x \in A \rightarrow y \in A)$,

(ii) $xOy \leftrightarrow (\exists A \in GF(W))(x \in A$ *and* $y \in A)$,

(iii) $x\widehat{O}y \leftrightarrow (\exists A \in GF(W))(x \notin A$ *and* $y \notin A)$.

Proof. (i) $(\rightarrow)$ – obvious.

($\leftarrow$) We will reason by contraposition. Let $x \not\leq y$. Then $[x)\cap(y] = \varnothing$. By $(\overline{O} \leq)$, and $(\overline{\widehat{O}} \leq)$ and $x \not\leq y$ we obtain xOx and $y\widehat{O}y$. Then by lemma 13 (iv) $[x)$ is a filter and $(y]$ is an ideal. Since $[x) \cap (y] = \varnothing$, by the Separation Lemma there exist a good filter F and a good ideal I such that $[x) \subseteq F$, $(y] \subseteq I$ and $F \cap I = \varnothing$. It follows from these conditions that $x \in F$ and $y \notin F$.

(ii) ($\leftarrow$) – obvious.

($\rightarrow$) Suppose xOy. Then by Lemma 13 (vi) the set $[x) \cup [y)$ is a filter. Since $\varnothing$ is an ideal, then $([x) \cup [y)) \cap \varnothing = \varnothing$, and by the Separation Lemma there exist a good filter F such that $[x) \cup [y) \subseteq F$, which implies that $x, y \in F$.

(iii) ($\leftarrow$) is the obvious part.

($\rightarrow$) Suppose $x\widehat{O}y$. Then, as in (ii) but reasoning in a dual way, we can obtain a good ideal I such that $x, y \in I$. Then putting $F = -I$ we find a good filter F such that $x, y \notin F$. ■

A characterization of mereotopological relations $C, \widehat{C}$ and $\ll$.

DEFINITION 18. Let $\underline{W}$ be a mereotopological structure and A, B be subsets of W and R be any of the relations $\not\ll, C$ and $\widehat{C}$. We define the following three relations between such subsets:

$A\rho_R B$ iff $(\forall x \in A, \forall y \in B)(xRy)$.

We define the following relation ρ in the set of all filter-ideal pairs:

$\Gamma\rho\Delta$ iff $\Gamma_1\rho_C\Delta_1$ and $\Gamma_2\rho_{\widehat{C}}\Delta_2$ and $\Gamma_1\rho_{\not\ll}\Delta_2$ and $\Delta_1\rho_{\not\ll}\Gamma_2$.

LEMMA 19. (*i*) *In the set of filters of $\underline{W}$, ρ_C is a reflexive and symmetric relation.*

(*ii*) *In the set of ideals of $\underline{W}$, $\rho_{\widehat{C}}$ is a reflexive and a symmetric relation.*

(*iii*) *If Γ is a filter-ideal pair, then $\Gamma_1\rho_{\not\ll}\Gamma_2$.*

(*iv*) *The relation ρ in the set of filter-ideal pairs is a reflexive and a symmetric relation.*

(*v*) *If Γ and Δ are filters and $\Gamma\rho_C\Delta$, then $(\Gamma, \varnothing)\rho(\Delta, \varnothing)$.*

(*vi*) *If Γ and Δ are ideals and $\Gamma\rho_{\widehat{C}}\Delta$, then $(\varnothing, \Gamma)\rho(\varnothing, \Delta)$.*

(*vii*) *If Γ is a filter, Δ is an ideal and $\Gamma\rho_{\not\ll}\Delta$, then $(\Gamma, \varnothing)\rho(\varnothing, \Delta)$ and $(\varnothing, \Delta)\rho(\Gamma, \varnothing)$.*

Proof. (i) The statement follows from axiom (CO1) and (C).

(ii) The statement follows from axiom $(\widehat{C}\widehat{O}1)$ and $(\widehat{C})$.

(iii) Suppose that Γ is a filter-ideal pair and that for some $x \in \Gamma_1$ and $y \in \Gamma_2$ we have $x \ll y$. Then by axiom $(\ll\leq 1)$ we have $x \leq y$, which implies that $y \in \Gamma_1$. This contradicts the fact that $\Gamma_1 \cap \Gamma_2 \neq \varnothing$.

(iv) follows from (i),(ii) and (iii).

(v), (vi) and (vii) follow just from the definition of the ρ-relation between filter-ideal pairs. ■

LEMMA 20. (*i*) *If xCy, then there exist filter-ideal pairs Γ, Δ such that $\Gamma\rho\Delta$ and $x \in \Gamma_1$ and $y \in \Delta_1$.*

(ii) If $x\widehat{C}y$, then there exist filter-ideal pairs Γ, Δ such that $\Gamma\rho\Delta$ and $x \in \Gamma_2$ and $y \in \Delta_2$.

(iii) If $x \not\ll y$, then there exists a filter-ideal pairs Γ, Δ such that $\Gamma\rho\Delta$, $x \in \Gamma_1$ and $y \in \Delta_2$.

Proof. (i) Let xCy. Then by Lemma 10 we have xOx and yOy, so by Lemma 13 $[x)$ and $[y)$ are filters. We shall show that $[x)\rho_C[y)$. Let $x' \in [x)$ and $y' \in [y)$. Then $x \leq x'$ and $y \leq y'$ and by xCy this implies by Lemma 10 that $x'Cy'$. Now by Lemma 19 we have $([x), \varnothing)\rho([y), \varnothing)$ which proves the statement.

(ii) The proof is similar (dual) to that of (i).

(iii) Let $x \not\ll y$. Then by axioms ($\ll$ O) and ($\ll \widehat{O}$) we obtain xOx and $y\widehat{O}y$ and by Lemma 13 $[x)$ is a filter and $(y]$ is an ideal. We will show that $[x)\rho_{\not\ll}(y]$. Let $x \leq x'$ and $y' \leq y$. Since $x \not\ll y$, we have by Lemma 10 that $x' \not\ll y'$ which proves $[x)\rho_{\not\ll}(y]$. Now by Lemma 19 we obtain $([x), \varnothing)\rho(\varnothing, (y])$. ■

LEMMA 21. **Point extension Lemma for filter-ideal pairs.** *Let Γ, Δ be a filter-ideal pairs and let $\Gamma\rho\Delta$. Then for any $x \in W$: either (1) $\Delta_1 \cup [x)$ is a filter and $\Gamma\rho(\Delta_1 \cup [x), \Delta_2)$ or (2) $\Delta_2 \cup (x]$ is an ideal and $\Gamma\rho(\Delta_1, \Delta_2 \cup (x])$.*

Proof. Suppose $\Gamma\rho\Delta$ and that we have $\neg(1)$ and $\neg(2)$. Due to the assumption $\Gamma\rho\Delta$ we obtain that $\neg(1)$ is equivalent to the disjunction of the following conditions:

$\neg(1) \equiv (\Delta_1 \cup [x)$ is a not a filter) or $(\Gamma_1\overline{\rho_C}(\Delta_1 \cup [x))$ or $(\Delta_1 \cup [x))\overline{\rho_{\not\ll}}\Gamma_2)$.

It is easy to see that $(\Gamma_1\overline{\rho_C}(\Delta_1 \cup [x))$ is equivalent to $(\exists z_1 \in \Gamma_1)(z_1\overline{C}x)$.

Similarly $(\Delta_1 \cup [x))\overline{\rho_{\not\ll}}\Gamma_2)$ is equivalent to $(\exists t_1 \in \Gamma_2)(x \ll t_1)$. Having in mind these equivalencies and Lemma 13 (v) we obtain that $\neg(1)$ is equivalent to the following disjunction:

$\neg(1) \equiv$ (11) $x\overline{O}x$ or (12) $(\exists y_1 \in \Delta_1)(x\overline{O}y_1)$ or (13) $(\exists z_1 \in \Gamma_1)(z_1\overline{C}x)$ or (14) $(\exists t_1 \in \Gamma_2)(x \ll t_1)$.

In a similar way we can see that $\neg(2)$ is equivalent to the following disjunction:

$\neg(2) \equiv$ (21) $x\overline{\widehat{O}}x$ or (22) $(\exists y_2 \in \Delta_2)(x\overline{\widehat{O}}y_2)$ or (23) $(\exists z_2 \in \Gamma_1)(z_2 \ll x)$ or (24) $(\exists t_2 \in \Gamma_2)(t_2\overline{\widehat{C}}x)$.

We have to combine all conditions (1i) with (2j) for $i, j = 1, 2, 3, 4$ and in all 16 cases to obtain a contradiction.

Case (11)(21): $x\overline{O}x$ and $x\widehat{O}x$ – this contradicts axiom $(O\overline{\widehat{O}})$.

Case (11)(22): $x\overline{O}x$ and $(\exists y_2 \in \Delta_2)(x\overline{\widehat{O}}y_2)$. From $x\overline{O}x$ we get by $(\overline{O} \leq)$ that $x \leq y_2$. From here and $y_2 \in \Delta_2$ we obtain $x \in \Delta_2$, because Δ_2 is an ideal. Also from $x, y_2 \in \Delta_2$ we obtain $x\widehat{O}y_2$ – a contradiction.

In a similar way we can treat the **cases (11)(23)** and **(11)(24)** and reasoning by duality – the cases **(12)(21), (13)(21), (14)(21)**.

Case (12)(22): $(\exists y_1 \in \Delta_1)(x\overline{O}y_1)$, $(\exists y_2 \in \Delta_2)(x\overline{\widehat{O}}y_2)$. From $x\overline{O}y_1$ and $x\overline{\widehat{O}}y_2$ we get by axiom $(\leq O\widehat{O})$ that $y_1 \leq y_2$. From this and $y_1 \in \Delta_1$ we obtain $y_2 \in \Delta_1$. Since $y_2 \in \Delta_2$ we obtain that $\Delta_1 \cap \Delta_2 \neq \varnothing$ - a contradiction.

Case (12)(23): $(\exists y_1 \in \Delta_1)(x\overline{O}y_1)$, $(\exists z_2 \in \Gamma_1)(z_2 \ll x)$. From $y_1 \in \Delta_1$ and $z_2 \in \Gamma_1$ we get $y_1 C z_2$. From here and $z_2 \ll x$ we obtain xOy_1 which contradicts $x\overline{O}y_1$.

Case (12)(24) $(\exists y_1 \in \Delta_1)(x\overline{O}y_1)$, $(\exists t_2 \in \Gamma_2)(t_2\overline{\widehat{C}}x)$. From $\Gamma\rho\Delta$ we get $\Delta_1\rho_{\not\ll}\Gamma_2$ and from here that $y_1 \not\ll t_2$. From $x\overline{O}y_1$ and $t_2\overline{\widehat{C}}x$, by $(\widehat{C})$ and $(\ll \widehat{C}O)$ we obtain $y_1 \ll t_2$ - a contradiction.

Case (13)(22): $(\exists z_1 \in \Gamma_1)(z_1\overline{C}x)$, $(\exists y_2 \in \Delta_2)(x\overline{\widehat{O}}y_2)$. From $\Gamma\rho\Delta$ we get $\Gamma_1\rho_{\not\ll}\Delta_2$ and from here – $z_1 \not\ll y_2$. From $z_1\overline{C}x$ and $x\overline{\widehat{O}}y_2$ we obtain $z_1 \ll y_2$ – a contradiction.

Case (13)(23): $(\exists z_1 \in \Gamma_1)(z_1\overline{C}x)$, $(\exists z_2 \in \Gamma_1)(z_2 \ll x)$. From $z_1 \in \Gamma_1$ and $z_2 \in \Gamma_1$ we get $z_1 O z_2$. $z_2 \ll x$ implies $z_2 \leq x$. Conditions $z_1 O z_2$ and $z_2 \leq x$ imply $z_1 O x$ which implies $z_1 C x$ - a contradiction.

Case (13)(24): $(\exists z_1 \in \Gamma_1)(z_1\overline{C}x)$, $(\exists t_2 \in \Gamma_2)(t_2\overline{\widehat{C}}x)$. From $z_1 \in \Gamma_1$ and $t_2 \in \Gamma_2$ we get by Lemma 19 (iii) that $z_1 \not\ll t_2$. From $z_1\overline{C}x$ and $t_2\overline{\widehat{C}}x$ we obtain by $(CO1)$ and $(\ll \widehat{C}O)$ that $z_1 \ll t_2$ – a contradiction.

Case (14)(22) $(\exists t_1 \in \Gamma_2)(x \ll t_1)$, $(\exists y_2 \in \Delta_2)(x\overline{\widehat{O}}y_2)$. From $t_1 \in \Gamma_2$ and $y_2 \in \Delta_2$ we get $t_1\widehat{C}y_2$. This with $x \ll t_1$ implies $x\widehat{O}y_2$ which contradicts $x\overline{\widehat{O}}y_2$.

Case (14)(23): $(\exists t_1 \in \Gamma_2)(x \ll t_1)$, $(\exists z_2 \in \Gamma_1)(z_2 \ll x)$. From $x \ll t_1$ and $z_2 \ll x$ we obtain $z_2 \leq t_1$ and consequently – $t_1 \in \Gamma_1$. This contradicts the fact that $\Gamma_1 \cap \Gamma_2 = \varnothing$.

Case (14)(24): $(\exists t_1 \in \Gamma_2)(x \ll t_1)$, $(\exists t_2 \in \Gamma_2)(t_2\overline{\widehat{C}}x)$. From $t_1, t_2 \in \Gamma_2$ we obtain $t_2\widehat{C}t_1$. From $x \ll t_1$ and $t_2\overline{\widehat{C}}x$ we obtain $t_2\overline{\widehat{C}}t_1$ - a contradiction. ■

LEMMA 22. **ρ-extension Lemma.** *Let Γ_0, Δ_0 be filter-ideal pairs and let $\Gamma_0\rho\Delta_0$. Then Γ_0 and Δ_0 can be extended correspondingly into complete pairs Γ and Δ such that $\Gamma\rho\Delta$.*

Proof. Let $\Gamma_0\rho\Delta_0$. By an application of the Zorn Lemma and Lemma 21 we can find a complete pair Δ such that $\Delta_0 \subseteq \Delta$ and $\Gamma_0\rho\Delta$. By the symmetry of ρ we obtain $\Delta\rho\Gamma_0$. Then in the same way we can find a complete pair Γ such that $\Gamma_0 \subseteq \Gamma$ and $\Delta\rho\Gamma$. By symmetry of ρ we obtain $\Gamma\rho\Delta$ and the proof is finished. ■

DEFINITION 23. Let $\underline{W}$ be a mereotopological structure. We define the following relation R in the set $GF(\underline{W})$ of good filters:

$\Gamma R \Delta$ iff $(\Gamma, -\Gamma)\rho(\Delta, -\Delta)$ where $-\Gamma = W \setminus \Gamma$ and $-\Delta = W \setminus \Delta$.

The relational system $(GF(\underline{W}), R)$ will be called the canonical system of $\underline{W}$.

LEMMA 24. *If Γ, Δ are complete pairs then $\Gamma\rho\Delta$ iff $\Gamma_1 R \Delta_1$. R is a reflexive and symmetric relation.*

Proof. The proof follows from the definition of R and the fact that for a complete pair Γ we have $\Gamma_2 = -\Gamma_1$. ■

LEMMA 25. **Good-filter characterization of C, $\widehat{C}$ and $\ll$.** *Let $\underline{W}$ be a mereotopological structure and let $GF(\underline{W})$ be the set of good filters of $\underline{W}$. Then for any $x, y \in W$ we have:*

(i) xCy iff $(\exists\Gamma, \Delta \in GF(\underline{W}))(\Gamma R\Delta,\ x \in \Gamma$ and $y \in \Delta)$.

(ii) $x\widehat{C}y$ iff $(\exists\Gamma, \Delta \in GF(\underline{W}))(\Gamma R\Delta,\ x \notin \Gamma$ and $y \notin \Delta)$.

(iii) $x \not\ll y$ iff $(\exists\Gamma, \Delta \in GF(\underline{W}))(\Gamma R\Delta,\ x \in \Gamma$ and $y \notin \Delta)$.

Proof. (i) $(\rightarrow)$ Suppose xCy. Then by Lemma 20 (i) there exist filter-ideal pairs Γ', Δ' such that $\Gamma'\rho\Delta'$, $x \in \Gamma'_1$ and $y \in \Delta'_1$. Then by the ρ-extension Lemma 22 we can extend Γ' and Δ' into complete pairs Γ'' and Δ'' such that $\Gamma''\rho\Delta''$. Let $\Gamma = \Gamma''_1$, $\Delta = \Delta''_1$. Then we have $x \in \Gamma$, $y \in \Delta$ and by Lemma 24 that $\Gamma R\Delta$.

$(\leftarrow)$ Let $\Gamma R\Delta$, $x \in \Gamma$ and $y \in \Delta$. Then by the definition of R we have $(\Gamma, -\Gamma)\rho(\Delta, -\Delta)$. From here we obtain $\Gamma\rho_C\Delta$ which implies xCy.

(ii) The proof of (ii) is similar (dual) to that of (i).

(iii) $(\rightarrow)$ Suppose $x \not\ll y$. Then by lemma 20 (iii) there exist filter-ideal pairs Γ', Δ' such that $\Gamma'\rho\Delta'$, $x \in \Gamma'_1$ and $y \in \Delta'_2$. Then by the ρ-extension Lemma 22 we can extend Γ' and Δ' into complete pairs Γ'' and Δ'' such that $\Gamma''\rho\Delta''$. Let $\Gamma = \Gamma''_1$, $\Delta = \Delta''_1$. Then we have $x \in \Gamma$, $y \notin \Delta$ (because $y \in \Delta''_2 = -\Delta''_1 = -\Delta$) and by Lemma 24 that $\Gamma R\Delta$.

$(\leftarrow)$ The proof is similar to the corresponding proof of (i). ■

Now we are ready to prove a representation theorem for mereotopological structures. To each mereotopological structure $\underline{W}$ we associate its canonical system $(GF(\underline{W}), R)$. Since R is a reflexive and symmetric relation in $GF(\underline{W})$, then by the construction of non-topological example of contact algebra in Examples 2 (2) we associate to $(GF(\underline{W}), R)$ a contact algebra consisting of all subsets of $GF(\underline{W})$ with the standard Boolean operations and a contact C_R between any subsets $a, b \subseteq GF(\underline{W})$ defined by: aC_Rb iff $(\exists\Gamma \in a, \exists\Delta \in b)(\Gamma R\Delta)$.

THEOREM 26. **Representation Theorem for mereotopological structures.** *Let $\underline{W}$ be a mereotopological structure, $(GF(\underline{W}), R)$ be the corresponding canonical structure and $(B(GF(\underline{W})), C_R)$ be the contact algebra over $(GF(\underline{W}), R)$. For $x \in W$ define $h(x) = \{\Gamma \in GF(\underline{W} : x \in \Gamma\}$. Then h is an isomorphic embedding of $\underline{W}$ into the contact algebra $(B(GF(\underline{W})), C_R)$.*

Proof. The proof follows from the following equivalencies.

- $x \leq y$ iff (by Lemma 17 (i)) $(\forall\Gamma \in GF(W))(x \in \Gamma \rightarrow y \in \Gamma)$ iff $(\forall\Gamma \in GF(W))(\Gamma \in h(x) \rightarrow \Gamma \in h(y))$ iff $h(x) \subseteq h(y)$ iff $h(x) \leq h(y)$.
- xOy iff (by Lemma 17 (ii) $(\exists\Gamma \in GF(W))(x \in \Gamma$ and $y \in \Gamma)$ iff $(\exists\Gamma \in GF(W))(\Gamma \in h(x)$ and $\Gamma \in h(y))$ iff $h(x) \cap h(y) \neq \varnothing$ iff $h(x)Oh(y)$.
- $x\widehat{O}y$ iff (by Lemma 17 (iii) $(\exists\Gamma \in GF(W))(x \notin \Gamma$ and $y \notin \Gamma)$ iff $(\exists\Gamma \in GF(W))(\Gamma \notin h(x)$ and $\Gamma \notin h(y))$ iff $h(x) \cup h(y) \neq GF(W)$ iff $h(x)\widehat{O}h(y)$.
- xCy iff (by Lemma 25 (i)) $(\exists\Gamma, \Delta \in GF(\underline{W}))(\Gamma R\Delta, x \in \Gamma$ and $y \in \Delta)$ iff $(\exists\Gamma, \Delta \in GF(\underline{W}))(\Gamma R\Delta,\ \Gamma \in h(x)$ and $\Delta \in h(y))$ iff $h(x)C_Rh(y)$.

- $x\widehat{C}y$ iff (by Lemma 25 (ii)) $(\exists\Gamma,\Delta \in GF(\underline{W}))(\Gamma R\Delta,\ x \notin \Gamma)$ and $y \notin \Delta$ iff $(\exists\Gamma,\Delta \in GF(\underline{W}))(\Gamma R\Delta,\ \Gamma \notin h(x)$ and $\Delta \notin h(y))$ iff $-h(x)C_R - h(y)$ iff $h(x)\widehat{C}h(y)$.
- $x \not\ll y$ iff (by Lemma 25 (iii)) $(\exists\Gamma,\Delta \in GF(\underline{W}))(\Gamma R\Delta,\ x \in \Gamma$ and $y \notin \Delta)$ iff $(\exists\Gamma,\Delta \in GF(\underline{W}))(\Gamma R\Delta, \Gamma \in h(x)$ and $\Delta \notin h(y))$ iff $h(x)C_R{-}h(y)$ iff $h(x) \not\ll h(y)$. ■

COROLLARY 27. *Every mereotopological structure is completely standard.*

Proof. By Theorem 26 every mereotopological structure $\underline{W}$ is a standard one and by Lemma 8 $\underline{W}$ is also a completely standard. ■

REMARK 28. Note that Theorem 26 generalizes considerably Theorem 4 from [12] and Corollary 27 extends the topological representation theory of contact algebras by regular closed sets from [10]. If we consider the axiomatic definition of mereotopological structures as their *point-free* formulation, then the representation process can be considered as the *Whiteheadian process* of defining points. The first kind of points are the *good filters*, but they are not enough, because they allow only a non-topological representation in which regions are arbitrary sets and the mereotopological relations between them are defined by a binary relation between points. The second kind of points are introduced in the second phase of the representation, where we apply the topological representation theorem (Theorem 3) to the obtained discrete contact algebra. The definition and the theory of this second kind of points (called clans) can be found in [10]. Note also that the first phase of the representation introduces not only the first kind of points, but also extends the mereotopological structures with the Boolean operations between regions, which are necessary for introducing the second kind of points.

1.4 RCC-8 and mereotopological structures

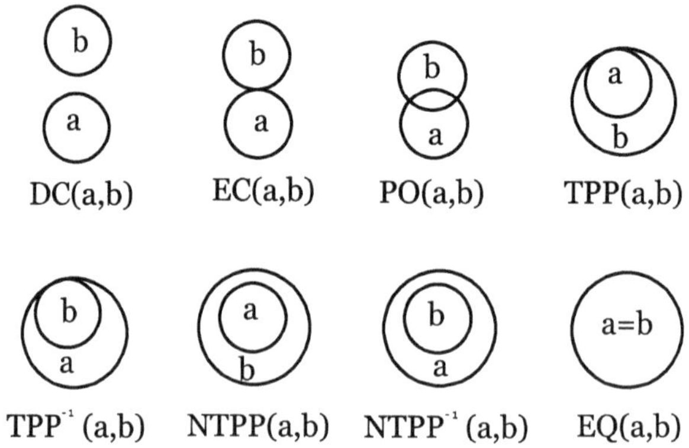

RCC-8 relations

One of the most popular systems of topological relations in the community of QSR is RCC-8. Probably this was one of the main motivations this system to be taken by Lutz and Wolter as a semantical base of the modal logic of topological relations [17]. The system RCC-8 was introduced for the first time by Egenhofer and Franzosa in [13]. It consists of 8 JEPD relations between non-empty regular closed subsets of arbitrary topological space. Having in mind the topological representation of contact algebras, it was given in [25] an equivalent definition of RCC-8 in the language of contact algebras:

DEFINITION 29. The system **RCC-8.**

- **disconnected** – **DC**(a,b)**:** $a\overline{C}b$,
- **external contact** – **EC**(a,b)**:** aCb and $a\overline{O}b$,
- **partial overlap** – **PO**(a,b)**:** aOb and $a \not\leq b$ and $b \not\leq a$,
- **tangential proper part** – **TPP**(a,b)**:** $a \leq b$ and $a \not\ll b$ and $b \not\leq a$,
- **tangential proper part**$^{-1}$ – **TPP**$^{-1}(a,b)$**:** $b \leq a$ and $b \not\ll a$ and $a \not\leq b$,
- **nontangential proper part NTPP**(a,b)**:** $a \ll b$ and $a \neq b$,
- **nontangential proper part**$^{-1}$ – **NTPP**$^{-1}(a,b)$**:** $b \ll a$ and $a \neq b$,
- **equal** – **EQ**(a,b)**:** $a = b$.

Looking at the above definitions we see that they can be repeated in the language of mereotopological structures. The following lemma represents some relationships between RCC-8 and mereotopological structures.

LEMMA 30. *Let* $\underline{W} = (W, \leq, O, \widehat{O}, C, \widehat{C}, \ll)$ *be a mereotopological structure and let* $W^- = \{a \in W : aOa\}$. *Then:*

(i) The system of relations in W^- *given as in Definition 29 represents an equivalent definition of RCC-8 relations. So RCC-8 is definable in the system of mereotopological relations.*

(ii) The following equivalencies are true in W^-*:*

$a \leq b$ *iff* $TPP(a,b) \vee NTPP(a,b) \vee a = b$,
$a\overline{O}b$ *iff* $DC(a,b) \vee EC(a,b)$
aCb *iff* $\neg DC(a,b)$,

Hence the relations $\leq, \geq, O$ *and* C *are definable in RCC-8.*

(iii) The relation $\widehat{O}$ *is not definable in RCC-8. Hence the system of mereotopological structures is more rich than RCC-8.*

Proof. (i) follows from the topological representation of mereotopological structures and Definition 29. (ii) follows from (i) and the axioms of mereotopological relations. (iii) follows from a result in [11] where a system, called RCC-10 is introduced as an extension of RCC-8. The definitions of the new relations in RCC-10 are given by means of the relation $\widehat{O}$. It follows from this fact that the relation $\widehat{O}$ is not definable in RCC-8. ■

2 A modal logic for mereotopological structures

In this section we introduce a poly-modal logic based on mereotopological structures, denoted by MTML (Mereo-Topological Modal Logic). MTML has the following modal box operators: $[\leq], [\geq], [\ll], [\gg], [O], [\widehat{O}], [C], [\widehat{C}]$, $[U]$, where $[U]$ is the universal modality. The corresponding diamond modality is denoted by $\langle R\rangle$ and defined as $\neg[R]\neg$. We adopt standard notations for Boolean connectives. The semantics of this language is the Kripke semantic over mereotopological structures. If $\underline{W}$ is a mereotopological structure and v is a valuation of the propositional variables in W, then the pair $M = (\underline{W}, v)$ is called, as usual, a model over $\underline{W}$. The fact that a formula A is true (false) at a point $x \in W$ will be denoted by $v(x, A) = 1$ ($v(x, A) = 0$). We adopt the standard semantical definitions of truth of a formula in a model, in a Kripke structure , etc. Let us note that all conditions of mereotopological structure except (≤ 0) are modally definable in this language by Sahlqvist formulas which then can be taken as axioms of the corresponding axiomatic system. So, in order to obtain a complete axiomatization of MTML we introduce another, non-standard semantics, which consists of a class of relational structures in which the non-definable axiom (≤ 0) is replaced by several modally definable consequences. This class admits an easy and straightforward modal axiomatization by means of generated canonical models. By using p-morphism techniques, we prove that generalized models of MTML are equivalent to the standard ones, which yields the completeness with respect to the standard semantics of the logic.

Comparing MTML and the modal logic of topological relations introduced by Lutz and Wolter in [17] (LW-logic), we can see that, on the base of Lemma 30, our modalities $[\leq], [\geq], [O]$ and $[C]$ are definable in LW-logic while the modality $[\widehat{O}]$ is not definable. Conversely, all basic modalities of LW-logic are not definable in MTML. If however we extend the language of MTML including modalities corresponding to Boolean combinations of the basic relations, then all modalities of LW-logic will be definable in this extended version of MTML, which will imply its undecidability.

2.1 Generalized mereotopological structures and the bulldozer construction

DEFINITION 31. A generalized mereotopological structure is a generalization of the notion of a mereotopological structure by dropping the axiom (≤ 0) and by adding the following additional axioms:

$(=_1)$ $a\overline{O}a \wedge b \leq a \rightarrow a = b,$ $\qquad(=_2)$ $a\overline{\widehat{O}}a \wedge a \leq b \rightarrow a = b,$

$(=_3)$ $a\overline{O}c \wedge b\overline{\widehat{O}}c \wedge b \leq a \rightarrow a = b.$

It can easily be seen that the above three conditions hold in mereotopological structures, so we have the following lemma.

LEMMA 32. *Each mereotopological structure is a generalized mereotopological structure.*

Now we shall show that each generalized mereotopological structure is a p-morphic image of a mereotopological structure. The construction is similar to the given one in [24] for a similar logic and is an adaptation of the Segerberg's bulldozer construction from [20]. To this end we first introduce in a given generalized mereotopological structure $\underline{W}$ the following equivalence relation. For $x, y \in W$, $x \equiv y$ iff $x \leq y$ and $y \leq x$. We denote by $\equiv (x) = \{y : x \equiv y\}$ the equivalence class generated by x and call such sets clusters. If $\equiv (x) = \{x\}$, then $\equiv (x)$ is called degenerated cluster. The following lemma states some easy properties of degenerated clusters.

LEMMA 33. *Let $\underline{W}$ be a generalized mereotopological structure. Then:*

(*i*) *If $x\overline{O}x$, then $\equiv (x)$ is a degenerated cluster.*

(*ii*) *If $x\overline{\widehat{O}}x$, then $\equiv (x)$ is a degenerated cluster.*

(*iii*) *If $x\overline{O}z$ and $x\overline{\widehat{O}}z$, then $\equiv (x)$ is a degenerate cluster.*

Proof. We will give a proof of (iii). Let $y \in \equiv (x)$. Then we have $x \leq y$ and $y \leq x$. From $x\overline{\widehat{O}}z$ and $x \leq y$ we get $y\overline{\widehat{O}}z$. Then $x\overline{O}z$, $y\overline{\widehat{O}}z$ and $y \leq x$ imply by $(=_3)$ that $x = y$, which shows that $\equiv (x)$ is a degenerate cluster. In a similar way, making use of the conditions $(=_1)$ and $(=_2)$, one can prove (i) and (ii). ■

DEFINITION 34. Let $\underline{W} = (W, \leq, O, \widehat{O}, \ll, C, \widehat{C})$ be a generalized mereotopological structure. We say that the structure $\underline{W'} = (W', \leq', O', \widehat{O}', \ll', C', \widehat{C}')$ is obtained from the structure $\underline{W}$ by the **bulldozer construction** if the following constructions hold.

Let $Z = \{\ldots, -2, -1, 0, 1, 2, \ldots\}$ be the set of integers and ∞ be a symbol such that $\infty \notin W \cup Z$. For $f \in Z$ and $x \in W$ define

$$f(x) = \begin{cases} (x, \infty) & \text{if } \equiv (x) \text{ is a degenerate cluster} \\ (x, f) & \text{otherwise.} \end{cases}$$

Define $W' = \{f(x) : x \in W, f \in Z\}$. For $R \in \{O, \widehat{O}, C, \widehat{C}\}$ define $f(x)R'g(y)$ iff xRy. For $\leq'$ and $\ll'$ we have the following definitions:

$$f(x) \leq' g(y) \leftrightarrow \begin{cases} x \leq y & \text{if } x \not\equiv y \text{ or } f(x) = (x, \infty) \\ f < g \text{ or } (f = g \text{ and } x = y) & \text{otherwise.} \end{cases}$$

$f(x) \geq' g(y)$ iff $g(y) \leq' f(x)$, $f(x) \ll' g(y)$ iff $x \ll y$ and $f(x) \leq' g(y)$, and $f(x) \gg' g(y)$ iff $g(y) \ll' f(x)$. Define the mapping $P : W' \to W$ as follows: $P(f(x)) = x$, for every $f(x) \in W'$.

LEMMA 35. **Bulldozer Lemma.** *Let $\underline{W}$ be a generalized mereotopological structure and let $\underline{W'}$ be obtained from $\underline{W}$ by the bulldozer construction. Then:*

(*i*) *$\underline{W'}$ is a mereotopological structure.*

(*ii*) *The mapping P is a p-morphism from $\underline{W'}$ onto $\underline{W}$.*

Proof. The proof that P is a p-morphism from $\underline{W'}$ onto $\underline{W}$ is straightforward. The proof that $\underline{W'}$ is a mereotopological structure is long because requires verification of a great number of axioms. For the most of the axioms this is a quite easy exercise. We will illustrate this giving several proofs only for the more difficult axioms. Note that for all relations R we have the following: if $f(x)R'g(y)$, then xRy, for all $x, y \in W$, and we will use this without explicit reference.

- **Axiom** (≤ 0) $f(x) \leq' g(y)$ and $g(y) \leq' f(x) \rightarrow f(x) = g(y)$.

Suppose $f(x) \leq' g(y)$. Then $x \leq y$ and $y \leq x$ and hence $x \equiv y$ and $\equiv (x) = \equiv (y)$.

Case 1: $\equiv (x)$ is a degenerate cluster. Then $\equiv (x) = \{x\}$, $\equiv (y) = \{y\}$ and consequently $x = y$. We have in this case $f(x) = (x, \infty)$, $g(y) = (y, \infty)$ and hence $f(x) = g(y)$.

Case 2: $\equiv (x)$ is not a degenerate cluster. Then we have $f(x) = (x, f)$ and $g(y) = (y, g)$ and we are in the second case of the definition of $\leq'$. Then we have: ($f < g$ or $f = g \& x = y$) and ($g < f$ or $g = f \& y = x$). This implies $f = g$ and $x = y$ which again gives $f(x) = g(y)$.

- **Axiom** (≤ 2)**:** Transitivity of $\leq'$. Suppose $f(x) \leq' g(y)$, $g(y) \leq' h(z)$. then we have $x \leq y$ and $y \leq z$ which implies $x \leq z$. We have to show that $f(x) \leq' h(z)$.

Case 1: $\equiv (x)$ is a degenerate cluster or $x \not\equiv z$. Since $x \leq z$ we obtain $f(x) \leq' h(z)$.

Case 2: $\equiv (x)$ is a not a degenerate cluster and $x \equiv z$. Then we obtain $x \leq z$ and $z \leq x$. Then from $z \leq x$ and $x \leq y$ we obtain $z \leq y$. From $y \leq z$ and $z \leq y$ we get $\equiv (y) = \equiv (z)$, and hence $\equiv (x) = \equiv (y) = \equiv (z)$. From here we obtain that $\equiv (y)$ and $\equiv (z)$ are not degenerate clusters. So for $f(x) \leq' g(y)$ and $g(y) \leq' h(z)$ we are in the second case of the definition of $\leq'$. This yields: ($f < g$ or $f = g \& x = y$) and ($g < h$ or $g = h \& y = z$). From here we obtain ($f < h$ or $f = h \& x = z$) which gives $f(x) \leq' h(z)$.

- **Axiom** $(\leq O\widehat{O})$**:** $h(z)\overline{O'}f(x)$ and $h(z)\overline{\widehat{O}'}g(y) \rightarrow f(x) \leq' g(y)$.

Suppose $h(z)\overline{O'}f(x)$ and $h(z)\overline{\widehat{O}'}g(y)$ Then we have $z\overline{O}x$ and $z\overline{\widehat{O}}y$ which implies $x \leq y$.

Case 1: $x \not\equiv y$ or $\equiv (x)$ is a degenerate cluster. In this case we have (by $x \leq y$) that $f(x) \leq' g(y)$.

Case 2: $x \equiv y$ and $\equiv (x)$ is not a degenerate cluster. From $z\overline{\widehat{O}}y$ and $x \equiv y$ we obtain $z\overline{\widehat{O}}x$. Then from $z\overline{O}x$, $z\overline{\widehat{O}}x$ and Lemma 33 (iii) we get that $\equiv (x)$ is a degenerate cluster, which shows that this case is impossible.

- **Axiom** $(\ll C\widehat{O})$**:** $h(z)\overline{C'}f(x)$ and $h(z)\overline{\widehat{O}'}g(y) \rightarrow f(x) \ll' g(y)$.

Suppose $h(z)\overline{C'}f(x)$ and $h(z)\overline{\widehat{O}'}g(y)$. This implies $z\overline{C}x$ and $z\overline{\widehat{O}}y$, which yield $x \ll y$. Condition $h(z)\overline{C'}f(x)$ implies $h(z)\overline{O'}f(x)$. This, together with $h(z)\overline{\widehat{O}'}g(y)$ implies (as we have just proved) $f(x) \leq' g(y)$. Conditions $x \ll y$ and $f(x) \leq' g(y)$ imply $f(x) \ll' g(y)$.

- **Axiom** $(=_1)$ $f(x)\overline{O'}f(x)$ and $g(y) \leq' f(x) \rightarrow f(x) = g(y)$.

Suppose $f(x)\overline{O'}f(x)$ and $g(y) \leq' f(x)$. This implies $x\overline{O}x$ and $y \leq x$. From $x\overline{O}x$ we obtain $x \leq y$ which with $y \leq x$ implies $x \equiv y$ and hence $\equiv (x) = \equiv (y)$. Condition $x\overline{O}x$ implies by Lemma 33 (i) that $\equiv (x)$ is a degenerate cluster. Then also $\equiv (y)$ is a degenerate cluster and hence $\equiv (x) = \{x\}$ and $\equiv (y) = \{y\}$ and hence $x = y$. In this case we have $f(x) = (x, \infty)$ and $g(y) = (y, \infty)$. Consequently $f(x) = g(y)$.

We expect that the above examples will show the reader how to verify the remaining axioms of generalized mereotopological structures. ■

2.2 Axiomatization and completeness theorem

We adopt the following system of axiom schemes and rules for MTML. All axioms are just the Sahlqvist modal equivalents of the axioms of generalized mereotopological structures.

Axiom Schemes

$(Bool)$ All boolean tautologies
(K) $[R](A \Rightarrow B) \Rightarrow ([R]A \Rightarrow [R]B)$,
(A_0) $\langle\leq\rangle[\geq]A \Rightarrow A$, $\langle\geq\rangle[\leq]A \Rightarrow A$, $\langle\ll\rangle[\gg]A \Rightarrow A$, $\langle\gg\rangle[\ll]A \Rightarrow A$,
$[U]A \Rightarrow A, \langle U\rangle[U]A \Rightarrow A$, $[U]A \Rightarrow [U][U]A$, $[R]A \Rightarrow [U]A$,
$(A_{\leq 1})$ $[\leq]A \Rightarrow A$, $(A_{\leq 2})$ $[\leq]A \Rightarrow [\leq][\leq]A$, (A_{O1}) $\langle O\rangle[O]A \Rightarrow A$,
$(A_{\widehat{O}1})$ $\langle\widehat{O}\rangle[\widehat{O}]A \Rightarrow A$, $(A_{O\leq})$ $[O]A \Rightarrow [O][\leq]A$,
$(A_{\widehat{O}\leq})$ $[\widehat{O}]A \Rightarrow [\widehat{O}][\geq]A$, $(A_{O\widehat{O}})$ $([O]A \Rightarrow A) \vee ([\widehat{O}]B \Rightarrow B)$,
$(A_{\leq O\widehat{O}})$ $[O]A \wedge [\widehat{O}]B \wedge \langle U\rangle([\leq]C \wedge \neg A) \Rightarrow [U](B \vee C)$,
(A_C) $\langle C\rangle[C]A \Rightarrow A$, $(A_{\widehat{C}})$ $\langle\widehat{C}\rangle[\widehat{C}]A \Rightarrow A$,
(A_{CO1}) $[C]A \Rightarrow [O]A$, $(A_{\widehat{C}\widehat{O}1})$ $[\widehat{C}]A \Rightarrow [\widehat{O}]A$,
(A_{CO2}) $\langle C\rangle\top \wedge [O]A \Rightarrow A$, $(A_{\widehat{C}\widehat{O}2})$ $\langle\widehat{C}\rangle\top \wedge [\widehat{O}]A \Rightarrow A$,
$(A_{C\leq})$ $[C]A \Rightarrow [C][\leq]A$, $(A_{\widehat{C}\leq})$ $[\widehat{C}]A \Rightarrow [\widehat{C}][\geq]A$,
$(A_{\ll\leq 1})$ $[\leq]A \Rightarrow [\ll]A$, $(A_{\ll\leq 2})$ $[\ll]A \Rightarrow [\leq][\ll]A$,
$(A_{\ll\leq 3})$ $[\ll]A \Rightarrow [\ll][\leq]A$, $(A_{\ll O})$ $\neg A \wedge [O]A \wedge [\ll]B \Rightarrow [U]B$,
$(A_{\ll\widehat{O}})$ $\neg A \wedge [\widehat{O}]A \wedge [\gg]B \Rightarrow [U]B$,
$(A_{\ll CO})$ $[O]A \Rightarrow [C][\ll]A$, $(A_{\ll\widehat{C}\widehat{O}})$ $[\widehat{O}]A \Rightarrow [\widehat{C}][\gg]A$,
$(A_{\ll C\widehat{O}})$ $[C]A \wedge [\widehat{O}]B \wedge \langle U\rangle([\ll]C \wedge \neg A) \Rightarrow [U](B \vee C)$,
$(A_{\ll\widehat{C}O})$ $[O]A \wedge [\widehat{C}]B \wedge \langle U\rangle([\ll]C \wedge \neg A) \Rightarrow [U](B \vee C)$,
$(A_{=1})$ $\langle\leq\rangle([O]A \wedge \neg A \wedge B) \Rightarrow B$, $(A_{=_2})$ $\langle\geq\rangle([\widehat{O}]A \wedge \neg A \wedge B) \Rightarrow B$,
$(A_{=_3})$ $\langle U\rangle(B \wedge \neg C \wedge \langle\leq\rangle(A \wedge C)) \Rightarrow \langle O\rangle A \vee \langle\widehat{O}\rangle B$.

Rules of inference:
Modus Ponens(MP) $A, A \Rightarrow B \vdash B$,
Necessitation (N) $A \vdash [R]A$ for $R \in \{\leq, \geq, \ll, \gg, O, \widehat{O}, C, \widehat{C}, U\}$.

THEOREM 36. **Completeness theorem for MTML.** *The following conditions are equivalent for any formula A of MTML:*
(i) A is a theorem of MTML,
(ii) A is true in all generalized mereotopological structures,
(iii) A is true in all mereotopological structures,
(iv) A is true in all standard and completely standard mereotopological structures.

Proof. The implications $(i) \rightarrow (ii) \rightarrow (iii) \rightarrow (iv)$ form the soundness part of the theorem and are straightforward. The implication $(iv) \rightarrow (iii)$ follows by Corollary 27. $(iii) \rightarrow (ii)$ is true by the Bulldozer Lemma 35. And finally the implication $(ii) \rightarrow (i)$ can be proved by using the standard techniques of generated canonical models (see [4, 5]). ∎

2.3 Filtration

LEMMA 37. *MTML do not possess fmp with respect to its standard semantics.*

Proof. It is easy to see that the Grzegorczyk formula

$$[\leq]([\leq](p \Rightarrow [\leq]p) \Rightarrow p) \Rightarrow p$$

is true in all finite mereotopological structures (because they are finite partial orderings with respect to $\leq$) but that it is falsified in the generalized mereotopological structure $W = \{a, b\}$ in which the relations $\leq, O, \widehat{O}, C, \widehat{C}$ and $\ll$ coincide with W^2, which proves the lemma. ∎

LEMMA 38. **Filtration Lemma for MTML.** *MTML admits filtration with respect to its nonstandard semantics and hence is decidable.*

Proof. The next definition presents the relevant constructions of the filtration.

DEFINITION 39. **Filtration for MTML.** Let $M = (\underline{W}, v)$ be a model over a generalized mereotopological structure and A_0 be a formula. Let Γ be the smallest set of formulas closed under sub-formulas, containing A_0 and satisfying the following closure conditions:

(Γ1) $\langle O\rangle\top$ and $\langle\widehat{O}\rangle\top$ are in Γ,

(Γ2) if $[R]A \in \Gamma$ for some $R \in \{O, \widehat{O}, \leq, \geq, \ll, \gg, C, \widehat{C}\}$, then $[R]A \in \Gamma$ for all $R \in \{O, \widehat{O}, \leq, \geq, \ll, \gg, C, \widehat{C}\}$.

We define an equivalence relation $\sim$ in W as follows:

$(\forall x, y \in W)(x \sim y \leftrightarrow (\forall A \in \Gamma)(v(x, A) = v(y, A)))$.

Further we define $|x| = \{y : x \sim y\}$ and $W' = \{|x| : x \in W\}$.

The valuation v' in W' is defined as follows: for $|x| \in W'$ and for propositional variable p we put $v'(|x|, p) = 1$ iff $v(x, p) = 1$.

We define the relational structure $\underline{W'} = (W', O', \widehat{O}', \leq', \geq', \ll', \gg', C', \widehat{C}')$ over $\underline{W}$ by specifying the relations $O', \widehat{O}', \leq', \geq', \ll', \gg', C', \widehat{C}'$ as follows. For any $|x|, |y| \in W'$ we define:

- $|x| \leq' |y|$ iff $(\forall[\leq]A \in \Gamma)$ $((v(x,[\leq]A) = 1 \rightarrow v(y,[\leq]A) = 1)$ & $(v(y,[\geq]A) = 1 \rightarrow v(x,[\geq]A) = 1)$ & $(v(x,[\ll]A) = 1 \rightarrow v(y,[\ll]A) = 1)$ & $(v(y,[\gg]A) = 1 \rightarrow v(x,[\gg]A) = 1)$ & $(v(y,[O]A) = 1 \rightarrow v(x,[O]A) = 1)$ & $(v(x,[\widehat{O}]A) = 1 \rightarrow v(y,[\widehat{O}]A) = 1)$ & $(v(y,[C]A) = 1 \rightarrow v(x,[C]A) = 1)$ & $(v(x,[\widehat{C}]A) = 1 \rightarrow v(y,[\widehat{C}]A) = 1)$ & $(v(x,\langle O\rangle\top) = 1 \rightarrow v(y,\langle O\rangle\top) = 1)$ & $(v(y,\langle\widehat{O}\rangle\top) = 1 \rightarrow v(x,\langle\widehat{O}\rangle\top) = 1))$,
- $|x| \geq' |y|$ iff $|y| \leq |x|$,
- $|x| \ll |y|$ iff $(\forall[\ll]A \in \Gamma)$ $((v(x,[\ll]A) = 1 \rightarrow v(y,[\leq]A) = 1)$ & $(v(y,[\gg]A) = 1 \rightarrow v(x,[\geq]A) = 1)$ & $(v(y,[O]A) = 1 \rightarrow v(x,[C]A) = 1)$ & $(v(x,[\widehat{O}]A) = 1 \rightarrow v(y,[\widehat{C}]A) = 1)$ & $(v(x,\langle O\rangle\top) = 1 \rightarrow v(y,\langle O\rangle\top) = 1)$ & $(v(y,\langle\widehat{O}\rangle\top) = 1 \rightarrow v(x,\langle\widehat{O}\rangle\top) = 1))$,
- $|x| \gg' |y|$ iff $|y| \ll' |x|$,
- $|x|O'|y|$ iff $(\forall[O]A \in \Gamma)$ $((v(x,[O]A) = 1 \rightarrow v(y,[\leq]A) = 1)$ & $(v(y,[O]A) = 1 \rightarrow v(x,[\leq]A) = 1)$ & $(v(x,\langle O\rangle\top) = 1$ & $v(y,\langle O\rangle\top) = 1))$,
- $|x|\widehat{O}'|y|$ iff $(\forall[\widehat{O}]A \in \Gamma)$ $((v(x,[\widehat{O}]A) = 1 \rightarrow v(y,[\geq]A) = 1)$ & $(v(y,[\widehat{O}]A) = 1 \rightarrow v(x,[\geq]A) = 1)$ & $(v(x,\langle\widehat{O}\rangle\top) = 1$ & $v(y,\langle\widehat{O}\rangle\top) = 1))$,
- $|x|C'|y|$ iff $(\forall[C]A \in \Gamma)$ $((v(x,[C]A) = 1 \rightarrow v(y,[\leq]A) = 1)$ & $(v(y,[C]A) = 1 \rightarrow v(x,[\leq]A) = 1)$ & $(v(x,[O]A) = 1 \rightarrow v(y,[\ll]A) = 1)$ & $(v(y,[O]A) = 1 \rightarrow v(x,[\ll]A) = 1)$ & $(v(x,\langle O\rangle\top) = 1$ & $v(y,\langle O\rangle\top) = 1))$,
- $|x|\widehat{C}'|y|$ iff $(\forall[\widehat{C}]A \in \Gamma)$ $((v(x,[\widehat{C}]A) = 1 \rightarrow v(y,[\geq]A) = 1)$ & $(v(y,[\widehat{C}]A) = 1 \rightarrow v(x,[\geq]A) = 1)$ & $(v(x,[\widehat{O}]A) = 1 \rightarrow v(y,[\gg]A) = 1)$ & $(v(y,[\widehat{O}]A) = 1 \rightarrow v(x,[\gg]A) = 1)$ & $(v(x,\langle\widehat{O}\rangle\top) = 1$ & $v(y,\langle\widehat{O}\rangle\top) = 1))$.

We have to prove two things. First that the new model $(\underline{W}', v')$ satisfies the two conditions of filtration for each relation R, namely for all $x, y \in W$

(F1) If xRy, then $|x|R'|y|$, and

(F2) If $|x|R'|y|$, then $(\forall[R]A \in \Gamma)(v(x,[R]A) = 1 \rightarrow v(y,A) = 1)$.

And second, to show that the new structure $\underline{W}'$ is a finite generalized mereotopological structure.

The finiteness of $\underline{W}'$ follows by the fact that Γ is a finite set – the closure conditions for Γ do not make it infinite.

The most tedious part of the proof is the verification of the conditions (F1) and (F2) – it is quite long but in each case easy. As an example we will verify the conditions (F1) and (F2) for the relation O'.

(F1,O) If xOy then $|x|O'|y|$.

Suppose xOy and let $[O]A \in \Gamma$. We have to verify the following conditions corresponding to the clauses of the definition of O':

(a) $v(x, [O]A) = 1 \rightarrow v(y, [\leq]A) = 1$,

(b) $v(y, [O]A) = 1 \rightarrow v(x, [\leq]A) = 1$,

(c) $v(x, \langle O \rangle \top) = 1$,

(d) $v(y, \langle O \rangle \top) = 1$.

Proof of (a). Suppose $v(x, [O]A) = 1$. To prove $v(y, [\leq]A) = 1$ suppose $y \leq z$. Then xOy and $y \leq z$ imply xOz and since $v(x, [O]A) = 1$ we obtain $v(z, A) = 1$. In a similar way we prove (b).

Proof of (c). From xOy we get xOx and since $v(x, \top) = 1$ we obtain $v(x, \langle O \rangle \top) = 1$. In the same way we verify (d).

Condition (F2) for O' can be verified rather easy. Suppose $|x|O'|y|$, $[O]A \in \Gamma$ and $v(x, [O]A) = 1$. By the first line of the definition of O' we obtain $v(y, [\leq]A) = 1$, Since $y \leq y$ we get $v(y, A) = 1$.

We left to the reader the verification of the conditions (F1) and (F2) for the other relations.

The verification of the axioms of generalized mereotopological structure is also quite long but in each case it is easy. We will demonstrate proofs only for some examples.

- **Axiom** (≤ 1) $|x| \leq' |x|$. By (≤ 1) we have $x \leq x$. Then by (F1) we obtain $|x| \leq' |x|$.
- **Axiom** (≤ 2) $|x| \leq' |y|$ and $|y| \leq' |z| \rightarrow |x| \leq' |z|$.

Suppose $|x| \leq' |y|$ and $|y| \leq' |z|$ and proceed to show $|x| \leq' |z|$. We have to verify the 10 clauses of the definition of $\leq'$ for $|x| \leq' |z|$. Let us demonstrate the clause for O:

$v(z, [O]A) = 1 \rightarrow v(x, [O]A) = 1$.

Suppose $v(z, [O]A) = 1$. Since $|y| \leq' |z|$ we get $v(y, [O]A) = 1$. This and $|x| \leq' |y|$ imply $v(x, [O]A) = 1$.

- **Axiom** ($O1$) $|x|O'|y| \rightarrow |y|O'|x|$. The axiom follows from the fact that the definition of O' is symmetric with respect to its arguments.
- **Axiom** ($O2$) $|x|O'|y| \rightarrow |x|O'|x|$.

Suppose $|x|O'|y|$ and proceed to verify $|x|O'|x|$. The first two conditions of the definition of O' for $|x|O'|x|$ are equal and easy to proof. The third condition $v(x, \langle O \rangle \top) = 1$ follows from the assumption $|x|O'|y|$. The forth condition is equal to the third one.

Most of the other axioms can be treated in a similar way. Since the axioms ($=_1$), ($=_2$) and ($=_3$) present some difficulties we will consider one of them, say ($=_3$) (the other two can be treated similarly).

- **Axiom** ($=_3$) $|x|\overline{O'}|z|$, $|y|\overline{\widehat{O}'}|z|$ and $|y| \leq' |x| \rightarrow |x| = |y|$.

Suppose $|x|\overline{O'}|z|$, $|y|\overline{\widehat{O'}}|z|$ and $|y| \leq' |x|$. By (F1) we get $x\overline{O}z$ and $y\overline{\widehat{O}}z$. We shall show that $x = y$ which automatically implies $|x| = |y|$. Suppose that $x \neq y$. Then By axiom $(=_3)$ (and $x\overline{O}z$ and $y\overline{\widehat{O}}z$) we obtain $y \not\leq x$. Then by axiom $(\leq O\widehat{O})$ we obtain yOz or $x\widehat{O}z$. By (F1) we obtain $|y|O'|z|$ or $|x|\widehat{O}|z|$. We shall show that both alternatives yield a contradiction.

(a) $|y|O'|z|$ and $|y| \leq' |x|$ imply $|x|O'|z|$ which contradicts $|x|\overline{O'}|z|$.

(b) $|x|\widehat{O'}|z|$ and $|y| \leq' |x|$ imply $|y|\widehat{O'}|z|$ which contradicts $|y|\overline{\widehat{O'}}|z|$.

■

3 Concluding remarks

We conclude the paper by formulating some open problems.

The first open problem concerns the completeness theorem of an extension of MTML over mereotopological structures satisfying the axiom of connectedness (see Remarks 9 (4)). The standard models of this extension are over connected topological spaces, for instance, models over R^n. Let us note that the techniques of the representation theorem for mereotopological structures, used in this paper, do not hold in the presence of this axiom. So one has to invent some new techniques. Another reasonable problem is to look for possible extensions of MTML with some new modalities, preserving decidability. And the last problem is the complexity of the satisfiability of MTML.

Acknowledgments. Thanks are due to the three anonymous referees for their very helpful and professional remarks helping us to improve the quality of the presentation. The work of the second author was supported by the project MI 1510 "Applied Logic and Topological Structures" of the Bulgarian Ministry of Science and Education.

BIBLIOGRAPHY

[1] R. Balbes and Ph. Dvinger, *Distributive lattices*, University of Missoury press, 1974.

[2] Ph. Balbiani, T. Tinchev and D. Vakarelov, Modal logics for region-based theory of space. *Fundamenta Informaticae*, vol. **81**(1–3):29-82, 2007.

[3] B. Bennett and I. Düntsch, Axioms, Algebras and Topology. In: *Handbook of Spatial Logics*, M. Aiello, I. Pratt, and J. van Benthem (Eds.), Springer, 2007, 99-160.

[4] P. Blackburn, M. de Rijke, and Y. Venema, *Modal Logic*, Cambridge Univ. Press, 2001.

[5] A. Chagrov and M. Zakharyaschev. *Modal Logic.* Oxford Univ. Press, 1997.

[6] A. Cohn and S. Hazarika. Qualitative spatial representation and reasoning: An overview. *Fuandamenta informaticae* **46**:1–20, 2001.

[7] A. Cohn and J. Renz. Qualitative spatial representation and reasoning. In: F. van Hermelen, V. Lifschitz and B. Porter (Eds.) *Handbook of Knowledge Representation*, Elsevier, 2008, 551-596.

[8] De Laguna, T. Point, line and surface as sets of solids. *The Journal of Philosophy* **19** :449–461, 1922.

[9] A. Deneva and D. Vakarelov, Modal Logics for Local and Global Similarity Relations. *Fundamenta Informaticae*, **31**(3-4):295-304, 1997.

[10] G. Dimov and D. Vakarelov, Contact Algebras and Region-based Theory of Space. A proximity approach. I and II. *Fundamenta Informaticae,* **74**(2-3):209-249, 251-282, 2006.

[11] I. Düntsch, G. Schmidt and M. Winter, A Necessary Relation Algebra for Mereotopology. *Studia Logica* **69**:381-409, 2001.

[12] I. Düntsch and D. Vakarelov, Region-based theory of discrette spaces: A proximity approach. In: Nadif, M., Napoli, A., SanJuan, E., and Sigayret, A. EDS, *Proceedings of Fourth International Conference Journées de l'informatique Messine*, 123-129, Metz, France, 2003. Journal version in: *Annals of Mathematics and Artificial Intelligence*, **49**(1-4):5-14, 2007.

[13] M. Egenhofer, R. Franzosa, Point-set topological spatial relations. *Int. J. Geogr. Inform. Systems* **5**:161–174, 1991.

[14] R. Engelking, *General Topology*, PWN, Warszawa, 1977.

[15] Valentin Goranko, Dimiter Vakarelov, Hyperboolean Algebras and Hyperboolean Modal Logic. *Journal of Applied Non-Classical Logics*, **9**(2-3):345-368, 1999.

[16] J. Y. Halpern and Y. Shoham, A propositional modal logic of time intervals. *Journal of the ACM*, **38**(4):935-962, 1991.

[17] C. Lutz and F. Wolter, Modal logics for topological relations. *Logical Meth. Computer Sci.*, **2**(2-5): 1-41, 2006.

[18] Y. Nenov,*A deciadable modal logic for topological relations*. Master thesis, Sofia University, Faculty of mathematics and informatics, Dept. of matematical logic. Sofia, 2008. (in Bulgarian).

[19] I. Pratt-Hartmann, First-order region-based theories of space, In: *Handbook of Spatial Logics* , M. Aiello, I. Pratt and J. van Benthem (Eds.), Springer, 2007, 13-97.

[20] K. Segerberg, *An Essay in Classical Modal Logic*. Uppsala 1971.

[21] P. Simons, *PARTS. A Study in Ontology*, Oxford, Clarendon Press, 1987.

[22] D. Vakarelov, Logical analysis of positive and negative similarity relations in property systems. *In: WOKFAI'91, First World Conference on the Fundamentals of Artificial Intelligence, 1-5 July 1991, Paris, France, Proceedings* ed. Mishel De Glas and Dov Gabbay, 491-499.

[23] D. Vakarelov, A modal logic for set relations. *10-th International Congress of Logic, Methodology and Philosophy of Science*, 1995, Florence, Italy, Abstracts p. 183.

[24] D. Vakarelov, A Modal Characterization of Indiscernibility and Similarity Relations in Pawlak's Information Systems. Invite paper in:*Rough Sets, Fuzzy Sets, Data Mining, and Granular Computing, 10th International Conference RSFDGrC-2005, Regina, Canada, August/September 2005, Proceedings, Part I.* LNAI No 3641, 12-22, Springer.

[25] D. Vakarelov, Region-Based Theory of space: Algebras of Regions, Representation Theory, and Logics. In: Dov Gabbay et al. (Eds.) *Mathematical Problems from Applied Logic II. Logics for the XXIst Century*, Springer, 2007, 267-348.

[26] A. N. Whitehead, *Process and Reality*, New York, MacMillan, 1929.

[27] F. Wolter. and M. Zakharyaschev, Spatial representation and reasoning in RCC-8 with Boolean region terms, In: *Proceedings of the 14th European Conference on Artificial Intelligence (ECAI 2000)*, Horn W. (Ed.), IOS Press, pp. 244–248.

Yavor Nenov
Dep. of Mathematical Logic, Faculty of Mathematics and Informatics,
Sofia University,
Blvd James Bourchier 5, 1164 Sofia, Bulgaria
yavor_nenov@yahoo.com

Dimiter Vakarelov
Dep. of Mathematical Logic, Faculty of Mathematics and Informatics,
Sofia University,
Blvd James Bourchier 5, 1164 Sofia, Bulgaria
dvak@fmi.uni-sofia.bg

A Lindström characterisation of the guarded fragment and of modal logic with a global modality [1]

Martin Otto and Robert Piro

ABSTRACT. We establish a Lindström type characterisation of the extension of basic modal logic by a global modality (ML[∀]) and of the guarded fragment of first-order logic (GF) as maximal among compact logics with the corresponding bisimulation invariance and the Tarski Union Property.

Keywords: Lindström theorems, modal logic, bisimulation invariance, global modality, guarded fragment, Tarski union property.

1 Introduction

This investigation is motivated by a recent Lindström theorem for basic modal logic (ML) by van Benthem [4] and related investigations in [5]. It is shown in [4] that no logic that is compact, bisimulation invariant and has the relativisation property can properly extend ML. This characterisation itself may be seen as a methodological improvement on an earlier Lindström characterisation of ML by de Rijke [13], which explicitly stipulated a finite depth (or locality) condition as a crucial criterion. [A formula φ (over pointed Kripke structures, say) is called r-local if whether or not a pointed τ-structure $(\mathfrak{M}, w)$ satisfies φ only depends on the substructure induced on the r-neighbourhood of w (the set of elements accessible from w in at most r steps).]

The proof of van Benthem's characterisation in [4] does not carry over to the interesting case of the guarded fragment GF, indeed not even to the extension of basic modal logic by a global (or universal) modality ML[∀]. Crucially, the finite depth criterion is still instrumental in that proof, though instead of being stipulated as a condition it is shown to be a consequence of the combination of compactness and relativisation for any logic invariant under ordinary bisimulation. But locality, or the finite depth criterion, fail for GF and even for ML[∀]. Neither global nor guarded bisimulation invariance implies locality. We therefore switch to an alternative characterisation crucially based on the Tarski Union Property (TUP), which is another natural model theoretic criterion that has been studied in abstract model theory [2]. Just as a variant characterisation of FO can be based on compactness,

[1]This paper summarises results from the second author's diploma thesis [12], which was supervised by the first author.

TUP and invariance under partial isomorphy, we here characterise ML[∀] and GF as maximally expressive among compact logics that are invariant under the appropriate notion of bisimulation (global or guarded bisimulation, respectively) and satisfy TUP. Some discussion of the role of TUP can be found in the concluding section 4. An analogous characterisation of basic modal logic ML itself is of course also available.

Part of the point of such investigations, as expounded in particular in [5], is a new interest in the abstract model theory of logics well below first-order logic. Many of the techniques and constructions that are available in the more classical investigations into abstract model theory, which is aimed at levels above first-order, [2], are no longer available or meaningful for corresponding investigations at levels below FO. Much of the usual coding machinery relies on first-order interpretations of, for instance, embedded substructures or systems of partial isomorphism, etc., which are not generally available at the level of logics of a typically modal character.

We point out that it remains open whether for instance ML[∀] is also maximal in the class of compact logics with the relativisation property that are invariant under global bisimulation.

In the following we presuppose some familiarity with basic model theoretic notions from modal logic (syntax and semantics of basic modal logic, Kripke structures, bisimulation relations and the basic bisimulation game, etc.) as presented in various textbooks and, for instance, in [6]. Corresponding variations for ML[∀] and GF will be reviewed where they arise.

2 Characterisation of ML[∀]

We summarise some standard notions that are important throughout the paper and for which the reader may also want to compare the classical setting for abstract model theory in [2]. A *logic* $\mathcal{L}$ is a pair $(L, \models_{\mathcal{L}})$, where L is a function that maps signatures σ to the sets $L(\sigma)$ of $\mathcal{L}$-formulae over σ. $\models_{\mathcal{L}}$ is a relation between structures and formulae of $\mathcal{L}$. We tacitly assume that structures are of an appropriate type also w.r.t. accommodating any 'free variables' as appropriate for $\mathcal{L}$: speaking of 'structures' we allow structures with parameters, like pointed structures; this latitude is included without explicit mention in (1) below. Any logic $\mathcal{L}$ is assumed to satisfy the following:

1. If $\mathfrak{M} \models_{\mathcal{L}} \varphi$, then $\mathfrak{M}$ is a σ-structure such that $\varphi \in L(\sigma)$.
2. If $\sigma \subseteq \sigma'$, then $L(\sigma) \subseteq L(\sigma')$.
3. If $\sigma \subseteq \sigma'$, $\varphi \in L(\sigma)$ and $\mathfrak{M}$ a σ'-structure, then $\mathfrak{M} \restriction \sigma \models \varphi$ iff $\mathfrak{M} \models \varphi$.
4. If $\mathfrak{M}$ is isomorphic to $\mathfrak{N}$, then $\mathfrak{M} \models \varphi$ iff $\mathfrak{N} \models \varphi$.

For the sake of simplicity we denote the set of $\mathcal{L}$-formulae over σ by $\mathcal{L}(\sigma)$, and mostly write just $\models$ for $\models_{\mathcal{L}}$.

A logic $\mathcal{L}'$ is at *least as expressive* as a logic $\mathcal{L}$, if for every signature σ and every formula $\varphi \in \mathcal{L}(\sigma)$ there is a formula $\varphi' \in \mathcal{L}'(\sigma)$ such that $\mathfrak{M} \models \varphi$

iff $\mathfrak{M} \models \varphi'$ for every σ-structure $\mathfrak{M}$. If $\mathcal{L}$ is at least as expressive as $\mathcal{L}'$ and vice versa, then $\mathcal{L}$ and $\mathcal{L}'$ are *equi-expressive* or *equivalent*. Since syntactic variations are immaterial for our purposes, we regard equivalent logics as equal. For the same reason, if $\mathcal{L}'$ is at least expressive as $\mathcal{L}$, we may assume that $\mathcal{L}(\sigma) \subseteq \mathcal{L}'(\sigma)$ for every signature σ. We thus write just $\mathcal{L} \subseteq \mathcal{L}'$ and $\mathcal{L} = \mathcal{L}'$ for the corresponding relations between logics.

We call a logic $\mathcal{L}$ *compact*, if it satisfies the following, for every set $\Psi \subseteq \mathcal{L}(\sigma)$: existence of models for every finite $\Psi_0 \subseteq \Psi$ implies the existence of a model of Ψ (any finitely satisfiable set of $\mathcal{L}$-formulae is satisfiable). Note that we make no restriction on the cardinality of the sets Ψ under consideration (full compactness).

A modal signature (for ML or ML[∀] and similar extensions) consists of a pair of sets (τ, Φ), where the $\alpha \in \tau$ label the binary accessibility relations R_α and the $P \in \Phi$ correspond to unary predicates interpreting basic propositions in Kripke structures of type (τ, Φ). The class of Kripke structures of this type is denoted $\mathrm{Mod}(\tau, \Phi)$. In the sense of the general stipulation above, the relevant signatures σ are thus of the form $\sigma = (\tau, \Phi)$ and we stick to notation like $\mathcal{L}(\tau, \Phi)$. Also, σ-structures are pointed (τ, Φ)-structures, according to the natural semantics of modal logics like ML and ML[∀]. Where appropriate, we shall make this implicit as usual, in notation as in $(\mathfrak{M}, w) \models \varphi$.

ML[∀] is the extension of basic modal logic ML by a global modality (corresponding to the full accessibility relation), which we denote as $\forall$.

We are interested in logics that extend ML[∀] in the following sense.

DEFINITION 1. A logic $\mathcal{L}$ extends ML[∀] if, for every (τ, Φ), $\mathrm{ML}[\forall](\tau, \Phi) \subseteq \mathcal{L}(\tau, \Phi)$ and $\mathcal{L}(\tau, \Phi)$ is closed under $\wedge, \neg$ as well as $\exists$ and $\langle\alpha\rangle$ for every $\alpha \in \tau$.

The bisimulation game of basic modal logic ML may be extended in a natural manner to cover moves that capture the power of the global accessibility relation associated with the extension of ML by $\forall$. For this, one allows the first player to call 'global rounds' in which both players are allowed to freely relocate pebbles within the respective structure unconstrained by the accessibility relations R_α. For the second player to have a winning strategy in this modified infinite bisimulation game starting from configurations $(\mathfrak{M}, w); (\mathfrak{N}, v)$, which we denote by $(\mathfrak{M}, w) \overset{\forall}{\longleftrightarrow} (\mathfrak{N}, v)$, is the same as to require an ordinary bisimulation relation between $\mathfrak{M}$ and $\mathfrak{N}$ that is global (covers all of $\mathfrak{M}$ and all of $\mathfrak{N}$) and contains the pair (w, v). We speak of global bisimulation equivalence. It is easy to see that global bisimulation equivalence is exactly the right analogue of ordinary bisimulation equivalence that is appropriate for ML[∀]. In particular ML[∀] is invariant under global bisimulation equivalence in the following sense.

DEFINITION 2. A logic $\mathcal{L}$ is invariant under global bisimulation equivalence, or $\overset{\forall}{\longleftrightarrow}$ *invariant*, if for any two globally bisimilar pointed structures $(\mathfrak{M}, w) \overset{\forall}{\longleftrightarrow} (\mathfrak{N}, v)$ of type (τ, Φ) and any formula $\varphi \in \mathcal{L}(\tau, \Phi)$: $(\mathfrak{M}, w) \models \varphi$ iff $(\mathfrak{N}, v) \models \varphi$.

We write $\mathit{Th}(\mathfrak{M}, w)$ for the $\mathrm{ML}[\forall](\tau, \Phi)$-theory and $\mathit{Th}_{\mathcal{L}}(\mathfrak{M}, w)$ for the $\mathcal{L}(\tau, \Phi)$-theory of $(\mathfrak{M}, w)$.

DEFINITION 3. Let $\mathfrak{M}, \mathfrak{N} \in \mathrm{Mod}(\tau, \Phi)$. $\mathfrak{N}$ is an *$\mathcal{L}$-elementary extension* of $\mathfrak{M}$, $\mathfrak{M} \preccurlyeq_{\mathcal{L}} \mathfrak{N}$, if $\mathfrak{M}$ is an induced substructure of $\mathfrak{N}$ and $\mathit{Th}_{\mathcal{L}(\tau,\Phi)}(\mathfrak{M}, w) = \mathit{Th}_{\mathcal{L}(\tau,\Phi)}(\mathfrak{N}, w)$ for all w in $\mathfrak{M}$.

DEFINITION 4. A logic $\mathcal{L}$ is said to have the *Tarski Union Property (TUP)* if for every $\mathcal{L}$-elementary chain, $(\mathfrak{M}_i)_{i\in\mathbb{N}}$: $\mathfrak{M}_0 \preccurlyeq_{\mathcal{L}} \mathfrak{M}_1 \preccurlyeq_{\mathcal{L}} \mathfrak{M}_2 \preccurlyeq_{\mathcal{L}} \ldots$, the union $\bigcup_{i\in\mathbb{N}} \mathfrak{M}_i$ is an $\mathcal{L}$-elementary extension of each $\mathfrak{M}_j$.

OBSERVATION 5. $\mathrm{ML}[\forall]$ has the Tarski Union Property.

Proof. Let $\mathfrak{M}^* := \bigcup_{i\in\mathbb{N}} \mathfrak{M}_i$. As $\mathfrak{M}_j$ is a substructure of $\mathfrak{M}^*$, atomic formulae are preserved at all w in $\mathfrak{M}_j$. The claim is trivially compatible with boolean operations. It remains to give inductive arguments for the $\langle\alpha\rangle$- and $\exists$-steps in formula formation.

$\langle\alpha\rangle$. Let $(\mathfrak{M}_j, w) \models \langle\alpha\rangle\varphi$, i.e., $(\mathfrak{M}_j, w') \models \varphi$ for some $(w, w') \in R_\alpha^{\mathfrak{M}_j}$. By the inductive hypothesis for φ, $(\mathfrak{M}^*, w') \models \varphi$, and therefore $(\mathfrak{M}^*, w) \models \langle\alpha\rangle\varphi$, as also $(w, w') \in R_\alpha^{\mathfrak{M}^*}$.

Conversely, if $(\mathfrak{M}^*, w) \models \langle\alpha\rangle\varphi$ through some $(w, w') \in R_\alpha^{\mathfrak{M}^*}$ such that $(\mathfrak{M}^*, w') \models \varphi$, then $w' \in M_k$ and $(w, w') \in R_\alpha^{\mathfrak{M}_k}$ for some $k \in \mathbb{N}$. By the inductive hypothesis for φ, $(\mathfrak{M}_k, w') \models \varphi$, whence $(\mathfrak{M}_k, w) \models \langle\alpha\rangle\varphi$. By the $\mathcal{L}$-elementary nature of the chain, $(\mathfrak{M}_j, w) \models \langle\alpha\rangle\varphi$ for all j such that $w \in M_j$.

$\exists$ is treated analogously. ■

Our goal is the following characterisation of $\mathrm{ML}[\forall]$ as maximally expressive among a natural class of $\overset{\forall}{\longleftrightarrow}$ invariant logics.

THEOREM 6. *Any compact $\overset{\forall}{\longleftrightarrow}$ invariant logic $\mathcal{L}$ with the Tarski Union Property that extends* $\mathrm{ML}[\forall]$ *is equivalent to* $\mathrm{ML}[\forall]$ *itself.*

We define several natural notions and provide some lemmas towards the proof.

DEFINITION 7. A set of $\mathrm{ML}(\tau, \Phi)$-formulae Γ is called an *α-type of* $(\mathfrak{M}, w)$ (or of $\mathit{Th}(\mathfrak{M}, w)$) if $(\mathfrak{M}, w) \models \langle\alpha\rangle \bigwedge \Gamma_0$ for all finite $\Gamma_0 \subseteq \Gamma$.

$\exists$-types of $\mathfrak{M}$ (or of $\mathit{Th}(\mathfrak{M})$) are similarly defined: $\mathfrak{M} \models \exists \bigwedge \Gamma_0$ must apply for all finite subsets Γ_0 of Γ.

DEFINITION 8. An α-type Γ of $(\mathfrak{M}, w)$ is *realised in* $(\mathfrak{M}, w)$ if there is some w' in $\mathfrak{M}$ such that $(w, w') \in R_\alpha^{\mathfrak{M}}$ and $(\mathfrak{M}, w') \models \Gamma$. An $\exists$-type Γ of $\mathfrak{M}$ is *realised in* $\mathfrak{M}$ if $(\mathfrak{M}, w') \models \Gamma$ for some w' in $\mathfrak{M}$.

A structure is called *saturated* if for all w in $\mathfrak{M}$ and all $\alpha \in \tau$ every α-type of $(\mathfrak{M}, w)$ is realised in $(\mathfrak{M}, w)$ and if every $\exists$-type of $\mathfrak{M}$ is realised in $\mathfrak{M}$.

The following is the natural variant of the Hennessy–Milner theorem for global bisimulation equivalence and $\mathrm{ML}[\forall]$ over the class of saturated Kripke structures. In fact it is easily seen via the game that, over saturated Kripke structures, $\mathrm{ML}[\forall]$-equivalence induces a global bismulation, see, e.g., [6].

THEOREM 9 (Hennessy–Milner). *If* $\mathfrak{M}, \mathfrak{N} \in \mathrm{Mod}(\tau, \Phi)$ *are saturated, then* $\mathit{Th}(\mathfrak{M}, w) = \mathit{Th}(\mathfrak{N}, v)$ *implies* $(\mathfrak{M}, w) \overset{\forall}{\longleftrightarrow} (\mathfrak{N}, v)$.

A forest-unfolding $(\mathfrak{M}^F, w)$ of $(\mathfrak{M}, w)$ is the disjoint union of all tree-unravelings in every element of $(\mathfrak{M}, w)$. Since any forest-unfolding $(\mathfrak{M}^F, w)$ is globally bisimilar to its underlying model $(\mathfrak{M}, w)$, they have the same $\mathcal{L}$-theory. Below, we shall use them as a normalised representation of models that allow us to embed one into another.

LEMMA 10. *Let* $(\mathfrak{M}, w)$ *be a forest model with a uniquely assigned propositional letter* $P_{w'}$ *for each element* w' *in* $(\mathfrak{M}, w)$. *Then any* $(\mathfrak{N}, v) \models \mathit{Th}(\mathfrak{M}, w)$ *admits an isomorphic embedding* $\iota : (\mathfrak{M}, w) \hookrightarrow (\mathfrak{N}, v)$.

Proof. $(\mathfrak{M}, w)$ can be embedded in $(\mathfrak{N}, v)$ by an injection $\iota : (\mathfrak{M}, w) \hookrightarrow (\mathfrak{N}, v)$, which is inductively defined (w.r.t. distance from the roots in the component trees of the forest model $\mathfrak{M}$) such that $(\mathfrak{N}, \iota(w')) \models P_{w'}$ for every w' in $\mathfrak{M}$. ■

PROPOSITION 11. *Let* $\mathcal{L}$ *be a compact logic extending* ML[∀]. *Then every forest model* $\mathfrak{M}$ *admits an* $\mathcal{L}$*-elementary extension* $\mathfrak{M}'$ *that realises all* α*-types of* $(\mathfrak{M}, w)$ *(as* α*-types of* $(\mathfrak{M}', w)$*) for all* w *in* $\mathfrak{M}$ *and realises all* $\exists$*-types of* $\mathfrak{M}$ *(as* $\exists$*-types of* $\mathfrak{M}'$*).*

Proof. We introduce new propositional letters to Φ by setting

$$\begin{array}{rcll} \Psi & := & \Phi & \cup \ \{P_w \mid w \in \mathfrak{M}\} \\ & & & \cup \ \{P^{\alpha}_{w,\Gamma} \mid w \text{ in } \mathfrak{M},\ \Gamma \text{ an } \alpha\text{-type of } (\mathfrak{M}, w)\} \\ & & & \cup \ \{P_\Gamma \mid \Gamma \text{ an } \exists\text{-type of } \mathfrak{M}\} \end{array}$$

for disjoint sets of new unary predicates. Let T' be the following $\mathcal{L}(\tau, \Psi)$-theory (towards an axiomatisation of the (τ, Ψ)-expansion of the desired $\mathfrak{M}'$). T' comprises, for all w in $\mathfrak{M}$, the following $\mathcal{L}(\tau, \Psi)$-formulae:

1. $\exists P_w$.
2. $\forall(P_w \longrightarrow \neg P_{w'})$, for all $w' \neq w$ in $\mathfrak{M}$.
3. $\forall(P_w \longrightarrow \langle\alpha\rangle P_{w'})$, for all $(w, w') \in R^{\mathfrak{M}}_{\alpha}$, $\alpha \in \tau$.
4. $\forall(P_w \longrightarrow \neg\langle\alpha\rangle P_{w'})$, for all $(w, w') \notin R^{\mathfrak{M}}_{\alpha}$, $\alpha \in \tau$.
5. $\forall(P_w \longrightarrow \xi)$, for every $\xi \in \mathit{Th}_{\mathcal{L}}(\mathfrak{M}, w)$.
6. $\forall(P_w \longrightarrow \langle\alpha\rangle P^{\alpha}_{w,\Gamma})$, for all $\alpha \in \tau$ and every α-type Γ of $(\mathfrak{M}, w)$.
7. $\exists P_\Gamma$, for every $\exists$-type Γ of $\mathfrak{M}$.
8. $\forall(Q \longrightarrow \xi)$, for all $Q = P^{\alpha}_{w,\Gamma}, P_\Gamma$ in Ψ and every $\xi \in \Gamma$.

T' is finitely satisfiable (in expansions of $\mathfrak{M}$), hence satisfiable by compactness of $\mathcal{L}$. Let $\mathfrak{M}' \models T'$. As $\mathcal{L}$ is invariant under global bisimulation $\overset{\forall}{\longleftrightarrow}$, we may assume w.l.o.g. that $\mathfrak{M}'$ is a forest model. By construction, the (τ, Φ)-reduct of $\mathfrak{M}'$ is isomorphic to an $\mathcal{L}$-elementary extension of the forest model $\mathfrak{M}$: the isomorphism as in the proof of lemma 10 here yields an $\mathcal{L}$-elementary embedding, due to the formulae in (5). W.l.o.g., the forest model $\mathfrak{M}'$ is an $\mathcal{L}$-elementary extension of $\mathfrak{M}$. Moreover, $\mathfrak{M}'$ realises all required types, by (6)–(8). ■

COROLLARY 12. *Let $\mathcal{L}$ be a compact logic extending* ML[∀] *with TUP. Then every forest model $(\mathfrak{M}, w)$ possesses a saturated $\mathcal{L}$-elementary extension.*

Proof. Starting with the given model $(\mathfrak{M}, w)$, a repeated application of proposition 11 yields a chain of forest models, in which the (τ, Φ)-reduct of each model is an $\mathcal{L}$-elementary extension of its predecessor (restricted to (τ, Φ)). Since $\mathcal{L}$ has the Tarski Union Property, the limit of this chain is an $\mathcal{L}$-elementary extension for all members of the chain. Every type of $\mathfrak{M}_i$ is realised in $\mathfrak{M}_{i+1}$; so the limit is saturated. ■

The following indicates how to complete an ML[∀](τ, Φ)-theory while maintaining ML[∀]-inexpressibility of a given $\varphi \in \mathcal{L}(\tau, \Phi)$.

LEMMA 13. *Let $\mathcal{L}$ be a logic extending* ML[∀], *$\varphi \in \mathcal{L}(\tau, \Phi)$, $T \subseteq$ ML[∀](τ, Φ) and $\psi \in$ ML[∀](τ, Φ). If there is no $\chi \in$ ML[∀](τ, Φ) such that $T \models \varphi \longleftrightarrow \chi$, then the same is true of at least one of $T \cup \{\psi\}$ or $T \cup \{\neg\psi\}$.*

Proof. Assume for both ψ and $\neg\psi$ there were formulae χ and χ' such that $T \models \psi \longrightarrow (\varphi \longleftrightarrow \chi)$ and $T \models \neg\psi \longrightarrow (\varphi \longleftrightarrow \chi')$. Then $T \models \varphi \longleftrightarrow ((\chi \wedge \psi) \vee (\chi' \wedge \neg\psi))$ contradicts our assumptions, since $(\chi \wedge \psi) \vee (\chi' \wedge \neg\psi)$ is in ML[∀](τ, Φ). ■

Compactness of $\mathcal{L}$ guarantees that, in the situation of the lemma, the set of ML[∀](τ, Φ)-theories under which φ is not equivalent to any ML[∀](τ, Φ)-formula is closed under unions of $\subseteq$-chains. By Zorn's lemma, we thus obtain a $\subseteq$-maximal such $T \subseteq$ ML[∀](τ, Φ). By the lemma, such T is a complete ML[∀](τ, Φ)-theory. Moreover, both $T \cup \{\varphi\}$ and $T \cup \{\neg\varphi\}$ are satisfiable, as otherwise φ would be equivalent to $\bot$ or $\top$ under T. We thus get the following.

PROPOSITION 14. *If $\mathcal{L}$ is a compact logic extending* ML[∀] *and for some signature (τ, Φ) there is a formula $\varphi \in \mathcal{L}(\tau, \Phi)$ not equivalent to any $\chi \in$* ML[∀](τ, Φ)*, then there are two (τ, Φ)-models $\mathfrak{M}$ and $\mathfrak{N}$ such that $Th(\mathfrak{M}, w) = Th(\mathfrak{N}, v)$ and $(\mathfrak{M}, w) \models \varphi$ while $(\mathfrak{N}, v) \models \neg\varphi$.*

Proof of theorem 6. Assume, for some (τ, Φ) there is a formula $\varphi \in \mathcal{L}(\tau, \Phi)$ which is not equivalent to any formula in ML[∀](τ, Φ). According to proposition 14 there are two (τ, Φ)-models $\mathfrak{M}$ and $\mathfrak{N}$ with $Th(\mathfrak{M}, w) = Th(\mathfrak{N}, v)$ and $(\mathfrak{M}, w) \models \varphi$ and $(\mathfrak{N}, v) \models \neg\varphi$. Since $\mathcal{L}$ is invariant under global bisimulation $\overset{\forall}{\longleftrightarrow}$, we may assume w.l.o.g. that $\mathfrak{M}$ and $\mathfrak{N}$ are forest models. By compactness, corollary 12 yields saturated (τ, Φ)-models $(\mathfrak{M}^*, w)$ and $(\mathfrak{N}^*, v)$ which have the same $\mathcal{L}(\tau, \Phi)$-theories as the originals. Therefore $\mathit{Th}(\mathfrak{M}^*, w) = \mathit{Th}(\mathfrak{N}^*, v)$ and, by theorem 9, $(\mathfrak{M}^*, w) \overset{\forall}{\longleftrightarrow} (\mathfrak{N}^*, v)$. But then $(\mathfrak{M}^*, w) \models \varphi$ and $(\mathfrak{N}^*, v) \models \neg\varphi$ shows that φ is not invariant under global bisimulation, contradicting the assumptions on $\mathcal{L}$. ■

3 Characterisation of GF

First we introduce the relevant basic notions for GF, in analogy with those used for ML[∀] above. For guarded bisimulation invariant candidate logics $\mathcal{L}$, suitable and natural notions of elementary extensions and the Tarski Union Property are presented, before we state the main theorem. For the following we work with arbitrary relational signatures τ.

The guarded fragment $\mathrm{GF}(\tau) \subseteq \mathrm{FO}(\tau)$ is introduced as the restriction of FO that only allows quantification of the following *guarded* format. For a relation symbol $R \in \tau$ (or $=$), we write $R(\bar{x}\bar{y})$ for an R-atom containing all the displayed variables (but not necessarily in this order, and repetitions are also allowed). A quantification $\exists \bar{y}.R(\bar{x}\bar{y}) \wedge \varphi$ is guarded if, and only if, $\mathrm{free}(\varphi) \subseteq \mathrm{free}(R(\bar{x}\bar{y}))$; the R-atom $R(\bar{x}\bar{y})$ is a *guard* in this first-order quantification. Since equality atoms may also serve as guards, $\exists y.y = y \wedge \varphi$ is a formula of $\mathrm{GF}(\tau)$ whenever y is the only free variable in $\varphi \in \mathrm{GF}(\tau)$.

As a fragment of FO, GF is compact.

In a τ-structure $\mathfrak{M}$, a subset $X \subseteq M$ is called *guarded* if it is a singleton set or for some $R \in \tau$ there is a tuple $\bar{a} \in R^{\mathfrak{M}}$ comprising all the elements of X. A tuple $\bar{a}$ is *guarded* in $\mathfrak{M}$ if there is a guarded set in $\mathfrak{M}$ that includes all components of $\bar{a}$. A tuple is called *strictly guarded* if its set of components is precisely the set of components of some $R^{\mathfrak{M}}$-atom, or a singleton set.

We also introduce the following terminology that is suitable for our purposes. A tuple $\bar{a}'$ is an $\exists \bar{y}.R(\bar{x}\bar{y})$-successor of $\bar{a}$ in $\mathfrak{M}$ if the assignment $\bar{x}\bar{y} \mapsto \bar{a}'$ is an extension of the assignment $\bar{x} \mapsto \bar{a}$ such that $\mathfrak{M}, \bar{a}' \models R(\bar{x}\bar{y})$. Clearly guarded quantifications correspond to modal quantifications w.r.t. transitions to $\exists \bar{y}.R(\bar{x}\bar{y})$-successors (which are strictly guarded tuples). This analogy is at the root of the appropriate notion of guarded bisimulation and of guarded tree-unfoldings to be discussed below.

Let $\mathfrak{M}, \mathfrak{N}$ be two τ-structures, possibly with tuples of distinguished parameters $\bar{a}$ and $\bar{b}$ of matching lengths. The guarded bisimulation game on $\mathfrak{M}$ and $\mathfrak{N}$ is played by two players **I** and **II**. A generic configuration of the game consists of designated strictly guarded tuples of the same length, one in each structure, denoted $(\mathfrak{M}, \bar{a}); (\mathfrak{N}, \bar{b})$.[1] **II** will have lost unless the componentwise mapping $\bar{a} \longmapsto \bar{b}$ is a partial isomorphism between $\mathfrak{M}$ and $\mathfrak{N}$.

In each new round player **I** chooses to play in one of the two structures, say $\mathfrak{M}$, and chooses a (possibly empty) subtuple $\bar{a}_0$ of the current tuple $\bar{a}$

[1] In the initial configuration of the game, the given, not necessarily guarded tuples of distinguished parameters are admitted.

that stays fixed, and a completion of $\bar{a}_0$ to some strictly guarded tuple $\bar{a}'$. **II** has to choose a strictly guarded tuple $\bar{b}'$ extending the corresponding $\bar{b}_0$ such that the componentwise mapping $\bar{a}' \longmapsto \bar{b}'$ is again a partial isomorphism between $\mathfrak{M}$ and $\mathfrak{N}$. Note that, if **I** chose an $\exists\bar{y}.R(\bar{x}\bar{y})$-successor of $\bar{a}_0$, then the rules force **II** to do likewise.

II looses the game if she cannot provide an answer that satisfies these constraints. We say that **II** wins the game, if she has a winning strategy which allows her to respond to all challenges of **I** indefinitely. Two τ-structures with designated tuples $\mathfrak{M}, \bar{a}$ and $\mathfrak{N}, \bar{b}$ are *guarded bisimilar*, $\mathfrak{M}, \bar{a} \overset{g}{\longleftrightarrow} \mathfrak{N}, \bar{b}$, if **II** wins the game from $(\mathfrak{M}, \bar{a}); (\mathfrak{N}, \bar{b})$.

Guarded bisimulation equivalence is the natural variant of Ehrenfeucht–Fraïssé equivalence associated with GF $\subseteq$ FO. In particular, the semantics of GF is invariant under this equivalence: if $\mathfrak{M}, \bar{a} \overset{g}{\longleftrightarrow} \mathfrak{N}, \bar{b}$, then $\mathfrak{M}, \bar{a}$ and $\mathfrak{N}, \bar{b}$ are GF-equivalent, or have the same GF-theories (cf. the following definition).

DEFINITION 15. The GF-*theory* of a tuple $\bar{a}$ in a τ-structure $\mathfrak{M}$ is the set of GF-formulae satisfied by $\bar{a}$ in $\mathfrak{M}$: $\mathit{Th}_{\mathrm{GF}(\tau)}(\mathfrak{M}, \bar{a}) := \{\varphi \in \mathrm{GF}(\tau) \mid (\mathfrak{M}, \bar{a}) \models \varphi\}$.

DEFINITION 16. Let $\mathfrak{M}$ be a τ-structure. A set $\Gamma \subseteq \mathrm{GF}(\tau)$ is called an $\exists\bar{y}.R(\bar{x}\bar{y})$-*type* of $(\mathfrak{M}, \bar{a})$ if for every finite subset $\Gamma_0 \subseteq \Gamma$ $(\mathfrak{M}, \bar{a}) \models \exists\bar{y}.R(\bar{x}\bar{y}) \wedge \bigwedge \Gamma_0$.

This type is *realised at* $(\mathfrak{M}, \bar{a})$ if there is an $\exists\bar{y}.R(\bar{x}\bar{y})$-successor $\bar{a}'$ of $\bar{a}$ in $\mathfrak{M}$ such that $(\mathfrak{M}, \bar{a}') \models \Gamma$.

A τ-structure $\mathfrak{M}$ is GF-*saturated* if, for every guarded tuple $\bar{a}$ and every $\exists\bar{y}.R(\bar{x}\bar{y})$, all $\exists\bar{y}.R(\bar{x}\bar{y})$-types of $(\mathfrak{M}, \bar{a})$ are realised at $(\mathfrak{M}, \bar{a})$.

The following analogue of the Hennessy–Milner theorem is then immediate.

PROPOSITION 17 (Hennessy–Milner). *Let $\mathfrak{M}, \bar{a}$ and $\mathfrak{N}, \bar{b}$ be two* GF-*saturated τ-structures with parameter tuples. If $\mathit{Th}_{\mathrm{GF}(\tau)}(\mathfrak{M}, \bar{a}) = \mathit{Th}_{\mathrm{GF}(\tau)}(\mathfrak{N}, \bar{b})$, then $\mathfrak{M}, \bar{a} \overset{g}{\longleftrightarrow} \mathfrak{N}, \bar{b}$.*

Proof. Indeed, GF-equivalence between (strictly guarded) tuples can be maintained by **II** and thus provides a winning strategy. Assume w.l.o.g. that **I** chooses an $\exists\bar{y}.R(\bar{x}\bar{y})$-successor $\bar{a}'$ of some subtuple $\bar{a}_0$ of $\bar{a}$ in $\mathfrak{M}$. Since $(\mathfrak{M}, \bar{a})$ and $(\mathfrak{N}, \bar{b})$ have the same theory,

$$(\mathfrak{M}, \bar{a}_0) \models \exists\bar{y}.R(\bar{x}\bar{y}) \wedge \bigwedge \Gamma_0 \quad \text{iff} \quad (\mathfrak{N}, \bar{b}_0) \models \exists\bar{y}.R(\bar{x}\bar{y}) \wedge \bigwedge \Gamma_0$$

for all finite $\Gamma_0 \subseteq \mathit{Th}_{\mathrm{GF}(\tau)}(\mathfrak{M}, \bar{a}')$. So $\mathit{Th}_{\mathrm{GF}(\tau)}(\mathfrak{M}, \bar{a}')$ is an $\exists\bar{y}.R(\bar{x}\bar{y})$-type of $\mathit{Th}_{\mathrm{GF}(\tau)}(\mathfrak{N}, \bar{b}_0)$. As $\mathfrak{N}$ is GF-saturated, there is an $\exists\bar{y}.R(\bar{x}\bar{y})$-successor $\bar{b}'$ of $\bar{b}_0$ such that $(\mathfrak{N}, \bar{b}') \models \mathit{Th}_{\mathrm{GF}(\tau)}(\mathfrak{M}, \bar{a}')$ for **II** to play. ■

DEFINITION 18. A logic $\mathcal{L}$ extends GF if, for every signature τ, $\mathrm{GF}(\tau) \subseteq \mathcal{L}(\tau)$ and $\mathcal{L}$ is closed under boolean operations $\wedge, \neg$ and guarded quantification. Closure under guarded quantification means that for any τ-atom

$R(\bar{x}\bar{y})$ and any $\varphi \in \mathcal{L}(\tau)$ with free variables[2] amongst $\bar{x}\bar{y}$, there is an $\mathcal{L}$-formula φ' with the semantics of $\exists \bar{y}.R(\bar{x}\bar{y}) \wedge \varphi$.

DEFINITION 19. Let $\mathfrak{M}, \mathfrak{N}$ be two τ-structures. $\mathfrak{N}$ is an *$\mathcal{L}$-elementary extension of* $\mathfrak{M}$ (in a guarded sense), abbreviated by $\mathfrak{M} \preccurlyeq_{\mathcal{L}} \mathfrak{N}$, if $\mathfrak{M}$ is an induced substructure of $\mathfrak{N}$ and $Th_{\mathcal{L}(\tau)}(\mathfrak{M}, \bar{a}) = Th_{\mathcal{L}(\tau)}(\mathfrak{N}, \bar{a})$ for all guarded tuples $\bar{a}$ in $\mathfrak{M}$. $\mathfrak{M}, \bar{a} \preccurlyeq_{\mathcal{L}} \mathfrak{N}, \bar{a}$ can be similarly defined for a not necessarily guarded tuple of distinguished parameters.

Recall from definition 4 that $\mathcal{L}$ has the Tarski Union Property (TUP) if the limit of every $\mathcal{L}$-elementary chain $(\mathfrak{M}_i)_{i \in \mathbb{N}}$ is an $\mathcal{L}$-elementary extension of each $\mathfrak{M}_i$.

OBSERVATION 20. GF has TUP.

Proof. Let $\mathfrak{M}_0 \preccurlyeq_{\mathrm{GF}} \mathfrak{M}_1 \preccurlyeq_{\mathrm{GF}} \mathfrak{M}_2 \ldots$ be a chain of GF-elementary extensions and $\mathfrak{M}^* := \bigcup_{i \in \mathbb{N}} \mathfrak{M}_i$. Let furthermore $\mathfrak{M}_j$ be a member of the chain and $\bar{a}$ a guarded tuple in $\mathfrak{M}_j$. It has to be shown that $Th_{\mathrm{GF}(\tau)}(\mathfrak{M}_j, \bar{a}) = Th_{\mathrm{GF}(\tau)}(\mathfrak{M}^*, \bar{a})$. The only interesting step in the syntactic induction is that of guarded quantification.

Assume $(\mathfrak{M}_j, \bar{a}) \models \exists \bar{y}.R(\bar{x}\bar{y}) \wedge \varphi$. For some $\exists \bar{y}.R(\bar{x}\bar{y})$-successor $\bar{a}'$ we have $(\mathfrak{M}_j, \bar{a}') \models \varphi$, which entails $(\mathfrak{M}^*, \bar{a}') \models \varphi$ by the induction hypothesis, and hence $(\mathfrak{M}^*, \bar{a}) \models \exists \bar{y}.R(\bar{x}\bar{y}) \wedge \varphi$.

Conversely, if $(\mathfrak{M}^*, \bar{a}) \models \exists \bar{y}.R(\bar{x}\bar{y}) \wedge \varphi$, then there is some $k \in \mathbb{N}$ such that the $\exists \bar{y}.R(\bar{x}\bar{y})$-successor $\bar{a}'$ satisfying φ lives in $\mathfrak{M}_k$. For $m := \max\{j, k\}$ this $\bar{a}'$ is an $\exists \bar{y}.R(\bar{x}\bar{y})$-successor of $\bar{a}$ in $\mathfrak{M}_m$. By the induction hypothesis $(\mathfrak{M}_m, \bar{a}') \models \varphi$ and hence $(\mathfrak{M}_m, \bar{a}) \models \exists \bar{y}.R(\bar{x}\bar{y}) \wedge \varphi$. The latter entails $(\mathfrak{M}_j, \bar{a}) \models \exists \bar{y}.R(\bar{x}\bar{y}) \wedge \varphi$, since $\mathfrak{M}_m$ is an GF-elementary extension of $\mathfrak{M}_j$. ■

In analogy to tree-unravellings of Kripke-structures, every τ-structure is guarded bisimilar to a structure that is tree-like w.r.t. the accessibility relations induced by the $\exists \bar{y}.R(\bar{x}\bar{y})$-successor relations. Technically, these *tree-like structures* are tree-decomposable into substructures consisting of (strictly) guarded tuples, or *guarded tree-decomposable*, cf. the generalised tree model property of [7], and see [8] for details of the analogy with ordinary unravellings. Intuitively, these guarded tree unfoldings are obtained by introducing new disjoint copies of elements along every path of overlapping guarded subsets leading to them. For a logic that is invariant under guarded bisimulation, one may w.l.o.g. restrict attention to the tree-like models thus obtained.

REMARK 21. Let $\mathfrak{M}$ be a tree-like (i.e., guarded tree-decomposable) τ-structure such that for every guarded tuple $\bar{a}$ in $\mathfrak{M}$ there is a predicate $P_{\bar{a}} \in \tau$ with $P_{\bar{a}}^{\mathfrak{M}} = \{\bar{a}\}$. Then $\mathfrak{M}$ can be embedded in any structure $\mathfrak{N}$ that satisfies the following GF-formulae for all $P_{\bar{a}}, P_{\bar{a}'} \in \tau$:

---

[2] Instead of explicitly referring to a notion of free variable for formulae in an abstract logic $\mathcal{L}$, a semantic description can be given: for every $\varphi \in \mathcal{L}(\tau)$ there is $\varphi' \in \mathcal{L}(\tau)$, s.t. $(\mathfrak{M}, \bar{a}) \models \varphi'$ iff there is an $\exists \bar{y}.R(\bar{x}\bar{y})$-successor $\bar{a}'$ of $\bar{a}$ with $(\mathfrak{M}, \bar{a}') \models \varphi$.

1. $\exists \bar{x}.P_{\bar{a}}\bar{x}$.

2. $\forall \bar{x}.P_{\bar{a}}\bar{x} \longrightarrow \bigwedge_{1\leq i\leq k} P_{a_i}x_i$ where $\bar{a} = (a_1, \ldots, a_k)$.

3. $\forall x.P_a x \longrightarrow \neg P_{a'}x$ for all $a \neq a'$ in $\mathfrak{M}$.

4. $\forall \bar{x}.P_{\bar{a}}\bar{x} \longrightarrow \exists \bar{y}.P_{\bar{a}'}(\bar{x}\bar{y})$, where $\bar{y}$ represents the components of $\bar{a}' \setminus \bar{a}$.

5. $\forall \bar{x}.P_{\bar{a}}\bar{x} \longrightarrow R(\bar{x})$ for all atomic formulae $R(\bar{x}) \in \mathit{Th}_{\mathrm{GF}(\tau)}(\mathfrak{M}, \bar{a})$.

6. $\forall \bar{x}.R\bar{x} \longrightarrow \neg(\bigwedge_{i\leq k} P_{a_i}x_i)$
for all $(a_1, \ldots, a_k) \notin R^{\mathfrak{M}}$, $R \in \tau$ of arity k and $(a_1, \ldots, a_k) \in M^k$.

Since $\mathfrak{M}$ is tree-like, the embedding can be defined inductively w.r.t. distance from the root in a guarded tree-decomposition (i.e., successively proceeding to $\exists \bar{y}.R(\bar{x}\bar{y})$-successors).

PROPOSITION 22. *Let $\mathcal{L}$ be a compact, guarded bisimulation invariant logic that has TUP. Then every tree-like τ-structure $\mathfrak{M}$ has an $\mathcal{L}$-elementary extension that is* GF*-saturated.*

The proof is based on a chain limit (TUP) of a chain obtained through the following process.

PROPOSITION 23. *Let $\mathfrak{M}$ be a tree-like τ-structure. Then there is a tree-like $\mathcal{L}$-elementary extension $\mathfrak{N}$ of $\mathfrak{M}$ such that, for every guarded tuple $\bar{a}$ in $\mathfrak{M}$ and for every $\exists \bar{y}.R(\bar{x}\bar{y}) \in \mathrm{GF}(\tau)$, every $\exists \bar{y}.R(\bar{x}\bar{y})$-type of $\mathfrak{M}, \bar{a}$ is realised at $(\mathfrak{N}, \bar{a})$.*

Proof. Let $G(\mathfrak{M})$ be the set of all guarded tuples in $\mathfrak{M}$. We extend τ to $\sigma := \tau \,\dot{\cup}\, \{P_{\bar{a}} \mid \bar{a} \in G(\mathfrak{M})\}$ and σ further to

$$\rho := \sigma \,\dot{\cup} \bigcup_{\bar{a}\in G(\mathfrak{M})} \{P^{\alpha}_{\bar{a},\Gamma} \mid \alpha = \exists \bar{y}.R(\bar{x}\bar{y}), \Gamma \subseteq \mathrm{GF}(\tau) \text{ an } \alpha\text{-type of } (\mathfrak{M}, \bar{a})\}.$$

The set of formulae defined in remark 21 can now be formulated in $\mathrm{GF}(\sigma)$. We extend this set to $T \subseteq \mathcal{L}(\rho)$ by adding the following formulae for every $\bar{a} \in G(\mathfrak{M})$.

7. $\forall \bar{x}.P_{\bar{a}}\bar{x} \longrightarrow \varphi$ for all $\varphi \in \mathit{Th}_{\mathcal{L}(\tau)}(\mathfrak{M}, \bar{a})$.

8. $\forall \bar{x}.P_{\bar{a}}\bar{x} \longrightarrow \exists \bar{y}.R(\bar{x}\bar{y}) \wedge P^{\alpha}_{\bar{a},\Gamma}(\bar{x}\bar{y})$
for every $\alpha = \exists \bar{y}.R(\bar{x}\bar{y})$ and every $\exists \bar{y}.R(\bar{x}\bar{y})$-type Γ of $(\mathfrak{M}, \bar{a})$.

9. $\forall \bar{x}.P^{\alpha}_{\bar{a},\Gamma}\bar{x} \longrightarrow \varphi$ for every $\varphi \in \Gamma$ and $P^{\alpha}_{\bar{a},\Gamma} \in \rho$.

A simple compactness argument shows that T is satisfiable; indeed any finite subset of T is satisfiable in an expansion of $\mathfrak{M}$.

Let $\mathfrak{N}$ be a ρ-structure satisfying T. Since $\mathcal{L}$ is bisimulation invariant we may assume that $\mathfrak{N}$ is tree-like. By setting $P^{\mathfrak{M}}_{\bar{a}} := \{\bar{a}\}$ the τ-structure $\mathfrak{M}$ can be extended to a σ-structure, and is embeddable into the σ-reduct $\mathfrak{N}\restriction\sigma$

according to remark 21. Hence we may assume, that $\mathfrak{M}$ is a substructure of $\mathfrak{N}\restriction\tau$.

For every $\bar{a} \in G(\mathfrak{M})$ the formulae in (7) guarantee $(\mathfrak{N}, \bar{a}) \models \mathit{Th}_{\mathcal{L}(\tau)}(\mathfrak{M}, \bar{a})$ and therefore $\mathfrak{N}\restriction\tau$ is an $\mathcal{L}$-elementary extension of $\mathfrak{M}$.

It remains to show, that every $\exists\bar{y}.R(\bar{x}\bar{y})$-Type of $(\mathfrak{M}, \bar{a})$ is realised at $(\mathfrak{N}, \bar{a})$. This is clear from (8) and (9). ■

Proof of proposition 22. Let $\mathfrak{M}_0 := \mathfrak{M}$ and inductively let $\mathfrak{M}_{i+1}$ be a tree-like $\mathcal{L}$-elementary extension that realises every GF-type of $\mathfrak{M}_i$ as obtained through proposition 23. We thus get an $\mathcal{L}$-elementary chain $\mathfrak{M}_0 \preccurlyeq_{\mathcal{L}} \mathfrak{M}_1 \preccurlyeq_{\mathcal{L}} \mathfrak{M}_2 \preccurlyeq_{\mathcal{L}} \ldots$, and since $\mathcal{L}$ has TUP, the limit $\mathfrak{M}^*$ is an $\mathcal{L}$-elementary extension of every member $\mathfrak{M}_i$ and in particular of $\mathfrak{M}$.

$\mathfrak{M}^*$ is GF-saturated: let Γ be an $\exists\bar{y}.R(\bar{x}\bar{y})$-type for some guarded tuple $\bar{a}$ in $\mathfrak{M}^*$. There is some $i \in \mathbb{N}$ such that $\bar{a}$ is a guarded tuple of $\mathfrak{M}_i$. Since $\mathfrak{M}^*$ is an $\mathcal{L}$-elementary extension for $\mathfrak{M}_i$, in particular $\mathit{Th}_{\mathrm{GF}(\tau)}(\mathfrak{M}_i, \bar{a}) = \mathit{Th}_{\mathrm{GF}(\tau)}(\mathfrak{M}^*, \bar{a})$. Therefore $(\mathfrak{M}_i, \bar{a})$ has exactly the same $\exists\bar{y}.R(\bar{x}\bar{y})$-types as $(\mathfrak{M}^*, \bar{a})$. All those types are realised in $\mathfrak{M}_{i+1}$ and hence in $\mathfrak{M}^*$ by the $\mathcal{L}$-elementary nature of the extension.

Hence $\mathfrak{M}^*$ is a GF-saturated $\mathcal{L}$-elementary extension of $\mathfrak{M}$. ■

COROLLARY 24. *For every τ-structure $\mathfrak{M}$ with distinguished parameters $\tilde{a}$ there is a* GF*-saturated $\mathcal{L}$-elementary extension $\mathfrak{M}^*$ for $\mathfrak{M}$ such that also $\mathit{Th}_{\mathcal{L}(\tau)}(\mathfrak{M}, \tilde{a}) = \mathit{Th}_{\mathcal{L}(\tau)}(\mathfrak{M}^*, \tilde{a})$.*

Proof. We expand the signature τ to $\dot{\tau} := \tau \cup \{P_{\tilde{a}}\}$ and set $P_{\tilde{a}}^{\mathfrak{M}} := \{\tilde{a}\}$. Using proposition 22 for the $\dot{\tau}$-structure $\mathfrak{M}$, we obtain a GF-saturated $\dot{\tau}$-structure $\mathfrak{M}^*$ such that $\mathfrak{M} \preccurlyeq_{\mathcal{L}} \mathfrak{M}^*$ w.r.t. $\dot{\tau}$. So $\mathit{Th}_{\mathcal{L}(\dot{\tau})}(\mathfrak{M}, \bar{a}) = \mathit{Th}_{\mathcal{L}(\dot{\tau})}(\mathfrak{M}^*, \bar{a})$ for every guarded tuple $\bar{a}$ in $\mathfrak{M}$ and especially for $\tilde{a}$, which was guarded by $P_{\tilde{a}} \in \dot{\tau}$. Since $\mathcal{L}$ is compatible with reducts (cf. condition (3) on abstract logics in section 2), we get $\mathit{Th}_{\mathcal{L}(\tau)}(\mathfrak{M}, \bar{a}) = \mathit{Th}_{\mathcal{L}(\tau)}(\mathfrak{M}^*, \bar{a})$ for every guarded tuple, including $\tilde{a}$. Therefore $\mathfrak{M}\restriction\tau \preccurlyeq_{\mathcal{L}} \mathfrak{M}^*\restriction\tau$ and $\mathit{Th}_{\mathcal{L}(\tau)}(\mathfrak{M}, \tilde{a}) = \mathit{Th}_{\mathcal{L}(\tau)}(\mathfrak{M}^*, \tilde{a})$. Since the τ-reduct of $\mathfrak{M}^*$ remains GF-saturated, $\mathfrak{M}^*\restriction\tau$ is the model we are looking for. ■

THEOREM 25. *Any compact $\stackrel{g}{\Longleftrightarrow}$ invariant logic $\mathcal{L}$ with the Tarski Union Property that extends GF is equivalent to GF itself.*

Proof. Assume $\mathcal{L}$ were more expressive than GF. Then there would be a signature τ and a formula $\varphi \in \mathcal{L}(\tau)$ that is not equivalent to any formula in $\mathrm{GF}(\tau)$. Since $\mathcal{L}$ is compact, there are two τ-structures $\mathfrak{M}, \bar{a}$ and $\mathfrak{N}, \bar{b}$ with parameters such that $\mathit{Th}_{\mathrm{GF}(\tau)}(\mathfrak{M}, \bar{a}) = \mathit{Th}_{\mathrm{GF}(\tau)}(\mathfrak{N}, \bar{b})$ yet $\mathfrak{M}, \bar{a} \models \varphi$ and $\mathfrak{N}, \bar{b} \models \neg\varphi$ (cf. proposition 14).

According to corollary 24 there are two GF-saturated structures $\mathfrak{M}^*$ and $\mathfrak{N}^*$ such that $\mathit{Th}_{\mathcal{L}(\tau)}(\mathfrak{M}, \bar{a}) = \mathit{Th}_{\mathcal{L}(\tau)}(\mathfrak{M}^*, \bar{a})$ and $\mathit{Th}_{\mathcal{L}(\tau)}(\mathfrak{N}, \bar{b}) = \mathit{Th}_{\mathcal{L}(\tau)}(\mathfrak{N}^*, \bar{b})$. It follows that $\mathit{Th}_{\mathrm{GF}(\tau)}(\mathfrak{M}^*, \bar{a}) = \mathit{Th}_{\mathrm{GF}(\tau)}(\mathfrak{N}^*, \bar{b})$. As both structures are GF-saturated, $\mathfrak{M}^*, \bar{a}$ and $\mathfrak{N}^*, \bar{b}$ are guarded bisimilar, by the Hennessy–Milner theorem. But by the $\mathcal{L}$-elementary nature of the extensions, still $\mathfrak{M}^*, \bar{a} \models \varphi$ while $\mathfrak{N}^*, \bar{b} \models \neg\varphi$, contradicting invariance under guarded bisimulation. ■

4 Concluding remarks

We have shown that ML[∀] and GF are maximally expressive logics whose semantics is invariant under the corresponding notion of bisimulation among compact logics with the Tarski Union Property (TUP).

The choice of TUP as a leading criterion may deserve some comment. Any choice of model theoretic criteria in a characterisation of the proposed kind has to be argued in the light of the question "*What constitutes a good Lindström characterisation?*". Clearly Lindström characterisations are very sensitive to the particular conditions imposed; the setting of the stage involves a critical choice as to *which competing logics* are admitted. While this may partly be a matter of taste or of tradition, it is also clear that a proposal is the more creditable, the wider the class of competitors is a priori, and the more fundamental the individual constraints are in the broader context of abstract model theory. There is no claim that the choices we made here are optimal in any sense. The following discussion is merely meant to indicate the setting in which this choice is being made, and thus points to some considerations that led us to favour the Tarski Union Property as a reasonably natural choice in a situation where compactness and basic semantic invariance conditions alone are at least not known to suffice to pin down the logics in question.

As pointed out in [4, 5], Lindström characterisations are closely related to semantic characterisation theorems in the tradition of classical preservation results. For basic modal logic, this companion/precursor is van Benthem's classical characterisation of ML as the bisimulation invariant fragment of first-order logic [3]; see [10, 9, 6] for a locality based account of, for instance, ML and ML[∀] as bisimulation invariant fragments of FO. The key difference is, of course, that in a Lindström characterisation we usually do not want to assume any a priori inclusion in some background logic, certainly not inclusion in FO. Modal logics like basic modal logic ML itself, or its extension with a global modality ML[∀], or the guarded fragment GF, are to be characterised not just *as fragments* of FO, but rather within the family of all logics that respect the same fundamental semantic invariance condition (bisimulation, global bisimulation, or guarded bisimulation invariance), including in particular candidate logics that are incomparable with FO. In this context, it is useful to recall Karp's theorem. All the semantic invariances considered concern equivalences that are

(a) bounded by partial isomorphism $\simeq_{\mathrm{part}}$.[3]

(b) game based in the sense that equivalence corresponds to the existence of a winning strategy in infinite plays of some Ehrenfeucht–Fraïssé type model theoretic game.[4]

By Karp's theorem, $\simeq_{\mathrm{part}}$ coincides with $\equiv_{\infty\omega}$, i.e., with equivalence in the

[3]We say that one equivalence relation is bounded by another, if the classes of the former are unions of classes of the latter, i.e., if the former can only be coarser than the latter.

[4]In the cases at hand: the ordinary bisimulation game, its variant with global moves, or the guarded bisimulation game.

infinitary logic $L_{\infty\omega}$, the extension of FO that allows infinite disjunctions and conjunctions rather than just finite ones. Therefore, (a) implies that equivalence $\equiv_{\mathcal{L}}$ for the candidate logics $\mathcal{L}$ is bounded by $\equiv_{\infty\omega}$.

Any infinitary game based equivalence as in (b), on the other hand, has natural finite approximations induced by existence of strategies in truncated games with a fixed finite number of rounds. By the corresponding variant of the classical Ehrenfeucht–Fraïssé theorem, these finitary approximate levels correspond – for finite vocabularies at least – to equivalences in fragments of the target logic that are finite up to logical equivalence and definable in FO. In the examples mentioned, these are the fragments of fixed finite nesting depths of the logics under consideration. In such a situation, a Hennessy–Milner–Karp connection between equivalence w.r.t. the common refinement of the finite approximation levels of the infinitary equivalence and full infinitary equivalence comes into focus. For basic modal logic, for instance, the classical Hennessy–Milner theorem tells us that equivalence in basic modal logic, i.e., equivalence w.r.t. all finite levels of n-bisimulation, guarantees full bisimulation equivalence for instance over the class of all finitely branching Kripke structures, and more generally over modally saturated Kripke structures.

Following an approach outlined in [11], one could sum up the key parameters in this setting as follows:

We consider a target logic $L = \bigcup_{\ell\in\omega} L_\ell \subseteq \mathrm{FO}$, stratified into syntactic levels L_ℓ such that equivalence w.r.t. L_ℓ has finite index and is captured by the ℓ-round game equivalence $\leftrightharpoons^\ell$, which is in FO for every fixed finite vocabulary. It follows that equivalence w.r.t. L, $\equiv_L$, is captured by the common refinement of all its finite approximations, $\leftrightharpoons^\omega := \bigcap_\ell \leftrightharpoons^\ell$. Moreover, $\leftrightharpoons^\omega$ coincides with $\leftrightharpoons$ at least over ω-saturated structures, since here 'good responses in every finite game' constitute (partial) types, whose realisations yield 'good responses for the infinite game'. (This reasoning is pursued in [11] in terms of first-order interpretations translating between the games for $\leftrightharpoons$ and the ordinary bisimulation game, also at the level of their finite approximations.)

We want to characterise L as maximal among certain well-behaved logics $\mathcal{L}$ whose semantics is invariant under $\leftrightharpoons$, the full infinitary game equivalence. Setting aside the issue of finite vocabularies (a finite occurrence property may be stipulated explicitly, or may be derivable from compactness assumptions), there are the following two essential hurdles in showing that L is maximally expressive among all $\leftrightharpoons$ invariant logics $\mathcal{L} \supseteq L$ satisfying some additional model theoretic criteria:

(i) the gap between $\leftrightharpoons^\omega$ (which is the same as $\equiv_L$) and $\leftrightharpoons$ (which we know to be a refinement of $\equiv_{\mathcal{L}}$, by the fundamental assumption of $\leftrightharpoons$ invariance of $\mathcal{L}$). Here Hennessy–Milner–Karp is useful: $\equiv_L$ ($\leftrightharpoons^\omega$) coincides with $\leftrightharpoons$ and hence with $\equiv_{\mathcal{L}}$ at least for ω-saturated structures.

(ii) the gap between showing that $\equiv_{\mathcal{L}}$ is bounded by $\equiv_L$ and showing that $\mathcal{L} \subseteq L$: this gap can typically be bridged by a compactness argument. Clearly compactness is a most natural criterion in the context of a

Lindström characterisation of a compact logic.

It follows that, in the light of (i), any additional assumption on $\mathcal{L}$ that guarantees the existence of $\mathcal{L}$-equivalent companions that are ω-saturated allows us to upgrade $\equiv_L$ to $\equiv_{\mathcal{L}}$. If F is some model transformation that preserves $\mathcal{L}$ and hence in particular $\equiv_L$ and produces ω-saturated companions, then $\mathfrak{M} \equiv_L \mathfrak{N}$ implies $F(\mathfrak{M}) \equiv_L F(\mathfrak{N})$, and hence $F(\mathfrak{M}) \equiv_{\mathcal{L}} F(\mathfrak{N})$, since $\leftrightarrows^{\omega}$ coincides with $\leftrightarrows$ in ω-saturated structures. So $\mathfrak{M} \equiv_{\mathcal{L}} F(\mathfrak{M}) \equiv_{\mathcal{L}} F(\mathfrak{N}) \equiv_{\mathcal{L}} \mathfrak{N}$ shows that $\mathfrak{M} \equiv_{\mathcal{L}} \mathfrak{N}$ for any $\mathfrak{M} \equiv_L \mathfrak{N}$.

$$
\begin{array}{ccccc}
\mathfrak{M} & \text{———} & \begin{array}{c}\leftrightarrows^{\omega}\\ \equiv_L\end{array} & \text{———} & \mathfrak{N} \\
\Big| & & & & \Big| \\
\equiv_{\mathcal{L}} & & & & \equiv_{\mathcal{L}} \\
\downarrow & & & & \downarrow \\
F(\mathfrak{M}) & \text{———} & \begin{array}{c}\leftrightarrows^{\omega}\\ \leftrightarrows\\ \equiv_{\mathcal{L}}\end{array} & \text{———} & F(\mathfrak{N})
\end{array}
$$

Clearly, ω-saturation can, in typical concrete instances, be replaced by weaker, specifically adapted notions of saturation (for instance, modal saturation suffices in the case of basic modal logic and bisimulation). Indeed, we here used the Tarski Union Property and compactness to establish the availability of not necessarily ω-saturated companions, but companions that are 'sufficiently saturated in the sense of $\leftrightarrows$', as limits of suitable elementary chains. Here 'sufficiently saturated in the sense of $\leftrightarrows$' really was sufficiency for the purpose of a Hennessy–Milner argument for the passage from $\leftrightarrows^{\omega}$ to $\leftrightarrows$.

Of course, other model theoretic criteria can serve the same purpose. In particular, in the spirit of de Rijke [14], preservation of $\mathcal{L}$ under (countable) ultrapowers produces ω-saturated (even ω_1-saturated) companions and hence immediately clinches the argument. If moreover, preservation under ultraproducts is assumed, then even compactness follows and needs not be stipulated separately. But even just preservation under countable ultrapowers immediately shows that $\equiv_{\mathcal{L}}$ is bounded by elementary equivalence. Since passage to ultrapowers in this case preserves both $\mathcal{L}$ and FO, it also gives us an upgrading of $\equiv$ to $\simeq_{\text{part}}$ and hence to $\equiv_{\mathcal{L}}$. In a sense, therefore, preservation assumptions of this calibre go some way in reducing Lindström characterisations to the semantic characterisations of fragments of FO in the style of a classical preservation theorem.

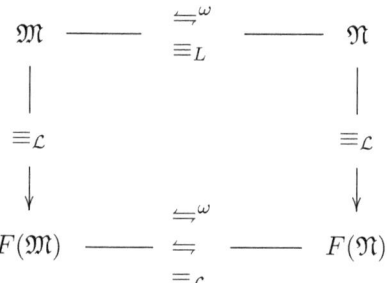

Our Lindström characterisations are meant to be different in this respect: we did not want to assume any a priori guarantees for the logics $\mathcal{L}$ under consideration to be fragments of FO or even for $\equiv_{\mathcal{L}}$ to be bounded by elementary equivalence. (As pointed out above, though, all equivalences considered are bounded by $\simeq_{\mathrm{part}}$ and hence by $\equiv_{\infty\omega}$.) Nevertheless, our arguments are also based on an upgrading of $\equiv_L$ ($\leftrightarrows^{\omega}$) to $\equiv_{\mathcal{L}}$ ($\leftrightarrows$) in suitably saturated companion structures $F(\mathfrak{M}) \equiv_{\mathcal{L}} \mathfrak{M}$, but since our model transformations, unlike ultrapower constructions, do not guarantee $F(\mathfrak{M}) \equiv \mathfrak{M}$, the same argument does not upgrade elementary equivalence $\equiv$ to $\equiv_{\mathcal{L}}$. In other words, $\mathcal{L}$-equivalence is only seen to be coarser than elementary equivalence a posteriori, because the target logic L happens to be a fragment of FO.

As already mentioned, it remains open whether a characterisation in the style of van Benthem's [4] is also available for ML[∀] or GF. We do not even know whether compactness, the appropriate notion of bisimulation invariance and possibly the relativisation property (and/or some other innocuous condition) might imply TUP.

BIBLIOGRAPHY

[1] H. Andréka, J. van Benthem, and I. Németi: Modal languages and bounded fragments of predicate logic. Journal of Philosophical Logic, 27(3), pp. 217-274, 1998.

[2] J. Barwise and S. Feferman (eds.): Model-Theoretic Logics. Springer, 1985.

[3] J. van Benthem: Modal Logic and Classical Logic. Bibliopolis, Napoli, 1983.

[4] J. van Benthem: A new modal Lindström theorem. Logica Universalis, 1(1), pp. 125-138, 2007.

[5] J. van Benthem, B. ten Cate, and J. Väänänen: Lindström theorems for fragments of first-order logic. Proceedings of the 22nd IEEE Symp. on Logic in Computer Science (LICS 07), pp. 280-292, 2007.

[6] V. Goranko and M. Otto: Model theory of modal logic, in Handbook of Modal Logic, P. Blackburn et al. (eds.), Elsevier, 2006.

[7] E. Grädel: On the restraining power of guards. Journal of Symbolic Logic, 64, pp. 1719-1742, 1999.

[8] E. Grädel, C. Hirsch, and M. Otto: Back and forth between guarded and modal logics. ACM Transactions on Computational Logic, 3, pp. 418-463, 2002.

[9] M. Otto: Modal and guarded characterisation theorems over finite transition systems. Annals of Pure and Applied Logic, 130, pp. 173-205, 2004.

[10] M. Otto: Elementary proof of the van Benthem–Rosen characterisation theorem. Technical report 2342, Fachbereich Mathematik, Technische Universität Darmstadt, 2004.

[11] M. Otto: Model theoretic methods for fragments of FO and special classes of (finite) structures. Draft of survey on the basis of a tutorial given at the 2006 Durham workshop on Finite and Algorithmic Model Theory (Isaac Newton Institute, Logic and Algorithms, 2006), 2007.

[12] R. Piro: Lindströmsche Sätze für modale Logiken. Diploma thesis, Technische Universität Darmstadt, 2008.

[13] M. de Rijke: Extending Modal Logic. Dissertation, Institute for Logic, Language and Computation, University of Amsterdam, 1993.

[14] M. de Rijke: A Lindström theorem for modal logic. In: Modal Logic and Process Algebra – A Bisimulation Perspective, A. Ponse, M. de Rijke and Y. Venema (eds), CSLI Publications, 1995, pp. 217–230.

Martin Otto and Robert Piro
Department of Mathematics
Technische Universität Darmstadt
64289 Damstadt, Germany
otto@mathematik.tu-darmstadt.de and robert.piro@gmx.de

PSPACE-decidability of Japaridze's polymodal logic

Ilya Shapirovsky

ABSTRACT. In this paper we prove that Japaridze's Polymodal Logic is PSPACE-decidable. To show this, we describe a decision procedure for satisfiability on *hereditarily ordered frames* that can be applied to obtain upper complexity bounds for various modal logics.

Keywords: Japaridze's polymodal Logic, computational complexity, conditional satisfiability

1 Introduction

In this paper we investigate the complexity of well-known *propositional polymodal provability logic* GLP. This logic was introduced in [8], and now plays a significant role in proof theory (see [4, 1]). In [8, 7], the logic GLP was proved to be decidable, the question about its complexity was left open. The PSPACE-decidability of GLP was conjectured in the recent paper [2].

The logic GLP is known to be Kripke-incomplete. In [2], it was shown that GLP is polynomial-time reducible to a logic J with an explicit Kripke semantics: J is characterized by a class of finite *hereditary orders*. This class is defined as follows: a strict partial order is a hereditary order; a strictly ordered (by a new relation, which is also a strict partial order) set of hereditary orders is a hereditary order.

This paper proves the PSPACE-decidability of J. We propose a technique allowing us to check modal satisfiability on frames obtained by "hereditarily ordering". This approach seems to be applicable to a large class of transitive logics: in section 4 we give semantical conditions, sufficient for PSPACE-decidability (Theorems 21 and 22 for the monomodal case, Theorem 35 for the multi-modal case).

The paper is organized as follows. Section 2 introduces some standard notions and notations. In section 3 we describe some truth-preserving transformations for ordered sets of frames. In section 4, we introduce a notion of *conditional satisfiability*, and show, how it can be applied to obtain decision procedures for satisfiability on hereditarily ordered frames. First we formulate it for the monomodal case and then generalize for the multi-modal case. In section 5, we apply the described technique to obtain a PSPACE-decision procedure for J, and thus for GLP.

© 2008, the author(s).

2 Preliminaries

We consider *propositional normal modal logics* with finitely or countably many modalities. Modal formulas are built using the connectives $\bot$ (*false*), $\to$ (*implication*), a countable set of unary connectives $\Diamond_1, \Diamond_2, \dots$ (*diamonds*) and a countable set of *propositional variables* $PV = \{p_1, p_2, \dots\}$. All other connectives are defined in the standard way, in particular $\Box_i\psi = \neg\Diamond_i\neg\psi$. An N-formula is a formula that contains only connectives $\Diamond_1, \dots, \Diamond_N, \to, \bot$.

An N-frame F is a tuple $(W, R_1, \dots, R_N)$, where $W \neq \varnothing$, $R_1, \dots, R_N \subseteq W \times W$; $R_1, \dots, R_N$ are called *accessability relations of* F.

In this paper we assume that all considered accessability relations are *transitive*.

An N-model M over a frame F is a pair (F, θ), where $\theta : PV \to 2^W$. The notations $w \in \mathsf{M}$, $w \in \mathsf{F}$ mean $w \in W$.

A *weak submodel* M' of M is a model $((W', R'_1, \dots, R'_N), \eta)$, such that $W' \subseteq W$, $R'_1 \subseteq R_1, \dots, R'_N \subseteq R_N$, and $\eta(p) = \theta(p) \cap W'$ for any $p \in PV$.

For $R \subseteq W \times W$, $V \subseteq W$, by $R|V$ we denote the restriction R to V: $R|V = R \cap (V \times V)$. For an N-frame $\mathsf{F} = (W, R_1, \dots, R_N)$, by $\mathsf{F}|V$ we denote the *restriction* F *to* V: $\mathsf{F}|V = (V, R_1|V, \dots, R_N|V)$. If M is a model over F, and G is the restriction F to V, then the submodel of M over G is called the *restriction* M *to* V *(to* G*)*, in symbols, $\mathsf{M}|\mathsf{G}$ or $\mathsf{M}|V$.

The *trues of a formula at a point in a model*, and also the *validity of a formula in a frame* (*in a class of frames*) are defined in the standard way, see e.g. [3]; in symbols, $\mathsf{M}, w \vDash \varphi$ means that φ is true at w in M, $\mathsf{F} \vDash \varphi$ means that φ is valid in F. Also, for a set of formulas Ψ, $\mathsf{F} \vDash \Psi$ means $\mathsf{F} \vDash \varphi$ for any $\varphi \in \Psi$.

For an N-frame F, an N-formula φ is *satisfiable in* F (or F-*satisfiable*), if φ is true at some point of a model over F. For a class of frames $\mathcal{F}$, φ is *satisfiable in* $\mathcal{F}$ (or $\mathcal{F}$-*satisfiable*), if φ is F-satisfiable for some $\mathsf{F} \in \mathcal{F}$. For a logic L, φ is L-*satisfiable*, if φ is F-satisfiable for some $\mathsf{F} \vDash \mathrm{L}$.

As usual, a *cluster* in a frame (W, R) is an $\sim_R$-equivalence class, where $\sim_R \; = (R \cap R^{-1}) \cup \{(w, w) \mid w \in W\}$. Also, by a *cluster* we mean a frame $\mathsf{F} = (W, W \times W)$, or a frame $(\{w\}, \varnothing)$ (*degenerate cluster*). For $n \geq 1$, C_n denotes the n-element cluster $(W_n, W_n \times W_n)$, where $W_n = \{1, \dots, n\}$; $\mathsf{C}_0 = (\{0\}, \varnothing)$.

For a frame $\mathsf{F} = (W, R)$, $w \in W$, put $R(w) = \{w' \mid wRw'\}$. If $W = R(w) \cup \{w\}$, then F is a *cone* (or *rooted frame*), and w is called a *root* of F.

For N-frames F and G, the notation $g : \mathsf{F} \twoheadrightarrow \mathsf{G}$ means that g is a p-morphism from F onto G; $\mathsf{F} \twoheadrightarrow \mathsf{G}$ means that $g : \mathsf{F} \twoheadrightarrow \mathsf{G}$ for some g. Recall that if $\mathsf{F} \twoheadrightarrow \mathsf{G}$ then any G-satisfiable formula is F-satisfiable (see e.g. [3]).

Let us recall the notion of *selective filtration.*

DEFINITION 1. Let M be an N-model, Ψ a set of N-formulas closed under subformulas. A weak submodel M' of M is called a *selective filtration of* M *through* Ψ, if for any $w \in \mathsf{M}'$, for any formula ψ, for all $i = 1, \dots, N$, we

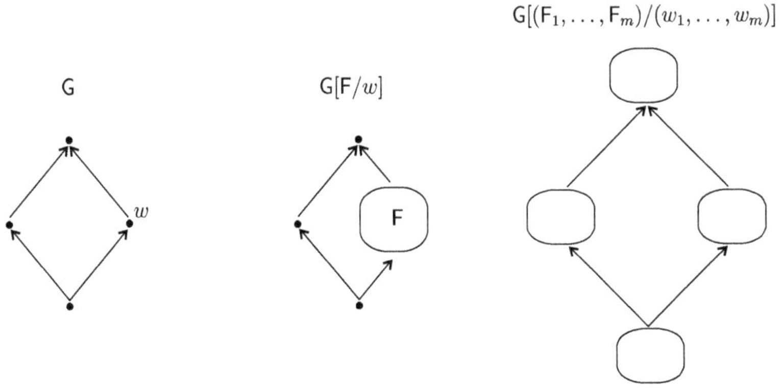

Figure 1.

have

$$\Diamond_i \psi \in \Psi \ \& \ \mathsf{M}, w \vDash \Diamond_i \psi \ \Rightarrow \ \exists u \in R'_i(x) \ \mathsf{M}, u \vDash \psi,$$

where $R'_1, \dots, R'_N$ are the the accessability relations of M'.

LEMMA 2. *If* M' *is a selective filtration of* M *through* Ψ*, then for any* $w \in \mathsf{M}'$*, for any* $\psi \in \Psi$*, we have*

$$\mathsf{M}, w \vDash \psi \Leftrightarrow \mathsf{M}', w \vDash \psi.$$

For a modal formula φ, $Sub(\varphi)$ denotes the set of all subformulas of φ, $\langle \varphi \rangle$ denotes the cardinality of $Sub(\varphi)$.

On the set of all modal formulas we fix a linear order $\lessdot$, such that for any ϕ, ψ, if $\phi \in Sub(\psi)$ then $\phi \lessdot \psi$. For a set of formulas Ψ, let $\Psi_{\lessdot}$ denote the list of elements of Ψ ordered by $\lessdot$. If $\Psi_{\lessdot} = (\psi_1, \dots, \psi_n)$, then for a boolean vector $\mathbf{v} = (v_1, \dots, v_n) \in \{0,1\}^n$ we put

$$\Psi_{\mathbf{v}} = \{\psi_i \mid v_i = 1, \ 1 \leq i \leq n\}, \quad \Psi^{\mathbf{v}} = \bigwedge_{1 \leq i \leq n} \psi_i^{v_i},$$

where $\psi^1 = \psi$ and $\psi^0 = \neg\psi$.

3 Partially ordered sets of frames

It is well-known that any transitive frame can be viewed as a set of clusters ordered by a transitive and antisymmetric relation (*skeleton*, see e.g. [6]). The following construction allows us to consider arbitrary frames instead of clusters.

DEFINITION 3. Let $\mathsf{G} = (W, R)$ be a finite (strict or non-strict) partial order, $m = |W|$, $W = \{w_1, \dots, w_m\}$.

For frames $\mathsf{F}_1 = (V_1, S_1), \dots, \mathsf{F}_m = (V_m, S_m)$, we define the frame $\mathsf{G}[(\mathsf{F}_1, \dots, \mathsf{F}_m)/(w_1, \dots, w_m)] = (\overline{W}, \overline{R})$ obtained by *replacing points*

$w_1, \ldots, w_m$ *with frames* $\mathsf{F}_1, \ldots, \mathsf{F}_m$ (Fig. 1):

$$\overline{W} = (\{w_1\} \times V_1) \cup \cdots \cup (\{w_m\} \times V_m),$$

$$(w', v')\overline{R}(w'', v'') \Leftrightarrow (w' \neq w'' \ \&\ w'Rw'') \ \text{ or } \ (w' = w'' = w_i \ \&\ v'S_iv'').$$

Also, for $w \in W$ and a frame $\mathsf{F} = (V, S)$, we define the frame $\mathsf{G}[\mathsf{F}/w] = (W', R')$ obtained by *replacing* w *with* F (Fig. 1):

$$W' = (W - \{w\}) \cup V', \text{ where } V' = \{w\} \times V,$$

$$R' = R|(W - \{w\}) \cup \{((w, u'), (w, u'')) \mid u'Su''\} \cup$$
$$\cup (V' \times (R(w) - \{w\})) \cup ((R^{-1}(w) - \{w\}) \times V').$$

For a class $\mathcal{F}$ of monomodal frames, we put

$$\mathsf{G}[\mathcal{F}] = \{\mathsf{G}[(\mathsf{F}_1, \ldots, \mathsf{F}_m)/(w_1, \ldots, w_m)] \mid \mathsf{F}_1, \ldots, \mathsf{F}_m \in \mathcal{F}\}.$$

Finally, for a class $\mathcal{G}$ of finite partial orders, we put

$$\mathcal{G}[\mathcal{F}] = \bigcup \{\mathsf{G}[\mathcal{F}] \mid \mathsf{G} \in \mathcal{G}\},$$

i.e. $\mathcal{G}[\mathcal{F}]$ is the class of frames, obtained from frames that belong to $\mathcal{G}$ by replacing all their points with frames from $\mathcal{F}$.

REMARK 4. By a straightforward argument,

$$\mathsf{G}[(\mathsf{F}_1, \ldots, \mathsf{F}_m)/(w_1, \ldots, w_m)] = \mathsf{G}[\mathsf{F}_1/w_1] \ldots [\mathsf{F}_m/w_m].$$

Note that the frame $\mathsf{G}[\mathsf{F}/w]$ is transitive due to the transitivity of the relations R and S. Thus all frames that described in the above definition are transitive.

Note also that if G is strict, and G' is the corresponding non-strict partial order, then

$$\mathsf{G}[(\mathsf{F}_1, \ldots, \mathsf{F}_m)/(w_1, \ldots, w_m)] = \mathsf{G}'[(\mathsf{F}_1, \ldots, \mathsf{F}_m)/(w_1, \ldots, w_m)].$$

EXAMPLE 5. Let $\mathcal{PO}$ denote the class of all finite non-strict partial orders.

If $\mathcal{F} = \{\mathsf{C}_0\}$, then $\mathcal{PO}[\mathcal{F}]$ is the class of all finite strict partial orders, up to isomorphisms.

If $\mathcal{F}$ is the class of all finite (non-degenerate) clusters, then $\mathcal{PO}[\mathcal{F}]$ is the class of all finite transitive (and reflexive) frames, up to isomorphisms.

Let us generalize the above construction for the multi-modal case.

DEFINITION 6. Let $\mathsf{G} = (W, R) \in \mathcal{PO}$, $m = |W|$, $W = \{w_1, \ldots, w_m\}$.

For an N-frame $\mathsf{F} = (V, S_1, \ldots, S_N)$, $1 \leq k \leq N$, we define the frame $\mathsf{G}[k; \mathsf{F}/w] = (W', R'_1, \ldots R'_N)$ as follows. To define W' and R'_k, we put $(W', R'_k) = \mathsf{G}[(V, S_k)/w]$; for $l \neq k$ we put

$$R'_l = \{((w, u'), (w, u'')) \mid u', u'' \in V, \ u'S_lu''\}.$$

For a class of N-frames $\mathcal{F}$, $1 \leq k \leq N$, we put

$$\mathsf{G}[k;\mathcal{F}] = \{\mathsf{G}[k;\mathsf{F}_1/w_1]\dots[k;\mathsf{F}_m/w_m] \mid \mathsf{F}_1,\dots,\mathsf{F}_m \in \mathcal{F}\},$$

and for a class $\mathcal{G}$ of finite partial orders,

$$\mathcal{G}[k;\mathcal{F}] = \bigcup\{\mathsf{G}[k;\mathcal{F}] \mid \mathsf{G} \in \mathcal{G}\}.$$

PROPOSITION 7. *Let* $\mathsf{G} \in \mathcal{PO}$, $w \in \mathsf{G}$, F *and* F' *be* N*-frames,* $1 \leq k \leq N$. *Then* $\mathsf{F}' \twoheadrightarrow \mathsf{F}$ *implies* $\mathsf{G}[k;\mathsf{F}'/w] \twoheadrightarrow \mathsf{G}[k;\mathsf{F}/w]$.

Proof. Let $g : \mathsf{F}' \twoheadrightarrow \mathsf{F}$. The required p-morphism g' is defined as follows. For $v \in \mathsf{F}'$ put $g'(w,v) = (w, g(v))$; for $w' \in \mathsf{G} - \{w'\}$ put $g'(w') = w'$. ■

Let $\mathsf{F} \in \mathcal{PO}$. By $Ht(\mathsf{F})$ we denote the height of F, i.e., the maximal length of strictly ascending chains in F (by length of a chain we mean the number of its elements); by branching of a point $w \in \mathsf{F}$ we mean the number of immediate successors of w in F; $Br(\mathsf{F})$ denotes the branching of F, i.e., the maximal branching of its points.

By a *tree* we mean a rooted non-strict partial order (W,R) such that $R^{-1}(w)$ is a chain for every w. By $\mathcal{T}$ we denote the class of all finite trees. By $\mathcal{T}_{n,b}$ we denote the class of trees with the height not more then h and the branching not more then b:

$$\mathcal{T}_{h,b} = \{\mathsf{T} \in \mathcal{T} \mid Ht(\mathsf{T}) \leq h,\ Br(\mathsf{T}) \leq b\}.$$

Let us recall the notions of *disjoint sum* (or *disjoint union*) and *ordinal sum* of frames. Suppose that frames $\mathsf{F}_1 = (W_1,R_1)$ and $\mathsf{F}_2 = (W_2,R_2)$ have no common points. Put

$\mathsf{F}_1 \sqcup \mathsf{F}_2 = (W_1 \cup W_2, R_1 \cup R_2)$	*disjoint sum of* F_1 *and* F_2,
$\mathsf{F}_1 + \mathsf{F}_2 = (W_1 \cup W_2, R_1 \cup (W_1 \times W_2) \cup R_2)$	*ordinal sum of* F_1 *and* F_2.

If M is a model over the frame $\mathsf{F}_1 \sqcup \mathsf{F}_2$ (over the frame $\mathsf{F}_1 + \mathsf{F}_2$), and $\mathsf{M}_1 = \mathsf{M}|W_1$, $\mathsf{M}_2 = \mathsf{M}|W_2$, then M is called the *disjoint (ordinal) sum of models* M_1 and M_2, in symbols: $\mathsf{M} = \mathsf{M}_1 \sqcup \mathsf{M}_2$ ($\mathsf{M} = \mathsf{M}_1 + \mathsf{M}_2$).

REMARK 8. By sum of frames that have common points, we mean sum of their isomorphic copies:

$$\mathsf{F}_1 \sqcup \mathsf{F}_2 = \mathsf{G}_\sqcup[(\mathsf{F}_1,\mathsf{F}_2)/(1,2)], \text{ where } \mathsf{G}_\sqcup = (\{1,2\},\varnothing);$$

$$\mathsf{F}_1 + \mathsf{F}_2 = \mathsf{G}_+[(\mathsf{F}_1,\mathsf{F}_2)/(1,2)], \text{ where } \mathsf{G}_+ = (\{1,2\},\{(1,2)\}).$$

Disjoint sum of N-frames is defined analogously (see e.g. [3]). Let us modify the notion of ordinal sum for the multi-modal case.

DEFINITION 9. Let $\mathsf{F} = (W,R_1,\dots,R_N)$ and $\mathsf{F}' = (W',R'_1,\dots,R'_N)$ be N-frames, $1 \leq k \leq N$. Put $\mathsf{F} +_k \mathsf{F}' = (V,S_1,\dots,S_N)$, where $(V,S_k) = (W,R_k) + (W',R'_k)$, and $(V,S_l) = (W,R_l) \sqcup (W',R'_l)$ for $l \neq k$.

The above definitions imply the following

PROPOSITION 10. *Let $\mathcal{F}$ be a class of N-frames, $1 \leq k \leq N$, and let $\mathsf{G} \in \mathcal{T}_{h+1,b}[k;\mathcal{F}]$ for some $h, b \geq 1$. Then G is either isomorphic to a frame $\mathsf{F} \in \mathcal{F}$ or isomorphic to a frame $\mathsf{F} +_k (\mathsf{G}_1 \sqcup \cdots \sqcup \mathsf{G}_{b'})$, where $1 \leq b' \leq b$, $\mathsf{F} \in \mathcal{F}$, $\mathsf{G}_1, \ldots \mathsf{G}_{b'} \in \mathcal{T}_{h,b}[k;\mathcal{F}]$.*

Proof. For some $\mathsf{T} \in \mathcal{T}_{h+1,b}$, we have $\mathsf{G} \in \mathsf{T}[k;\mathcal{F}]$. Then either T is a singleton (when $Ht(\mathsf{T}) = 1$), and in this case G is isomorphic to a frame $\mathsf{F} \in \mathcal{F}$, or T is isomorphic to a frame $\mathsf{C}_1 + (\mathsf{T}_1 \sqcup \cdots \sqcup \mathsf{T}_{b'})$, where b' is the branching at the root of T and $\mathsf{T}_1, \ldots \mathsf{T}_{b'} \in \mathcal{T}_{h,b}$. ∎

LEMMA 11. *Let $\mathcal{F}$ be a class of N-frames, $1 \leq k \leq N$, G be a finite rooted partial order. Then for any $\mathsf{H} \in \mathsf{G}[k;\mathcal{F}]$ there exists a tree $\mathsf{T} \in \mathcal{T}$ such that for some $\mathsf{H}' \in \mathcal{T}[k;\mathcal{F}]$ we have $\mathsf{H}' \twoheadrightarrow \mathsf{H}$.*

Proof. By the standard unravelling argument. Let w_0 be the root of G. To define $\mathsf{T} = (W, R)$, put

$$W = \{(w_0, \ldots, w_k) \mid w_0, \ldots, w_k \in W,\ \ w_{i+1} \text{ is an immediate successor of } w_i \text{ for all } i = 0, \ldots, k-1\};$$

$$(w_0, \ldots, w_k) R (w_0, \ldots, w_l) \Leftrightarrow (w_0, \ldots, w_k) \text{ is a prefix of } (w_0, \ldots, w_l).$$ ∎

It is well-known that any K4-satisfiable formula φ is satisfiable in some finite frame with the height and the branching of its skeleton not more then $\langle\varphi\rangle$ (see e.g. [6]). The following lemma generalizes this observation.

LEMMA 12. *Let $\mathcal{F}$ be a class of N-frames, $1 \leq k \leq N$. If an N-formula φ is $\mathcal{PO}[k;\mathcal{F}]$-satisfiable, then φ is $\mathcal{T}_{\langle\varphi\rangle,\langle\varphi\rangle}[k;\mathcal{F}]$-satisfiable.*

Proof. By Lemma 11, φ is satisfiable in a frame $\mathsf{H} \in \mathsf{T}[k;\mathcal{F}]$, where $\mathsf{T} \in \mathcal{T}$. Then $\mathsf{M}, (w_0, v) \vDash \varphi$, where M is a model over H, $w_0 \in \mathsf{T}$, $(w_0, v) \in \mathsf{H}$.

Let $\mathsf{T} = (W, R)$. For a point $w \in W$, put

$$\Psi_w = \{\Diamond_k \psi \in Sub(\varphi) \mid \mathsf{M}, (w, v) \vDash \Diamond_k \psi \text{ for some } (w, v) \in \mathsf{H}\}.$$

Inductively we define a set W_i. Put $W_0 = \{w_0\}$, $\Psi_i = \bigcup\{\Psi_w \mid w \in W_i\}$.

If $\Psi_i \neq \varnothing$, we define W_{i+1}. First, for every $w \in W_i$ we define a set U_w: if $\Psi_w = \varnothing$, we put $U_w = \varnothing$; for $\Psi_w = \{\Diamond_k \psi_1, \ldots, \Diamond_k \psi_l\}$, put $U_w = \{u_1, \ldots, u_l\}$, where u_i is an R-maximal point in the set

$$\{u \mid u \in R(w),\ \mathsf{M}, (u, v) \vDash \psi_i \text{ for some } (u, v) \in \mathsf{M}\}.$$

Put $W_{i+1} = \bigcup\{U_w \mid w \in W_i\}$.

Note that $|\Psi_{i+1}| < |\Psi_i|$, thus for some $l < \langle\varphi\rangle$ we obtain $\Psi_l = \varnothing$. Then we put $W' = W_0 \cup \cdots \cup W_l$, $V' = \{(w, v) \in \mathsf{H} \mid w \in W'\}$, and put $\mathsf{T}' = \mathsf{T}|W'$, $\mathsf{M}' = \mathsf{M}|V'$, $\mathsf{H}' = \mathsf{H}|V'$. Due to the construction, $\mathsf{T}' \in \mathcal{T}_{\langle\varphi\rangle,\langle\varphi\rangle}$ and $\mathsf{H}' \in \mathcal{T}_{\langle\varphi\rangle,\langle\varphi\rangle}[k;\mathcal{F}]$. Also, M' is a selective filtration of M through $Sub(\varphi)$, thus φ is $\mathcal{T}_{\langle\varphi\rangle,\langle\varphi\rangle}[k;\mathcal{F}]$-satisfiable. ∎

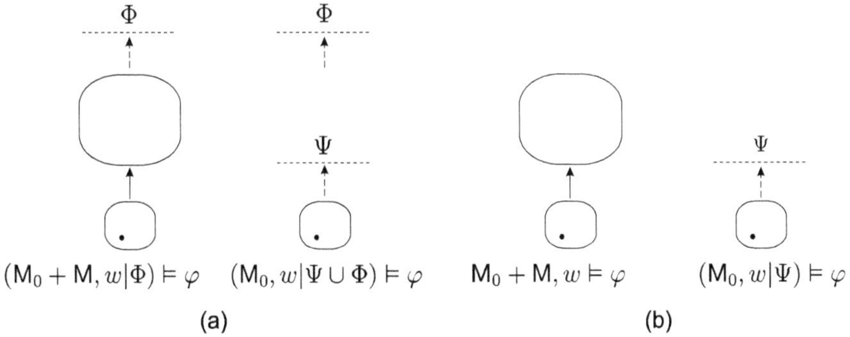

Figure 2.

4 Conditional satisfiability

4.1 Monomodal case

Is this subsection we assume that all frames, models and formulas are monomodal.

DEFINITION 13. Let M be a model, Ψ be a set of formulas. For a formula φ and a point $w \in \mathsf{M}$, we define the truth-relation $(\mathsf{M}, w|\Psi) \models \varphi$:

$$\begin{array}{lcl} (\mathsf{M}, w|\Psi) \models p & \Leftrightarrow & \mathsf{M}, w \models p \\ (\mathsf{M}, w|\Psi) \not\models \bot & & \\ (\mathsf{M}, w|\Psi) \models \varphi \to \psi & \Leftrightarrow & (\mathsf{M}, w|\Psi) \not\models \varphi \text{ or } (\mathsf{M}, w|\Psi) \models \psi \\ (\mathsf{M}, w|\Psi) \models \Diamond\varphi & \Leftrightarrow & \varphi \in \Psi \text{ or } \Diamond\varphi \in \Psi \text{ or} \\ & & \text{for some } v \in R(w) \text{ we have } (\mathsf{M}, v|\Psi) \models \varphi, \end{array}$$

where R is the accessability relation in M.

We read $(\mathsf{M}, w|\Psi) \models \varphi$ as "*φ is true at w in M under the condition Ψ*".

Note that $(\mathsf{M}, w|\varnothing) \models \varphi \Leftrightarrow \mathsf{M}, w \models \varphi$.

PROPOSITION 14. *Consider models M_0, M, their ordinal sum $\mathsf{M}_0 + \mathsf{M}$, and a set of formulas Φ (Fig. 2,a). If*

$$\Psi = \{\psi \in Sub(\varphi) \mid (\mathsf{M}, v|\Phi) \models \psi \textit{ for some } v\},$$

then for any formula φ, $w \in \mathsf{M}_0$,

$$(\mathsf{M}_0 + \mathsf{M}, w|\Phi) \models \varphi \Leftrightarrow (\mathsf{M}_0, w|\Psi \cup \Phi) \models \varphi.$$

Proof. The proof is straightforward, by induction on the length of φ. Consider only the case $\varphi = \Diamond\psi$, $\psi \notin \Phi$, $\Diamond\psi \notin \Phi$.

Suppose $(\mathsf{M}_0 + \mathsf{M}, w|\Phi) \models \Diamond\psi$. Then $(\mathsf{M}_0 + \mathsf{M}, v|\Phi) \models \psi$ for some $v \in R(w)$, where R is the accessability relation of $\mathsf{M}_0 + \mathsf{M}$. If $v \in \mathsf{M}$, then $\psi \in \Psi$; if $v \in \mathsf{M}_0$, then $(\mathsf{M}_0, v|\Psi \cup \Phi) \models \psi$ by the induction hypothesis; in both cases $(\mathsf{M}_0, w|\Psi \cup \Phi) \models \Diamond\psi$.

Suppose $(\mathsf{M}_0, w|\Psi\cup\Phi) \vDash \Diamond\psi$. If $\psi \in \Psi$ or $\Diamond\psi \in \Psi$, then $(\mathsf{M}, v|\Phi) \vDash \psi\vee\Diamond\psi$ for some $v \in \mathsf{M}$. Thus $(\mathsf{M}_0 + \mathsf{M}, w|\Phi) \vDash \Diamond\psi$. ∎

COROLLARY 15. *Consider models* M_0 *and* M. *For any formula* φ, $w \in \mathsf{M}_0$, *we have (Fig. 2,b)*

$$\mathsf{M}_0 + \mathsf{M}, w \vDash \varphi \Leftrightarrow (\mathsf{M}_0, w|\{\psi \in Sub(\varphi) \mid \mathsf{M}, v \vDash \psi \textit{ for some } v\}) \vDash \varphi.$$

DEFINITION 16. Let F be a cone, Ψ be a set of formulas. We say that φ is F-*satisfiable under the condition* Ψ, if φ is true at a root of F in some model over F under the condition Ψ. For a formula φ and vectors $\mathbf{v}, \mathbf{u} \in \{0,1\}^{\langle\varphi\rangle}$, the notation

$$\mathsf{F} \mid \mathbf{u} \Vdash_\varphi \mathbf{v}$$

means that the formula $Sub(\varphi)^{\mathbf{v}}$ is F-satisfiable under the condition $Sub(\varphi)_{\mathbf{u}}$. For a class $\mathcal{F}$ of cones, $\mathcal{F} \mid \mathbf{u} \Vdash_\varphi \mathbf{v}$ means that $\mathsf{F} \mid \mathbf{u} \Vdash_\varphi \mathbf{v}$ for some $\mathsf{F} \in \mathcal{F}$.

The following constructions are generalization of the construction proposed in [10].

DEFINITION 17. For a positive integer d, a sequence $(\mathcal{F}_n)_{n\in\mathbb{N}}$ of sets of cones is called d-*moderate*, if there exists an algorithm such that for any formula φ and any vectors $\mathbf{u}, \mathbf{v} \in \{0,1\}^{\langle\varphi\rangle}$ it decides whether

$$\mathcal{F}_{\langle\varphi\rangle} \mid \mathbf{u} \Vdash_\varphi \mathbf{v}$$

in space $O(\langle\varphi\rangle^d)$. A sequence $(\mathcal{F}_n)_{n\in\mathbb{N}}$ is *moderate*, if it is d-moderate for some integer d.

EXAMPLE 18. It is clear, that if a sequence $(\mathcal{F}_n)_{n\in\mathbb{N}}$ can be effectively described in polynomial of n space, then it is moderate. In particular, if $(\mathcal{F}_n)_{n\in\mathbb{N}}$ is a sequence of finite sets of finite cones, such that for some k $\mathcal{F}_k = \mathcal{F}_{k+1} = \dots$, then $(\mathcal{F}_n)_{n\in\mathbb{N}}$ is moderate. For instance, $(\mathcal{F}_n)_{n\in\mathbb{N}}$ is moderate, if:

- $\mathcal{F}_n$ is the set of all (non-degenerate) clusters with cardinality not more then n: for all n $\mathcal{F}_n = \{\mathsf{C}_0, \dots, \mathsf{C}_n\}$ or for all n $\mathcal{F}_n = \{\mathsf{C}_1, \dots, \mathsf{C}_n\}$;
- $\mathcal{F}_n$ consists of a single frame which is a singleton: for all n $\mathcal{F}_n = \{\mathsf{C}_0\}$ or for all n $\mathcal{F}_n = \{\mathsf{C}_1\}$.

Next we show that tree-like structures "constructed" from moderate sequences are also moderate.

For boolean vectors $\mathbf{u}, \mathbf{v}$ of the same length, let $\mathbf{u}\vee\mathbf{v}$ denote their bitwise disjunction.

PROPOSITION 19. *Let* $\mathcal{F}$ *be a class of cones. Then for any formula* φ, *for any* $\mathbf{u}, \mathbf{v} \in \{0,1\}^{\langle\varphi\rangle}$, *for any integers* $h, b \geq 1$, *the following two conditions are equivalent.*

1. $\mathcal{T}_{h+1,b}[\mathcal{F}] \mid \mathbf{u} \Vdash_\varphi \mathbf{v}$.

2. *Either* $\mathcal{F} \mid \mathbf{u} \Vdash_\varphi \mathbf{v}$, *or for some vectors* $\mathbf{v}_1, \dots, \mathbf{v}_{b'} \in \{0,1\}^{\langle\varphi\rangle}$, *where* $1 \leq b' \leq b$, *we have:* $\mathcal{T}_{h,b}[\mathcal{F}] \mid \mathbf{u} \Vdash_\varphi \mathbf{v}_j$ *for all* $j = 1, \dots, b'$, *and* $\mathcal{F} \mid \mathbf{u} \vee \mathbf{v}_1 \cdots \vee \mathbf{v}_{b'} \Vdash_\varphi \mathbf{v}$.

Proof. Put $\Phi = Sub(\varphi)$.

(1 $\Rightarrow$ 2) Suppose that $\mathcal{T}_{h+1,b}[\mathcal{F}] \mid \mathbf{u} \Vdash_\varphi \mathbf{v}$. Then $\mathsf{G} \mid \mathbf{u} \Vdash_\varphi \mathbf{v}$ for some $\mathsf{G} \in \mathsf{T}[\mathcal{F}]$, where $\mathsf{T} \in \mathcal{T}$, $Ht(\mathsf{T}) = h' \leq h+1$ and $Br(\mathsf{T}) \leq b$.

The case $h' = 1$ is trivial: here G is isomorphic to some frame from $\mathcal{F}$, thus $\mathcal{F} \mid \mathbf{u} \Vdash_\varphi \mathbf{v}$.

Suppose $h' > 1$. Let b' be the branching at the root of T. Then G is isomorphic to a frame $\mathsf{F} + (\mathsf{G}_1 \sqcup \dots \sqcup \mathsf{G}_{b'})$, where $\mathsf{F} \in \mathcal{F}$, $1 \leq b' \leq b$, $\mathsf{G}_1, \dots \mathsf{G}_{b'} \in \mathcal{T}_{h,b}[\mathcal{F}]$.

For some model M over G we have $(\mathsf{M}, w|\Phi_{\mathbf{u}}) \vDash \Phi^{\mathbf{v}}$, where w is a root of G. For $1 \leq j \leq b'$, let w_j be a root of G_j,

$$\Phi_j = \{\psi \in \Phi \mid (\mathsf{M}, w_j|\Phi_{\mathbf{u}}) \vDash \psi\}.$$

Then $\mathsf{G}_j \mid \mathbf{u} \Vdash_\varphi \mathbf{v}_j$, where $\mathbf{v}_j$ is determined by the equation $\Phi_j = \Phi_{\mathbf{v}_j}$.

For a formula $\psi \in \Phi$, we have:

$$(\mathsf{M}, w'|\Psi_{\mathbf{u}}) \vDash \psi \text{ for some } w' \in \mathsf{G}_1 \sqcup \dots \sqcup \mathsf{G}_{b'} \text{ iff } \psi \in \Phi_j \text{ for some } j.$$

Let M' be the restriction M to F. By Proposition 14, we obtain $(\mathsf{M}', w|\Psi_{\mathbf{u}} \cup \Phi_1 \cup \dots \cup \Phi_{b'}) \vDash \Phi^{\mathbf{v}}$. Thus $\mathcal{F} \mid \mathbf{u} \vee \mathbf{v}_1 \cdots \vee \mathbf{v}_{b'} \Vdash_\varphi \mathbf{v}$.

(2 $\Rightarrow$ 1) If $\mathsf{F} \mid \mathbf{u} \Vdash_\varphi \mathbf{v}$ for some $\mathsf{F} \in \mathcal{F}$, then $\mathcal{T}_{h+1,b}[\mathcal{F}] \mid \mathbf{u} \Vdash_\varphi \mathbf{v}$, since F is isomorphic to some frame from $\mathcal{T}_{1,1}[\mathcal{F}]$.

In the second case, $(\mathsf{M}', w|\Psi_{\mathbf{u}} \cup \Phi_{\mathbf{v}_1} \cup \dots \cup \Phi_{\mathbf{v}_{b'}}) \vDash \Phi^{\mathbf{v}}$ for some model M' over a frame $\mathsf{F} \in \mathcal{F}$, and $(\mathsf{M}_j, w_j|\Phi_u) \vDash \Psi^{\mathbf{v}_j}$ for some models M_j over frames from $\mathcal{T}_{h,b}[\mathcal{F}]$, where w is a root of M, w_j is a root of M_j, $j = 1, \dots, b'$. Put $\mathsf{M} = \mathsf{M}' + (\mathsf{M}_1 \sqcup \dots \sqcup \mathsf{M}_{b'})$. By Proposition 14, we have $(\mathsf{M}, w|\Psi_{\mathbf{u}}) \vDash \Phi^{\mathbf{v}}$. Thus $\mathcal{T}_{h+1,b}[\mathcal{F}] \mid \mathbf{u} \Vdash_\varphi \mathbf{v}$. ■

COROLLARY 20. *Let* $\mathcal{F}$ *be a class of cones. Suppose that* SatModerate *is an algorithm such that for any formula* φ, *for any* $\mathbf{u}, \mathbf{v} \in \{0,1\}^{\langle\varphi\rangle}$, *it decides whether*

$$\mathcal{F} \mid \mathbf{u} \Vdash_\varphi \mathbf{v}.$$

Then SatTree *(see Table 1) is an algorithm such that for any formula* φ, *for any* $\mathbf{u}, \mathbf{v} \in \{0,1\}^{\langle\varphi\rangle}$, *for any integers* $h, b \geq 1$, *it decides whether*

$$\mathcal{T}_{h,b}[\mathcal{F}] \mid \mathbf{u} \Vdash_\varphi \mathbf{v}.$$

THEOREM 21. *If* $(\mathcal{F}_n)_{n\in\mathbb{N}}$ *is d-moderate sequence of sets of cones, and* P *is a polynomial of degree* d', *then the sequence* $(\mathcal{T}_{P(n),P(n)}[\mathcal{F}_n])_{n\in\mathbb{N}}$ *is* $\max\{2+d', d\}$*-moderate.*

Table 1. Algorithm SatTree

Function SatTree(*formula* φ; *boolean vectors* $\mathbf{v}, \mathbf{u}$; *integers* h, b) returns boolean;

* SatTree *decides whether* $\mathcal{T}_{h,b}[\mathcal{F}] \mid \mathbf{u} \Vdash_\varphi \mathbf{v}$ *\

Begin
  if SatModerate($\varphi, \mathbf{v}, \mathbf{u}$) then
    * $\mathcal{F} \mid \mathbf{u} \Vdash_\varphi \mathbf{v}$ *\
    return(true);
  if $h > 1$ then
    for every integer b' such that $1 \leq b' \leq b$
      * b' *is the branching* *\
      for every boolean vectors $\mathbf{v}_1, \ldots, \mathbf{v}_{b'} \in \{0,1\}^{\langle\varphi\rangle}$
        if SatModerate($\varphi, \mathbf{v}, \mathbf{u} \vee \mathbf{v}_1 \cdots \vee \mathbf{v}_{b'}$) then
          * $\mathcal{F} \mid \mathbf{u} \vee \mathbf{v}_1 \cdots \vee \mathbf{v}_{b'} \Vdash_\varphi \mathbf{v}$ *\
          if $\bigwedge_{1 \leq j \leq b'}$ SatTree($\varphi, \mathbf{v}_j, \mathbf{u}, h-1, b$) then
            * $\mathcal{T}_{h,b}[\mathcal{F}] \mid \mathbf{u} \Vdash_\varphi \mathbf{v}_j$ *for all* j *\
            return(true);
  return(false);
End.

Proof. At every step of recursion, the algorithm SatTree uses $O(n^2)$ amount of space for a formula φ, where $n = \langle\varphi\rangle$. We also need $O(n^d)$ amount of space that used by SatModerate. The depth of recursion is $P(n)$, thus we need $O(n^2 P(n) + n^d)$ amount of space. ■

The above fact implies the following

THEOREM 22. *Suppose that a logic* L *is characterized by* $\mathcal{PO}[\mathcal{F}]$ *for some class* $\mathcal{F}$. *If there exists a moderate sequence* $(\mathcal{F}_n)_{n\in\mathbb{N}}$ *such that* $\mathcal{F}_n \subseteq \mathcal{F}$ *for all* $n \in \mathbb{N}$, *and any* L*-satisfiable formula* φ *is* $\mathcal{PO}[\mathcal{F}_{\langle\varphi\rangle}]$*-satisfiable, then* L *is in* PSPACE.

Proof. Consider an L-satisfiable formula φ with $\langle\varphi\rangle = n$. Then φ is satisfiable at a root of some $\mathsf{G} \in \mathcal{PO}[\mathcal{F}_n]$. By Lemma 12, φ is satisfiable at a root of some $\mathsf{G}' \in \mathcal{T}_{n,n}[\mathcal{F}_n]$. Thus, by Corollary 20, φ is L-satisfiable iff for some $\mathbf{v} = \{v_1, \ldots, v_{n-1}, 1\} \in \{0,1\}^n$ we have SatTree($\varphi, \mathbf{v}, (0, \ldots, 0), n, n$) $=$ *true*. ■

COROLLARY 23. *If* $\mathrm{L} = \mathrm{L}(\mathcal{PO}[\mathcal{F}])$ *for some finite class of finite cones* $\mathcal{F}$, *then* L *is in PSPACE.*

Proof. Put $\mathcal{F}_n = \mathcal{F}$ for all n. ■

EXAMPLE 24. As an example, consider the logics K4, S4 , Gödel-Löb logic GL, and Grzegorczyk logic GRZ. They are well-known to be PSPACE-decidable, see e.g [9, 11, 5]. Let us illustrate, how this fact follows from Theorem 22.

GRZ (GL) is the logic of all finite non-strict (strict) partial orders, see e.g. [6]: GRZ $= \mathrm{L}(\mathcal{PO}[\{C_1\}])$, GL $= \mathrm{L}(\mathcal{PO}[\{C_0\}])$. By corollary 23, GRZ and GL are in PSPACE.

Note that any K4-satisfiable formula is satisfiable at some finite transitive frame F such that the cardinality of any cluster in F does not exceed $\langle\varphi\rangle$. Put

$$\mathcal{F}_n^{\mathrm{K4}} = \{\mathsf{C}_0, \dots, \mathsf{C}_n\}, \quad \mathcal{F}_n^{\mathrm{S4}} = \{\mathsf{C}_1, \dots, \mathsf{C}_n\}.$$

Then for any φ we have:

$$\varphi \text{ is K4-satisfiable iff } \varphi \text{ is } \mathcal{PO}[\mathcal{F}^{\mathrm{K4}}_{\langle\varphi\rangle}]\text{-satisfiable,}$$

$$\varphi \text{ is S4-satisfiable iff } \varphi \text{ is } \mathcal{PO}[\mathcal{F}^{\mathrm{S4}}_{\langle\varphi\rangle}]\text{-satisfiable.}$$

Since the sequences $(\mathcal{F}_n^{\mathrm{K4}})_{n\in\mathbb{N}}$ and $(\mathcal{F}_n^{\mathrm{S4}})_{n\in\mathbb{N}}$ are moderate, then by Theorem 22, K4 and S4 are in PSPACE.

REMARK 25. Using the standard method of *translating of QBF-formula into modal logics* [9], it is not difficult to obtain PSPACE-hardness for logics of classes $\mathcal{PO}[\mathcal{F}]$, thus described in Theorem 22 logics are PSPACE-complete (for non-empty $\mathcal{F}$).

4.2 Multi-modal case

DEFINITION 26. Let M be an N-model. A *condition for* M is a tuple $\overline{\Psi} = (\Psi_1, \dots, \Psi_N)$ of sets of N-formulas. For an N-formula φ and a point $w \in \mathsf{M}$, we define the truth-relation $(\mathsf{M}, w|\overline{\Psi}) \vDash \varphi$ ("*φ is true at w in* M *under the condition* $\overline{\Psi}$"):

$$\begin{array}{lll}
(\mathsf{M}, w|\overline{\Psi}) \vDash p & \Leftrightarrow & \mathsf{M}, w \vDash p \\
(\mathsf{M}, w|\overline{\Psi}) \nvDash \bot & & \\
(\mathsf{M}, w|\overline{\Psi}) \vDash \varphi \to \psi & \Leftrightarrow & (\mathsf{M}, w|\Psi) \nvDash \varphi \text{ or } (\mathsf{M}, w|\Psi) \vDash \psi \\
(\mathsf{M}, w|\overline{\Psi}) \vDash \Diamond_k\varphi & \Leftrightarrow & \varphi \in \Psi_k \text{ or } \Diamond_k\varphi \in \Psi_k \text{ or} \\
 & & \text{for some } v \in R_k(w) \text{ we have } (\mathsf{M}, v|\overline{\Psi}) \vDash \varphi,
\end{array}$$

where $R_1, \dots, R_N$ are the accessability relations in M.

PROPOSITION 27. *Consider N-frames F and G. Let $1 \le k \le N$, M be a model over the frame $\mathsf{F} +_k \mathsf{G}$, and let M' be the restriction M to F. Let φ be an N-formula φ, $\overline{\Phi}$ be a condition.*

Then for any $w \in \mathsf{M}'$ we have

$$(\mathsf{M}, w|\overline{\Phi}) \vDash \varphi \Leftrightarrow (\mathsf{M}', w|(\Phi'_1, \dots, \Psi'_N)) \vDash \varphi,$$

where $\Phi'_i = \Phi_i$ for $i \neq k$, and

$$\Phi'_k = \Phi_k \cup \{\psi \in Sub(\varphi) \mid (\mathsf{M}, w|\overline{\Phi}) \vDash \psi \textit{ for some } w \in \mathsf{G}\}.$$

Proof. By induction on the length of φ, analogously to the proof of Proposition 14. ■

DEFINITION 28. We call an N-frame $\mathsf{G} = (W, R_1, \ldots, R_N)$ *rooted*, if for some $w \in W$ we have $\{w\} \cup R_1(w) \cup \cdots \cup R_N(w) = W$; w is called a *root* of G.

The following propositions are straightforward consequences of the above definition.

PROPOSITION 29. *Suppose that in the condition of Proposition 27 we also have* $\mathsf{G} = \mathsf{G}_1 \sqcup \cdots \sqcup \mathsf{G}_b$ *for some rooted N-frames* $\mathsf{G}_1, \ldots, \mathsf{G}_b$. *Let w_i denote a root of G_i, $i = 1, \ldots, b$. Then for any $w \in \mathsf{M}'$ we have*

$$(\mathsf{M}, w|\overline{\Phi}) \vDash \varphi \Leftrightarrow (\mathsf{M}', w|(\Phi'_1, \ldots, \Psi'_N)) \vDash \varphi,$$

where $\Phi'_i = \Phi_i$ for $i \neq k$, and

$$\Phi'_k = \Phi_k \cup \bigcup_{1 \leq i \leq b} \{\psi \in Sub(\varphi) \mid (\mathsf{M}, w_i|\overline{\Phi}) \vDash \psi \vee \Diamond_1 \psi \vee \cdots \vee \Diamond_N \psi\}.$$

PROPOSITION 30. *Let F be a class of rooted N-frames, $1 \leq k \leq N$, $\mathsf{G} \in \mathcal{PO}$. If G is rooted, $\mathsf{H} \in \mathsf{G}[k; \mathcal{F}]$, then H is rooted.*

DEFINITION 31. As well as in the monomodal case, we say that an N-formula φ is F-*satisfiable under the condition* $\overline{\Psi} = (\Psi_1, \ldots, \Psi_N)$, if φ is true at some root of F in some model over F under the condition $\overline{\Psi}$.

For an N-formula φ, put

$$Sub^*(\varphi) = Sub(\varphi) \cup \{\Diamond_i \psi \mid 1 \leq i, j \leq N, \quad \Diamond_j \psi \in Sub(\varphi)\}.$$

Consider an N-formula φ with $Sub^*(\varphi)_{\lessdot} = (\psi_1, \ldots, \psi_n)$. For vectors $\mathbf{v}, \mathbf{u}_1, \ldots, \mathbf{u}_N \in \{0,1\}^n$, the notation

$$\mathsf{F} \mid (\mathbf{u}_1, \ldots, \mathbf{u}_N) \Vdash_\varphi \mathbf{v}$$

means that $Sub^*(\varphi)^{\mathbf{v}}$ is F-satisfiable under the condition

$$(Sub^*(\varphi)_{\mathbf{u}_1}, \ldots, Sub^*(\varphi)_{\mathbf{u}_N}).$$

(Note that if $N = 1$ then $Sub^*(\varphi) = Sub(\varphi)$, so this notation does not contradict the monomodal case). For a class $\mathcal{F}$ of rooted N-frames, $\mathcal{F} \mid (\mathbf{u}_1, \ldots, \mathbf{u}_N) \Vdash_\varphi \mathbf{v}$ means that $\mathsf{F} \mid (\mathbf{u}_1, \ldots, \mathbf{u}_N) \Vdash_\varphi \mathbf{v}$ for some $\mathsf{F} \in \mathcal{F}$.

Also, for $1 \leq k \leq N$, we define auxiliary function f_k. For boolean vector $\mathbf{v} = (v_1, \ldots, v_n)$, we put $f_k(\mathbf{v}) = (v'_1, \ldots, v'_n)$, where $v'_i = 1$ iff $v_i = 1$ or for some j, l, k' we have $\psi_i = \Diamond_k \psi_l$, $\psi_j = \Diamond_{k'} \psi_l$, and $v_j = 1$.

DEFINITION 32. For a positive integer d, a sequence $(\mathcal{F}_n)_{n \in \mathbb{N}}$ of sets of rooted N-frames is called *d-moderate*, if there exists an algorithm such that for any N-formula φ, for any vectors $\mathbf{v}, \mathbf{u}_1, \ldots, \mathbf{u}_N \in \{0,1\}^{|Sub^*(\varphi)|}$, it decides whether

$$\mathcal{F}_{|Sub^*(\varphi)|} \mid (\mathbf{u}_1, \ldots, \mathbf{u}_N) \Vdash_\varphi \mathbf{v}$$

Table 2. Algorithm $\mathrm{SatTree_N}$

Function $\mathrm{SatTree_N}$(*formula* φ; *boolean vectors* $\mathbf{v}, \mathbf{u}_1, \ldots, \mathbf{u}_N$; *integers* h, b, k) returns boolean;

Begin
 if $\mathrm{SatModerate_N}(\varphi, \mathbf{v}, \mathbf{u}_1, \ldots, \mathbf{u}_N)$ then return(true);
 if $h > 1$ then
  for every integer b' such that $1 \le b' \le b$
   for every boolean vectors $\mathbf{v}_1, \ldots, \mathbf{v}_{b'} \in \{0,1\}^{\langle\varphi\rangle}$
    if $\mathrm{SatModerate_N}(\varphi, \mathbf{v}, \mathbf{u}_1, \ldots, \mathbf{u}_k \vee f_k(\mathbf{v}_1) \cdots \vee f_k(\mathbf{v}_{b'}), \ldots, \mathbf{u}'_N)$ then
     if $\bigwedge_{1 \le j \le b'} \mathrm{SatTree_N}(\varphi, \mathbf{v}_j, \mathbf{u}_1, \ldots, \mathbf{u}_N, h-1, b, k)$ then
      return(true);
 return(false);
End.

in space $O(|Sub^*(\varphi)|^d)$.

The following statement is a straightforward generalization of Proposition 19.

PROPOSITION 33. *Let $\mathcal{F}$ ba a class or rooted N-frames, φ be an N-formula, $n = |Sub^*(\varphi)|$, $1 \le k \le N$. Then for any $\mathbf{u}_1, \ldots, \mathbf{u}_N, \mathbf{v} \in \{0,1\}^n$, for any integers $h, b \ge 1$, the following two conditions are equivalent.*

1. $\mathcal{T}_{h+1,b}[k; \mathcal{F}] \mid (\mathbf{u}_1, \ldots, \mathbf{u}_N) \Vdash_\varphi \mathbf{v}$.

2. *Either $\mathcal{F} \mid (\mathbf{u}_1, \ldots, \mathbf{u}_N) \Vdash_\varphi \mathbf{v}$ or there exist vectors $\mathbf{v}_1, \ldots, \mathbf{v}_{b'} \in \{0,1\}^n$, where $1 \le b' \le b$, such that*

$$\mathcal{T}_{h,b}[k; \mathcal{F}] \mid (\mathbf{u}_1, \ldots, \mathbf{u}_N) \Vdash_\varphi \mathbf{v}_j \text{ for all } j = 1, \ldots, b', \text{ and}$$

$$\mathcal{F} \mid (\mathbf{u}'_1, \ldots, \mathbf{u}'_N) \Vdash_\varphi \mathbf{v},$$

where $\mathbf{u}'_i = \mathbf{u}_i$ for $i \ne k$, $\mathbf{u}'_k = \mathbf{u}_k \vee f_k(\mathbf{v}_1) \cdots \vee f_k(\mathbf{v}_{b'})$.

Proof. By Propositions 29 and 30, analogously to the proof of Proposition 19. ∎

COROLLARY 34. *Let $\mathcal{F}$ be a class of rooted N-frames. Suppose that* $\mathrm{SatModerate_N}$ *is an algorithm such that for any N-formula φ, for any $\mathbf{u}_1, \ldots, \mathbf{u}_N, \mathbf{v} \in \{0,1\}^{|Sub^*(\varphi)|}$, it decides whether*

$$\mathcal{F} \mid (\mathbf{u}_1, \ldots, \mathbf{u}_N) \Vdash_\varphi \mathbf{v}.$$

Then $\mathrm{SatTree_N}$ *(see Table 2) is an algorithm such that for any formula φ, for any $\mathbf{u}_1, \ldots, \mathbf{u}_N, \mathbf{v} \in \{0,1\}^{|Sub^*(\varphi)|}$, for any integers $h, b \ge 1$, it decides whether*

$$\mathcal{T}_{h,b}[k; \mathcal{F}] \mid (\mathbf{u}_1, \ldots, \mathbf{u}_N) \Vdash_\varphi \mathbf{v}.$$

Since at every step of recursion the algorithm $\mathrm{SatTree}_{\mathrm{N}}$ uses $O(n^2)$ amount of memory for a formula φ with $|Sub^*(\varphi)| = n$, we obtain

THEOREM 35. *If* $(\mathcal{F}_n)_{n\in\mathbb{N}}$ *is d-moderate sequence of sets of rooted N-frames,* $1 \leq k \leq N$*, P is a polynomial of degree d′, then the sequence*

$$(\mathcal{T}_{P(n),P(n)}[k;\mathcal{F}_n])_{n\in\mathbb{N}}$$

is $\max\{2+d', d\}$*-moderate.*

5 PSPACE-decidability of GLP

Is this section we construct PSPACE-decision procedure for GLP.

First, we need to quote some results from [2].

For an N-frame $\mathsf{F} = (W, R_1, \ldots, R_N)$ let F_+ denote the $(N+1)$-frame $(W, \varnothing, R_1, \ldots, R_N)$, and let F_∞ denote the frame $(W, R_1, \ldots, R_N, \varnothing, \varnothing, \ldots)$ with countably many relations.

DEFINITION 36 ([2]). For $N \geq 1$, we inductively define a class $\mathcal{F}^{(N)}$ of N-frames. Let $\mathcal{F}^{(1)}$ be the class of all finite strict partial orders,

$$\mathcal{F}^{(N+1)} = \mathcal{PO}[1;\mathcal{G}^{(N)}], \text{ where } \mathcal{G}^N = \{\mathsf{F}_+ \mid \mathsf{F} \in \mathcal{F}^{(N)}\}.$$

Also put $\mathcal{F}^{\mathrm{J}}$=$\{\mathsf{F}_\infty \mid \mathsf{F} \in \mathcal{F}^{(N)}$ for some $N\}$.

Let J be the logic of the class $\mathcal{F}^{\mathrm{J}}$ (complete axiomatization of this logic is given in [2]). These frames are called *hereditary strict orders*, and were introduced in [2], where the following result was proved:

THEOREM 37 ([2]). *There exists a polynomial-time translation f such that for any formula* φ *we have*

$$\mathrm{GLP} \vdash \varphi \Leftrightarrow \mathrm{J} \vdash f(\varphi).$$

PROPOSITION 38 ([2]).
(1) If $(W, R_1, \ldots, R_N, R, S_1, \ldots, S_K) \in \mathcal{F}^{(N+K+1)}$ *then* $(W, R_1, \ldots, R_N, S_1, \ldots, S_K)$ *is isomorphic to some frame from* $\mathcal{F}^{(N+K)}$.
(2) If $(W, R_1, \ldots, R_N, S_1, \ldots, S_K) \in \mathcal{F}^{(N+K)}$ *then* $(W, R_1, \ldots, R_N, \varnothing, S_1, \ldots, S_K)$ *is isomorphic to some frame from* $\mathcal{F}^{(N+K+1)}$.

Proof. The proof is straightforward (by Definition 36).

Another proof is based on the first-order conditions that characterized the class of hereditary strict orders, see [2] for details. ■

LEMMA 39. *Let* φ *be an N-formula, and let* $\{\Diamond_{i_1}, \ldots, \Diamond_{i_K}\}$ *be the set of all diamonds that occur in* φ*, where* $i_1 < i_2 < \cdots < i_K$*. Let* φ' *be the K-modal formula that obtained from* φ *by replacing* $\Diamond_{i_j}$ *with* $\Diamond_j$ *for all* $j = 1 \ldots K$*. Then* φ *is* $\mathcal{F}^{(N)}$*-satisfiable iff* φ' *in* $\mathcal{F}^{(K)}$*-satisfiable.*

Proof. $\Rightarrow$). Suppose that φ is satisfiable in a frame $\mathsf{F} = (W, R_1, \dots, R_N) \in \mathcal{F}^{(N)}$. Clearly, φ' is satisfiable at the frame $\mathsf{G} = (W, R_{i_1}, \dots, R_{i_K})$. G is obtained from F by 'deleting' some relations, so by Proposition 38, an isomorphic copy of G belongs to $\mathcal{F}^{(K)}$. Thus φ' is $\mathcal{F}^{(K)}$-satisfiable.

$\Leftarrow$). Suppose that φ' is satisfiable in a frame $\mathsf{G} = (W, S_1, \dots, S_K) \in \mathcal{F}^{(K)}$. Then φ is satisfiable in $\mathsf{F} = (W, R_1, \dots, R_N)$, where $R_{i_j} = S_j$ for all $j = 1 \dots K$, and all other relations of F are empty. By Proposition 38, φ is $\mathcal{F}^{(N)}$-satisfiable. ■

For $N, h, b \geq 1$, by induction on N we define a class $\mathcal{T}_{h,b}^{(N)}$. Let $\mathcal{T}_{h,b}^{(1)} = \mathcal{T}_{h,b}[\{\mathsf{C}_0\}]$, i.e., $\mathcal{T}_{h,b}^{(1)}$ is the class (up to isomorphisms) of all finite transitive irreflexive trees with the height not more then h and the branching not more then b. Put

$$\mathcal{T}_{h,b}^{(N+1)} = \mathcal{T}_{h,b}[1; \{\mathsf{F}_+ \mid \mathsf{F} \in \mathcal{T}_{h,b}^{(N)}\}].$$

THEOREM 40. *The satisfiability problem for* J *is in* PSPACE.

Proof. Consider an N-formula φ. Let $n = \langle\varphi\rangle$. By Lemma 39, we may assume that $N < n$.

Suppose that φ is J-satisfiable. Then φ is satisfiable at some N-frame $\mathsf{F} \in \mathcal{F}^{(N)}$. Using Lemma 12, by induction on N, one can show that φ is satisfiable at the root of some N-frame $\mathsf{T} \in \mathcal{T}_{n,n}^{(N)}$. Thus

$$\varphi \text{ is J-satisfiable} \iff \varphi \text{ is } \mathcal{T}_{n,n}^{(N)}\text{-satisfiable.}$$

By induction on N we obtain that there exists d, such that for any N the sequence $(\mathcal{T}_{n,n}^{(N)})_{n\in\mathbb{N}}$ is d-moderate (Theorem 35), and, moreover, we obtain that it is possible to check whether φ is $\mathcal{T}_{\langle\varphi\rangle,\langle\varphi\rangle}^{(\langle\varphi\rangle)}$-satisfiable in polynomial of $\langle\varphi\rangle$ space. Thus satisfiability problem for J is in PSPACE. ■

THEOREM 41. GLP *is in* PSPACE.

Proof. Follows from Theorems 37 and 40. ■

Acknowledgements. I would like to thank Lev Beklemishev and Valentin Shehtman for their help. Also, I would like to thank anonymous referees for their very useful and very detailed comments on previous versions of this paper.

The work on this paper was supported by Poncelet Laboratory (UMI 2615 of CNRS and Independent University of Moscow) and by RFBR grant 06-01-72555.

BIBLIOGRAPHY

[1] S. Artemov, L. Beklemishev. Provability Logic. In: Handbook of Philosophical Logic, D. Gabbay and F. Guenthner, eds., vol. 13, 2nd ed. Kluwer, Dordrecht, 229–403, 2004.

[2] L. Beklemishev. Kripke semantics for provability logic GLP. Submitted to Annals of Pure and Applied Logic.
http://www.phil.uu.nl/preprints/preprints/PREPRINTS/preprint260.pdf

[3] P. Blackburn, M. de Rijke and Y. Venema. Modal Logic. Cambridge University Press, 2001
[4] G. Boolos. The Logic of Provability. Cambridge University Press, Cambridge, 1993.
[5] L. Farinas del Cerro, O. Gasquet. A general framework for pattern-driven modal tableaux. Logic Journal of the IGPL, 10(1):51–83, 2002.
[6] A. Chagrov, M. Zakharyaschev. Modal logic. Oxford University Press, 1997.
[7] K.N. Ignatiev. On strong provability predicates and the associated modal logics. The Journal of Symbolic Logic, 58:249–290, 1993.
[8] G. Japaridze. The polymodal logic of provability. In Intensional Logics and Logical Structure of Theories: Material from the fourth Soviet Finnish Symposium on Logic, 16–48, 1988. In Russian.
[9] R. Ladner. The computational complexity of provability in systems of modal propositional logic. SIAM Journal of Computing, 6:467–480, 1977.
[10] I. Shapirovsky. On PSPACE-decidability in transitive modal logic. Advances in Modal Logic, V. 5, 269–287, 2005.
[11] E. Spaan. Complexity of modal logics. PhD thesis, University of Amsterdam, 1993.

Ilya Shapirovsky
Institute for Information Transmission Problems
Russian Academy of Sciences
B.Karetny 19, Moscow, Russia, 127994
shapir@iitp.ru

On the intermediate logic of open subsets of metric spaces

Timofei Shatrov

ABSTRACT. In this paper we study the intermediate logic $ML_{O(\mathcal{X})}$ of open subsets of a metric space $\mathcal{X}$. This logic is closely related to Medvedev's logic of finite problems ML. We prove several facts about this logic: its inclusion in ML, impossibility of its finite axiomatization and indistinguishability from ML within some large class of propositional formulas.

Keywords: intermediate logic, Medvedev's logic, axiomatization, indistinguishability

1 Introduction

In this paper we introduce and study a new intermediate logic $ML_{O(\mathcal{X})}$, which we first define using Kripke semantics and then we will try to rationalize this definition using the concepts from prior research in this field.

An (intuitionistic) *Kripke frame* is a partially ordered set $(F, \leq)$. A *Kripke model* is a Kripke frame with a valuation θ (a function which maps propositional variables to upward-closed subsets of the Kripke frame). The notion of a propositional formula being true at some point $w \in F$ of a model $M = (F, \theta)$ is defined recursively as follows:

$$\text{For propositional letter } p_i,\ M, w \vDash p_i \text{ iff } w \in \theta(p_i)$$
$$M, w \vDash \psi \wedge \chi \text{ iff } M, w \vDash \psi \text{ and } M, w \vDash \chi$$
$$M, w \vDash \psi \vee \chi \text{ iff } M, w \vDash \psi \text{ or } M, w \vDash \chi$$
$$M, w \vDash \psi \to \chi \text{ iff for any } w' \geq w,\ M, w \vDash \psi \text{ implies } M, w' \vDash \chi$$
$$M, w \nvDash \bot$$

A formula is true in a model M if it is true in each point of M. A formula is valid in a frame F if it is true in all models based on that frame.

The set of formulas which are valid on every Kripke frame is *intuitionistic logic* **Int**. An *intermediate logic* is an extension of **Int** closed under modus ponens and substitution. Every consistent intermediate logic is contained in *classical logic* **CL**.

The set $L(F)$ of all formulas valid in a given frame F forms an intermediate logic, which allows us to define a logic just by constructing its Kripke frame (although not every intermediate logic can be constructed in this way).

For example, the logics ML and ML_1, which will be often mentioned in this paper are defined as follows:

DEFINITION 1. Let X be a set, then $P_1(X)$ is the Kripke frame $(2^X \setminus \emptyset, \supseteq)$ (non-empty subsets of X ordered by converse inclusion). *Medvedev's logic of finite problems* ML is $\bigcap\{L(P_1(X))|X \text{ is finite}\}$ [1] and the *logic of infinite problems* ML_1 is $L(P_1(\omega))$ (or $L(P_1(X))$ for any infinite X) [6].

It is well-known that $ML_1 \subseteq ML$, but it is an open problem whether $ML_1 = ML$ or not.

Let $\mathcal{X}$ be a topological space, and $O(\mathcal{X})$ a set of non-empty open sets in $\mathcal{X}$. Then $(O(\mathcal{X}), \supseteq)$ is a Kripke frame with the least element $\mathcal{X}$. Logic $ML_{O(\mathcal{X})}$ is defined as the logic of this frame:

DEFINITION 2. $ML_{O(\mathcal{X})} = L(O(\mathcal{X}), \supseteq)$. We will often use $O(\mathcal{X})$ to denote the frame $(O(\mathcal{X}), \supseteq)$ as a simple notation.

Finally, a few words on why this logic is interesting. We can understand *information* as a subset of some basic set Ω. An *information type* is an arbitrary set of informations (these notions were introduced by Yuri Medvedev in [2]). A type σ is called regular iff

$$\forall E_1 \forall E_2 \; E_1 \in \sigma \;\&\; E_2 \subseteq E_1 \Rightarrow E_2 \in \sigma$$

As described in [5] regular information types form a structure that is naturally connected to Medvedev's logic of finite problems ML (in case of finite Ω), or logic of infinite problems ML_1 in case of infinite Ω. Namely, the set of all regular types $I(\Omega)$ ordered by inclusion is a Heyting algebra, and it is isomorphic to the Heyting algebra of upward-closed subsets of $P_1(\Omega)$. Thus, the logic of Heyting algebra $I(\Omega)$ is $L(P_1(\Omega))$.

Now, for a topological space $\mathcal{X}$ it may be more natural to understand information as an open subset of $\mathcal{X}$, and not an arbitrary one. For example, every measurement in a physical experiment has a margin of error, so the only kind of information we can get about the measured value is that it is within some interval (open subset of $\mathbb{R}$). In this case regular information types correspond to upward-closed subsets of $O(\mathcal{X})$ and thus form a Heyting algebra of $ML_{O(\mathcal{X})}$.

2 Useful facts and definitions

In this section we briefly recall facts related to Kripke semantics of intermediate logics.

DEFINITION 3. Let $u \in F$ be a point of a Kripke frame $(F, \leq)$. Then the Kripke frame $F^u = \{v \in F | u \leq v\}$ with the ordering $\leq$ is called a *cone*.

DEFINITION 4. The surjective mapping h from frame F to frame G is called a *p-morphism* iff $\forall u \in F \; h(F^u) = G^{h(u)}$, that is

1. $\forall u, v \in F \; (u \leq v) \Rightarrow h(u) \leq h(v)$

2. $\forall u \in F, w \in G\ (h(u) \leq w) \Rightarrow \exists v \geq u : h(v) = w$

If there is a p-morphism from F to G, this is denoted by $F \twoheadrightarrow G$.

LEMMA 5.

1. *$L(F) \subseteq L(F^u)$.*
2. *If $F \twoheadrightarrow G$ then $L(F) \subseteq L(G)$.*

Let F be a finite Kripke frame with the least element (called the root and denoted by 0_F). Then it is possible to construct a propositional formula $X(F)$, called *Jankov (characteristic) formula* with the following properties:

LEMMA 6.

1. *For any propositional formula A, $X(F) \in (\mathbf{Int} + A) \iff F \not\models A$*
2. *For any frame G, $G \not\models X(F) \iff \exists u \in G : G^u \twoheadrightarrow F$*

3 $ML_{O(\mathcal{X})}$ as a subset of Medvedev's logic

In this section we construct a p-morphism from the Kripke frame of open subsets of $\mathcal{X}$ to the Kripke frame of ML_1, thus proving that $ML_{O(\mathcal{X})}$ is a subset of ML_1 and ML.

THEOREM 7. *Let $\mathcal{X}$ be an infinite metric space. Then $ML_{O(\mathcal{X})} \subseteq ML_1$.*

Proof. To prove this we construct a p-morphism $f : O(\mathcal{X}) \twoheadrightarrow P_1(\omega)$. $\mathcal{X}$ can be split into 3 disjoint subsets: $\mathcal{X} = \mathcal{X}_1 \cup \mathcal{X}_2 \cup \mathcal{X}_3$.
$\mathcal{X}_1 = \{x \in \mathcal{X} | \exists \varepsilon \forall y \in \mathcal{X}, y \neq x \Rightarrow \rho(x,y) > \varepsilon\}$ contains all isolated points.
$\mathcal{X}_2 = \{x \in \mathcal{X} | \exists \{x_n\}_{n=1}^{\infty} \subset \mathcal{X}_1, x_n \to x, n \to \infty\} \setminus \mathcal{X}_1$ contains the limits of isolated points.
$\mathcal{X}_3 = (\mathcal{X} \setminus \mathcal{X}_1) \setminus \mathcal{X}_2$ contains the limit points which are not the limits of isolated points. A point of $\mathcal{X}_3$ is always a limit of other points in $\mathcal{X}_3$, thus $\mathcal{X}_3$ is either empty or infinite.

If $\mathcal{X}_3$ is empty then $\mathcal{X}_1$ cannot be finite (since otherwise $\mathcal{X}_2$ is empty, and $\mathcal{X}$ is finite). We can choose a sequence $x_n \in \mathcal{X}_1$, $n = 0, 1, 2, \ldots$ where all x_n are different. Let $D_i = \{x_i\}$, $i = 1, 2, 3, \ldots$, $D_0 = \{x_0\} \cup \{x \in \mathcal{X}_1 | \forall n\ x \neq x_n\}$. Each of D_i is open, since each x_i is an isolated point. Let $G \subset \mathcal{X}$ be a nonempty open subset. If $G \cap D_i = \emptyset$, $i = 0, 1, 2, \ldots$ then $G \cap \mathcal{X}_1 = \emptyset$, so $G \subseteq \mathcal{X}_2$, since $\mathcal{X}_3$ is empty. But since each point in $\mathcal{X}_2$ is a limit of isolated points from $\mathcal{X}_1$, G has to contain some, which is a contradiction. Thus, there always exists D_n which intersects G. We define $f(G) = \{n \in \omega | G \cap D_n \neq \emptyset\}$, which is easily checked to be p-morphism.

Let $\mathcal{X}_3$ be nonempty. It is open, since its complement $\mathcal{X} \setminus \mathcal{X}_3 = \mathcal{X}_1 \cup \mathcal{X}_2$ is clearly closed (a limit of points in $\mathcal{X}_2$ is also in $\mathcal{X}_2$). Choose an arbitrary $x_0 \in \mathcal{X}_3$ and let $D_n = \{x \in \mathcal{X}_3 | \frac{1}{n+1} < \rho(x, x_0) < \frac{1}{n}\}$, $n = 2, 3, \ldots$. If we exclude empty D_n, there would be still an infinite number of them left (since

x_0 is a limit point of $\mathcal{X}_3$). Also define $D_1 = \{x \in \mathcal{X}_3 | \rho(x, x_0) > \frac{1}{2}\}$, $D_0 = \mathcal{X}_1$. Each open subset $G \subset \mathcal{X}$ intersects one of D_n. After renumbering D_n such that all of them are non-empty we can define $f(G) = \{n \in \omega | G \cap D_n \neq \emptyset\}$, which is a p-morphism. ■

It should be noted that by definition $ML_1 = ML_{O(\omega)}$. For a metric space $\mathcal{X}$, $ML_{O(\mathcal{X})}$ would be equal to ML_1 if it is possible to construct a p-morphism from $P_1(Y)$ to $O(\mathcal{X})$ for some infinite set Y.

For example, consider $\mathcal{X} = \{0\} \cup \{1/n | n \in \mathbb{N}\}$ with the usual metric and $Y = \mathbb{N} \times \{0, 1\}$. Open sets of $\mathcal{X}$ are either arbitrary subsets of $\mathcal{X} \setminus 0$, or cofinite subsets of $\mathcal{X}$. We define $h : P_1(Y) \to O(\mathcal{X})$ as follows:

- $h(E) = \{1/n | \exists j\ (n, j) \in E\}$ if $|\mathcal{X} \setminus E| = \omega$
- $h(E) = \{1/n | \exists j\ (n, j) \in E\} \cup \{0 | \exists n\ (n, 1) \in E\}$ if $|\mathcal{X} \setminus E| < \omega$

It is easy to check that h is a p-morphism. The following lemma is a generalization of this construction.

LEMMA 8. *Suppose $\mathcal{X}$ is an infinite metric space consisting only of isolated points and limits of isolated points ($\mathcal{X} = \mathcal{X}_1 \cup \mathcal{X}_2$ using the notation from the proof of Theorem 7). Then $ML_1 = ML_{O(\mathcal{X})}$.*

Proof. Let $Y = \mathcal{X}_1 \times (\mathcal{X}_2 \cup \{x_0\})$ ($x_0 \in \mathcal{X}_1$), and for $y = (y_1, y_2) \in Y$ $\phi(y) = y_1$, $\psi(y) = y_2$. Define $h : P_1(Y) \to O(\mathcal{X})$ as follows:

$$h(E) = \text{int}(\phi(E) \cup (\mathcal{X}_2 \cap \psi(E)))$$

$\phi(E)$ is always in $h(E)$, and we only add appropriate limit points from $\psi(E)$ so that the result is an open set. If $G \in O(\mathcal{X})$ then $h(E) = G$ for $E = H(G) = (G \cap \mathcal{X}_1) \times (\{x_0\} \cup (G \cap \mathcal{X}_2))$ (every open subset of $\mathcal{X}$ contains some points from $\mathcal{X}_1$, so $G \cap \mathcal{X}_1 \neq \emptyset$). So, h is surjective. If $E' \subseteq E$ then of course $h(E') \subseteq h(E)$. If $G \subseteq h(E)$ then $h(E') = G$ for $E' = E \cap H(G)$. So, h is a p-morphism, and that means that $ML_1 \subseteq ML_{O(\mathcal{X})}$. Combining this with the result of Theorem 7 we conclude that $ML_1 = ML_{O(\mathcal{X})}$. ■

On the other hand, if $\mathcal{X}_3 \neq \emptyset$, there is no p-morphism from $P_1(Y)$ to $O(\mathcal{X})$. Suppose h is such p-morphism and $h(Y') = \mathcal{X}_3$ ($\mathcal{X}_3$ is an open set, so such $Y' \subseteq Y$ must exist). Then for $y \in Y'$ $h(\{y\})$ must be a singleton (isolated point), but there are no such open subsets of $\mathcal{X}_3$.

It is unknown whether $ML_{O(\mathcal{X})}$ is always equal to ML_1, although we will prove later that they are quite close to each other.

4 $ML_{O(\mathcal{X})}$ is not finitely axiomatizable

Let $\mathcal{X}$ be a metric space, F is a finite Kripke frame and suppose there exists a p-morphism $h : O(\mathcal{X}) \twoheadrightarrow F$. We will construct a mapping $\overline{h}$, which is an extension of h to some other (non-open) subsets of $\mathcal{X}$.

Let $A \subset \mathcal{X}$ be a non-empty subset of $\mathcal{X}$, and let $\{A_n\}$ be a sequence of open subsets of $\mathcal{X}$ such that $A_{i+1} \subseteq A_i$ and $\bigcap_{i=1}^{\infty} A_i = A$. Then we define $\overline{h}(A) = \sup_{\{A_i\}} \lim_i h(A_i)$. Note that for every sequence A_i with these properties the limit does exist. This follows from the fact that F is finite and $h(A_i) \leq h(A_{i+1})$. If A_i and B_i are sequences which yield different limits, then the sequence $A_i \cap B_i$ yields a limit that is greater or equal to both of these limits ($h(A_i \cap B_i) \geq h(A_i)$). Thus, the supremum always exists as long as at least one such sequence exists.

Since $\mathcal{X}$ is a metric space, $\overline{h}$ is defined for every non-empty finite subset of $\mathcal{X}$ (for $\{x_1, \ldots, x_n\}$ there exists a sequence $A_i = \bigcup_{1 \leq j \leq n} \{x | \rho(x, x_j) < \frac{1}{i}\}$).

For $u \in F$ we denote the set of its immediate successors by $br(u)$; the maximum length of an increasing chain starting at u is denoted by $d(u)$.

LEMMA 9. *Let $\mathcal{X}$ be a metric space and let F be a finite frame with the least element, and $|br(u)| \neq 1$ for any $u \in F$. If there exists a p-morphism $h : O(\mathcal{X}) \twoheadrightarrow F$, then for any $u \in F$ there exists $E \subset \mathcal{X}$ such that $\overline{h}(E) = u$ and $|E| < 2^{d(u)}$.*

Proof. If $d(u) = 1$ then $\exists G \in O(\mathcal{X}) : h(G) = u$ and $\forall G' \subset G\ h(G') = u$. Thus any $E = \{g\}, g \in G$ satisfies $\overline{h}(E) = u$, as there is a sequence of neighborhoods of g which yields a limit u, but since u is a maximal point of F, $\overline{h}(E)$ is also u.

Now suppose $d(u) = n > 1$ and we have already proved the statement of the lemma for $d(u) < n$ (and for any appropriate $\mathcal{X}$ and F). Choose $G \in O(\mathcal{X}) : h(G) = u$. There exist at least 2 immediate successors of u in F, which we denote by v_1 and v_2. $\exists G_1, G_2 :\ G' = G_1 \cup G_2 \subset G, \overline{h}(G_1) = v_1$, $\overline{h}(G_2) = v_2$, $|G_1| < 2^{d(u)-1}$, $|G_2| < 2^{d(u)-1}$. This follows from the induction hypothesis for $\mathcal{X} := G, F := F^u$, $h = h|_{O(G)}$. Then $|G'| < 2^{d(u)}$ and $\overline{h}(G') = u$. This completes the induction step. ■

We will call a finite subset $A = \{a_1, a_2, \ldots, a_n\}$ *stable* under $\overline{h}$ iff

$$\exists \varepsilon \forall A' = \{a'_1, a'_2, \ldots, a'_n\}\ \forall i \leq n\ \rho(a_i, a'_i) < \varepsilon \Rightarrow \overline{h}(A') = \overline{h}(A)$$

If every non-empty subset of A is also stable, we say that A is *hereditarily stable* under $\overline{h}$.

LEMMA 10. *Let $E = \{e_1, e_2, \ldots, e_n\}$ be a subset constructed by applying the previous lemma. Then it is stable under $\overline{h}$.*

Proof. Note that every point $e_i \in E$ is chosen at the induction base as an arbitrary point in an open set G_i. Since E is finite, we can choose ε such every $e_i \in E$ will remain in its corresponding set G_i if we move it no farther than ε. Thus E is stable. ■

LEMMA 11. *If a finite set $E = \{e_1, e_2, \ldots, e_n\}$ is stable under $\overline{h}$ then $\forall \varepsilon \exists E' = \{e'_1, e'_2, \ldots, e'_n\} : \forall i \leq n\ \rho(e_i, e'_i) < \varepsilon$ such that $\overline{h}(E') = \overline{h}(E)$ and E' is hereditarily stable.*

Proof. Let δ be a positive number which is smaller than ε, $\min_{i \neq j} \rho(e_i, e_j)$ and $\forall E' = \{e'_1, e'_2, \ldots, e'_n\}\ \forall i \leq n\ \rho(e_i, e'_i) < \delta \Rightarrow \overline{h}(E') = \overline{h}(E)$. Such δ exists since E is stable.

A point $x \in \mathcal{X}^n$ with distinct coordinates can be regarded as n-element subset of $\mathcal{X}$. $x^0 = (e_1, e_2, \dots, e_n)$ corresponds to E and points in δ-neighborhood of x^0 correspond to possible E' (by choice of δ, all such points have distinct coordinates and $\overline{h}(E') = \overline{h}(E)$).

$k = 2^n - 1$ is the number of non-empty subsets of $I = \{1, \dots, n\}$, and we will denote $J \subseteq I$ by $I(\sum_{j \in J} 2^{j-1})$ (so that every non-empty subset is numbered from 1 to k).

We define the mapping $H : B_\delta(x^0) \to F^k$ as follows: $H_i(x) = \overline{h}(E_i(x))$, $i = 1, \dots, k$, where $E_i(x) = \{x_j | j \in I(i)\}$. The frame F^k is ordered as follows:

$$(f_1, \dots, f_k) \leq (g_1, \dots, g_k) \rightleftharpoons f_1 \leq g_1 \& \dots \& f_k \leq g_k$$

It can be easily shown that under that ordering H has a local minimum in each point x from δ-neighborhood of x^0. Indeed, if $H_i(x) = u$ then there is open $G \supset E_i(x)$ such that $h(G) = u$ and

$$\forall x' \; E_i(x') \subset G \Rightarrow H_i(x') = \overline{h}(E_i(x')) \geq h(G) = u$$

(this follows from the definition of $\overline{h}$). Note that a hereditarily stable set corresponds to a point in $\mathcal{X}^n$ such that H is constant in some neighborhood of it. If H is not constant near x^0 then in $\frac{\delta}{2}$-neighborhood of x^0 $\exists x^1 : H(x^1) > H(x^0)$. Likewise, if H is not locally constant in x^1, then in $\frac{\delta}{4}$-neighborhood of x^1 $\exists x^2 : H(x^2) > H(x^1)$. F^k is finite and hence has no infinite increasing sequence, so we will stop at some point. That is, there exists x^m where H is locally constant and it is within δ-neighborhood of x^0. A set E' which corresponds to x^m is hereditarily stable, and satisfies the requirement of being close enough to E. ■

LEMMA 12. *Under the assumptions of Lemma 9,* $\forall u \in F \; |br(u)| < 2^{d(u)}$.

Proof. Choose a finite set E using Lemma 9 such that $\overline{h}(E) = u$. It is stable and thus by Lemma 11 we can construct E' which is hereditarily stable and $\overline{h}(E') = u$. Choose $E_0 \subseteq E'$ such that $\overline{h}(E_0) = u$ and $\forall E_1 \subset E_0 \; \overline{h}(E_1) > u$. Assume that $|E_0| = n$, $E_0 = \{e_1, e_2, \dots, e_n\}$. We also introduce the following open sets: $A_{ij} = \{x \in \mathcal{X} : \rho(x, e_j) < 1/i\}$, $A_i = \bigcup_{j=1}^n A_{ij}$. By the definition of $\overline{h}$, $\lim h(A_i) = u$. There exists i_0 such that for $i > i_0$ $A_{ij_1} \cap A_{ij_2} = \emptyset$ if $j_1 \neq j_2$. In the remaining part of the proof we assume that $i > i_0$.

We assume that $br(u)$ is non-empty since otherwise the statement is trivial. Select $v \in br(u)$ and for each A_i construct $G_i \subset A_i : |G_i| < 2^{d(u)-1}$, $\overline{h}(G_i) = v$ by Lemma 9. $K_i = \{j : G_i \cap A_{ij} \neq \emptyset\}$ is a sequence of subsets of $\{1, \dots, n\}$. Since there is only a finite number of such subsets, there is a subset K, which is equal to infinitely many of K_i, say $K = K_{i_m}$, $m \in \mathbb{N}$, $i_{m+1} > i_m$. We can now define the following sets: $A'_m = \bigcup_{j \in K} A_{i_m j}$, $E_1 = \{e_j | j \in K\}$. Note that $\overline{h}(E_1) = \lim_i h(A_i)$.

If $E_1 = E_0$ then $G_{i_m} \subseteq A'_m = A_{i_m}$ and $\forall j \; G_{i_m} \cap A_{i_m j} \neq \emptyset$. Since E' is hereditarily stable, E_0 is stable. Thus for a sufficiently large m

($\frac{1}{i_m} < \varepsilon$) we can select $E^* = \{e_j^* | e_j^* \in G_{i_m} \cap A_{i_m j}, j = 1, \ldots, n\}$ such that $\overline{h}(E^*) = \overline{h}(E_0) = u$. But since $E^* \subset G_{i_m}$, $\overline{h}(E^*) \geq v$, which leads to a contradiction ($u < v$).

So, E_1 is a proper subset of E_0. $\overline{h}(E_1) > u$ (by the choice of E_0) and $\overline{h}(E_1) \leq v$, since each A'_m contains G_{i_m} and $h(G_{i_m}) = v$, so $h(A'_m) \leq v$. Thus $h(E_1) = v$. Choose $E_2 : E_1 \subseteq E_2 \subset E_0$, $|E_2| = n - 1$. $\overline{h}(E_2) = v$ (again, by the choice of E_0). There are only n possible subsets of E_0 of this cardinality, thus $|br(u)| \leq n \leq |E'| = |E| < 2^{d(u)}$. QED. ■

LEMMA 13. *Under the assumptions of Lemma 9, if E is a hereditarily stable subset under $\overline{h}$, and $\overline{h}(E) = u \in F$ then $\overline{h}|_{P_1(E)}$ is a p-morphism from $P_1(E)$ to F^u.*

Proof. We will prove that $\overline{h}(P_1(E)) = F^u$ using induction by $|E|$. If $|E| = 1$, then $|br(u)| = 0$, and $F^u = \overline{h}(P_1(E)) = \{u\}$.

Now consider E such that $|E| > 1$. For any $E' \subseteq E$ $\overline{h}(E') \geq u$, so $\overline{h}(P_1(E)) \subseteq F^u$. For any $w \in F^u \setminus \{u\}$ there is $v \in br(u)$ such that $u \leq v \leq w$, and there is $E_1 \subseteq E$ such that $\overline{h}(E_1) = v$ (as demonstrated in the proof of Lemma 12). By induction hypothesis, $\overline{h}(P_1(E_1)) = F^{\overline{h}(E_1)}$, so $\exists E_2 \subseteq E_1$ $\overline{h}(E_2) = w$. Thus, $\overline{h}(P_1(E)) = F^u$. ■

The remaining part of the proof is almost identical to the proof of Theorem 5.5 in [5]. This method was introduced in [4], where it was used to prove that ML is not finitely axiomatizable.

We will use the family of finite frames $\Phi(k,m)$ $(k, m > 0)$. $\Phi(k,m)$ is the set of pairs (i,j) such that either $0 \leq i \leq k$ and $0 \leq j \leq 1$, or $i = k+1$ and $1 \leq j \leq m$, or $i = k+2$ and $j = 0$. The ordering is defined as follows: $(i,j) < (i',j')$ iff $(i > i')$. We will also use the frames $\Phi^i(k,m) \rightleftharpoons \Phi(k,m) \setminus \{(i,1)\}$, where $0 \leq i \leq k$.

LEMMA 14. *Let $\mathcal{X}$ be an infinite metric space.*

1. $X(\Phi(k, 2^{k+3})) \in ML_{O(\mathcal{X})}$
2. $X(\Phi^i(k,m) \notin ML_{O(\mathcal{X})}$

Proof.

1. By Lemma 6 we need to prove that there doesn't exist a p-morphism from some cone of $O(\mathcal{X})$ (that is $O(Y)$ where $Y \subseteq \mathcal{X}$) to $\Phi(k, 2^{k+3})$. But if such p-morphism exists, by Lemma 12 we have $|br(k+2,0)| < 2^{d(k+2,0)}$, which is false, since $|br(k+2,0)| = 2^{k+3}$ and $d(k+2,0) = k+3$.

2. To prove this we can either construct a p-morphism from $ML_{O(\mathcal{X})}$ to $\Phi^i(k,m)$, which is easy, or use a result from [5] (Lemma 5.3), stating that $X(\Phi^i(k,m) \notin ML$, and we already know that $ML_{O(\mathcal{X})} \subseteq ML$.

■

DEFINITION 15. A *k-formula* is a formula that does not contain propositional letters besides $p_1, \dots, p_k$.

LEMMA 16.

If A is a k-formula and $\Phi(k,m) \not\models A$ then there exists $i \leq k$ such that $\Phi^i(k,m) \not\models A$. Or, in terms of Jankov formulas, if $X(\Phi(k,m)) \in (\mathbf{Int} + A)$ then $X(\Phi^i(k,m)) \in (\mathbf{Int} + A)$ for some $i \leq k$.

Proof. Let $(\Phi(k,m), \theta)$ be a model refuting A. Since each $\theta(p_r)$ is an upwards closed set, for every i, except maybe a single one, it either contains all (i,j) or none. Since there are only k variables, and $k+1$ possible values for i, there has to be a level i^* on which $(i^*, 0)$ and $(i^*, 1)$ have the same valuations. Clearly, the model $(\Phi^{i^*}(k,m), \theta)$ also refutes A. ∎

THEOREM 17. *Let $\mathcal{X}$ be an infinite metric space. Then $ML_{O(\mathcal{X})}$ is not axiomatizable in finite number of variables, that is for any $k \in \mathbb{N}$ $ML_{O(\mathcal{X})}$ is not axiomatizable by any set of k-formulas.*

Proof. Let Σ be a set of k-formulas and suppose that Σ axiomatizes $ML_{O(\mathcal{X})}$. Then $X(\Phi(k, 2^{k+3}))$ could be derived from some finite number of axioms $A_1, \dots, A_n \in \Sigma$. But then Lemma 16 with $A = A_1 \wedge \dots \wedge A_n$ provides a contradiction, since by Lemma 14 $X(\Phi^i(k,m))$ is never in $ML_{O(\mathcal{X})}$. ∎

COROLLARY 18. *$ML_{O(\mathcal{X})}$ is not finitely axiomatizable.*

Finally, let us show that $ML_{O(\mathcal{X})}$ is not distinguishable from ML within the class of characteristic formulas of finite frames, to which Lemma 9 is applicable.

THEOREM 19. *Let $\mathcal{X}$ be an infinite metric space and F a finite frame with the least element, and $|br(u)| \neq 1$ for any $u \in F$. Then*

$$X(F) \in ML_{O(\mathcal{X})} \iff X(F) \in ML$$

Proof. If $X(F) \in ML_{O(\mathcal{X})}$ then it is also in ML because $ML_{O(\mathcal{X})} \subseteq ML$.

If $X(F) \notin ML_{O(\mathcal{X})}$ then by Lemma 6 $\exists U \subseteq \mathcal{X}\ \exists h : O(U) \twoheadrightarrow F$. By Lemmas 9 and 11 we can find a finite $E \subset U$ such that $\overline{h}(E) = 0_F$ and E is hereditarily stable. $\overline{h} : P_1(E) \to F$ is a p-morphism (see Lemma 13), hence $X(F) \notin L(P_1(E))$ and thus $X(F) \notin ML$. ∎

COROLLARY 20. *Let L be an intermediate logic, and $ML_{O(\mathcal{X})} \subseteq L \subseteq ML$ for some infinite metric space $\mathcal{X}$. Then L is not axiomatizable in finite number of variables.*

Proof. By Theorem 19 L has the same formulas of the form $X(F)$ as $ML_{O(\mathcal{X})}$ or ML. Therefore Lemma 14 applies to L, and so does the rest of the proof of Theorem 17. ∎

5 Conclusion

To summarise the results of this paper, we have discovered that logics $ML_{O(\mathcal{X})}$ (for different infinite metric spaces $\mathcal{X}$) are subsets of ML, are not finitely axiomatizable and we cannot distinguish them from ML or each other by using formulas of the form $X(F)$, where F is a finite frame with a particular property. Although these results were stated only for metric spaces, they could easily be extended to various topological spaces as well.

Many questions still remain open. Here are some of them:

1. Is (for example) $ML_{O(\mathbb{R})}$ different from ML_1 and ML?
2. Is $ML_{O(\mathbb{R})}$ recursively axiomatizable?
3. Is $ML_{O(\mathbb{R})}$ decidable?
4. What is the intersection of the logics $ML_{O(\mathcal{X})}$? Does there exist a metric space $\mathcal{X}_0$ such that $ML_{O(\mathcal{X}_0)} \subseteq ML_{O(\mathcal{X})}$ for all metric spaces $\mathcal{X}$?
5. [5] introduces a sequence of logics $ML_n = L(P_n(\omega))$, where $P_n(X) = (\{E \subseteq X \mid |E| \geq n\}, \supseteq)$. $ML_i \subseteq ML_j$ if $i \leq j$. Is it true that $ML_2 \subseteq ML_{O(\mathcal{X})}$?

Instead of open sets we can choose another class of sets to build a Kripke frame. For example closed sets generate a logic between ML_1 and ML [5]. [3] shows that many families of sets such as n-balls in $\mathbb{R}^n$, connected compacts and convex compacts generate intuitionistic logic **Int**.

BIBLIOGRAPHY

[1] Y.T. Medvedev *On interpretation of logic formulas by means of finite problems.* Doklady AN SSSR, 169, No 1, 20-24, 1966 (Russian)

[2] Y.T. Medvedev *Transformations of information and calculi that describe them: types of information and their possible transformations.* Semiotika i informatika, 13, 109-141, 1979 (Russian)

[3] I. Shapirovsky, V.Shehtman *Modal logics of regions and Minkowski spacetime.* Journal of Logic and Computation, 15(4), 559-574, Oxford Univ Press, 2005

[4] L.L. Maksimova, D.P. Skvortsov, V.B. Shehtman *The impossibility of a finite axiomatization of Medvedev's logic of finitary problems.* Doklady AN SSSR, 245, No 5, 1051-1054, 1979 (Russian)

[5] V.B. Shehtman, D.P. Skvortsov *Logics of Some Kripke Frames Connected with Medvedev Notion of Informational Types.* Studia Logica, 45, 101-118, 1986.

[6] D.P. Skvortsov *Logic of infinite problems and Kripke models on atomic semilattices of sets.* Doklady AN SSSR, 245, No 4, 798-801, 1979 (Russian)

Timofei Shatrov

119234, Russia, Moscow, ul. Leninskiye Gory, 1b-1421

timofei.shatrov@gmail.com

Locality and subsumption testing in $\mathcal{EL}$ and some of its extensions

VIORICA SOFRONIE-STOKKERMANS

ABSTRACT. In this paper we show that subsumption problems in many lightweight description logics (including $\mathcal{EL}$ and $\mathcal{EL}^+$) can be expressed as uniform word problems in classes of semilattices with monotone operators. We use possibilities of efficient local reasoning in such classes of algebras, to obtain uniform PTIME decision procedures for CBox subsumption in $\mathcal{EL}$ and extensions thereof. These locality considerations allow us to present a new family of (possibly many-sorted) logics which extend $\mathcal{EL}$ and $\mathcal{EL}^+$ with n-ary roles and/or numerical domains.

Keywords: description logics, deduction, hierarchical reasoning

1 Introduction

Description logics are logics for knowledge representation used in databases and ontologies. They provide a logical basis for modeling and reasoning about objects, classes (or concepts), and relationships (or links, or roles) between them. Recently, tractable description logics such as $\mathcal{EL}$ [2] have attracted much interest. Although they have relatively restricted expressivity, this expressivity is sufficient for formalizing the type of knowledge used in widely used ontologies such as the medical ontology SNOMED [19, 20]. Several papers were dedicated to studying the properties of $\mathcal{EL}$ and of its extensions (e.g. $\mathcal{EL}^+$ [4]), especially to understanding the limits of tractability in extensions of $\mathcal{EL}$. Undecidability results in extensions of $\mathcal{EL}$ are obtained in [1] using a reduction to the word problem for semi-Thue systems.

In this paper we show that the subsumption problem in $\mathcal{EL}$ and $\mathcal{EL}^+$ can be expressed as satisfiability problems for ground clauses w.r.t. so-called *local (extensions of) theories*, for which methods for efficient (PTIME) checking of satisfiability of ground clauses exist. General results on local theories allow us to uniformly present some extensions of $\mathcal{EL}$ and $\mathcal{EL}^+$ with n-ary roles and/or numerical domains. The main contributions of the paper are:

- We show that the subsumption problem in $\mathcal{EL}$ (resp. $\mathcal{EL}^+$) can be expressed as uniform word problem in classes of semilattices with monotone operators (possibly satisfying certain composition laws).
- We show that the corresponding classes of semilattices with operators have local presentations and use methods for efficient reasoning in local theories or in local theory extensions in order to obtain PTIME decision procedures for $\mathcal{EL}$ and $\mathcal{EL}^+$.

Table 1. Constructors and their semantics

Constructor name	Syntax	Semantics
negation	$\neg C$	$D^{\mathcal{I}} \backslash C^{\mathcal{I}}$
conjunction	$C_1 \sqcap C_2$	$C_1^{\mathcal{I}} \cap C_2^{\mathcal{I}}$
disjunction	$C_1 \sqcup C_2$	$C_1^{\mathcal{I}} \cup C_2^{\mathcal{I}}$
existential restriction	$\exists r.C$	$\{x \mid \exists y((x,y) \in r^{\mathcal{I}} \text{ and } y \in C^{\mathcal{I}})\}$
universal restriction	$\forall r.C$	$\{x \mid \forall y((x,y) \in r^{\mathcal{I}} \implies y \in C^{\mathcal{I}})\}$

- These locality considerations allow us to present new families of PTIME logics with n-ary roles which extend $\mathcal{EL}$ and $\mathcal{EL}^+$, and a PTIME extension of $\mathcal{EL}$ with two sorts, concept and num, where the concepts of sort num are interpreted as elements in the ORD-Horn, convex fragment of Allen's interval algebra.

Structure of the paper. In Sect. 2 we present generalities on description logic and introduce the description logics $\mathcal{EL}$ and $\mathcal{EL}^+$. In Sect. 3 we show that CBox subsumption can be expressed as a uniform word problem in the class of semilattices with monotone operators satisfying certain composition axioms. In Sect. 4 we present general definitions and results on local and stably local equational theories and in Sect. 5 we show that the algebraic models of $\mathcal{EL}$ and $\mathcal{EL}^+$ have local resp. stably local presentations, thus providing an alternative proof of the fact that CBox subsumption in $\mathcal{EL}$ and $\mathcal{EL}^+$ is decidable in PTIME. Locality results for more general classes of semilattice with operators are used in Sect. 6 for defining extensions of $\mathcal{EL}$ and $\mathcal{EL}^+$ with a subsumption problem decidable in PTIME.

2 Description logics: generalities

The central notions in description logics are concepts and roles. In any description logic a set N_C of *concept names* and a set N_R of *roles* is assumed to be given. Complex concepts are defined starting with the concept names in N_C, with the help of a set of *concept constructors*. The available constructors determine the expressive power of a description logic. The semantics of description logics is defined in terms of interpretations $\mathcal{I} = (D^{\mathcal{I}}, \cdot^{\mathcal{I}})$, where $D^{\mathcal{I}}$ is a non-empty set, and the function $\cdot^{\mathcal{I}}$ maps each concept name $C \in N_C$ to a set $C^{\mathcal{I}} \subseteq D^{\mathcal{I}}$ and each role name $r \in N_R$ to a binary relation $r^{\mathcal{I}} \subseteq D^{\mathcal{I}} \times D^{\mathcal{I}}$. Table 1 shows the constructor names used in $\mathcal{ALC}$ and their semantics. The extension of $\cdot^{\mathcal{I}}$ to concept descriptions is inductively defined using the semantics of the constructors.

Terminology. A *terminology* (or TBox, for short) is a finite set consisting of *primitive concept definitions* of the form $C \equiv D$, where C is a concept name and D a concept description; and *general concept inclusions* (GCI) of the form $C \sqsubseteq D$, where C and D are concept descriptions.

Interpretations. An interpretation $\mathcal{I}$ is a model of a TBox $\mathcal{T}$ if it satisfies:

- all concept definitions in $\mathcal{T}$, i.e. $C^{\mathcal{I}}=D^{\mathcal{I}}$ for all definitions $C\equiv D \in \mathcal{T}$;
- all general concept inclusions in $\mathcal{T}$, i.e. $C^{\mathcal{I}}\subseteq D^{\mathcal{I}}$ for every $C\sqsubseteq D \in \mathcal{T}$.

Since definitions can be expressed as double inclusions, in what follows we will only refer to TBoxes consisting of general concept inclusions (GCI) only.

DEFINITION 1. Let $\mathcal{T}$ be a TBox, and C_1, C_2 two concept descriptions. C_1 is subsumed by C_2 w.r.t. $\mathcal{T}$ (for short, $C_1 \sqsubseteq_{\mathcal{T}} C_2$) if and only if $C_1^{\mathcal{I}} \subseteq C_2^{\mathcal{I}}$ for every model $\mathcal{I}$ of $\mathcal{T}$.

2.1 The description logics $\mathcal{EL}$ and $\mathcal{EL}^+$

By restricting the type of allowed concept constructors less expressive but tractable description logics can be defined. If we only allow intersection and existential restriction as concept constructors, we obtain the description logic $\mathcal{EL}$ [2], a logic used in terminological reasoning in medicine [19, 20]. In [4], the extension $\mathcal{EL}^+$ of $\mathcal{EL}$ with role inclusion axioms is studied. Relationships between concepts and roles are described using CBoxes.

Constraint box. A CBox consists of a terminology $\mathcal{T}$ and a set RI of role inclusions of the form $r_1 \circ \cdots \circ r_n \sqsubseteq s$. (Since any terminology can be expressed as a set of general concept inclusions, in what follows we will view CBoxes as unions $GCI \cup RI$ of a set GCI of general concept inclusions and a set RI of role inclusions of the form $r_1 \circ \cdots \circ r_n \sqsubseteq s$.)

Interpretation. An interpretation $\mathcal{I}$ is a model of the CBox $\mathcal{C} = GCI \cup RI$ if it is a model of GCI and satisfies all role inclusions in $\mathcal{C}$, i.e. $r_1^{\mathcal{I}} \circ \cdots \circ r_n^{\mathcal{I}} \subseteq s^{\mathcal{I}}$ for all $r_1 \circ \cdots \circ r_n \subseteq s \in RI$. If $\mathcal{C}$ is a CBox, and C_1, C_2 are concept descriptions then $C_1 \sqsubseteq_{\mathcal{C}} C_2$ if and only if $C_1^{\mathcal{I}} \subseteq C_2^{\mathcal{I}}$ for every model $\mathcal{I}$ of $\mathcal{C}$.

In [4] it was shown that subsumption w.r.t. CBoxes in $\mathcal{EL}^+$ can be reduced in linear time to subsumption w.r.t. *normalized* CBoxes, in which all GCIs have one of the forms: $C \sqsubseteq D, C_1 \sqcap C_2 \sqsubseteq D, C \sqsubseteq \exists r.D, \exists r.C \sqsubseteq D$, where C, C_1, C_2, D are concept names, and all role inclusions are of the form $r \sqsubseteq s$ or $r_1 \circ r_2 \sqsubseteq r$. Therefore, in what follows, we consider w.l.o.g. that CBoxes only contain role inclusions of the form $r \sqsubseteq s$ and $r_1 \circ r_2 \sqsubseteq r$.

3 Algebraic semantics for $\mathcal{EL}$ and $\mathcal{EL}^+$

We show that CBox subsumption for $\mathcal{EL}$ and $\mathcal{EL}^+$ can be expressed as a uniform word problem for classes of semilattices with monotone operators.

3.1 Algebra: preliminaries

We assume known notions such as partially-ordered set and order filter/ideal in a partially-ordered set. For further information cf. [13]. A structure $(L, \wedge)$ consisting of a non-empty set L together with a binary operation $\wedge$ is called *semilattice* if $\wedge$ is associative, commutative and idempotent. A structure $(L, \vee, \wedge)$ consisting of a non-empty set L together with two binary operations $\vee$ and $\wedge$ on L is called *lattice* if $\vee$ and $\wedge$ are associative, commutative and idempotent and satisfy the absorption laws. A *distributive lattice* is a lattice that satisfies either of the distributive laws $(D_\wedge)$ or $(D_\vee)$, which are equivalent in a lattice.

$$(D_\wedge) \qquad \forall x, y, z \quad x \wedge (y \vee z) = (x \wedge y) \vee (x \wedge z)$$
$$(D_\vee) \qquad \forall x, y, z \quad x \vee (y \wedge z) = (x \vee y) \wedge (x \vee z).$$

A lattice having both a first and a last element is called *bounded.* A Boolean algebra is a structure $(B, \vee, \wedge, \neg, 0, 1)$, such that $(B, \vee, \wedge, 0, 1)$ is a bounded distributive lattice and $\neg$ is a unary operation that satisfies:

$$(\mathsf{Complement}) \qquad \forall x \quad \neg x \vee x = 1 \qquad \forall x \quad \neg x \wedge x = 0$$

Let $\mathcal{V}$ be a class of algebras. The *universal Horn theory* of $\mathcal{V}$ is the collection of those closed formulae valid in $\mathcal{V}$ which are of the form

$$(1) \qquad \forall x_1 \dots \forall x_n (\bigwedge_{i=1}^{n} s_{i1} = s_{i2} \rightarrow t_1 = t_2)$$

The formula (1) above is valid in $\mathcal{V}$ if for each algebra $\mathcal{A} \in \mathcal{V}$ and each assignment v of values in A to the variables, if $v(s_{i1}) = v(s_{i2})$ for all $i \in \{1, \dots, n\}$ then $v(t_1) = v(t_2)$.[1] The problem of deciding the validity of universal Horn sentences in a class $\mathcal{V}$ of algebras is also called the *uniform word problem* for $\mathcal{V}$. It is known that the uniform word problem is decidable for the classes: SL of semilattices (in PTIME), DL of distributive lattices (NP-complete), and Bool of Boolean algebras (NP-complete).

3.2 An algebraic semantics for description logics

A translation of concept descriptions into terms in a signature naturally associated with the set of constructors can be defined as follows. For every role name r, we introduce unary function symbols, $f_{\exists r}$ and $f_{\forall r}$. The renaming is inductively defined by:

- $\overline{C} = C$ for every concept name C;
- $\overline{\neg C} = \neg \overline{C}$; $\quad \overline{C_1 \sqcap C_2} = \overline{C}_1 \wedge \overline{C}_2$, $\quad \overline{C_1 \sqcup C_2} = \overline{C}_1 \vee \overline{C}_2$;
- $\overline{\exists r.C} = f_{\exists r}(\overline{C})$, $\quad \overline{\forall r.C} = f_{\forall r}(\overline{C})$.

Set theoretical semantics. There exists a one-to-one correspondence between interpretations of description logics, $\mathcal{I} = (D, \cdot^{\mathcal{I}})$ and Boolean algebras of sets $(\mathcal{P}(D), \cup, \cap, \neg, \emptyset, D, \{f_{\exists r}, f_{\forall r}\}_{r \in N_R})$, together with valuations $v : N_C \rightarrow \mathcal{P}(D)$, where $f_{\exists r}, f_{\forall r}$ are defined, for every $U \subseteq D$, by:

$$f_{\exists r}(U) \quad = \quad \{x \mid \exists y((x, y) \in r^{\mathcal{I}} \text{ and } y \in U)\}$$
$$f_{\forall r}(U) \quad = \quad \{x \mid \forall y((x, y) \in r^{\mathcal{I}} \Rightarrow y \in U)\}.$$

[1] If A is an algebra and $v : X \rightarrow A$ an assignment, then v extends in a canonical way to a homomorphism $\overline{v}$ from the algebra of terms with variables X to A. For every term t with variables in X we will, for the sake of simplicity, write $v(t)$ instead of $\overline{v}(t)$.

Let $v : N_C \to \mathcal{P}(D)$ with $v(A) = A^{\mathcal{I}}$ for all $A \in N_C$, and let $\overline{v}$ be the (unique) homomorphic extension of v to terms. Let C be a concept description and $\overline{C}$ be its associated term. Then $C^{\mathcal{I}} = \overline{v}(\overline{C})$ (denoted by $\overline{C}^{\mathcal{I}}$).

Boolean algebras with operators. Let BAO_{N_R} be the class of all Boolean algebras with operators $(B, \vee, \wedge, \neg, 0, 1, \{f_{\exists r}, f_{\forall r}\}_{r \in N_R})$, where

- $f_{\exists r}$ is a join hemimorphism, i.e. $f_{\exists r}(x \vee y) = f_{\exists r}(x) \vee f_{\exists r}(y)$, $f_{\exists r}(0) = 0$;
- $f_{\forall r}$ is a meet hemimorphism, i.e. $f_{\forall r}(x \wedge y) = f_{\forall r}(x) \wedge f_{\forall r}(y)$, $f_{\forall r}(1) = 1$;
- $f_{\forall r}(x) = \neg f_{\exists r}(\neg x)$ for every $x \in B$.

It is known that the TBox subsumption problem for $\mathcal{ALC}$ can be expressed as uniform word problem for Boolean algebras with suitable operators.

THEOREM 2. *If $\mathcal{T}$ is an $\mathcal{ALC}$ TBox consisting of general concept inclusions between concept terms formed from concept names $N_C = \{C_1, \dots, C_n\}$, and D_1, D_2 are concept descriptions, the following are equivalent:*

(1) $D_1 \sqsubseteq_{\mathcal{T}} D_2$.

(2) $\mathcal{P}(\mathbf{D}) \models \forall C_1 \dots C_n \left(\left(\bigwedge_{C \sqsubseteq D \in \mathcal{T}} \overline{C} \leq \overline{D} \right) \to \overline{D_1} \leq \overline{D_2} \right)$ *for all interpretations* $\mathcal{I} = (D, \cdot^{\mathcal{I}})$, *where* $\mathcal{P}(\mathbf{D}) = (\mathcal{P}(D), \cup, \cap, \neg, \emptyset, D, \{f_{\exists r}, f_{\forall r}\}_{r \in N_R})$.

(3) $\mathsf{BAO}_{N_R} \models \forall C_1 \dots C_n \left(\left(\bigwedge_{C \sqsubseteq D \in \mathcal{T}} \overline{C} \leq \overline{D} \right) \to \overline{D_1} \leq \overline{D_2} \right)$.

Proof: The equivalence of (1) and (2) follows from the definition of $D_1 \sqsubseteq_{\mathcal{T}} D_2$. (3) $\Rightarrow$ (2) is immediate. (2) $\Rightarrow$ (3) follows from the fact that every algebra in BAO_{N_R} homomorphically embeds into a Boolean algebra of sets. □

3.3 An algebraic semantic for $\mathcal{EL}^+$

In [15] we studied the link between TBox subsumption in $\mathcal{EL}$ and uniform word problems in the corresponding classes of semilattices with monotone functions. We now show that these results naturally extend to the description logic $\mathcal{EL}^+$. Consider the following classes of algebras:

- $\mathsf{BAO}^{\exists}_{N_R}$ the class of boolean algebras with operators $(B, \vee, \wedge, \neg, 0, 1, \{f_{\exists r}\}_{r \in N_R})$, such that $f_{\exists r}$ is a join hemimorphism;
- $\mathsf{DLO}^{\exists}_{N_R}$ the class of bounded distributive lattices with operators $(L, \vee, \wedge, 0, 1, \{f_{\exists r}\}_{r \in N_R})$, such that $f_{\exists r}$ is a join hemimorphism;
- $\mathsf{SLO}^{\exists}_{N_R}$ the class of all bounded $\wedge$-semilattices with operators $(S, \wedge, 0, 1, \{f_{\exists R}\}_{R \in N_R})$, such that $f_{\exists R}$ is monotone and $f_{\exists R}(0) = 0$.[2]

[2]For the sake of simplicity, in what follows we assume that the description logics $\mathcal{EL}$ and $\mathcal{EL}^+$ contain the additional constructors $\bot, \top$, which will be interpreted as 0 and 1. Similar considerations can be used to show that the algebraic semantics for variants of $\mathcal{E}L$ and $\mathcal{E}L^+$ having only $\top$ (or $\bot$) is given by semilattices with 1 (resp. 0).

Assume given a set RI of axioms of the form $r \sqsubseteq s$ and $r_1 \circ r_2 \sqsubseteq r$, with $r_1, r_2, r \in N_R$. We associate with RI the following set of axioms:

$$\begin{aligned} RI_a &= \{\forall x \ \ (f_{\exists r_2} \circ f_{\exists r_1})(x) \leq f_{\exists r}(x) \mid r_1 \circ r_2 \sqsubseteq r \in RI\} \cup \\ & \quad\ \{\forall x \ \ f_{\exists r}(x) \leq f_{\exists s}(x) \mid r \sqsubseteq s \in RI\}. \end{aligned}$$

Let $\mathsf{BAO}^{\exists}_{N_R}(RI)$ (resp.$\mathsf{DLO}^{\exists}_{N_R}(RI)$, $\mathsf{SLO}^{\exists}_{N_R}(RI)$) be the subclass of $\mathsf{BAO}^{\exists}_{N_R}$ (resp. $\mathsf{DLO}^{\exists}_{N_R}$, $\mathsf{SLO}^{\exists}_{N_R}$) consisting of those algebras which satisfy RI_a.

LEMMA 3. *Let $\mathcal{I} = (D, \cdot^{\mathcal{I}})$ be a model of an $\mathcal{EL}^+$ CBox, $\mathcal{C} = GCI \cup RI$. Then $(\mathcal{P}(D), \cap, \{f_{\exists r}\}_{r \in N_R}) \in \mathsf{SLO}^{\exists}_{N_R}(RI)$.*

Proof: Clearly, $(\mathcal{P}(D), \cap, \{f_{\exists r}\}_{r \in N_R}) \in \mathsf{SLO}^{\exists}_{N_R}$. Let $r_1, r_2, r \in N_R$ and $U \in \mathcal{P}(D)$.

$$\begin{aligned} f_{\exists r_1}(U) &= \{x \mid \exists y \in U \text{ s.t. } (x,y) \in r_1^{\mathcal{I}}\} \subseteq f_{\exists r}(U) \quad \text{if } r_1 \sqsubseteq r \in RI \\ f_{\exists r_2}(f_{\exists r_1}(U)) &= \{x \mid \exists y \text{ s.t. } (x,y) \in r_2^{\mathcal{I}} \text{ and } y \in f_{\exists r_1}(U)\} \\ &= \{x \mid \exists y \text{ s.t. } (x,y) \in r_2^{\mathcal{I}} \text{ and } \exists z \in U \text{ with } (y,z) \in r_1^{\mathcal{I}}\} \\ &= \{x \mid \exists z \in U \text{ s.t. } (x,z) \in (r_1 \circ r_2)^{\mathcal{I}}\} \\ &\subseteq f_{\exists r}(U) \qquad\qquad\qquad\qquad\qquad\qquad \text{if } r_1 \circ r_2 \sqsubseteq r \in RI. \end{aligned}$$

LEMMA 4. *Every $\mathbf{S} \in \mathsf{SLO}^{\exists}_{N_R}(RI)$ embeds into a lattice in $\mathsf{DLO}^{\exists}_{N_R}(RI)$. Every lattice in $\mathsf{DLO}^{\exists}_{N_R}(RI)$ embeds (as a lattice) into a lattice in $\mathsf{BAO}^{\exists}_{N_R}(RI)$.*

Proof: Let $\mathbf{S} = (S, \wedge, 0, 1, \{f_S\}_{f \in \Sigma})$ be a semilattice with 0, 1, and with monotone operators in Σ such that $f_S(0) = 0$. Consider the the lattice of all order-ideals of S, $\mathcal{OI}(\mathbf{S}) = (\mathcal{OI}(\mathbf{S}), \cap, \cup, \{0\}, S, \{\overline{f}_S\}_{f \in \Sigma})$, where join is set union, meet is set intersection, and the additional operators in Σ are defined, for every order ideal U of S, by $\overline{f}_S(U) = \downarrow f_S(U)$. Note that $\overline{f}(\{0\}) = \{0\}$ and $\overline{f}(U_1 \vee U_2) = \downarrow f(U_1 \vee U_2) = \downarrow (f(U_1) \cup f(U_2)) = \downarrow f(U_1) \cup \downarrow f(U_2)$. Thus, $\mathcal{OI}(\mathbf{S}) \in \mathsf{DLO}^{\exists}_{N_R}$.[3] Moreover, $\eta : \mathbf{S} \to \mathcal{OI}(\mathbf{S})$ defined by $\eta(x) := \downarrow x$ is an injective homomorphism w.r.t. the operations in SLO_{N_R}, i.e. $\eta(f_S(x)) = \downarrow f_S(x) = \overline{f}_S(\downarrow x)$. Let $r_1 \circ \dots \circ r_n \sqsubseteq r \in RI$, and let $U \in \mathcal{OI}(\mathbf{S})$. Then:

$$\begin{aligned} \overline{f}_{\exists r_1}(U) &= \downarrow f_{\exists r_1}(U), \\ \overline{f}_{\exists r_2}(\overline{f}_{\exists r_1}(U)) &= \overline{f}_{\exists r_2}(\downarrow f_{\exists r_1}(U)) = \downarrow f_{\exists r_2}(f_{\exists r_1}(U)). \end{aligned}$$

The second statement is a consequence of Priestley duality for distributive lattices. Let $\mathbf{L} \in \mathsf{DLO}^{\exists}_{N_R}(RI)$. Let $\mathcal{F}_p$ be the set of prime filters of L, and $B(\mathbf{L}) = (\mathcal{P}(\mathcal{F}_p), \cup, \cap, \{\overline{f}_{\exists r}\}_{r \in N_r})$, where for $r \in R$, $\overline{f}_{\exists r}$ is defined by

$$\overline{f}_{\exists r}(U) = \{F \in \mathcal{F}_p \mid \exists G \in U : f_{\exists r}(G) \subseteq F\}.$$

[3] A similar construction can be made starting from $\wedge$-semilattices with monotone operators which have only 1 (resp. 0) or neither 0 nor 1.

Let $i : \mathbf{L} \to B(\mathbf{L})$ be defined by $i(x) = \{F \in \mathcal{F}_p \mid x \in F\}$. Obviously, i is a lattice homomorphism. We show that $i(f_{\exists r}(x)) = \overline{f}_{\exists r}(i(x))$.

$$\begin{array}{rcl} \overline{f}_{\exists r}(i(x)) & = & \{F \in \mathcal{F}_p \mid \exists G \in i(x) : f_{\exists r}(G) \subseteq F\} \\ & = & \{F \in \mathcal{F}_p \mid \exists G : x \in G \text{ and } f_{\exists r}(G) \subseteq F\} \\ & \subseteq & \{F \in \mathcal{F}_p \mid f_{\exists r}(x) \subseteq F\} = i(f_{\exists r}(x)). \end{array}$$

To prove the converse inclusion, let $F \in i(f_{\exists r}(x))$. Then $F \in \mathcal{F}_p$ and $f_{\exists r}(x) \in F$. Then $x \in G = f_{\exists r}^{-1}(F)$. As F is a prime filter, and $f_{\exists r}$ is a join hemimorphism, $G = f_{\exists r}^{-1}(F)$ is a prime filter with $x \in G$ and $f_{\exists r}(G) \subseteq F$, so $F \in \overline{f}_{\exists r}(i(x))$. Finally, we show that $B(\mathbf{L})$ satisfies the axioms in RI_a. Let $U \in B(\mathbf{L})$. By definition,

$$\begin{array}{rcl} \overline{f}_{\exists r_1}(U) & = & \{F \in \mathcal{F}_p \mid \exists G_1 \in U : f_{\exists r_1}(G_1) \subseteq F\}, \\ \overline{f}_{\exists r_2}(\overline{f}_{\exists r_1}(U)) & = & \{F \in \mathcal{F}_p \mid \exists G_1 \in \overline{f}_{\exists r_1}(U) : f_{\exists r_2}(G_1) \subseteq F\} \\ & = & \{F \in \mathcal{F}_p \mid \exists G_1, \exists G_2 \in U : f_{\exists r_1}(G_2) \subseteq G_1 \\ & & \text{and } f_{\exists r_2}(G_1) \subseteq F\} \\ & \subseteq & \{F \in \mathcal{F}_p \mid \exists G_2 \in U : f_{\exists r_2}(f_{\exists r_1}(G_2)) \subseteq F\}. \end{array}$$

Assume that $r_1 \sqsubseteq r \in RI$. Then for all x, $f_{\exists r_1}(x) \leq f_{\exists r}(x)$. Let $F \in \overline{f}_{\exists r_1}(U)$. Then $f_{\exists r_1}(G_1) \subseteq F$ for some $G_1 \in U$, so also $f_{\exists r}(G_1) \subseteq F$. Hence, $\overline{f}_{\exists r_1}(U) \subseteq \overline{f}_{\exists r}(U)$. Similarly we can prove that if $r_1 \circ r_2 \sqsubseteq r \in RI$ then $\overline{f}_{\exists r_2}(\overline{f}_{\exists r_1}(U)) \subseteq \overline{f}_{\exists r}(U)$. □

THEOREM 5. *If the only concept constructors are intersection and existential restriction, then for all concept descriptions D_1, D_2 and every $\mathcal{EL}^+$ CBox $\mathcal{C}$=GCI∪RI, with concept names $N_C = \{C_1, \dots, C_n\}$ the following are equivalent:*

(1) $D_1 \sqsubseteq_{\mathcal{C}} D_2$.

(2) $\mathsf{SLO}^{\exists}_{N_R}(RI) \models \forall C_1 \dots C_n \left(\left(\bigwedge_{C \sqsubseteq D \in GCI} \overline{C} \leq \overline{D} \right) \to \overline{D_1} \leq \overline{D_2} \right)$.

Proof: We know that $C_1 \sqsubseteq_{\mathcal{C}} C_2$ iff $C_1^{\mathcal{I}} \subseteq C_2^{\mathcal{I}}$ for every model $\mathcal{I}$ of the CBox $\mathcal{C}$. Assume first that (2) holds. Let $\mathcal{I}=(D, \cdot^{\mathcal{I}})$ be an interpretation that satisfies $\mathcal{C}$. Then $(\mathcal{P}(D), \cap, \{f_{\exists r}\}_{r \in N_R}) \in \mathsf{SLO}^{\exists}_{N_R}(RI)$, hence $(\mathcal{P}(D), \cap, \{f_{\exists r}\}_{r \in N_R}) \models \left(\bigwedge_{C \sqsubseteq D \in GCI} \overline{C} \leq \overline{D} \right) \to \overline{D_1} \leq \overline{D_2}$. As $\mathcal{I}$ is a model of GCI, $\overline{C}^{\mathcal{I}} \subseteq \overline{D}^{\mathcal{I}}$ for all $C \sqsubseteq D \in GCI$, so $D_1^{\mathcal{I}} = \overline{D_1}^{\mathcal{I}} \subseteq \overline{D_2}^{\mathcal{I}} = D_2^{\mathcal{I}}$. To prove (1) ⇒ (2) note that, by Thm. 2, if $D_1 \sqsubseteq_{\mathcal{T}} D_2$ then $\mathsf{BAO}_{N_R} \models \left(\bigwedge_{C \sqsubseteq D \in \mathcal{C}} \overline{C} \leq \overline{D} \right) \to \overline{D_1} \leq \overline{D_2}$. Let $\mathbf{S} \in \mathsf{SLO}^{\exists}_{N_R}(RI)$. By Lemma 4, $\mathbf{S}$ embeds into an algebra in $\mathsf{BAO}^{\exists}_{N_R}$ which satisfies RI_a. Therefore, $\mathbf{S} \models \left(\bigwedge_{C \sqsubseteq D \in GCI} \overline{C} \leq \overline{D} \right) \to \overline{C_1} \leq \overline{C_2}$. □

We will show that the word problem for the class of algebras $\mathsf{SLO}^{\exists}_{N_R}(RI)$ is decidable in PTIME. For this we will prove that $\mathsf{SLO}^{\exists}_{N_R}(RI)$ has a "local" presentation. The general locality definitions, as well as methods for recognizing local presentations are given in Sect. 4. The application to the class of models for $\mathcal{EL}$ and $\mathcal{EL}^+$ are given in Sect. 5.

4 Local theories; local theory extensions

First-order theories are sets of formulae (closed under logical consequence), typically the set of all consequences of a set of axioms. Alternatively, we may consider the set of all models of a theory. In this paper we consider theories specified by their sets of axioms. (At places, however, we will refer to a theory, and mean the set of all its models.)

Before defining the notion of local theory and local theory extension we will introduce some preliminary notions on partial models of a theory.

Partial and total models. A partial model is a model in which some function symbols may be partial. In this paper the models we consider are partially ordered algebraic structures, i.e. the only predicates are $\leq$ and $=$.

A *weak* Π*-embedding* between the partial structures $A = (\{A_s\}_{s\in S}, \{f_A\}_{f\in\Sigma}, \{P_A\}_{P\in\mathsf{Pred}})$ and $B = (\{B_s\}_{s\in S}, \{f_B\}_{f\in\Sigma}, \{P_B\}_{P\in\mathsf{Pred}})$ is a (many-sorted) family $i = (i_s)_{s\in S}$ of total maps $i_s : A_s \to B_s$ such that

- if $f_A(a_1, \dots, a_n)$ is defined then also $f_B(i_{s_1}(a_1), \dots, i_{s_n}(a_n))$ is defined and $i_s(f_A(a_1, \dots, a_n)) = f_B(i_{s_1}(a_1), \dots, i_{s_n}(a_n))$;
- for each s, i_s is injective and an embedding w.r.t. Pred i.e. for every $P \in \mathsf{Pred}$ with arity $s_1 \dots s_n$ and every $a_1, \dots, a_n$ where $a_i \in A_{s_i}$, $P_A(a_1, \dots, a_n)$ if and only if $P_B(i_{s_1}(a_1), \dots, i_{s_n}(a_n))$.

In this case we say that A *weakly embeds* into B.

If A is a partial structure and $\beta : X \to A$ is a valuation we say that $(A, \beta) \models t_1 = t_2$ iff at least one of the following conditions holds:

(a) $\beta(t_1), \beta(t_2)$ are defined and $\beta(t_1) = \beta(t_2)$, or

(b) $\beta(t_1)$ and $\beta(t_2)$ are undefined, or

(c) $\beta(t_1)$ is defined, $t_2{=}f(s_1, \dots, s_n)$ and $\beta(s_i)$ is undefined for some i, or

(d) if $\beta(t_1)$ is defined, $t_2 = f(s_1, \dots, s_n)$ and $\beta(s_i)$ is defined for all i then $\beta(t_2)$ has to be defined and $\beta(t_1) = \beta(t_2)$.

$(A, \beta) \models t_1 \leq t_2$ is defined similarly, replacing "=" with "$\leq$" in (a)–(d). We say that $(A, \beta) \models t_1 \neq t_2$ if at least one of the following conditions holds:

(a') $\beta(t_1), \beta(t_2)$ are defined and $\beta(t_1) \neq \beta(t_2)$, or

(b') $\beta(t_1)$ or $\beta(t_2)$ are undefined.

(A, β) *satisfies a clause* C (notation: $(A, \beta) \models C$) if it satisfies at least one literal in C. A is an *(Evans) partial model* of a set of clauses $\mathcal{K}$ if $(A, \beta) \models C$ for every valuation β and every clause C in $\mathcal{K}$.

We say that $(A, \beta) \models_w (\neg)P(t_1, \dots, t_n)$, with $P \in \mathsf{Pred} \cup \{=\}$ if either $\beta(t_i)$ are all defined and $(\neg)P_A(\beta(t_1), \dots, \beta(t_n))$ is true in A, or $\beta(t_i)$ is not defined for some argument t_i of P. Weak satisfaction of clauses ($(A, \beta) \models_w C$) can then be defined in the usual way. We say that A is a *weak partial model* of a set of clauses $\mathcal{K}$ if $(A, \beta) \models_w C$ for every $\beta : X \to A$ and $C \in \mathcal{K}$.

4.1 Local theories

The notion of local theory was introduced by Givan and McAllester [9, 10]. They studied sets of Horn clauses $\mathcal{K}$ with the property that, for any ground Horn clause C, $\mathcal{K} \models C$ only if already $\mathcal{K}[C] \models C$ (where $\mathcal{K}[C]$ is the set of instances of $\mathcal{K}$ in which all terms are subterms of ground terms in either $\mathcal{K}$ or C). Since the size of $\mathcal{K}[C]$ is polynomial in the size of C for a fixed $\mathcal{K}$ and satisfiability of sets of ground Horn clauses can be checked in linear time [7], it follows that for local theories, validity of ground Horn clauses can be checked in polynomial time. Givan and McAllester proved that every problem which is decidable in PTIME can be encoded as an entailment problem of ground clauses w.r.t. a local theory [10]. The property above can be easily generalized to the notion of locality of a set of (Horn) clauses:

DEFINITION 6. A *local theory* is a set of Horn clauses $\mathcal{K}$ such that, for any set G of ground Horn clauses, $\mathcal{K} \wedge G \models \perp$ if and only if already $\mathcal{K}[G] \wedge G \models \perp$, where $\mathcal{K}[G]$ is the set of instances of $\mathcal{K}$ in which all terms are subterms of ground terms in either $\mathcal{K}$ or G.

The same considerations as above can be used to show that in any local theory satisfiability of sets of ground Horn clauses can be checked in polynomial time. In [8], Ganzinger established a link between proof theoretic and semantic concepts for polynomial time decidability of uniform word problems which had already been studied in algebra [14, 6]. In the course of this work he introduced and studied, besides locality, also the less restrictive notion of *stable locality* for equational Horn theories.

DEFINITION 7. A set $\mathcal{K}$ of Horn clauses is *stably local* if for every set G of ground clauses, if $\mathcal{K} \wedge G \models \perp$ then G can be refuted using the set $\mathcal{K}^{[G]}$ of all instances of $\mathcal{K}$ obtained by instantiating the variables with (ground) subterms of G, i.e. if

$$\mathcal{K} \wedge G \models \perp \text{ if and only if } \mathcal{K}^{[G]} \wedge G \models \perp .$$

The more general notion of Ψ-stably local theory (in which the instances to be considered are described by a closure operation Ψ) is introduced in [11]. Let $\mathcal{K}$ be a set of clauses. Let $\Psi_{\mathcal{K}}$ be a function associating with any set T of ground terms a set $\Psi_{\mathcal{K}}(T)$ of ground terms such that

(i) all ground subterms in $\mathcal{K}$ and T are in $\Psi_{\mathcal{K}}(T)$;

(ii) for all sets of ground terms T, T' if $T \subseteq T'$ then $\Psi_{\mathcal{K}}(T) \subseteq \Psi_{\mathcal{K}}(T')$;

(iii) for all sets of ground terms T, $\Psi_{\mathcal{K}}(\Psi_{\mathcal{K}}(T)) \subseteq \Psi_{\mathcal{K}}(T)$;

(iv) Ψ is compatible with any map h between constants, i.e. for any map $h : C \to C$, $\Psi_{\mathcal{K}}(\overline{h}(T)) = \overline{h}(\Psi_{\mathcal{K}}(T))$, where $\overline{h}$ is the unique extension of h to terms.

Let $\mathcal{K}^{[\Psi_\mathcal{K}(G)]}$ be the set of instances of $\mathcal{K}$ where the variables are instantiated with terms in $\Psi_\mathcal{K}(\mathsf{st}(\mathcal{K}, G))$ (set denoted in what follows by $\Psi_\mathcal{K}(G)$), where $\mathsf{st}(\mathcal{K}, G)$ is the set of all ground terms occurring in $\mathcal{K}$ or G. We say that $\mathcal{K}$ is Ψ-stably local if it satisfies:

(SLoc^Ψ) for every finite set G of ground clauses, $\mathcal{K} \cup G \models \perp$ iff $\mathcal{K}^{[\Psi_\mathcal{K}(G)]} \cup G$ has no partial model in which all terms in $\Psi_\mathcal{K}(G)$ are defined.

In the particular case when $\Psi_\mathcal{K}(G) = \mathsf{st}(\mathcal{K}, G)$ we refer to *stable locality* of the extension. The corresponding condition is denoted SLoc.

Complexity. If a set $\mathcal{K}$ of Horn clauses satisfies (SLoc^Ψ) then satisfiability of any set G of Horn clauses w.r.t. $\mathcal{K}$ is decidable in polynomial time in the size of $\Psi_\mathcal{K}(G)$. This follows from the fact that $\mathcal{K}^{[\Psi_\mathcal{K}(G)]} \cup G$ is a set of ground Horn clauses of size polynomial in the size of $\Psi_\mathcal{K}(G)$, and satisfiability of sets of ground Horn clauses (in a relational encoding, taking into account only suitable instances of the congruence axioms – which are also Horn and not more than $|\Psi_\mathcal{K}(G)|^2$) can be checked in linear time ([7], see also [8]).

Recognizing stably local theories. Locality can be recognized by proving embeddability of partial into total models [16, 18, 11]. Theories satisfying (SLoc^Ψ) can be recognized by showing that Evans partial models of $\mathcal{T}_1$ embed into total models.

THEOREM 8. *Let $\mathcal{K}$ be a set of clauses. Assume $\Psi_\mathcal{K}$ satisfies conditions (i)–(iv) above, and that every Evans partial model of $\mathcal{K}$ with the property that the set of defined terms is closed under $\Psi_\mathcal{K}$ weakly embeds into a total model of $\mathcal{K}$. Then $\mathcal{K}$ satisfies* SLoc^Ψ.

Proof: Let G be a set of ground clauses. We show that, under the given assumptions, if $\mathcal{K} \cup G \models \perp$ then $\mathcal{K}^{[\Psi_\mathcal{K}(G)]} \cup G$ has no partial algebra model in which all (ground) terms in $\Psi_\mathcal{K}(G)$ are defined. Assume that $\mathcal{K}^{[\Psi_\mathcal{K}(G)]} \cup G$ has a partial Evans model P in which all (ground) terms occurring in $\Psi_\mathcal{K}(G)$ are defined. We construct a partial model A of $\mathcal{K} \cup G$ as follows. Let $A = \{t_P \mid t \in \Psi_\mathcal{K}(G)\}$. As we want A to be a model of $\mathcal{K} \cup G$ in Evans' sense, we need to make sure that if f is an n-ary function and $t^1_P, \dots, t^n_P \in A$ and $(f(t^1, \dots, t^n))_P$ is defined and equal to, say, $t_P \in A$, then $f_A(t^1_P, \dots, t^n_P)$ is defined in A and equal to t_P. Thus, we impose that $f_A(t^1_P, \dots, t^n_P)$ is defined and yields t_P as a result iff $t_P = f(t^1, \dots, t^n)_P \in A$. We show that the set of defined terms in $\mathcal{A}$ is closed under $\Psi_\mathcal{K}$. Note first that, by definition of A, for any ground term t, t_A is defined if and only if there exists $t' \in \Psi_\mathcal{K}(G)$ such that $t_A = t'_A$. Thus,

$$\mathsf{Def}(A) = \{t \mid t \text{ ground term }, t_A \text{ defined}\} = \overline{h}(\Psi_\mathcal{K}(G)),$$

where $\overline{h}$ is the unique homomorphism which extends the map h with $h(c) = c_P$ for every constant c occurring in $\Psi_\mathcal{K}(G)$. Then:

$$\Psi_\mathcal{K}(\mathsf{Def}(A)) = \Psi_\mathcal{K}(\overline{h}(\Psi_\mathcal{K}(G))) = \overline{h}(\Psi_\mathcal{K}(\Psi_\mathcal{K}(G))) \subseteq \overline{h}(\Psi_\mathcal{K}(G)) = \mathsf{Def}(A).$$

By condition (i), all ground literals occurring in G are defined in P and (by construction) also in A. Therefore, A satisfies a ground literal L which occurs in G iff P satisfies L. Hence, A satisfies all clauses in G.

It remains to show that A satisfies $\mathcal{K}$. Let $D \in \mathcal{K}$, and $\beta : X \to A$. For every $x \in X$ there exists at least one $t \in \Psi_{\mathcal{K}}(G)$ with $\beta(x) = t_P$. Thus, there exists at least one substitution $\sigma : X \to \Psi_{\mathcal{K}}(G)$ such that $h(\sigma(t)) = \beta(t)$ for all terms t, where h is the canonical projection which associates with every term t its interpretation t_P in P. Then $\sigma(D)$ is an instance of D in $\mathcal{K}^{[\Psi_{\mathcal{K}}(G)]}$. We know that P is a model of $K^{[\Psi_{\mathcal{K}}(G)]}$, hence $(P, h) \models \sigma(D)$. Therefore $(A, \beta) \models D$.

Thus, A satisfies $\mathcal{K} \cup G$. Therefore, A weakly embeds into a total model B of $\mathcal{K}$. It is easy to see that B satisfies the same ground literals as A, so B satisfies all clauses in G. Thus, B is a model of $\mathcal{K} \cup G$, so $\mathcal{K} \cup G \not\models \perp$. □

4.2 Local theory extensions

We will also consider extensions of theories, in which the signature is extended by new *function symbols* (i.e. we assume that the set of predicate symbols remains unchanged in the extension). Let $\mathcal{T}_0$ be an arbitrary theory with signature $\Pi_0 = (S_0, \Sigma_0, \mathsf{Pred})$, where S_0 is a set of sorts, Σ_0 a set of function symbols, and Pred a set of predicate symbols. We consider extensions $\mathcal{T}_1$ of $\mathcal{T}_0$ with signature $\Pi = (S, \Sigma, \mathsf{Pred})$, where the set of sorts is $S = S_0 \cup S_1$ and the set of function symbols is $\Sigma = \Sigma_0 \cup \Sigma_1$ (i.e. the signature is extended by new sorts and function symbols). We assume that $\mathcal{T}_1$ is obtained from $\mathcal{T}_0$ by adding a set $\mathcal{K}$ of (universally quantified) clauses in the signature Π. Thus, $\mathsf{Mod}(\mathcal{T}_1)$ consists of all Π-structures which are models of $\mathcal{K}$ and whose reduct to Π_0 is a model of $\mathcal{T}_0$. In what follows, when referring to *(weak) partial models* of $\mathcal{T}_0 \cup \mathcal{K}'$, we mean (weak) partial models of $\mathcal{K}'$ whose reduct to Π_0 is a total model of $\mathcal{T}_0$.

Locality. In what follows, when we refer to sets G of ground clauses we assume that they are in the signature $\Pi^c = (S, \Sigma \cup \Sigma_c, \mathsf{Pred})$, where Σ_c is a set of new constants.

We will focus on the following type of locality of a theory extension $\mathcal{T}_0 \subseteq \mathcal{T}_1$, where $\mathcal{T}_1 = \mathcal{T}_0 \cup \mathcal{K}$ with $\mathcal{K}$ a set of (universally quantified) clauses:

(Loc) For every finite set G of ground clauses $\mathcal{T}_1 \cup G \models \perp$ iff $\mathcal{T}_0 \cup \mathcal{K}[G] \cup G$ has no weak partial model with all terms in $\mathsf{st}(\mathcal{K}, G)$ defined.

We say that an extension $\mathcal{T}_0 \subseteq \mathcal{T}_1$ is *local* if it satisfies condition (Loc). (Note that a local equational theory [8] is a local extension of the pure theory of equality (with no function symbols).) Notions of stable locality, and Ψ-(stable) locality can be defined as in the case of local theories [16, 11]. In Ψ-(stably) local theories and theory extensions hierarchical reasoning is possible. We present the ideas for the case of local theories.

Hierarchical reasoning. Consider a local theory extension $\mathcal{T}_0 \subseteq \mathcal{T}_0 \cup \mathcal{K}$. The locality conditions defined above require that, for every set G of ground clauses, $\mathcal{T}_1 \cup G$ is satisfiable if and only if $\mathcal{T}_0 \cup \mathcal{K}[G] \cup G$ has a weak partial model with additional properties. All clauses in $\mathcal{K}[G] \cup G$ have the property that the function symbols in Σ_1 have as arguments only ground terms. Therefore, $\mathcal{K}[G] \cup G$ can be flattened and purified (i.e. the function symbols in Σ_1 are separated from the other symbols) by introducing, in a bottom-up manner, new constants c_t for subterms $t = f(g_1, \dots, g_n)$ with $f \in \Sigma_1$, g_i ground $\Sigma_0 \cup \Sigma_c$-terms (where Σ_c is a set of constants which contains the constants introduced by flattening, resp. purification), together with corresponding definitions $c_t = t$. The set of clauses thus obtained has the form $\mathcal{K}_0 \cup G_0 \cup D$, where D is a set of ground unit clauses of the form $f(g_1, \dots, g_n) = c$, where $f \in \Sigma_1$, c is a constant, $g_1, \dots, g_n$ are ground terms without function symbols in Σ_1, and $\mathcal{K}_0$ and G_0 are clauses without function symbols in Σ_1. Flattening and purification preserve both satisfiability and unsatisfiability w.r.t. total algebras, and also w.r.t. partial algebras in which all ground subterms which are flattened are defined [16].

For the sake of simplicity in what follows we will always flatten and then purify $\mathcal{K}[G] \cup G$. Thus we ensure that D consists of ground unit clauses of the form $f(c_1, \dots, c_n) = c$, where $f \in \Sigma_1$, and $c_1, \dots, c_n, c$ are constants.

LEMMA 9 ([16]). *Let $\mathcal{K}$ be a set of clauses. Assume that $\mathcal{T}_0 \subseteq \mathcal{T}_0 \cup \mathcal{K}$ is a local theory extension. For any set G of ground clauses, let $\mathcal{K}_0 \cup G_0 \cup D$ be obtained from $\mathcal{K}[G] \cup G$ by flattening and purification, as explained above. Then the following are equivalent:*

(1) $\mathcal{T}_0 \cup \mathcal{K}[G] \cup G$ has a partial model with all terms in $\mathsf{st}(\mathcal{K}, G)$ *defined.*

(2) $\mathcal{T}_0 \cup \mathcal{K}_0 \cup G_0 \cup D$ has a partial model with all terms in $\mathsf{st}(\mathcal{K}, G)$ *defined.*

(3) $\mathcal{T}_0 \cup \mathcal{K}_0 \cup G_0 \cup N_0$ has a (total) model, where

$$N_0 = \{\bigwedge_{i=1}^{n} c_i = d_i \rightarrow c = d \mid f(c_1, \dots, c_n) = c, f(d_1, \dots, d_n) = d \in D\}.$$

THEOREM 10 ([16]). *Assume that the theory extension $\mathcal{T}_0 \subseteq \mathcal{T}_1$ satisfies condition* (Loc). *If all variables in the clauses in $\mathcal{K}$ occur below some function symbol from Σ_1 and if testing satisfiability of ground clauses in $\mathcal{T}_0$ is decidable, then testing satisfiability of ground clauses in $\mathcal{T}_1$ is decidable.*

Recognizing local theory extensions. The locality of an extension can be recognized by proving embeddability of partial into total models [16, 18, 11]. We will use the following notation:

$\mathsf{PMod}^{\mathsf{fd}}_{\mathsf{w}}(\Sigma_1, \mathcal{T}_1)$ is the class of all weak partial models of $\mathcal{T}_1$ in which the Σ_1-functions are partial and have a finite domain of definition and all the other function symbols are total.

For theory extensions $\mathcal{T}_0 \subseteq \mathcal{T}_1 = \mathcal{T}_0 \cup \mathcal{K}$, where $\mathcal{K}$ is a set of clauses, we consider the following condition:

$(\mathsf{Emb}^{\mathsf{fd}}_{\mathsf{w}})$ Every $A \in \mathsf{PMod}^{\mathsf{fd}}_{\mathsf{w}}(\Sigma_1, \mathcal{T}_1)$ weakly embeds into a total model of $\mathcal{T}_1$.

In what follows we say that a non-ground clause is Σ_1-*flat* if function symbols (including constants) do not occur as arguments of function symbols in Σ_1. A Σ_1-flat non-ground clause is called Σ_1-*linear* if whenever a variable occurs in two terms in the clause which start with function symbols in Σ_1, the two terms are identical, and if no term which starts with a function symbol in Σ_1 contains two occurrences of the same variable.

THEOREM 11 ([16, 18]). *Let $\mathcal{K}$ be a set of Σ-flat and Σ-linear clauses. If the extension $\mathcal{T}_0 \subseteq \mathcal{T}_1$ satisfies* $(\mathsf{Emb}_{\mathsf{w}}^{\mathsf{fd}})$ *then it satisfies* (Loc).

Similar results hold also for stable locality or Ψ-locality of an extension (cf. e.g. [16, 11]).

5 Locality and complexity of $\mathcal{EL}^+$ and $\mathcal{EL}$

We now show that the classes of algebraic models of $\mathcal{EL}^+$ and of $\mathcal{EL}$ have presentations which satisfy certain locality properties. This gives an alternative, algebraic explanation of the fact that CBox subsumption in these logics is decidable in PTIME and makes generalizations possible.

5.1 Locality and $\mathcal{EL}^+$

In this section we prove that the class $\mathsf{SLO}_\Sigma(RI)$ of semilattices with monotone operators in a set Σ satisfying a family RI_a of axioms of the form

$$\forall x\, (f_1 \circ \cdots \circ f_n)(x) \leq f(x)$$

has a local presentation, and therefore the uniform word problem w.r.t. this class can be decided in polynomial time. For the sake of simplicity we restrict, w.l.o.g., to axioms as above with $n \in \{1, 2\}$.

It is known that the theory of lattices allows a local Horn axiomatization (cf. e.g. [14, 6]). Let SL be such an axiomatization for the theory of lattices. We denote by $\mathsf{Mon}(\Sigma)$ the set $\{\mathsf{Mon}(f) \mid f \in \Sigma\}$, where

$$\mathsf{Mon}(f) \quad \forall x, y(x \leq y \rightarrow f(x) \leq f(y)).$$

THEOREM 12. *The set of Horn clauses $SL \cup \mathsf{Mon}(\Sigma) \cup RI_a$ has the property that every Evans partial model A with the properties:*

(i) for every $f \in \Sigma$, f_A is a partial function with finite definiton domain;

(ii) for each axiom in RI_a of the form $(f_1 \circ f_2)(x) \leq f(x)$, and every $a \in A$, if $f(a)$ is defined then $f_2(a)$ is defined in A;

(iii) $A \models SL \cup \mathsf{Mon}(\Sigma) \cup RI_a$;

weakly embeds into a total model of $SL \cup \mathsf{Mon}(\Sigma) \cup RI_a$.

Proof: Let A be an Evans partial model of $SL \cup \mathsf{Mon}(\Sigma) \cup RI_a$ with properties (i)–(iii). In particular, A is a poset, hence it embeds into a complete (semi)lattice S such that the meets that exist in A are preserved. (We will think of A as a subset of S.) For every $f \in \Sigma$ we define $\overline{f} : S \rightarrow S$ by

$$\overline{f}(a) = \bigwedge \{f(c) \mid a \leq c, c \in A, f_A(c) \text{ is defined}\}.$$

For every $f \in \Sigma$, $\overline{f}$ is monotone (see e.g. also [18]). We show that the axioms in RI_a are satisfied by these extensions. Let $f_1(x) \leq f_2(x) \in RI_a$ and $a \in S$. Then $\overline{f}_i(a) = \bigwedge\{f_i(c) \mid a \leq c, c \in A, f_i(c) \text{ is defined}\}$. Let $d \in A$ with $a \leq d$ and $f_2(d)$ defined. Then $f_1(d)$ is also defined and $f_1(d) \leq f_2(d)$. Thus, $\overline{f}_1(a) \leq f_2(d)$ for all $d \in A$ with $a \leq d$ and $f_2(d)$ defined, so $\overline{f}_1(a) \leq \overline{f}_2(a)$. Let now $(f_1 \circ f_2)(x) \leq f(x) \in RI_a$ and $a \in S$. Then $\overline{f}_2(a) = \bigwedge\{f_2(c) \mid a \leq c, c \in A, f_2(c) \text{ is defined}\}$. Then for every $a \leq c$, if $f_2(c)$ is defined then $\overline{f}_2(a) \leq f_2(c)$. We prove that $\overline{f}_1(\overline{f}_2(a)) \leq \overline{f}(a)$.

Note first that if $a \leq c$ and $f_1(f_2(c))$ is defined then $\overline{f}_2(a) \leq f_2(c)$. Therefore, $f_1(f_2(c)) \in \{f_1(c_1) \mid \overline{f}_2(a) \leq c_1, \text{ and } f_1(c_1) \text{ defined}\}$. Hence, $\{f_1(f_2(c)) \mid a \leq c, f_1(f_2(c)) \text{ defined}\} \subseteq \{f_1(c_1) \mid \overline{f}_2(a) \leq c_1, f_1(c_1) \text{ defined}\}$. Therefore, the infimum of the first set is larger than the infimum of the second set. Hence:

$$\begin{aligned}
\overline{f}_1(\overline{f}_2(a)) &= \bigwedge\{f_1(c_1) \mid \overline{f}_2(a) \leq c_1, f_1(c_1) \text{ is defined}\} \\
&\leq \bigwedge\{f_1(f_2(c)) \mid a \leq c \text{ and } f_1(f_2(c)) \text{ defined}\} \\
&\leq \bigwedge\{f(c) \mid a \leq c \text{ and } f(c) \text{ defined}\} = \overline{f}(a).
\end{aligned}$$

The last inequality is a consequence of the fact that if $f(d)$ is defined in A then $f_2(d)$ is defined in A, and since $A \models RI_a$, $f_1(f_2(d))$ is defined in A and $f_1(f_2(d)) \leq f(d)$. Hence, $\bigwedge\{f_1(f_2(c)) \mid a \leq c \text{ and } f_1(f_2(c)) \text{ defined}\} \leq f_1(f_2(d)) \leq f(d)$. □

COROLLARY 13. *The following are equivalent:*

(1) $SL \cup \mathsf{Mon}(\Sigma) \cup RI_a \models \forall \overline{x} \bigwedge_{i=1}^n s_i(\overline{x}) \leq s_i'(\overline{x}) \rightarrow s(\overline{x}) \leq s'(\overline{x})$;

(2) $SL \cup \mathsf{Mon}(\Sigma) \cup RI_a \wedge G \models \perp$, *where* $G = \bigwedge_{i=1}^n s_i(\overline{c}) \leq s_i'(\overline{c}) \wedge s(\overline{c}) \not\leq s'(\overline{c})$;

(3) $(SL \cup \mathsf{Mon}(\Sigma) \cup RI_a)^{[\Psi_{RI}(G)]} \wedge G \models \perp$ *where* $\Psi_{RI}(G) = \bigcup_{i \geq 0} \Psi_{RI}^i$, *with* $\Psi_{RI}^0 = \mathsf{st}(G)$, *and* $\Psi_{RI}^{i+1} = \{f_2(d) \mid f(d) \in \Psi_{RI}^i, (f_1 \circ f_2)(x) \leq f(x) \in RI_a\}$.

Here $\mathsf{st}(G)$ is the set of all (ground) subterms occurring in G. Note that $\Psi_{RI}(G)$ can have at most $|\mathsf{st}(G)| \cdot |N_R|$ elements. Thus, its size is polynomial in the size of G. On the other hand, the number of clauses in $(SL \cup \mathsf{Mon}(\Sigma) \cup RI_a)^{[\Psi_{RI}(G)]}$ is polynomial in $|\Psi_{RI}(G)|$, and satisfiability of any set of ground clauses can be tested in polynomial time. This shows that the uniform word problem for the class $\mathsf{SLO}_\Sigma(RI)$ (and thus also for $\mathsf{SLO}_{NR}^{\exists}(RI)$) is decidable in polynomial time.

EXAMPLE 14. We illustrate the ideas on an example presented in [4] (here slightly simplified). Consider the CBox C consisting of the following *GCI*:

$$\begin{aligned}
\mathsf{Endocard} &\sqsubseteq \mathsf{Tissue} \sqcap \exists \mathsf{cont\text{-}in.HeartWall} \sqcap \exists \mathsf{cont\text{-}in.HeartValve} \\
\mathsf{HeartWall} &\sqsubseteq \exists \mathsf{part\text{-}of.Heart} \\
\mathsf{HeartValve} &\sqsubseteq \exists \mathsf{part\text{-}of.Heart} \\
\mathsf{Endocarditis} &\sqsubseteq \mathsf{Inflammation} \sqcap \exists \mathsf{has\text{-}loc.Endocard} \\
\mathsf{Inflammation} &\sqsubseteq \mathsf{Disease} \\
\mathsf{Heartdisease} &= \mathsf{Disease} \sqcap \exists \mathsf{has\text{-}loc.Heart}
\end{aligned}$$

and the following role inclusions RI:

$$\begin{aligned} \mathsf{part\text{-}of} \circ \mathsf{part\text{-}of} &\sqsubseteq \mathsf{part\text{-}of} \\ \mathsf{part\text{-}of} &\sqsubseteq \mathsf{cont\text{-}in} \\ \mathsf{has\text{-}loc} \circ \mathsf{cont\text{-}in} &\sqsubseteq \mathsf{has\text{-}loc} \end{aligned}$$

We want to check whether Endocarditis $\sqsubseteq_{\mathcal{C}}$ Heartdisease. This is the case iff (with some abbreviations – e.g. f_{ci} stands for $f_{\exists\mathsf{cont\text{-}in}}$ and f_{po} for $f_{\exists\mathsf{part\text{-}of}}$, h_w and h_v for HeartWall resp. HeartValve, e for Endocard, h for Heart, etc.):

$$\begin{aligned} SL \;\cup\; & \mathsf{Mon}(f_{\mathsf{ci}}, f_{\mathsf{hl}}, f_{\mathsf{po}}) \;\cup\; \{\forall x\; f_{\mathsf{ci}}(f_{\mathsf{ci}}(x)) \leq f_{\mathsf{ci}}(x), \\ & \qquad\qquad \forall x\; f_{\mathsf{po}}(x) \leq f_{\mathsf{ci}}(x), \\ & \qquad\qquad \forall x\; f_{\mathsf{hl}}(f_{\mathsf{ci}}(x)) \leq f_{\mathsf{hl}}(x)\} \\ \cup\; & \{e \leq t \wedge f_{\mathsf{ci}}(h_w) \wedge f_{\mathsf{ci}}(h_v), h_w \leq f_{\mathsf{po}}(h), \quad h_v \leq f_{\mathsf{po}}(h), \\ & \mathsf{Endocarditis} \leq i \wedge f_{\mathsf{hl}}(e), \quad i \leq d, \quad \mathsf{Heartdisease} = d \wedge f_{\mathsf{hl}}(h), \\ & \mathsf{Endocarditis} \not\leq \mathsf{Heartdisease}\} \;\models\; \perp . \end{aligned}$$

Then $\mathsf{st}(\mathcal{K}, G) = \{f_{\mathsf{ci}}(h_w), f_{\mathsf{ci}}(h_v), f_{\mathsf{po}}(h), f_{\mathsf{hl}}(e), f_{\mathsf{hl}}(h)\}$. To compute $\Psi_{\mathcal{K}}(G)$, note that $\Psi^0_{RI} = \mathsf{st}(\mathcal{K}, G)$, $\Psi^1_{RI} = \{f_{\mathsf{ci}}(e), f_{\mathsf{ci}}(h)\}$, and $\Psi^2_{RI} = \Psi^1_{RI}$.

Thus, $\Psi_{\mathcal{K}}(G) = \{f_{\mathsf{ci}}(h_w), f_{\mathsf{ci}}(h_v), f_{\mathsf{ci}}(e), f_{\mathsf{ci}}(h), f_{\mathsf{po}}(h), f_{\mathsf{hl}}(e), f_{\mathsf{hl}}(h)\}$. After computing $(RI_a \cup \mathsf{Mon}(f_{\mathsf{ci}}, f_{\mathsf{hl}}, f_{\mathsf{po}}) \cup \mathsf{Con})^{[\Psi(G)]}$ and $SL^{[\Psi(G)]}$ we obtain the following conjunction of (Horn) ground clauses:

G	$(RI_a \wedge \mathsf{Mon} \wedge \mathsf{Con})^{[\Psi(G)]} \wedge SL^{[\Psi(G)]}$	
$e \leq t \wedge f_{\mathsf{ci}}(h_w) \wedge f_{\mathsf{ci}}(h_v)$	$f_{\mathsf{ci}}(f_{\mathsf{ci}}(x)) \leq f_{\mathsf{ci}}(x)$	for $x \in \Psi_{\mathcal{K}}(G)$
$h_w \leq f_{\mathsf{po}}(h)$	$f_{\mathsf{po}}(x) \leq f_{\mathsf{ci}}(x)$	for $x \in \Psi_{\mathcal{K}}(G)$
$h_v \leq f_{\mathsf{po}}(h)$	$f_{\mathsf{hl}}(f_{\mathsf{ci}}(x)) \leq f_{\mathsf{hl}}(x)$	for $x \in \Psi_{\mathcal{K}}(G)$
$\mathsf{Endocarditis} \leq i \wedge f_{\mathsf{hl}}(e)$		
$i \leq d$	$xRy \rightarrow f_{\mathsf{ci}}(x) R f_{\mathsf{ci}}(y)$	for $x, y \in \Psi_{\mathcal{K}}(G)$
$\mathsf{Heartdisease} = d \wedge f_{\mathsf{hl}}(h)$	$xRy \rightarrow f_{\mathsf{po}}(x) R f_{\mathsf{po}}(y)$	for $x, y \in \Psi_{\mathcal{K}}(G)$
$\mathsf{Endocarditis} \not\leq \mathsf{Heartdisease}$	$xRy \rightarrow f_{\mathsf{hl}}(x) R f_{\mathsf{hl}}(y)$	for $x, y \in \Psi_{\mathcal{K}}(G)$
		$R \in \{\leq, \geq, =\}$
	$SL^{[\Psi(G)]}$	

By Corollary 13, Endocarditis $\sqsubseteq_{\mathcal{C}}$ Heartdisease iff $\phi = G \wedge (RI_a \wedge \mathsf{Mon} \wedge \mathsf{Con})^{[\Psi(G)]} \wedge SL^{[\Psi(G)]}$ is unsatisfiable. Note that ϕ is a set of ground clauses in first-order logic with equality, containing all instances of the congruence axioms corresponding to the (ground) terms which occur in ϕ. A translation to Datalog can easily be obtained by replacing the function symbols with binary predicate symbols. Alternatively, we can process the instances in ϕ by replacing, in a bottom-up fashion, all the terms starting with function symbols (which are all ground) with new constants (and adding, separately, the corresponding definitions) (cf. e.g. the remarks in [8, 6]). The satisfiability of ϕ can therefore be checked automatically in polynomial time in the size of ϕ which in its turn is polynomial in the size of $\Psi_{\mathcal{K}}(G)$. Hence, in this case, the size of ϕ is polynomial in the size of G.

Unsatisfiability can also be proved directly: G entails the inequalities:

(1) $\mathsf{Endocarditis} \leq (d \wedge f_{\mathsf{hl}}(e))$; (2) $e \leq (f_{\mathsf{ci}}(h_w) \wedge f_{ci}(h_v))$;
(3) $(h_w \leq f_{\mathsf{po}}(h))$; (4) $(h_v \leq f_{\mathsf{po}}(h))$.

Hence $G \wedge (RI_a \wedge \mathsf{Mon} \wedge \mathsf{Con})^{[\Psi(G)]} \models e \leq f_{\mathsf{ci}}(f_{\mathsf{po}}(h)) \leq f_{\mathsf{ci}}(f_{\mathsf{ci}}(h)) \leq f_{\mathsf{ci}}(h)$. Thus, $G \wedge (RI_a \wedge \mathsf{Mon} \wedge \mathsf{Con})^{[\Psi(G)]} \models f_{\mathsf{hl}}(e) \leq f_{\mathsf{hl}}(f_{\mathsf{ci}}(h)) \leq f_{\mathsf{hl}}(h)$, so $G \wedge (RI_a \wedge \mathsf{Mon} \wedge \mathsf{Con})^{[\Psi(G)]} \models \mathsf{Endocarditis} \leq d \wedge f_{\mathsf{hl}}(h)$, which together with $d \wedge f_{\mathsf{hl}}(h) = \mathsf{Heartdisease}$ and $\mathsf{Endocarditis} \not\leq \mathsf{Heartdisease}$ leads to a contradiction.

5.2 Locality and $\mathcal{EL}$

In [15] we proved that the algebraic counterpart of the description logic $\mathcal{EL}$, namely the class of semilattices with monotone operators – axiomatized by $SL \cup \mathsf{Mon}(\Sigma)$ – has an even stronger locality property, namely for every set G of ground clauses

$$SL \cup \mathsf{Mon}(\Sigma) \wedge G \models \perp \quad \text{if and only if} \quad (SL \cup \mathsf{Mon}(\Sigma))[G] \wedge G \models \perp$$

where $\mathcal{K}[G]$ is the set of instances of $\mathcal{K}$ containing only ground terms occurring in G. In fact, we showed that the extension of the theory SL of semilattices with a family of monotone functions is local in the sense defined in [16].

THEOREM 15 ([18]). *Let G be a set of ground clauses. The following are equivalent:*

(1) $SL \cup \mathsf{Mon}(\Sigma) \wedge G \models \perp$.

(2) $SL \cup \mathsf{Mon}(\Sigma)[G] \wedge G$ *has no partial model A such that its $\{\wedge\}$-reduct is a (total) semilattice and the functions in Σ are partially defined, their domain of definition is finite and all terms in G are defined in A.*

Let $\mathsf{Mon}(\Sigma)[G]_0 \wedge G_0 \wedge \mathsf{Def}$ *be obtained from* $\mathsf{Mon}(\Sigma)[G] \wedge G$ *by purification, i.e. by replacing, in a bottom-up manner, all subterms $f(g)$ with $f \in \Sigma$, with newly introduced constants $c_{f(g)}$ and adding the definitions $f(g) = c_t$ to the set* Def. *The following are equivalent (and equivalent to (1) and (2)):*

(3) $\mathsf{Mon}(\Sigma)[G]_0 \wedge G_0 \wedge \mathsf{Def}$ *has no partial model $(A, \wedge, \{f_A\}_{f \in \Sigma})$ such that $(A, \wedge)$ is a semilattice and for all $f \in \Sigma$, f_A is partially defined, its domain of definition is finite and all terms in* Def *are defined in A;*

(4) $\mathsf{Mon}(\Sigma)[G]_0 \wedge G_0$ *is unsatisfiable in SL.*

(Note that in the presence of $\mathsf{Mon}(\Sigma)$ *the instances* $\mathsf{Con}[G]_0$ *of the congruence axioms for the functions in Σ are not necessary.)*

$$\mathsf{Con}[G]_0 = \{g{=}g' \rightarrow c_{f(g)}{=}c_{f(g')} \mid f(g){=}c_{f(g)}, f(g'){=}c_{f(g')} \in \mathsf{Def}\}.$$

This equivalence allows us to hierarchically reduce, in polynomial time, proof tasks in $SL \cup \mathsf{Mon}(\Sigma)$ to proof tasks in SL (cf. e.g. [18]) which can then be solved in polynomial time. [4]

EXAMPLE 16. We illustrate the method on an example first considered in [2]. Consider the $\mathcal{EL}$ TBox $\mathcal{T}$ consisting of the following definitions:

$$\begin{array}{rcl} A_1 & = & P_1 \sqcap A_2 \sqcap \exists r_1.\exists r_2.A_3 \\ A_2 & = & P_2 \sqcap A_3 \sqcap \exists r_2.\exists r_1.A_1 \\ A_3 & = & P_3 \sqcap A_2 \sqcap \exists r_1.(P_1 \sqcap P_2) \end{array}$$

We want to prove that $P_3 \sqcap A_2 \sqcap \exists r_1.(A_1 \sqcap A_2) \sqsubseteq_{\mathcal{T}} A_3$. We translate this subsumption problem to the following satisfiability problem:

$$\begin{array}{rl} \mathsf{SL} \cup \mathsf{Mon}(f_1, f_2) \ \cup & \{a_1 = (p_1 \wedge a_2 \wedge f_1(f_2(a_3))), \\ & a_2 = (p_2 \wedge a_3 \wedge f_2(f_1(a_1))), \\ & a_3 = (p_3 \wedge a_2 \wedge f_1(p_1 \wedge p_2)), \\ & \neg(p_3 \wedge a_2 \wedge f_1(a_1 \wedge a_2) \leq a_3)\} \models \perp . \end{array}$$

We proceed as follows: We flatten and purify the set G of ground clauses by introducing new names for the terms starting with the function symbols f_1 or f_2. Let Def be the corresponding set of definitions. We then take into account only those instances of the monotonicity and congruence axioms for f_1 and f_2 which correspond to the instances in Def, and purify them as well, by replacing the terms themselves with the constants which denote them. We obtain the following separated set of formulae:

Def	$G_0 \wedge$	$(\mathsf{Mon}(f_1, f_2)[G])_0 \wedge \mathsf{Con}[G]_0$
$f_2(a_3) = c_1$	$(a_1 = p_1 \wedge a_2 \wedge c_2)$	$a_1 R c_1 \rightarrow c_3 R c_2, \ R \in \{\leq, \geq, =\}$
$f_1(c_1) = c_2$	$(a_2 = p_2 \wedge a_3 \wedge c_4)$	$a_3 R c_3 \rightarrow c_1 R c_4, \ R \in \{\leq, \geq, =\}$
$f_1(a_1) = c_3$	$(a_3 = p_3 \wedge a_2 \wedge d_1)$	$a_1 R e_1 \rightarrow c_3 R d_1, \ R \in \{\leq, \geq, =\}$
$f_2(c_3) = c_4$	$(p_3 \wedge a_2 \wedge d_2 \not\leq a_3)$	$a_1 R e_2 \rightarrow c_3 R d_2, \ R \in \{\leq, \geq, =\}$
$f_1(e_1) = d_1$	$p_1 \wedge p_2 = e_1$	$c_1 R e_1 \rightarrow c_2 R d_1, \ R \in \{\leq, \geq, =\}$
$f_1(e_2) = d_2$	$a_1 \wedge a_2 = e_2$	$c_1 R e_2 \rightarrow c_2 R d_2, \ R \in \{\leq, \geq, =\}$
		$e_1 R e_2 \rightarrow d_1 R d_2, \ R \in \{\leq, \geq, =\}$

The subsumption is true iff $G_0 \wedge (\mathsf{Mon}(f_1, f_2)[G])_0 \wedge \mathsf{Con}[G]_0$ is unsatisfiable in the theory of semilattices. We can see this as follows: note that $a_1 \wedge a_2 \leq p_1 \wedge p_2$, i.e. $e_2 \leq e_1$. Then (using an instance of monotonicity) $d_2 \leq d_1$, so $p_3 \wedge a_2 \wedge d_2 \leq p_3 \wedge a_2 \wedge d_1 = a_3$.

This can also be checked automatically in PTIME either by using the fact that there exists a local presentation of SL or using the fact that $\mathsf{SL} = ISP(S_2)$ (i.e. every semilattice is isomorphic with a sublattice of a power

[4] We could prove a similar theorem in the presence of role inclusion axioms for certain types of role inclusions. An extension to general role inclusions – which would provide more efficient instantiations, and therefore more efficient algorithms than those provided by Corollary 13 – is subject of work in progress.

Table 2. Constructors for $\mathcal{EL}$ with n-ary roles and their semantics

Constructor	Syntax	Semantics
conjunction	$C_1 \sqcap C_2$	$C_1^{\mathcal{I}} \cap C_2^{\mathcal{I}}$
existential	$\exists R.(C_1, \dots C_n)$	$\{x \mid \exists y_1, \dots, y_n\, (x, y_1, \dots, y_n) \in R^{\mathcal{I}}$ and $y_i \in C_i^{\mathcal{I}}\}$

of S_2), where S_2 is the semilattice with two elements, hence SL and S_2 satisfy the same Horn clauses. Since the theory of semilattices is convex, satisfiability of ground clauses w.r.t. SL can be reduced to SAT solving.

6 Extensions of $\mathcal{EL}$ and $\mathcal{EL}^+$

The results described in Section 5 can easily be generalized to semilattices with n-ary monotone functions satisfying composition axioms. This allows us to define natural generalizations of $\mathcal{EL}$ and $\mathcal{EL}^+$. We start by presenting a generalization of $\mathcal{EL}$ in which n-ary roles are allowed. We then sketch possible extensions in which role inclusions are also taken into account.

6.1 Extensions of $\mathcal{EL}$

We consider extensions of $\mathcal{EL}$ with n-ary roles. The semantics is defined in terms of interpretations $\mathcal{I} = (D^{\mathcal{I}}, \cdot^{\mathcal{I}})$, where $D^{\mathcal{I}}$ is a non-empty set, concepts are interpreted as usual, and each n-ary role $R \in N_R$ is interpreted as an n-ary relation $R^{\mathcal{I}} \subseteq (D^{\mathcal{I}})^n$ (cf. Table 2). A further extension is obtained by allowing for certain concrete sorts – having the same support in all interpretations; or additionally assuming that there exist specific concrete concepts which have a fixed semantics (or additional fixed properties) in all interpretations. The extensions we consider are different from the extensions with concrete domains and those with n-ary quantifiers studied in the description logic literature (cf. e.g. [5, 3].

EXAMPLE 17. Consider a description logic having a usual ($\mathsf{concept}$) sort and a 'concrete' sort num with fixed domain $\mathbb{N}$. We may be interested in general concrete concepts of sort num (interpreted as subsets of $\mathbb{R}$) or in special concepts of sort num such as $\uparrow n$, $\downarrow n$, or $[n, m]$ for $m, n \in \mathbb{R}$. For any interpretation $\mathcal{I}$, $\uparrow n^{\mathcal{I}} = \{x \in \mathbb{R} \mid x \geq n\}$, $\downarrow n^{\mathcal{I}} = \{x \in \mathbb{R} \mid x \leq n\}$, and $[n, m]^{\mathcal{I}} = \{x \in \mathbb{R} \mid n \leq x \leq m\}$. We will denote the arities of roles using a many-sorted framework. Let $(D, \mathbb{R}, \cdot^{\mathcal{I}})$ be an interpretation with two sorts $\mathsf{concept}$ and num. A role with arity $(s_1, \dots, s_n)$ is interpreted as a subset of $D_{s_1} \times \dots \times D_{s_n}$, where $D_{\mathsf{concept}} = D$ and $D_{\mathsf{num}} = \mathbb{R}$.

1. Let price be a binary role or arity $(\mathsf{concept}, \mathsf{num})$, which associates with every element of sort $\mathsf{concept}$ its possible prices. The concept

$$\exists \mathsf{price}.\uparrow n = \{x \mid \exists k \geq n : \mathsf{price}(x, k)\}$$

represents the class of all individuals with some price greater than n.

2. Let has-weight-price be a role of arity (concept, num, num). The concept

$$\exists\ \text{has-weight-price}.(\uparrow\mathsf{y}, \downarrow\mathsf{p}) = \{x \mid \exists y' \geq \mathsf{y}, \exists p' \leq \mathsf{p} \text{ and has-weight-price}(x, y', p')\}$$

denotes the family of individuals for which a weight above y and a price below p exist.

The example below can be generalized by allowing a set of concrete sorts. We discuss the algebraic semantics of this type of extensions of $\mathcal{EL}$.

Let $\mathsf{SLO}^{\exists}_{N_R,S}$ denote the class of all structures $(S, \mathcal{P}(A_1), \dots, \mathcal{P}(A_n), \{f_{\exists r} \mid r \in N_R\})$, where S is a semilattice, $A_1, \dots, A_n$ are concrete domains, and $\{f_{\exists r} \mid r \in N_R\}$ are n-ary monotone operators. We may allow constants of concrete sort, interpreted as sets in $\mathcal{P}(A_i)$. The classes $\mathsf{DLO}^{\exists}_{N_R,S}$ and $\mathsf{BAO}^{\exists}_{N_R,S}$ of all distributive lattices resp. Boolean algebras with concrete supports and n-ary join hemimorphisms $\{f_{\exists r} | r \in N_R\}$ are defined similarly.

THEOREM 18. *If the only concept constructors are intersection and existential restriction, then for all concept descriptions D_1, D_2, and every TBox $\mathcal{T}$ consisting of general concept inclusions GCI the following are equivalent:*

(1) $D_1 \sqsubseteq_{\mathcal{T}} D_2$.

(2) $\mathsf{SLO}^{\exists}_{N_R,S} \models \forall C_1, \dots, C_n \left(\left(\bigwedge_{C \sqsubseteq D \in GCI} \overline{C} \leq \overline{D} \right) \rightarrow \overline{D_1} \leq \overline{D_2} \right)$.

Proof: Analogous to the proof of Theorem 5. □

Let SL_S be the class of all structures $\mathcal{A} = (A, \mathcal{P}(A_1), \dots, \mathcal{P}(A_n))$, with signature $\Pi = (S, \{\wedge\} \cup \Sigma, \mathsf{Pred})$ with $S = \{\mathsf{concept}, \mathsf{s}_1, \dots, \mathsf{s}_n\}$, $\mathsf{Pred} = \{\leq\} \cup \{\subseteq_i \mid 1 \leq i \leq n\}$, where $A \in SL$, the support of sort concept of $\mathcal{A}$ is A, and for all i the support sort s_i of $\mathcal{A}$ is $\mathcal{P}(A_i)$.

THEOREM 19 ([18]). *Every structure $(A, \mathcal{P}(A_1), \dots, \mathcal{P}(A_n), \{f_A\}_{f \in \Sigma})$, where*

(i) $(A, \mathcal{P}(A_1), \dots, \mathcal{P}(A_n)) \in SL_S$, *and*

(ii) for every $f \in \Sigma$ of arity $s_1 \dots s_n \rightarrow s$, f_A is a partial function from $\prod_{i=1}^{n} U_{s_i}$ to U_s with a finite definition domain on which it is monotone,

weakly embeds into a total model of $SLO_{\Sigma,S}$ (axiomatized by $SL_S \cup \mathsf{Mon}(\Sigma)$).

COROLLARY 20. *Let $G = \bigwedge_{i=1}^{n} s_i(\overline{c}) \leq s'_i(\overline{c}) \wedge s(\overline{c}) \not\leq s'(\overline{c})$ be a set of ground unit clauses in the extension Π^c of Π with new constants Σ_c. The following are equivalent:*

(1) $SL_S \cup \mathsf{Mon}(\Sigma) \wedge G \models \perp$.

(2) $SL_S \cup \mathsf{Mon}(\Sigma)[G] \wedge G$ has no partial model with a total $\{\wedge_{SL}\}$-reduct in which all terms in G are defined.

Let $\bigcup_{i=0}^{n} \mathsf{Mon}(\Sigma)[G]_i \wedge G_i \wedge \mathsf{Def}$ *be obtained from* $\mathsf{Mon}(\Sigma)[G] \wedge G$ *by purification, i.e. by replacing, in a bottom-up manner, all subterms* $f(g)$ *of sort* s *with* $f \in \Sigma$, *with newly introduced constants* $c_{f(g)}$ *of sort* s *and adding the definitions* $f(g) = c_t$ *to the set* Def. *We thus separate* $\mathsf{Mon}(\Sigma)[G] \wedge G$ *into a conjunction of constraints* $\Gamma_i = \mathsf{Mon}(\Sigma)[G]_i \wedge G_i$, *where* Γ_0 *is a constraint of sort* semilattice *and for* $1 \leq i \leq n$, Γ_i *is a set of constraints over terms of sort* i *(*i *being the concrete sort with fixed support* $\mathcal{P}(A_i)$*). Then the following are equivalent (and are also equivalent to (1) and (2)):*

(3) $\bigcup_{i=0}^{n} \mathsf{Mon}(\Sigma)[G]_i \wedge G_i \wedge \mathsf{Def}$ *has no partial model with a total* $\{\wedge_{SL}\}$*-reduct in which all terms in* Def *are defined.*

(4) $\bigcup_{i=0}^{n} \mathsf{Mon}(\Sigma)[G]_i \wedge G_i$ *is unsatisfiable in the many-sorted disjoint combination of* SL *and the concrete theories of* $\mathcal{P}(A_i)$, $1 \leq i \leq n$.

The complexity of the uniform word problem of $SL_S \cup \mathsf{Mon}(\Sigma)$ depends on the complexity of the problem of testing the satisfiability — in the many-sorted disjoint combination of SL with the concrete theories of $\mathcal{P}(A_i)$, $1 \leq i \leq n$ — of sets of clauses $C_{\mathsf{concept}} \cup \bigcup_{i=1}^{n} C_i \cup \mathsf{Mon}$, where C_{concept} and C_i are unit clauses of sort concept resp. s_i, and Mon consists of possibly mixed ground Horn clauses.

Specific extensions of the logic $\mathcal{EL}$ can be obtained by imposing additional restrictions on the interpretation of the "concrete"-type concepts within $\mathcal{P}(A_i)$. (For instance, we can require that numerical concepts are always interpreted as intervals, as in Example 17.)

THEOREM 21. *Consider the following extensions of* $\mathcal{EL}$ *with* n*-ary roles:*

(1) *The one-sorted extension of* $\mathcal{EL}$ *with* n*-ary roles.*

(2) *The extension of* $\mathcal{EL}$ *with two sorts,* concept *and* num, *where the semantics of classical concepts is the usual one, and the concepts of sort* num *are interpreted as elements in the ORD-Horn, convex fragment of Allen's interval algebra [12], where any CBox can contain many-sorted GCI's over concepts, as well as constraints over the numerical data expressible in the ORD-Horn fragment.*

In both cases, CBox subsumption is decidable in PTIME.

Proof: (1) is an immediate consequence of results in [18]. We prove (2) as follows. The assumption on the semantics of the extension of $\mathcal{EL}$ we made ensures that all algebraic models are two-sorted structures of the form $\mathcal{A} = ((A, \wedge), (\mathcal{I}nt(\mathbb{R}, O), \{f_{\mathcal{A}}\}_{f \in \Sigma})$, with sorts $\{\mathsf{concept}, \mathsf{num}\}$, such that $(A, \wedge)$ is a semilattice, $\mathcal{I}nt(\mathbb{R}, O)$ is an interval algebra in the Ord-Horn fragment of Allen's interval arithmetic [12], and for all $f \in \Sigma$, f_A is a monotone (many-sorted) function. We will denote the class of all these structures by SL_{OrdHorn}.

Note that the Ord-Horn fragment of Allen's interval arithmetic has the property that all operations and relations between intervals can be represented by Ord-Horn clauses, i.e. clauses over atoms $x \leq y, x = y$, containing at most one positive literal ($x \leq y$ or $x = y$) and arbitrarily many negative literals (of the form $x \neq y$). Nebel and Bürkert [12] proved that a finite set of Ord-Horn clauses is satisfiable over the real numbers iff it is satisfiable over posets. As the theory of partial orders is convex, this means that although the theory of reals is not convex w.r.t. $\leq$, we can always assume that the theory of Ord-Horn clauses is convex. The main result in Corollary 20 can be adapted without problems to show that if $G = \bigwedge_{i=1}^{n} s_i(\overline{c}) \leq s'_i(\overline{c}) \wedge s(\overline{c}) \not\leq s'(\overline{c})$ is a set of ground unit clauses in the extension Π^c of Π with new constants Σ_c, and if $\mathsf{Mon}(\Sigma)[G]_{\mathsf{c}} \wedge \mathsf{Mon}(\Sigma)[G]_{\mathsf{num}} \wedge G_{\mathsf{c}} \wedge G_{\mathsf{num}} \wedge \mathsf{Def}$ are obtained from $\mathsf{Mon}(\Sigma)[G] \wedge G$ by purification, the following are equivalent:

- $SL_{\mathsf{OrdHorn}} \cup \mathsf{Mon}(\Sigma) \wedge G \models \perp$;
- $\mathsf{Mon}(\Sigma)[G]_0 \wedge G_0 \wedge \mathsf{Con}[\mathsf{Def}]_0$ is unsatisfiable in the combination of SL and the Ord-Horn fragment of Allen's interval arithmetic.

In order to test the unsatisfiability of the latter problem we proceed as follows. We first note that, due to the convexity of the theories involved and to the fact that all constraints in $G_0 \wedge \mathsf{Mon}(\Sigma)[G]_0 \wedge \mathsf{Con}[\mathsf{Def}]_0$ are separated (in the sense that there are no mixed atoms) if

(1) $G_0 \wedge \mathsf{Mon}(\Sigma)[G]_0 \wedge \mathsf{Con}[\mathsf{Def}]_0 \models \perp$, then:

(2) there exists a clause $C = (\bigwedge c_i = d_i \rightarrow c = d)$ in $\mathsf{Mon}(\Sigma)[G]_0 \cup \mathsf{Con}[\mathsf{Def}]_0$ such that $G_0 \models \bigwedge c_i = d_i$ and $G_0 \wedge \{c = d\} \wedge (\mathsf{Mon}(\Sigma)[G]_0 \wedge \mathsf{Con}[\mathsf{Def}]_0) \backslash \{\mathsf{C}\} \models \perp$.

In order to prove this, let $\mathcal{D}$ be the set of all atoms $c_i R_i d_i$ occurring in premises of clauses in $\mathsf{Mon}(\Sigma)[G]_0 \cup \mathsf{Con}[\mathsf{Def}]_0$. As every model of $G_0 \wedge \bigwedge_{(cRd) \in \mathcal{D}} \neg(cRd)$ is also a model of $G_0 \wedge \mathsf{Mon}(\Sigma)[G]_0 \cup \mathsf{Con}[\mathsf{Def}]_0$, and the last formula is by (1) unsatisfiable, $G_0 \wedge \bigwedge_{(cRd) \in \mathcal{D}} \neg(cRd) \models \perp$ in the combination of the Ord-Horn fragment over posets with the theory of semilattices. Let G_0^+ be the conjunction of all atoms in G_0, and G_0^- be the set of all negative literals in G_0. Then $G_0^+ \models \bigvee_{(cRd) \in \mathcal{D}} (cRd) \vee \bigvee_{\neg L \in (G_0)^-} L$. Since the constraints are sort-separated and both theories involved are convex, it follows that either $G_0 \models \perp$ or else $G_0 \models cRd$ for some $(cRd) \in \mathcal{D}$. We can repeat the process until all the premises of some clause in $\mathsf{Mon}(\Sigma)[G]_0 \cup \mathsf{Con}[\mathsf{Def}]_0$ are proved to be entailed by G_0. Thus, (2) holds.

By iterating the argument above we can always – if (1) holds – successively entail sufficiently many premises of monotonicity and congruence axioms in order to ensure that, in the end,

(3) there exists a set $\{C_1, \dots, C_n\}$ of clauses in $\mathsf{Mon}(\Sigma)[G]_0 \cup \mathsf{Con}[\mathsf{Def}]_0$ with $C_j = (\bigwedge c_i^j = d_i^j \rightarrow c^j = d^j)$, such that for all $k \in \{0, \dots, n-1\}$,

$$G_0 \wedge \bigwedge_{j=1}^{k} (c^j = d^j) \models \bigwedge c_i^{k+1} = d_i^{k+1} \text{ and } G_0 \wedge \bigwedge_{j=1}^{n} (c^j = d^j) \models \perp .$$

Note that (3) implies (1), since the conditions in (3) imply that $G_0 \wedge \bigwedge_{j=1}^{n}(c^j = d^j)$ is logically equivalent with $G_0 \wedge C_1 \wedge \dots C_n$, which (as set of clauses) is contained in the set of clauses $G_0 \wedge \mathsf{Mon}(\Sigma)[G]_0 \wedge \mathsf{Con}[\mathsf{Def}]_0$.

This means that in order to test satisfiability of $G_0 \wedge \mathsf{Mon}(\Sigma)[G]_0 \cup \mathsf{Con}[\mathsf{Def}]_0$ we need to test entailment of the premises of $\mathsf{Mon}(\Sigma)[G]_0 \cup \mathsf{Con}[\mathsf{Def}]_0$ from G_0; when all premises of some clause are provably true we delete the clause and add its conclusion to G_0. The PTIME assumptions for concept subsumption and for the Ord-Horn fragment ensure that this process terminates in PTIME. □

EXAMPLE 22. Consider the special case described in Example 17. Assume that the concepts of sort num used in any TBox are of the form $\uparrow n, \downarrow m$ and $[n, m]$. Consider the TBox $\mathcal{T}$ consisting of the following GCIs:

$$\{\exists \mathsf{price}(\downarrow n_1) \sqsubseteq \mathsf{affordable}, \quad \exists \mathsf{weight}(\uparrow m_1) \sqcap \mathsf{car} \sqsubseteq \mathsf{truck}, \\ \mathsf{has\text{-}weight\text{-}price}(\uparrow m, \downarrow n) \sqsubseteq \exists \mathsf{price}(\downarrow n) \sqcap \exists \mathsf{weight}(\uparrow m), \\ \downarrow n \sqsubseteq \downarrow n_1, \quad \uparrow m \sqsubseteq \uparrow m_1, \quad C \sqsubseteq \mathsf{car}, \quad C \sqsubseteq \exists\ \mathsf{has\text{-}weight\text{-}price}(\uparrow m, \downarrow n)\ \}$$

In order to prove that $C \sqsubseteq_{\mathcal{T}} \mathsf{affordable} \sqcap \mathsf{truck}$ we proceed as follows. We refute $\bigwedge_{D \sqsubseteq D' \in \mathcal{T}} \overline{D} \leq \overline{D}' \wedge \overline{C} \not\leq \mathsf{affordable} \wedge \mathsf{truck}$. We purify the problem introducing definitions for the terms starting with existential restrictions, and express the interval constraints using constraints over $\mathbb{Q}$ and obtain the following set of constraints:

Def	C_{num}	C_{concept}	Mon
$f_{\mathsf{price}}(\downarrow n_1) = c_1$	$n \leq n_1$	$c_1 \leq \mathsf{affordable}$	$n_1 \leq n \rightarrow c_1 \leq c$
$f_{\mathsf{price}}(\downarrow n) = c$	$m \geq m_1$	$d_1 \wedge \mathsf{car} \leq \mathsf{truck}$	$n_1 \geq n \rightarrow c_1 \geq c$
$f_{\mathsf{weight}}(\uparrow m_1) = d_1$		$e \leq c \wedge d$	$m_1 \geq m \rightarrow d_1 \leq d$
$f_{\mathsf{weight}}(\uparrow m) = d$		$C \leq \mathsf{car}$	$m_1 \leq m \rightarrow d_1 \geq d$
$f_{\text{h-w-p}}(\uparrow m, \downarrow n) = e$		$C \leq e$	
		$C \not\leq \mathsf{affordable} \wedge \mathsf{truck}$	

The task of proving $C \sqsubseteq_{\mathcal{T}} \mathsf{affordable} \sqcap \mathsf{truck}$ can therefore be reduced to checking if $C_{\mathsf{num}} \wedge C_{\mathsf{concept}} \wedge \mathsf{Mon}$ is satisfiable w.r.t. the combination of SL (sort $\mathsf{concept}$) with $LI(\mathbb{Q})$ (sort num). For this, we note that C_{num} entails the premises of the first, second, and fourth monotonicity rules. Thus, we can add $c \leq c_1$ and $d \leq d_1$ to C_{concept}. Thus, we deduce that $C \leq e \wedge \mathsf{car} \leq (c \wedge d) \wedge \mathsf{car} \leq c_1 \wedge (d_1 \wedge \mathsf{car}) \leq \mathsf{affordable} \wedge \mathsf{truck}$, which contradicts the last clause in C_{concept}.

A similar procedure can be used in general for testing (in PTIME) the satisfiability of mixed constraints in the many-sorted combination of SL with concrete domains of sort num, assuming that all concepts of sort num are interpreted as intervals and the constraints C_{num} are expressible in a PTIME, convex fragment of Allen's interval algebra.

6.2 Extensions of $\mathcal{EL}^+$

For roles with arbitrary arity we also consider role inclusion constraints of the form $r_1 \circ r_2 \sqsubseteq r$. This means that, for every interpretation

$\mathcal{I} = (D, A_1, \dots, A_n)$, if $(x_1, \dots, x_n) \in r_1^{\mathcal{I}}$ and $(x_n, \dots, x_{n+k}) \in r_2^{\mathcal{I}}$ then the tuple $(x_1, \dots, x_{n-1}, x_{n+1}, \dots, x_{n+k}) \in r^{\mathcal{I}}$. The monotone functions associated with r_1, r_2 are:

$$f_{\exists r_2}(U_{n+1}, \dots, U_{n+k}) = \{y_n \mid \exists y_i {\in} U_i, n+1 {\leq} i {\leq} n+k, (y_n, y_{n+1}, \dots, y_{n+k}) {\in} r_2\},$$
$$f_{\exists r_1}(U_2, \dots, U_n) = \{y_1 \mid \exists y_i \in U_i, 2 \leq i \leq n, (y_1, y_2, \dots, y_n) \in r_1\}.$$

The corresponding composition rule at algebraic level is:

$$\begin{array}{l} f_{\exists r_1}(U_2, \dots, U_{n-1}, f_{\exists r_2}(U_{n+1}, \dots, U_{n+k})) = \\ \{y_1 \mid \exists y_i {\in} U_i, 2 {\leq} i {\leq} n-1, \exists y_n {\in} f_{\exists r_2}(U_{n+1}, \dots, U_{n+k}), (y_1, y_2, \dots, y_n) {\in} r_1\} = \\ \{y_1 \mid \exists y_i \in U_i, 2 \leq i \leq n-1, \exists y_i \in U_i, n+1 \leq i \leq n+k, \\ \qquad (y_n, y_{n+1}, \dots, y_{n+k}) \in r_2^{\mathcal{I}} \text{ and } (y_1, y_2, \dots, y_n) \in r_1^{\mathcal{I}}\} = \\ \{y_1 \mid \exists y_i {\in} U_i, \text{ such that for } 2 \leq i \leq n+k, i \neq n, \\ \qquad (y_1, y_2, \dots, y_{n-1}, y_{n+1}, \dots, y_{n+k}) {\in} r_2^{\mathcal{I}} \text{o} r_1^{\mathcal{I}}\} \subseteq \\ \{y_1 \mid \exists y_i \in U_i, \text{ such that for } 2 \leq i \leq n+k, i \neq n, \\ \qquad (y_1, y_2, \dots, y_{n-1}, y_{n+1}, \dots, y_{n+k}) \in r^{\mathcal{I}}\} = \\ = f_{\exists r}(U_2, \dots, U_{n-1}, U_{n+1}, \dots, U_{n+k}). \end{array}$$

THEOREM 23. *The set of Horn clauses $SL \cup \mathsf{Mon}(\Sigma) \cup RI_a$, where the functions in Σ may be n-ary, has the property that every Evans partial model A with the properties:*

(i) for every $f {\in} \Sigma$, f_A is a partial function with a finite definition domain;

(ii) for each axiom $\forall x_1, \dots, x_{n+k}(f_1(x_1, \dots, x_{n-1}, f_2(x_{n+1}, \dots, x_{n+k})) \leq f(x_1, \dots, x_{n-1}, x_{n+1}, \dots, x_{n+k})) \in RI_a$, and all $a_1, \dots, a_{n+k} \in A$, if $f_A(a_1, \dots, a_{n-1}, a_{n+1}, \dots, a_{n+k})$ is defined then $f_{2A}(a_{n+1}, \dots, a_{n+k})$ is defined in A;

(iii) $A \models SL \cup \mathsf{Mon}(\Sigma) \cup RI_a$;

weakly embeds into a total model of $SL \cup \mathsf{Mon}(\Sigma) \cup RI_a$.

Proof: Similar to the proof of Theorem 12. □

7 Conclusions

In this paper we have shown that subsumption problems in $\mathcal{EL}$ can be expressed as uniform word problems in classes of semilattices with monotone operators, and that subsumption problems in $\mathcal{EL}^+$ can be expressed as uniform word problems in classes of semilattices with monotone operators satisfying certain composition laws. This allowed us to obtain, in a uniform way, PTIME decision procedures for $\mathcal{EL}$, $\mathcal{EL}^+$, and extensions thereof. These locality considerations allow us to present a new family of PTIME (many-sorted) logics which extend $\mathcal{EL}$ with n-ary roles and/or with numerical domains. These extensions are different from other types of extensions studied in the description logic literature such as extensions with n-ary existential quantifiers (cf. e.g. [3]) or with concrete domains [5].

The results in [17] show that the class of semilattices with monotone operations allows ground (equational) interpolation. We plan to use the results presented in this paper for studying interpolation properties in extensions of $\mathcal{EL}$ and for analyzing possibilities of efficient (modular) reasoning in combinations of ontologies based on extensions of $\mathcal{EL}$.

Acknowledgments. This work was partly supported by the German Research Council (DFG) as part of the Transregional Collaborative Research Center "Automatic Verification and Analysis of Complex Systems" (SFB/TR 14 AVACS). See `www.avacs.org` for more information.

BIBLIOGRAPHY

[1] F. Baader. Restricted role-value-maps in a description logic with existential restrictions and terminological cycles. In *Proc. of the 2003 International Workshop on Description Logics* (DL2003), CEUR-WS, 2003.

[2] F. Baader. Terminological cycles in a description logic with existential restrictions. In: G. Gottlob and T. Walsh, editors, *Proc. of the 18th International Joint Conference in Artificial Intelligence*, pages 325–330, Morgan Kaufmann, 2003.

[3] F. Baader, C. Lutz, E. Karabaev, and M. Theißen. A new n-ary existential quantifier in description logics. In *Proc. 28th Annual German Conference on Artificial Intelligence* (KI 2005), *LNAI 3698*, pages 18-033, Springer, 2005.

[4] F. Baader, C. Lutz, B. Suntisrivaraporn. Is tractable reasoning in extensions of the description logic EL useful in practice? *Journal of Logic, Language and Information* (M4M special issue). To appear.

[5] F. Baader, S. Brandt, and C. Lutz. Pushing the EL Envelope. In *Proceedings of the Nineteenth International Joint Conference on Artificial Intelligence* IJCAI-05, Morgan-Kaufmann Publishers, 2005.

[6] S. Burris. Polynomial time uniform word problems. *Mathematical Logic Quarterly*, 41:173–182, 1995.

[7] W.F. Dowling and J.H. Gallier. Linear time algorithms for testing the satisfiability of propositional Horn formulae. *J. Logic Programming* 77, 192–217, 1984.

[8] H. Ganzinger. Relating semantic and proof-theoretic concepts for polynomial time decidability of uniform word problems. In *Proc. 16th IEEE Symposium on Logic in Computer Science (LICS'01)*, pages 81–92. IEEE Computer Society Press, 2001.

[9] R. Givan and D. McAllester. New results on local inference relations. In *Principles of Knowledge Representation and reasoning: Proceedings of the Third International Conference (KR'92)*, pages 403–412. Morgan Kaufmann Press, 1992.

[10] R. Givan and D.A. McAllester. Polynomial-time computation via local inference relations. *ACM Transactions on Computational Logic*, 3(4):521–541, 2002.

[11] C. Ihlemann, S. Jacobs, and V. Sofronie-Stokkermans. On local reasoning in verification. In *Proc. TACAS 2008, LNCS 4963*, pages 265–281, Springer 2008.

[12] B.Nebel and H.-J.Bürckert. Reasoning about temporal relations: A maximal tractable subclass of Allen's interval algebra. *Journal of the ACM*, 42 (1): 43–66, 1995.

[13] B.A. Davey and H.A. Priestley. *Introduction to Lattices and Order*. Cambridge University Press, 1990.

[14] T. Skolem. (1920). Logisch-kombinatorische Untersuchungen über die Erfüllbarkeit und Beweisbarkeit mathematischer Sätze nebst einem Theorem über dichte Mengen. *Skrifter utgit av Videnskabsselskapet i Kristiania, I. Matematisk-naturvidenskabelig klasse, 4*, pages 1–36.

[15] V. Sofronie-Stokkermans. Automated theorem proving by resolution in non-classical logics. *Annals of Mathematics and Artificial Intelligence* (Special issue "Knowledge Discovery and Discrete Mathematics: Dedicated to the Memory of Peter L. Hammer"), 49 (1-4): 221-252, 2007.

[16] V. Sofronie-Stokkermans. Hierarchic reasoning in local theory extensions. In R. Nieuwenhuis, editor, *20th International Conference on Automated Deduction (CADE-20), LNAI 3632*, pages 219–234. Springer, 2005.
[17] V. Sofronie-Stokkermans. Interpolation in local theory extensions. In *IJCAR'2006: Int. Joint Conf. on Automated Reasoning, LNCS 4130*, pages 235–250. Springer, 2006.
[18] V. Sofronie-Stokkermans and C. Ihlemann. Automated reasoning in some local extensions of ordered structures. *Journal of Multiple-Valued Logics and Soft Computing* (Special issue dedicated to ISMVL'07), 13 (4-6), 397-414, 2007.
[19] K.A. Spackman, K.E. Campbell, R.A. Cote. SNOMED RT: A reference terminology for health care. *Journal of the Americal Medical Informatics Association*, pages 640-644, 1997. Fall Symposium Supplement.
[20] K.A. Spackman. Normal forms for description logic expression of clinical concepts in SNOMED RT. *Journal of the Americal Medical Informatics Association*, pages 627–631, 2001. Symposium Supplement.

Viorica Sofronie-Stokkermans
Max-Planck-Institut für Informatik,
Campus E1.4, Saarbrücken,
Germany
sofronie@mpi-inf.mpg.de

A decision procedure for alternation-free modal μ-calculi

YOSHINORI TANABE, KOICHI TAKAHASHI AND MASAMI HAGIYA

ABSTRACT. The alternation-free fragment of the propositional modal μ-calculus (AFμ) allows less complex decision procedure for satisfiability judgment than the full μ-calculus, yet it still has strong expressive power. In this paper, we present a concrete decision procedure with its complexity for AFμ enriched by features of nominals, backward modalities, and functional modalities. While AFμ with all three features is undecidable, AFμ with two out of the three features are decidable and the procedure is sound and complete for these combinations. The procedure is suitable for implementation with BDDs. An application of the decision procedure for program analysis is also reported.

Keywords: modal μ-calculus, satisfiability, decision procedure

1 Introduction

The propositional modal μ-calculus has been studied extensively, and has been applied to verification problems. The authors have proposed to apply its variants for analyzing graph transformation systems [1, 2]. One of the key operations for such analysis is determine whether a given formula is satisfiable.

Known decision procedures for the propositional modal μ-calculus are complex. To reduce the complexity, we have restricted ourselves to its alternation-free fragment. Although this restriction does not reduce the theoretical complexity (both are EXPTIME-complete), it does lead to efficient implementation using binary decision diagrams (BDDs). Meanwhile, the restricted fragments are still sufficiently powerful, and we can apply them to the analysis. In previous research, we established decision procedures for the alternation-free modal μ-calculus (AFμ) [3] and its extension with backward modalities [4]. They are used to analyze programs that handle XML documents. To apply them to other areas such as shape analysis [5], we require logics with more features, nominals and functional modalities, as well as backward modalities.

A nominal is a type of atomic formula, which is satisfied by one and only one node in a Kripke structure. A functional modality is interpreted in the Kripke structures as a (partial) function on the set of states, whereas an ordinary modality is interpreted as a relation on it. A backward modality

m^{-1}, where m is an ordinary (forward) modality, follows the transition relation of a Kripke structure in the reverse direction.

In this paper, we extend the above-mentioned decision procedure to AFμ that has (1) nominals, (2) backward modalities, and (3) functional modalities. Unfortunately, the logic with all three features is undecidable [6]. However, the logics with any two features out of the three are decidable, and we prove that the procedure is sound and complete for these combinations. Moreover, the procedure is sound for the logic with all three features. The complexity of the procedure is $2^{\mathcal{O}(n \log n)}$ for AFμ + (2) + (3) and $2^{\mathcal{O}(n^2 \log n)}$ for the other two combinations. We have an application of the decision procedure for shape analysis. We developed an experimental tool for analyzing programs that manipulate pointers. The procedure was implemented in the tool using JavaBDD [7] with a small modification to fulfill requirements of the tool. Some properties of programs were successfully verified, including the partial correctness of the Deutsch-Schorr-Waite marking algorithm.

The propositional modal μ-calculus was introduced by Kozen [8], which he proved to be decidable. Emerson and Jutla proved that the complexity of its satisfiability problem is EXPTIME-complete [9]. Bonatti and Peron showed that the satisfiability of the modal μ-calculus with nominals, backward modalities, and graded modalities is undecidable [6]. By checking their proof, one can see that AFμ + (1) + (2) + (3) is also undecidable. Bonatti *et al.* [10] proved that the satisfiability problems of the modal μ-calculus extended with any two features out of nominals, backward modalities, and graded modalities are decidable, and their complexity is EXPTIME-complete. Since a functional modality is a type of graded modality, and the satisfiability problem of AFμ is already EXPTIME-hard, it is apparent from their results that "AFμ + any two of (1), (2), and (3)" are also EXPTIME-complete.

The decision procedure given in [10] is based on alternating tree automata. From an application point of view, it is extremely complex and no running implementations are reported. On the other hand, our procedure consists of set operations, which are easily encoded in BDDs for efficient implementation owing to the restriction to the alternation-free fragment. To the best of our knowledge, our decision procedure is the first one that covers all the above-mentioned features, and has been actually applied to solve some concrete problems. Its applicability is also guaranteed by the accurate computational complexity obtained by our analysis.

Various implementations exist for variants of modal logics. Emerson provided a decision procedure for CTL based on the tableau method [11]. Pan *et al.* provided efficient implementations of the decision procedure using BDDs for the minimal modal logic **K** [12, 13]. MONA [14] is a famous tool that implements decision procedures for WS1S and WS2S. This tool also utilizes BDDs for its implementation. Kupferman and Vardi [15] showed an efficient decision procedure for tree automata which leads to implementation for the modal μ-calculus. Eijck developed a theorem prover for hybrid logics [16] based on the tableau method. The decision procedure presented in this paper cannot be replaced with any of the above-mentioned studies. The first four do not contain nominals, and the last does not contain fixed-point operators.

The remainder of the paper is organized as follows. In Section 2, we define syntax and semantics of the logics. In Section 3, we describe the decision procedure. In Section 4, its correctness is proved. The complexity is discussed in Section 5. In Section 6, we present an application that uses a variant of the procedure. Finally, future work is described in Section 7.

2 Preliminaries

2.1 Syntax

Let PS, Nom, PV, GMS, and FMS be countable sets of propositional symbols, nominals, propositional variables, general modality symbols, and functional modality symbols, respectively. The set Mod of modalities and the set Form of formulas are defined as follows.

$\mathrm{Mod} \ni m ::= g \mid f \mid g^{-1} \mid f^{-1}$

$\mathrm{Form} \ni \varphi ::= p \mid x \mid X \mid \neg\varphi \mid \varphi \vee \varphi \mid \langle m\rangle\varphi \mid \mu X\varphi \mid @x\,\varphi$

where $p \in \mathrm{PS}$, $x \in \mathrm{Nom}$, $X \in \mathrm{PV}$, $g \in \mathrm{GMS}$, and $f \in \mathrm{FMS}$. In $\mu X\varphi$, any free occurrence of X in φ (i.e., an occurrence of X in φ that is not bound by another μX or νX) must be positive (i.e., the number of negation symbols whose scope contains the occurrence is even). A modality in the form of m^{-1} is called a *backward modality*. We denote the set of formulas by $\mathcal{L}$. We define $\mathrm{Atom} = \mathrm{PS} \cup \mathrm{Nom}$, and $\mathrm{MS} = \mathrm{GMS} \cup \mathrm{FMS}$. For $m \in \mathrm{Mod}$, the set of the formulas in the form of $\langle m\rangle\varphi$ and $[m]\varphi$ are denoted by $\mathcal{L}_{\langle m\rangle}$ and $\mathcal{L}_{[m]}$, respectively. When $\varphi = \langle m\rangle\varphi'$ or $\varphi = [m]\varphi'$, we denote φ' by $\vec{\varphi}$

We define $(m^{-1})^{-1} = m$ for $m \in \mathrm{MS}$, therefore, $(m^{-1})^{-1} = m$ for any $m \in \mathrm{Mod}$.

The following standard abbreviations are used: $\mathbf{false} = p \wedge \neg p$ for some fixed $p \in \mathrm{PS}$, $\mathbf{true} = \neg\mathbf{false}$, $\varphi_1 \wedge \varphi_2 = \neg(\neg\varphi_1 \vee \neg\varphi_2)$, $\varphi_1 \rightarrow \varphi_2 = \neg\varphi_1 \vee \varphi_2$, $[m]\varphi = \neg\langle m\rangle\neg\varphi$, $\nu X\varphi = \neg\mu\neg\varphi[\neg X/X]$.

2.2 Semantics

A *Kripke structure* for $\mathcal{L}$ is a tuple $\mathcal{K} = (S, R, L)$ that satisfies the following conditions. We denote the powerset of S by $\mathcal{P}(S)$.

- S is a set. An element of S is called a *state*.
- $R : \mathrm{MS} \rightarrow \mathcal{P}(S \times S)$. $R(f)$ is a (graph of partial) function if $f \in \mathrm{FMS}$.
- $L : \mathrm{Atom} \rightarrow \mathcal{P}(S)$. $L(x)$ is a singleton if $x \in \mathrm{Nom}$.

For $x \in \mathrm{Nom}$, we denote the unique element of $L(x)$ by $L'(x)$, i.e., $L(x) = \{L'(x)\}$. If $f \in \mathrm{FMS}$ and $s \in \mathrm{dom}(f)$, we express $R(f, s)$ for the unique $s' \in S$ such that $(s, s') \in R(f)$. We also consider that $\mathrm{dom}(R) = \mathrm{Mod}$ by defining $R(m^{-1}) = (R(m))^{-1}$ for $m \in \mathrm{MS}$.

A function $\rho : \mathrm{PV} \rightarrow \mathcal{P}(S)$ is called a *valuation* for $\mathcal{K}$. The *interpretation* $[\![\varphi]\!]^{\mathcal{K},\rho} \subseteq S$ of $\varphi \in \mathcal{L}$ is defined in the standard manner as follows. Symbols

$\mathcal{K}$ and/or ρ are omitted if there is no possibility of confusion. For a function F, we denote by $F[a \mapsto b]$ a function G defined by $\mathrm{dom}(G) = \mathrm{dom}(F) \cup \{a\}$, $G(a) = b$, and $G(x) = F(x)$ for $x \in \mathrm{dom}(F) \setminus \{a\}$.

$$
\begin{array}{ll}
[\![a]\!] = L(a) \quad \text{for } a \in \text{Atom} & [\![X]\!]^{\rho} = \rho(X) \quad \text{for } X \in \text{PV} \\
[\![\neg\varphi]\!] = S \setminus [\![\varphi]\!] & [\![\varphi_1 \vee \varphi_2]\!] = [\![\varphi_1]\!] \cup [\![\varphi_2]\!] \\
\multicolumn{2}{l}{[\![\langle m\rangle\varphi]\!] = \{s \in S \mid \exists s' \in S.\, (s, s') \in R(m) \text{ and } s' \in [\![\varphi]\!]\}} \\
\multicolumn{2}{l}{[\![\mu X \varphi]\!]^{\rho} = \bigcap\{T \subseteq S \mid [\![\varphi]\!]^{\rho[X \mapsto T]} \subseteq T\}} \\
[\![@x\, \varphi]\!] = S \quad \text{if } L'(x) \in [\![\varphi]\!] & [\![@x\, \varphi]\!] = \varnothing \quad \text{if } L'(x) \notin [\![\varphi]\!]
\end{array}
$$

We write $\mathcal{K}, \rho, s \models \varphi$ if $s \in [\![\varphi]\!]^{\mathcal{K},\rho}$. Again, $\mathcal{K}$ and/or ρ are often omitted. We write $\mathcal{K} \models \varphi$ if $\mathcal{K}, \rho, s \models \varphi$ holds for any valuation ρ and state s. Formulas φ and φ' are *equivalent* ($\varphi \equiv \varphi'$) if $[\![\varphi]\!]^{\mathcal{K},\rho} = [\![\varphi']\!]^{\mathcal{K},\rho}$ for any Kripke structure $\mathcal{K}$ and valuation ρ. A formula φ is *valid* if it is equivalent to **true**. It is *satisfiable* if its negation is not valid.

2.3 Closures

A formula φ is in *positive normal form (PNF)*, if φ satisfies the following conditions:

- All negation operators ($\neg$) in φ appear immediately before propositional symbols, nominals, or propositional variables.
- All propositional variables in φ are bound at most once.

It is easy to see that any formula is equivalent to a formula in PNF.

Symbol λ is used to express either μ or ν. Thus $\lambda X \varphi$ is either $\mu X \varphi$ or $\nu X \varphi$. For the formula $\lambda X \varphi$, we denote by $\exp(\lambda X \varphi)$ the formula $\varphi[\lambda X \varphi / X]$, which is obtained from φ by replacing all free occurrences of X with $\lambda X \varphi$, and call it the *expansion* of $\lambda X \varphi$. For example, $\exp(\mu X(p \vee \langle m\rangle X)) = p \vee \langle m\rangle(\mu X(p \vee \langle m\rangle X))$. It is easy to see that $\exp(\lambda X \varphi) \equiv \lambda X \varphi$.

We define relation F on the set of all formulas in PNF that satisfies the following:

$$
\begin{array}{ll}
\multicolumn{2}{l}{(\varphi_1 \vee \varphi_2, \varphi_j) \in F \text{ and } (\varphi_1 \wedge \varphi_2, \varphi_j) \in F \quad \text{for } j = 1, 2} \\
(\langle m\rangle\varphi, \varphi) \in F & ([m]\varphi, \varphi) \in F \\
(\lambda X \varphi, \exp(\lambda X \varphi)) \in F & (@n\, \varphi, \varphi) \in F
\end{array}
$$

and that no other pairs belong to F.

For a formula φ in PNF, the *closure* of φ is the least set of formulas that contains φ and is closed under the relation F, i.e., if φ is in the closure and $(\varphi, \psi) \in F$ then ψ is in the closure. We denote the closure of φ by $\mathrm{cl}(\varphi)$. The *lean* of φ is a subset of $\mathrm{cl}(\varphi)$ defined by $\{\psi \in \mathrm{cl}(\varphi) \mid \psi \in$ Atom or ψ is in the form of $\langle m\rangle\chi, [m]\chi$, or $@n\, \chi\}$.

An occurrence of a propositional variable X in a formula φ is *guarded* if there exists a formula ψ such that (1) ψ contains the occurrence, (2) ψ is in the form of $\langle m\rangle\psi'$, $[m]\psi'$, or $@n\, \psi'$, and (3) ψ is a subformula of the formula $\lambda X \chi$ that binds the occurrence in φ. A formula is *guarded* if all occurrences of bound propositional variables are guarded. For example, $\mu X(p \vee \langle m \rangle X \vee @n\, X)$ is guarded, but $\mu X(p \vee X)$ is not.

PROPOSITION 1. (Kozen) *For every formula φ, there is a guarded formula ψ such that $\varphi \equiv \psi$.*

Refer to [17] for a proof of the proposition. With a little modification of the proof, we see that ψ can be taken so that $|\operatorname{cl}(\psi)| \leq |\operatorname{cl}(\varphi)|$ if φ is in PNF.

Hereafter in this paper, we assume that formulas are guarded and are in PNF unless explicitly stated otherwise.

2.4 Alternation-freeness

A formula φ is *alternation-free* if the following conditions are satisfied:

- For any subformula ψ of φ in the form of $\mu X \psi'$ and for any subformula χ of ψ' in the form of $\nu Y \chi'$, X does not occur freely in χ'.
- For any subformula ψ of φ in the form of $\nu Y \psi'$ and for any subformula χ of ψ' in the form of $\mu X \chi'$, Y does not occur freely in χ'.

For example, $\varphi_1 = \mu X(\nu Y(p \wedge \langle m \rangle Y) \vee [m]X)$ is an alternation-free formula, whereas $\varphi_2 = \mu X(\nu Y(p \wedge \langle m \rangle (X \wedge Y)) \vee [m]X)$ is not.

We denote the set of alternation-free formulas by $\mathcal{L}(\mu_{\mathrm{AF}}, \mathrm{nom}, \mathrm{back}, \mathrm{func})$. Also, $\mathcal{L}(\mu_{\mathrm{AF}}, \mathrm{nom}, \mathrm{back})$, $\mathcal{L}(\mu_{\mathrm{AF}}, \mathrm{nom}, \mathrm{func})$, and $\mathcal{L}(\mu_{\mathrm{AF}}, \mathrm{back}, \mathrm{func})$ are the sets of alternation-free formulas that do not contain functional modalities, backward modalities, and nominals, respectively.

We consider $(\operatorname{cl}(\varphi), F)$ as a graph and denote the set of strongly connected components (SCCs) of the graph by $\mathcal{D}(\varphi)$. Also, we denote by $\mathcal{D}_\mu(\varphi)$ and $\mathcal{D}_\nu(\varphi)$ the set of strongly connected components of the graph that contains a formula in the form of $\mu X \psi$ and $\nu X \psi$, respectively. If φ is clear from the context, they are written as $\mathcal{D}_\mu$ and $\mathcal{D}_\nu$, respectively. Also we write $D_\mu = \bigcup \mathcal{D}_\mu$ and $D_\nu = \bigcup \mathcal{D}_\nu$. If $\psi \in \operatorname{cl}(\varphi)$ is an element of $\bigcup \mathcal{D}(\varphi)$, we denote by $D(\psi)$ the strong connected component $D \in \mathcal{D}(\varphi)$ that contains ψ.

For example, consider the above formula φ_1. By letting $\psi_1 = \nu Y(p \wedge \langle m \rangle Y)$, $D_1 = \{\varphi_1, \psi_1 \vee [m]\varphi_1, [m]\varphi_1\}$, $D_2 = \{\psi_1, p \wedge \langle m \rangle \psi_1, \langle m \rangle \psi_1\}$, and $D_3 = \{p\}$, we have $\mathcal{D}(\varphi_1) = \{D_1, D_2, D_3\}$, $\mathcal{D}_\mu(\varphi_1) = \{D_1\}$, and $\mathcal{D}_\nu(\varphi_1) = \{D_2\}$.

It is easy to show the following lemma.

LEMMA 2. *$\mathcal{D}_\mu(\varphi) \cap \mathcal{D}_\nu(\varphi) = \varnothing$ if and only if φ is alternation-free.*

2.5 Rank and choice functions

In this section, we prepare lemmas to be used to show the correctness of the decision procedure. We denote the class of ordinal numbers by On.

Let φ_{I} be a closed alternation-free formula. Let $\mathcal{K} = (S, R, L)$ be a Kripke structure and $D \in \mathcal{D}_\mu(\varphi_{\mathrm{I}})$. We define $U_{D,\alpha} \subseteq \operatorname{cl}(\varphi_{\mathrm{I}}) \times S$ for $\alpha \in \mathrm{On}$ as the least set that satisfies the following conditions. We omit D and write U_α if no confusion occurs.

- $U_0 = \{(\varphi, s) \mid \varphi \in \mathrm{cl}(\varphi_\mathrm{I}) \setminus D,\ s \in S,\ s \models \varphi\}$.
- If $\varphi \in \mathrm{cl}(\varphi_\mathrm{I})$ and one of the following conditions holds, $(\varphi, s) \in U_\alpha$.
  - $\varphi = \varphi_1 \vee \varphi_2$ and either $(\varphi_1, s) \in U_\alpha$ or $(\varphi_2, s) \in U_\alpha$.
  - $\varphi = \varphi_1 \wedge \varphi_2$ and both $(\varphi_1, s) \in U_\alpha$ and $(\varphi_2, s) \in U_\alpha$.
  - $\varphi = \lambda X \varphi_1$ and $(\exp(\varphi), s) \in U_\alpha$.
  - $\varphi = \langle m \rangle \varphi_1$ and there is $s' \in S$ such that $(s, s') \in R(m)$ and $(\varphi_1, s') \in U_\beta$ for some $\beta < \alpha$.
  - $\varphi = [m]\varphi_1$ and for all $s' \in S$ if $(s, s') \in R(m)$ then there exists $\beta < \alpha$ such that $(\varphi_1, s') \in U_\beta$.
  - $\varphi = @n\, \varphi_1$ and $(\varphi_1, L'(n)) \in U_\beta$ for some $\beta < \alpha$.

LEMMA 3. *For any $D \in \mathcal{D}_\mu(\varphi_\mathrm{I})$, we have $\{(\varphi, s) \in \mathrm{cl}(\varphi_\mathrm{I}) \times S \mid s \models \varphi\} = \bigcup_{\alpha \in \mathrm{On}} U_{D,\alpha}$.*

Proof. We use the fact that $s \models \varphi$ holds if and only if (φ, s) belongs to the winning region of Player 0 in the corresponding parity game [18].

Let $U_\infty = \bigcup_{\alpha \in \mathrm{On}} U_\alpha$ and assume $(\varphi, s) \in \mathrm{cl}(\varphi_\mathrm{I}) \times (S \setminus U_\infty)$. It is easy to see that Player 1 can keep the vertex outside of U_∞. Since $\mathrm{cl}(\varphi_\mathrm{I}) \times S \setminus U_\infty \subseteq D \times S$, the winner of the trace is Player 1. Therefore $s \not\models \varphi$.

The other direction can be shown by induction on α. ∎

We define the rank function rank_D: its domain is $\{(\varphi, s) \in \mathrm{cl}(\varphi_\mathrm{I}) \times S \mid \mathcal{K}, s \models \varphi\}$, its range is On, and $\mathrm{rank}_D(\varphi, s) = \min\{\alpha \in \mathrm{On} \mid (\varphi, s) \in U_\alpha\}$. The following lemma clearly holds from the definition.

LEMMA 4. *Suppose $(\varphi, s) \in \mathrm{dom}(\mathrm{rank}_D)$. If $\varphi \notin D$, $\mathrm{rank}_D(\varphi, s) = 0$. Otherwise, the following holds:*

- $\mathrm{rank}_D(\varphi_1 \vee \varphi_2, s) = \min(\{\mathrm{rank}_D(\varphi_j, s) \mid s \models \varphi_j,\ j = 1, 2\})$.
- $\mathrm{rank}_D(\varphi_1 \wedge \varphi_2, s) = \max(\mathrm{rank}_D(\varphi_1, s), \mathrm{rank}_D(\varphi_2, s))$.
- $\mathrm{rank}_D(\langle m \rangle \varphi, s) = \min\{\mathrm{rank}_D(\varphi, s') + 1 \mid (s, s') \in R(m),\ s' \models \varphi\}$.
- $\mathrm{rank}_D([m]\varphi, s) = \sup\{\mathrm{rank}_D(\varphi, s') + 1 \mid (s, s') \in R(m)\}$.
- $\mathrm{rank}_D(\lambda X \varphi, s) = \mathrm{rank}_D(\exp(\lambda X \varphi), s)$.
- $\mathrm{rank}_D(@n\, \varphi, s) = \mathrm{rank}_D(\varphi, L'(n)) + 1$.

A *choice function* c for φ_I and $\mathcal{K}$ is a function that satisfies the following conditions:

- $\mathrm{dom}(c) = \Phi \times S$, where Φ is the set of formulas in $\mathrm{cl}(\varphi_\mathrm{I})$ that is in the form of either $\varphi_1 \vee \varphi_2$ or $\langle m \rangle \varphi$ where $m \in \mathrm{Mod}$.
- For $\varphi_1 \vee \varphi_2 \in \mathrm{cl}(\varphi_\mathrm{I})$ and $s \in S$, $c(\varphi_1 \vee \varphi_2, s)$ is either φ_1 or φ_2. If $s \models \varphi_1 \vee \varphi_2$, $s \models c(\varphi_1 \vee \varphi_2, s)$.

- For $\langle m\rangle\varphi \in \mathrm{cl}(\varphi_{\mathrm{I}})$ and $s \in S$, $s' = c(\langle m\rangle\varphi, s) \in S$. If $s \models \langle m\rangle\varphi$, $(s, s') \in R(m)$ and $s' \models \varphi$.

A *trace* τ is a finite or an infinite sequence $((\varphi_i, s_i) \mid i < \alpha)$, where $\alpha < \omega + 1$, $\varphi_i \in \mathrm{cl}(\varphi_{\mathrm{I}})$, and $s_i \in S$, satisfying the following conditions:

(1) $(\varphi_i, \varphi_{i+1}) \in F$ for $i + 1 < \alpha$.

(2) If $\varphi_i = \langle m\rangle\varphi_{i+1}$ or $\varphi_i = [m]\varphi_{i+1}$, $(s_i, s_{i+1}) \in R(m)$.

(3) If $\varphi_i = @n\,\varphi_{i+1}$, $s_{i+1} = L'(n)$.

(4) In cases other than (2) and (3), $s_{i+1} = s_i$.

Trace $\tau = ((\varphi_i, s_i) \mid i < \alpha)$ *conforms with* choice function c if the following conditions are satisfied for $i + 1 < \alpha$:

- If $\varphi_i = \xi \vee \eta$, $\varphi_{i+1} = c(\varphi_i, s_i)$.
- If $\varphi_i = \langle m\rangle\varphi_{i+1}$, $s_{i+1} = c(\varphi_i, s_i)$.

LEMMA 5. *Let c be a choice function for φ_{I} and $\mathcal{K}$ and assume $Z \subseteq \mathrm{cl}(\varphi_{\mathrm{I}}) \times S$. If the following conditions are satisfied, $\mathcal{K}, s \models \varphi$ holds for all $(\varphi, s) \in Z$.*

(1) If a is an atom or its negation, $(a, s) \in Z$ implies $\mathcal{K}, s \models a$.

(2) If $((\varphi, s), (\varphi', s'))$ is a trace that conforms with c and $(\varphi, s) \in Z$, then $(\varphi', s') \in Z$.

(3) Assume $((\varphi_i, s_i) \mid i < \omega)$ is an infinite trace that conforms with c, $(\varphi_i, s_i) \in Z$ for all $i < \omega$, $D \in \mathcal{D}_\mu(\varphi_{\mathrm{I}})$, and $\varphi_0 \in D$. Then, there is $k < \omega$ such that $\varphi_k \notin D$.

Proof. Using the choice function, a strategy σ for Player 0 is defined in an obvious manner. We show that σ is a winning strategy on any $(\varphi, s) \in Z$. By using condition (2), one can show that for any trace τ beginning at (φ, s) and conforming σ and for any $n \in \mathrm{dom}(\tau)$, $\tau(n) = (\varphi_n, s_n) \in Z$. If $\mathrm{dom}(\tau)$ is finite, the winner of τ is Player 0 by condition (1). If $\mathrm{dom}(\tau)$ is infinite, there is $N \in \omega$ and $D \in \mathcal{D}$ such that $\varphi_n \in D$ for all $n \geq N$. By condition (3), $D \in \mathcal{D}_\nu$. This means that all priorities (greater than zero) appearing infinitely often are even and therefore the winner of τ in this case is also Player 0. ■

3 Decision procedure

Now, we describe the decision procedure. In Section 3.1, we define the necessary concepts, and the procedure is defined in Section 3.2.

Let φ_{I} be the given formula of which we judge satisfiability. We denote the lean of φ_{I} by Lean. When φ_{I} has free variables, we replace each of them with a distinct fresh propositional symbol (i.e., one that does not appear in $\varphi_{\rm I}$), and denote it by $\varphi'_{\rm I}$. It is clear that $\varphi_{\rm I}$ and $\varphi'_{\rm I}$ are equi-satisfiable. Therefore, hereafter, we assume that $\varphi_{\rm I}$ is closed. In what follows, we denote by PS the set of propositional symbols that appear in $\varphi_{\rm I}$. Thus, PS is a finite set. The same convention is also applied to Nom, PV, GMS and FMS.

3.1 Tableau

For $D \in \mathcal{D}_\mu$, we denote by BMod_D the set of modality symbol m such that there are formulas $\varphi \in D$ in the form of $\langle m\rangle\varphi'$ or $[m]\varphi'$ and $\psi \in D$ in the form of $\langle m^{-1}\rangle\psi'$ or $[m^{-1}]\psi'$. We denote by BForm_D the subset of D that consists of the formulas in the form of $\langle m\rangle\varphi$, $[m]\varphi$, $\langle m^{-1}\rangle\varphi$, or $[m^{-1}]\varphi$ for some $m \in \mathrm{BMod}_D$.

The decision procedure is a variant of the tableau method. A node of the tableau is a pair (x, y) that satisfies the following conditions:

- x is a function and its domain is Lean. For $\varphi \in$ Lean, $x(\varphi)$ is either a natural number or value "∞". If $\varphi \in D_\mu$ and $\mathrm{BForm}_{D(\varphi)} \neq \varnothing$, then either $x(\varphi) \leq |\mathrm{BForm}_{D(\varphi)}| + 1$ or $x(\varphi) = \infty$; otherwise $x(\varphi)$ is either 0 or ∞. We regard $0 < 1 < \cdots < \infty$.

- y is a function and its domain is $\{\varphi \in D_\mu \cap \mathrm{Lean} \mid x(\varphi) < \infty\}$. For $\varphi \in \mathrm{dom}(y)$, $y(\varphi)$ is a natural number and $y(\varphi) \leq |\mathrm{Nom}| \cdot |D(\varphi)|$.

- If there is $D \in \mathcal{D}_\mu$ such that $\varphi_1, \varphi_2 \in D$ and $x(\varphi_1) < x(\varphi_2) < \infty$, then $y(\varphi_1) \leq y(\varphi_2)$.

Note that there are only finitely many pairs (x, y) that satisfy the conditions. We denote by Tab the set of all pairs that satisfy the above conditions. When $t = (x, y) \in \mathrm{Tab}$, x is denoted by x_t and y is denoted by y_t.

The intention is that at node $t = (x, y)$, $\varphi \in$ Lean is satisfied if $x(\varphi) < \infty$ and is not satisfied if $x(\varphi) = \infty$. In the case $\varphi \in D \in \mathcal{D}_\mu$, due to (3) of Lemma 5, if we can appropriately define a choice function c, there should be a finite trace that conforms with c, starts with (φ, t), and goes outside D. The values $x(\varphi)$ and $y(\varphi)$ both relate to the trace. If $x(\psi) \leq x(\varphi)$, we regard that (ψ, t) can appear in the trace that starts with (φ, t). The value $y(\varphi)$ expresses the number of nominals that appear in the trace. For precise meanings, refer to Section 4.

A function g from Nom to Tab is called a *naming function* when it satisfies the following conditions:

- $x_{g(n)}(n) = 0$ for all $n \in \mathrm{Nom}$.

- $x_{g(n_1)}(n_2) = 0 \implies g(n_1) = g(n_2)$ for any $n_1, n_2 \in \mathrm{Nom}$.

We denote the set of naming functions by NF. Note that NF is a finite set. A naming function designates a node at which each nominal is satisfied. By recalling that $x_t(\varphi) = 0$ means that t satisfies φ, the first condition means that a nominal should be satisfied at the node where the nominal is satisfied. The second condition means that if n_2 is satisfied at the node where n_1 is satisfied, then the node where n_1 is satisfied should be equal to the node where n_2 is satisfied.

Values $x(\varphi)$ are defined only on Lean but we extend its domain to $\mathrm{cl}(\varphi_I)$ by induction in the following manner. Assume $t = (x, y)$. For $a \in \mathrm{Atom}$, $x(\neg a) = 0$ if $x(a) = \infty$ and $x(\neg a) = \infty$ if $x(a) = 0$. For disjunctions, conjunctions, and fixed-points, we define the x-value basically as the minimum values of its subformulas, the maximum values of its subformulas, and the value of its expansion, respectively; however, if the formula goes outside the SCC, the value goes to 0 or ∞. More precisely, $x(\varphi_1 \vee \varphi_2) = \min(\tilde{x}(\varphi_1, D(\varphi_1 \vee \varphi_2)), \tilde{x}(\varphi_2, D(\varphi_1 \vee \varphi_2)))$, $x(\varphi_1 \wedge \varphi_2) = \max(\tilde{x}(\varphi_1, D(\varphi_1 \wedge \varphi_2)), \tilde{x}(\varphi_2, D(\varphi_1 \wedge \varphi_2)))$, and $x(\lambda X \varphi) = x(\exp(\lambda X \varphi))$, where $\tilde{x}(\varphi, D) = 0$ if $x(\varphi) < \infty$ and $\varphi \notin D$ and $\tilde{x}(\varphi, D) = x(\varphi)$ otherwise. Note that the induction is sound since the formulas are guarded.

The domain of y is also extended to $\{\varphi \in D_\mu \mid x(\varphi) < \infty\}$ in a similar manner: $y(\varphi_1 \vee \varphi_2) = \min(\tilde{y}(\varphi_1, D(\varphi_1 \vee \varphi_2)), \tilde{y}(\varphi_2, D(\varphi_1 \vee \varphi_2)))$, $y(\varphi_1 \wedge \varphi_2) = \max(\tilde{y}(\varphi_1, D(\varphi_1 \wedge \varphi_2)), \tilde{y}(\varphi_2, D(\varphi_1 \wedge \varphi_2)))$, and $y(\lambda X \varphi) = y(\exp(\lambda X \varphi))$. If $x(\varphi) = \infty$, then $\tilde{y}(\varphi, D) = \infty$, else if $\varphi \notin D$, then $\tilde{y}(\varphi, D) = 0$, else $\tilde{y}(\varphi, D) = y(\varphi)$.

We write $t \Vdash \varphi$ when $x_t(\varphi) < \infty$. The set $\{\varphi \in \mathrm{Lean} \mid t \Vdash \varphi\}$ is denoted by $\mathrm{sat}(t)$. A node $t \in \mathrm{Tab}$ *has a name* if there is $n \in \mathrm{Nom}$ such that $t \Vdash n$.

For each $m \in \mathrm{MS}$, we define the *transition relation* $\mathrm{Tr}(m)$ on Tab as the conjunction of the following five conditions: $\mathrm{ConBox}(t, t', m)$, $\mathrm{ConBox}(t', t, m^{-1})$, $\mathrm{ConNom}(t, t', m)$, $\mathrm{ConNom}(t', t, m^{-1})$, and $\mathrm{LoopFree}(t, t', m)$, each of which is defined below. For $m \in \mathrm{MS}$, $\mathrm{Tr}(m^{-1})$ is defined as $\mathrm{Tr}(m^{-1}) = \mathrm{Tr}(m)^{-1}$.

For $t \in \mathrm{Tab}$ and $m \in \mathrm{Mod}$, we define $\mathrm{Box}(t, m) = \mathrm{sat}(t) \cap \mathcal{L}_{[m]}$ if $m \in \mathrm{Mod} \setminus \mathrm{FMS}$ and $\mathrm{Box}(t, m) = \mathrm{sat}(t) \cap (\mathcal{L}_{[m]} \cup \mathcal{L}_{\langle m \rangle})$ if $m \in \mathrm{FMS}$. Then, $\mathrm{ConBox}(t, t', m)$ is defined as "for all $\varphi \in \mathrm{Box}(t, m)$, $t' \Vdash \vec{\varphi}$". The intention of this definition should be clear if we consider $\mathrm{Tr}(m)$ as a corresponding relation of $R(m)$ of a Kripke structure. Note that if $m \in \mathrm{FMS}$, diamond formulas behave similarly to box formulas since there is at most one successor.

For $t, t' \in \mathrm{Tab}$ and $\varphi \in \mathrm{dom}(y_t) \cap \mathrm{Lean}$, $\mathrm{ConNom}(t, t', \varphi)$ is defined as follows: "$\vec{\varphi} \in \mathrm{dom}(y_{t'})$ and $y_t(\varphi) \geq y_{t'}(\vec{\varphi})$ hold. Moreover if t' has a name, $y_t(\varphi) > y_{t'}(\vec{\varphi})$ holds." This is also a natural requirement by considering the intuitive meaning of the function y. For $t, t' \in \mathrm{Tab}$ and $m \in \mathrm{Mod}$, we write $\mathrm{ConNom}(t, t', m)$ if $\mathrm{ConNom}(t, t', \varphi)$ holds for any $\varphi \in \mathrm{dom}(y_t) \cap \mathrm{Box}(t, m)$.

For $t, t' \in \mathrm{Tab}$, $m \in \mathrm{Mod}$, and $\varphi \in \mathrm{Box}(t, m)$, $\mathrm{LoopFree}(t, t', \varphi)$ is defined as follows: "there exists *no* formula $\psi \in \mathcal{L}_{[m]} \cap D(\varphi)$ such that both $x_t(\varphi) \leq x_t(\vec{\psi}) < \infty$ and $x_{t'}(\psi) \leq x_{t'}(\vec{\varphi}) < \infty$ hold." We write $\mathrm{LoopFree}(t, t', m)$ if $\mathrm{LoopFree}(t, t', \varphi)$ holds for any $\varphi \in \mathrm{Box}(t, m) \cap D_\mu$. This completes the definition of $\mathrm{Tr}(m)$.

The intuitive reason why we require the loop-freeness for $\mathrm{Tr}(m)$ is as follows: roughly speaking, we will create a Kripke structure from a tableau in the manner that $t \in \mathrm{Tab}$ is a state, the transition relation of modality m is defined by $\mathrm{Tr}(m)$, and φ is satisfied at state t if and only if $t \Vdash \varphi$. Suppose

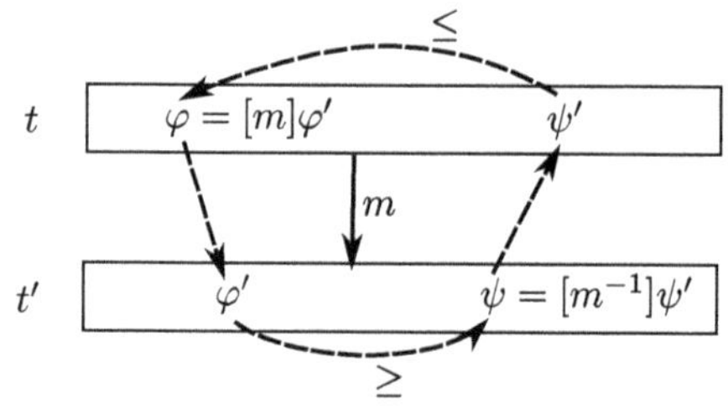

Figure 1. loop-freeness

that there is a transition of modality m from t to t' and it is not loop-free with regard to $\varphi = [m]\varphi'$. Take $\psi = [m^{-1}]\psi'$ as in the definition. Then, (φ, t) may appear in a trace from (ψ', t), and (ψ, t') may appear in a trace from (φ', t'). Furthermore, both $((\varphi, t), (\varphi', t'))$ and $((\psi, t'), (\psi', t))$ can be adjacent pairs in traces. Combining these four parts (see Figure 1), one can construct an infinite trace within D, which contradicts to Lemma 5.

For $m \in \mathrm{Mod}$ and $\varphi \in \mathrm{Lean} \cap \mathcal{L}_{\langle m \rangle}$, we also define the relation $\mathrm{Tr}(\varphi)$ on Tab as the set of tuples (t, t') that satisfies the following three conditions. (1) $(t, t') \in \mathrm{Tr}(m)$. (2) $t \Vdash \varphi$. (3) If $\varphi \in D_\mu$, $\mathrm{ConNom}(t, t', \varphi)$ and $\mathrm{LoopFree}(t, t', \varphi)$ hold.

3.2 Procedure

For given φ_{I}, the decision procedure will try to find a suitable subset T of Tab such that a model for φ_{I} is constructed from T. One of the difficulty is to decide where nominals should be satisfied in T. For example if we know that t should be in T and $t \Vdash \langle m \rangle (n \wedge (\varphi \vee \psi))$. If n is a mere propositional symbol, we can request t_1 or t_2 be a member of T, where t_1 and t_2 are appropriate element of Tab such that $t_1 \Vdash n \wedge \varphi$ and $t_2 \Vdash n \wedge \psi$ and $(t, t_i) \in \mathrm{Tr}(m)$. However, if n is a nominal, and especially if φ and ψ are mutually inconsistent, we need to ask exactly one of t_1 and t_2 exists in T, but which of them should we choose?

Naming functions are used to solve the problem. We fix a naming function g, and describe a subprocedure for g. In the subprocedure, we assume that nominal n is satisfied at $g(n) \in \mathrm{Tab}$. The entire procedure consists of a loop over NF. If there is a $g \in \mathrm{NF}$ for which the subprocedure succeeds, then we consider that φ_{I} is satisfiable. If the procedure fails for all $g \in \mathrm{NF}$, we judge that φ_{I} is not satisfiable. In the rest of this section, we define the subprocedure for g.

We construct a sequence $(T_k)_{k \le K}$ of subsets of Tab so that $T_0 \supseteq T_1 \supseteq \cdots$. The construction is repeated until $T_k = T_{k+1}$ holds. Since Tab is a finite set, there exists such $k \in \omega$. We write this k as K. The procedure succeeds if there is $t \in T_K$ such that $t \Vdash \varphi_{\mathrm{I}}$, and for all $n \in \mathrm{Nom}$, $g(n) \in T_K$.

The initial set T_0 consists of the elements t of Tab that satisfy the following conditions:

- For all $n \in \mathrm{Nom}$, $t \Vdash n \implies t = g(n)$.
- For all $\varphi \in \mathrm{Lean}$ in the form of $@n\,\varphi'$, if $t \Vdash \varphi$, then $g(n) \Vdash \varphi'$. Moreover, if $\varphi \in D_\mu$, $y_t(\varphi) > y_{g(n)}(\varphi')$.

T_{k+1} is obtained from T_k by removing nodes that are $\Diamond$-inconsistent or μ-inconsistent in T_k. Let T be a subset of Tab. Node $t \in T$ is *$\Diamond$-consistent* in T, if:

- for any $m \in \mathrm{Mod} \setminus \mathrm{FMS}$ and $\varphi \in \mathrm{Lean} \cap \mathcal{L}_{\langle m \rangle}$, $t \Vdash \varphi$ implies that there exists $t' \in T$ such that $(t, t') \in \mathrm{Tr}(\varphi)$, and
- for any $m \in \mathrm{FMS}$, if there exists $\varphi \in \mathrm{Lean} \cap \mathcal{L}_{\langle m \rangle}$ such that $t \Vdash \varphi$, then there exists $t' \in T$ such that $(t, t') \in \mathrm{Tr}(m)$.

We say $t \in T$ is *$\Diamond$-inconsistent* if it is not $\Diamond$-consistent.

In order to define μ-consistency, we construct a sequence $(V_j)_{j \le J}$ of subsets of $T \times \mathcal{P}(D_\mu \cap \mathrm{Lean})$ so that $V_0 \subseteq V_1 \subseteq \cdots$. We repeat it until $V_j = V_{j+1}$ holds, and set $J = j$. (J depends on T.) Since $T \times \mathcal{P}(D_\mu \cap \mathrm{Lean})$ is a finite set, there must be such $j \in \omega$. A node $t \in T$ is *μ-consistent* in T if for all $D \in \mathcal{D}_\mu$, $(t, \{\varphi \in D \cap \mathrm{Lean} \mid t \Vdash \varphi\}) \in V_J$, and t is *μ-inconsistent* if it is not μ-consistent.

Let $t = (x, y) \in \mathrm{Tab}$ and $E \subseteq D_\mu$. For $\varphi \in \mathrm{Lean}$, we define $x^E(\varphi)$ by $x^E(\varphi) = \infty$ if $\varphi \in D_\mu \setminus E$, and $x^E(\varphi) = x(\varphi)$ otherwise. Then, we extend the domain of x^E to $\mathrm{cl}(\varphi_\mathrm{I})$ in the same manner as we extended the domain of x. We write $t^E \Vdash \varphi$ when $x^E(\varphi) < \infty$.

The initial set V_0 is defined as $V_0 = \{(t, \varnothing) \mid t \in T\}$. Assume we have V_j. Then V_{j+1} is defined so that $(t, E) \in V_{j+1}$ if and only if $(t, E) \in V_j$ or the following conditions are satisfied:

(a) The following holds for all $\varphi \in E$.

  - $t \Vdash \varphi$.
  - For all $\psi \in D(\varphi) \cap \mathrm{Lean}$, if $x_t(\varphi) > x_t(\psi)$ then $\psi \in E$.
  - For all $\psi \in D(\varphi) \cap \mathrm{Lean} \cap \mathrm{dom}(y_t)$, if $y_t(\varphi) > y_t(\psi)$ then $\psi \in E$.

(b) For any $m \in \mathrm{Mod} \setminus \mathrm{FMS}$ and $\varphi \in \mathcal{L}_{\langle m \rangle} \cap \mathrm{sat}(t)$, there is $(t', E') \in V_j$ such that:

  - $(t, t') \in \mathrm{Tr}(\varphi)$.
  - For any $\psi \in E \cap \mathcal{L}_{[m]}$, $t'^{E'} \Vdash \vec{\psi}$.
  - If $\varphi \in E$, $t'^{E'} \Vdash \vec{\varphi}$.

(c) For any $m \in \mathrm{FMS}$ that satisfies $\mathcal{L}_{\langle m \rangle} \cap \mathrm{sat}(t) \neq \varnothing$, there is $(t', E') \in V_j$ such that:

  - $(t, t') \in \mathrm{Tr}(m)$.
  - For any $\psi \in E \cap (\mathcal{L}_{\langle m \rangle} \cup \mathcal{L}_{[m]})$, $t'^{E'} \Vdash \vec{\psi}$.

This completes the description of the decision procedure.

4 Correctness

In this section we prove that the decision procedure presented in the previous section is sound and complete for $\mathcal{L}(\mu_{\mathrm{AF}}, \mathrm{nom}, \mathrm{back})$, $\mathcal{L}(\mu_{\mathrm{AF}}, \mathrm{nom}, \mathrm{func})$, and $\mathcal{L}(\mu_{\mathrm{AF}}, \mathrm{back}, \mathrm{func})$; and sound for $\mathcal{L}(\mu_{\mathrm{AF}}, \mathrm{nom}, \mathrm{back}, \mathrm{func})$. We first show the soundness (if there is a Kripke structure that satisfies φ_{I}, the procedure succeeds) in Section 4.1; then, we show the completeness, the other direction, in Section 4.2.

4.1 Soundness

Assume that there is a Kripke structure $\mathcal{K} = (S, R, L)$ and $s_{\mathrm{I}} \in S$ such that $\mathcal{K}, s_{\mathrm{I}} \models \varphi_{\mathrm{I}}$. We define a function $h : S \to \mathrm{Tab}$ and show that $h(s_{\mathrm{I}}) \Vdash \varphi_{\mathrm{I}}$ and all nodes in the range of h remain in T_k. For $\varphi \in D_\mu$ and $s \in S$, we denote the set $\{\mathrm{rank}_D(\psi, s) \mid \psi \in \mathrm{BForm}_D,\ \mathrm{rank}_D(\psi, s) \le \mathrm{rank}_D(\varphi, s)\}$ by $X(\varphi, s)$, where $D = D(\varphi)$. It is finite since BForm_D is finite. If we write $h(s) = (x, y)$, x and y are defined as follows. If $s \not\models \varphi$, $x(\varphi) = \infty$. Otherwise, if $\varphi \in D_\mu$ then $x(\varphi) = |X(\varphi, s)| + 1$, else $x(\varphi) = 0$. For $\varphi \in \mathrm{dom}(y)$, $y(\varphi) = |Y(\varphi, s)|$, where $Y(\varphi, s) = \{(n, \psi) \in \mathrm{Nom} \times D \mid \mathrm{rank}_D(\psi, L'(n)) < \mathrm{rank}_D(\varphi, s)\}$ and $D = D(\varphi)$.

The following three lemmas can be easily proved and we omit the proofs.

LEMMA 6. *For* $\varphi \in \mathrm{cl}(\varphi_{\mathrm{I}})$ *and* $s \in S$, $\quad \mathcal{K}, s \models \varphi \iff h(s) \Vdash \varphi$.

LEMMA 7. $(s, s') \in R(m) \implies (h(s), h(s')) \in \mathrm{Tr}(m)$ *holds for* $s, s' \in S$ *and* $m \in \mathrm{Mod}$.

LEMMA 8. *Suppose* $s \in S$, $(x, y) = h(s)$, $\varphi, \psi \in D_\mu$ *and* $D(\varphi) = D(\psi)$. *Let us write* $D = D(\varphi) = D(\psi)$.

(1) $x(\varphi) \le x(\psi) \iff \mathrm{rank}_D(\varphi, s) \le \mathrm{rank}_D(\psi, s)$.

(2) If $\varphi, \psi \in \mathrm{dom}(y)$, $y(\varphi) \le y(\psi) \iff \mathrm{rank}_D(\varphi, s) \le \mathrm{rank}_D(\psi, s)$.

For $D \in \mathcal{D}_\mu$, $\alpha \in \mathrm{On}$, and $s \in S$, let us denote by $F_D(\alpha, s)$ the set $\{\varphi \in D \cap \mathrm{Lean} \mid s \models \varphi,\ \mathrm{rank}_D(\varphi, s) < \alpha\}$.

LEMMA 9. *For any* $\alpha \in \mathrm{On}$ *and* $D \in \mathcal{D}_\mu$ *there exists* $j \in \omega$ *such that for any* $s \in S$, $(h(s), F_D(\alpha, s)) \in V_j$.

Proof. We prove the lemma by induction on α. The case $\alpha = 0$ is trivial since $F_D(0, s) = \varnothing$. The case in which α is limit is also clear since there exists $J \in \omega$ such that $V_j = V_J$ for all $j \ge J$.

For the remaining case of $\alpha + 1$, we assume that $(h(s), F_D(\alpha, s)) \in V_j$ for all $s \in S$ and prove that $(h(s), F_D(\alpha+1, s)) \in V_{j+1}$ for all $s \in S$. Take $s \in S$ and let $t = h(s)$ and $E = F_D(\alpha+1, s)$. We need to check the conditions (a), (b) and (c) of the definition of V_{j+1}. The condition (a) can be proved without difficulty by using Lemmas 6 and 8. For the condition (b), assume $m \in \mathrm{Mod} \setminus \mathrm{FMS}$, $\varphi = \langle m \rangle \varphi' \in \mathrm{Lean}$, and $t \Vdash \varphi$. Since $s \models \varphi$ by Lemma 6, we can take $s' \in S$ such that $s' \models \varphi$ and $(s, s') \in R(m)$. Moreover, if $\varphi \in D_\mu$, we take s' so that $\mathrm{rank}_D(\varphi', s') < \mathrm{rank}_D(\varphi, s)$ is satisfied. Let $t' = h(s') = (x', y')$ and $E' = F_D(\alpha, s')$. By the induction hypothesis, we have $(t', E') \in V_j$. The three items in the condition (b) can be checked for (t', E'). We skipped the details. For the condition (c): Assume $m \in$ FMS, $\varphi = \langle m \rangle \varphi' \in$ Lean, and $t \Vdash \varphi$. By Lemma 6, we have $s' \in S$ such that $(s, s') \in R(m)$ and $s' \models \varphi'$. Let $t' = h(s')$ and $E' = F_D(\alpha, s')$. The two items in the condition (c) can be checked for (t', E'). ■

LEMMA 10. *If* $h[S] \subseteq T \subseteq$ Tab*, A node* $t \in h[S]$ *is both* $\Diamond$*-consistent and* μ*-consistent in* T*.*

Proof. The $\Diamond$-consistency clearly holds by combining Lemmas 6 and 7. For the μ-consistency, note that $F_D(\alpha, s) = \{\varphi \in D \cap \text{Lean} \mid s \models \varphi\}$ if α is sufficiently large, for example if α is a larger cardinal than the cardinality of S. The conclusion follows from Lemmas 6 and 9. ■

Now we prove the soundness.

THEOREM 11. *If there is a Kripke structure* $\mathcal{K} = (S, R, L)$ *and* $s_{\mathrm{I}} \in S$ *such that* $\mathcal{K}, s_{\mathrm{I}} \models \varphi_{\mathrm{I}}$*, the procedure succeeds.*

Proof. Let us define $g(n) = h(L'(n))$ for $n \in$ Nom. It is easy to see that g is a naming function. We claim for this g that $h[S] \subseteq T_k$ holds for all $k \in \omega$. This claim, combined with Lemma 6, is sufficient for the proof.

We prove the claim by induction on k. For $k = 0$, $h[S] \subseteq T_0$ can be checked using Lemma 6 and the definition of the y-part of h.

Assume $h[S] \subseteq T_k$. Then by Lemma 10, $h[S] \subseteq T_{k+1}$. This completes the proof. ■

4.2 Completeness

To prove the completeness, we assume that the procedure succeeds. Let g be the naming function for which the procedure succeeds.

We fix an SCC $D_0 \in \mathcal{D}_\mu$ and a cyclic permutation τ on $\mathcal{D}_\mu$. In case $\mathcal{D}_\mu = \varnothing$, we consider $\mathcal{D}_\mu = \{\varnothing\}$; therefore $D_0 = \varnothing$ and τ is the identity.

Let Nom$'$ be a set of representatives of the equivalence class induced by the equivalence relation $\{(n, n') \in \text{Nom} \mid g(n') \Vdash n\}$. For $t \in$ Tab and $D \in \mathcal{D}_\mu$, the set $\text{sat}(t) \cap D$ is denoted by $\text{sat}_D(t)$.

We construct a (possibly infinite) forest (W, R^W), that is, a disjoint union of trees, where W is the underlying set and R^W is the forest relation on W, together with functions $t : W \to T_K$, $l : R^W \to$ Mod, $D : W \to \mathcal{D}_\mu$, $E : W \to \mathcal{P}(D_\mu \cap \text{Lean})$, and $j : W \to \omega$. During the construction, we will keep the following invariant: for $w, w' \in W$, $j(w) > 0 \implies (t(w), E(w)) \in V_{j(w)} \setminus V_{j(w)-1}$.

At the first stage of the construction, we create elements w_{I} and w_n for $n \in$ Nom$'$ and let $W = \{w_{\mathrm{I}}\} \cup \{w_n \mid n \in \text{Nom}'\}$ and $R^W = \varnothing$. Also, we define $t(w_{\mathrm{I}}) = t_{\mathrm{I}}$ and $t(w_n) = g(n)$ for $n \in$ Nom$'$. And for $w \in W$, we define $D(w) = D_0$, $E(w) = \text{sat}_{D_0}(t(w))$, and $j(w) = \min\{j \in \omega \mid (t(w), E(w)) \in V_j\}$. The invariant holds for the first stage: since $t(w) \in T_K$, $t(w)$ is μ-consistent, therefore, the set in the right hand side of the definition of $j(w)$ is not empty.

At the second and succeeding stages, we pick a leaf node w of W which has not yet been processed and is located at the shallowest level of the forest among such nodes. If there is $n \in \mathrm{Nom}'$ such that $t(w) = t(w_n)$ but $w \neq w_n$, nothing needs to be done. In this case, w is a leaf of the forest.

Otherwise, for each $m \in \mathrm{Mod} \setminus \mathrm{FMS}$ and $\varphi \in \mathrm{sat}(t(w)) \cap \mathcal{L}_{\langle m \rangle}$, we create a node w' and add it to W. Also for each $m \in \mathrm{FMS}$ that satisfies (1) $\mathrm{sat}(t(w)) \cap \mathcal{L}_{\langle m \rangle} \neq \varnothing$, and (2) there is no $\hat{w} \in W$ such that $(\hat{w}, w) \in R^W$ and $l(\hat{w}, w) = m^{-1}$, we create a node w' and add it to W. In both cases, a pair (w, w') is added to R^W and we define $l(w, w') = m$.

We define $t(w')$ depending on the value of $j(w)$. If $j(w) = 0$, since $t(w) \in T_K$, it is $\Diamond$-consistent in T_K. When $m \in \mathrm{Mod} \setminus \mathrm{FMS}$, take $t' \in T_K$ such that $(t(w), t') \in \mathrm{Tr}(\varphi)$, where φ is the formula that corresponds to w'. When $m \in \mathrm{FMS}$, take $t' \in T_K$ such that $(t(w), t') \in \mathrm{Tr}(m)$. In both cases we define $t(w') = t'$. Also, we define $D(w') = \tau(D(w))$, $E(w') = \mathrm{sat}_{D(w')}(t(w'))$, and $j(w') = \min\{j \in \omega \mid (t(w'), E(w')) \in V_j\}$. The invariant can be checked as before.

If $j(w) > 0$, from the invariant there is $(t', E') \in V_{j(w)} \setminus V_{j(w)-1}$ that satisfies the conditions in the description of the procedure. We define $t(w') = t'$, $D(w') = D(w)$, $E(w') = E'$ and $j(w') = \min\{j \in \omega \mid (t', E') \in V_j\}$. The invariant trivially holds. This completes the construction of the forest.

We define an equivalence relation $\sim$ on W: $w_1 \sim w_2$ if and only if $w_1 = w_2$ or there exists $n \in \mathrm{Nom}'$ such that $t(w_1) = t(w_2) = g(n)$. Note that in each equivalence class there is at most one element that has successors. Using this, we define a tuple $\mathcal{K} = (S, R, L)$ as follows:

- The underlying set S is W divided by the equivalence relation $\sim$. For brevity, we also write w for the equivalence class that contains w.
- $R(m) = \{(w_1, w_2) \in R^W \mid l(w_1, w_2) = m \text{ or } l(w_2, w_1) = m^{-1}\}$.
- $L(p) = \{w \in S \mid t(w) \Vdash p\}$.

Tuple $\mathcal{K}$ is almost a Kripke structure, but $R(m)$ might not be a function for $m \in \mathrm{FMS}$. For example if two new nodes are added with $\langle m^{-1} \rangle n$ for $n \in \mathrm{Nom}$ at two different nodes, they are equivalent and the equivalence class has two successors in $\mathcal{K}$. However, one can easily check that the following lemma holds.

LEMMA 12. *If* φ_{I} *is in either* $\mathcal{L}(\mu_{\mathrm{AF}}, \mathrm{nom}, \mathrm{back})$, $\mathcal{L}(\mu_{\mathrm{AF}}, \mathrm{back}, \mathrm{func})$, *or* $\mathcal{L}(\mu_{\mathrm{AF}}, \mathrm{nom}, \mathrm{func})$, *tuple* $\mathcal{K}$ *is a Kripke structure.*

To prove the completeness using Lemma 5, we take a choice function c as follows. Let $w \in S$, $(x, y) = t = t(w)$, $D = D(w)$, and $E = E(w)$. When $\varphi = \varphi_0 \vee \varphi_1 \in \mathrm{cl}(\varphi_{\mathrm{I}})$, $c(\varphi, w)$ is determined in the following order. (1) If $x(\varphi) = \infty$, either φ_0 or φ_1 can be chosen. (2) If only one of $x(\varphi_i) < \infty$ $(i = 0, 1)$ holds, choose φ_i. (3) If $\varphi \notin D$, either φ_0 or φ_1 can be chosen. (4) If only one of $\varphi_i \notin D$ $(i = 0, 1)$ holds, take φ_i. (5) If $y(\varphi_0)$ and $y(\varphi_1)$ differs, choose the smaller of the two. (6) If $x(\varphi_0)$ and $x(\varphi_1)$ differs, choose the smaller of the two. (7) If $x^E(\varphi_0)$ and $x^E(\varphi_1)$ differs, choose the smaller of the two. (8) Either φ_0 or φ_1 can be taken. When $\varphi = \langle m\rangle\varphi'$, (1) if $t \not\Vdash \varphi$, any $w' \in S$ can be chosen. (2) If $m \in \mathrm{FMS}$ and there is $w' \in S$ such that $l(w', w) = m^{-1}$, choose it. (3) In this case, a successor w' of w for φ was created during the construction. Choose it.

Let $Z = \{(\varphi, w) \in \mathrm{cl}(\varphi_\mathrm{I}) \times S \mid t(w) \Vdash \varphi\}$. We will show that the choice function c and Z satisfies the conditions of Lemma 5. We start with a preliminary lemma.

LEMMA 13. *Assume* $t = (x, y) \in \mathrm{Tab}$, $D \in \mathcal{D}_\mu$, $E \subseteq D \cap \mathrm{Lean}$, *and* E *is closed under downward order of* x, *that is,* $\psi \in D \cap \mathrm{Lean}$, $\psi' \in E$, *and* $x(\psi) < x(\psi')$ *implies* $\psi \in E$. *Then, for* φ_1, $\varphi_2 \in D$, *if* $x(\varphi_1) < x(\varphi_2)$ *then* $x^E(\varphi_1) < x^E(\varphi_2)$ *or* $x^E(\varphi_1) = x^E(\varphi_2) = \infty$.

Proof. This lemma can be shown by double induction, first on φ_1, then on φ_2. We omit details, which are tedious but not difficult. ■

Next, let us summarize the properties of the forest as a lemma.

LEMMA 14. *The following holds for all* $w \in W$:

(1) $j(w) > 0 \implies (t(w), E(w)) \in V_{j(w)} \setminus V_{j(w)-1}$.

(2) $E(w) \subseteq D(w)$

(3) Assume $m \in \mathrm{Mod}$, $\varphi \in \mathrm{sat}(t(w)) \cap \mathcal{L}_{\langle m\rangle}$, *and* $w' = c(\varphi)$.

- *(a) Either* $(w, w') \in R^W$ *or* $(w', w) \in R^W$ *holds. If* $m \notin \mathrm{FMS}$, *then* $(w, w') \in R^W$ *holds.*
- *(b)* $(t(w), t(w')) \in \mathrm{Tr}(\varphi)$ *holds.*
- *(c) If* $m \notin \mathrm{FMS}$ *and* $\varphi \in E(w)$, $t(w')^{E(w')} \Vdash \vec{\varphi}$ *holds.*

(4) For any $w' \in W$, $(w, w') \in R^W$ *implies* $(t(w), t(w')) \in \mathrm{Tr}(l(w, w'))$ *and* $(t(w'), t(w)) \in \mathrm{Tr}(l(w, w')^{-1})$.

(5) For any $w' \in W$, $(w, w') \in R^W$ *implies the following:*

- *(a) If* $j(w) > 0$, *then we have* $j(w) > j(w')$, $D(w') = D(w)$, *and* $t(w')^{E(w')} \Vdash \vec{\psi}$ *for any* $\psi \in E(w) \cap \mathcal{L}_{[l(w,w')]}$.
- *(b) If* $j(w) = 0$, *then we have* $E(w) = \varnothing$, $D(w') = \tau(D(w))$ *and* $E(w') = \mathrm{sat}_{D(w')}(t(w'))$.

(6) For $m \in \mathrm{FMS}$, w *has at most one* $w' \in W$ *such that either* $(w, w') \in R^W$ *and* $l(w, w') = m$ *or* $(w', w) \in R^W$ *and* $l(w', w) = m^{-1}$.

Proof. (1) is the invariant we keep through the construction and the others can also be checked easily. ■

LEMMA 15. *Suppose that* $(w_i \mid i \in \omega)$ *is a sequence of elements of* W *and* $(w_i, w_{i+1}) \in R^W$ *for all* $i \in \omega$. *Also suppose* $D \in \mathcal{D}_\mu$ *and* $l \in \omega$.

(1) There exists $k \in \omega$ such that $l \leq k$ and $E(w_k) = \varnothing$.

(2) There exists $k \in \omega$ such that $l < k$, $D(w_k) = D$, and $E(w_k) = \mathrm{sat}_D(t(w_k))$.

Proof. (1) is clear from Lemma 14 (5).
(2) By Lemma 14 (5), there is $k' \geq l$ such that $D(w_{k'}) = \tau(D(w_l))$. Since $\mathcal{D}_\mu$ is finite and τ is a cyclic permutation on $\mathcal{D}_\mu$, by repeating this finitely many times, we have $k > l$ such that $j(w_{k-1}) = 0$, $E(w_{k-1}) = 0$, $D(w_k) = \tau(D(w_{k-1})) = D$, and $E(w_k) = \mathrm{sat}_D(t(w_k))$. ∎

Now we prove the key lemma of this section.

LEMMA 16. *The choice function c and the set Z satisfies the conditions of Lemma 5. Therefore, $t(w) \Vdash \varphi$ implies $\mathcal{K}, w \models \varphi$.*

Proof. The conditions (1) and (2) of Lemma 5 can be checked easily from the definitions.

To show the condition (3), we use the reduction to absurdity and assume that there exist $D \in \mathcal{D}_\mu$ and an infinite trace $((\varphi_k, w_k) \mid k < \omega)$ that conforms with c such that $t(w_k) \Vdash \varphi_k$ and $\varphi_k \in D$ for all $k < \omega$.

CLAIM 17. For all $k \in \omega$, $y_{t(w_k)}(\varphi_k) \geq y_{t(w_{k+1})}(\varphi_{k+1})$. If $w_k \neq w_{k+1}$ and $t(w_{k+1})$ has a name, $y_{t(w_k)}(\varphi_k) > y_{t(w_{k+1})}(\varphi_{k+1})$.

Proof. When $\varphi_k = \varphi \vee \psi$ or $\varphi_k = \langle m \rangle \varphi$, the claim holds by the definition of c. (If $m \in \mathrm{FMS}$, we need Lemma 14 (3) as well.) When φ_k is in the form of $\varphi \wedge \psi$ or $\mu X \varphi$, the claim follows from the definition of y. For $\varphi_k = [m]\varphi$, use Lemma 14 (4) and for $\varphi_k = @n\, \varphi$, use the definition of T_0. And φ_k cannot be an atomic formula, its negation, or $\nu X \varphi$ since $\varphi_k \in D \in \mathcal{D}_\mu$. ∎

From this claim, it is clear that named nodes appear only finitely many times in the trace. (Note also that formulas are guarded. Therefore there cannot exist $K \in \omega$ such that $w_k = w_K$ for all $k \geq K$.) More precisely, we can take $k_0 \in \omega$ such that $t(w_k) \nVdash n$ for all $k \geq k_0$ and $n \in \mathrm{Nom}$.

CLAIM 18. If $K, L \in \omega$, $k_0 \leq K \leq L$, and $w_K = w_L$ (we denote it by w), then $x_{t(w)}(\varphi_K) \geq x_{t(w)}(\varphi_L)$. Moreover, if there is k' such that $K < k' < L$ and $w_{k'} \neq w$, then $x_{t(w)}(\varphi_K) > x_{t(w)}(\varphi_L)$.

Proof. We prove the claim by induction on $L - K$. The claim becomes trivial when $L - K = 0$, so we assume $K < L$.

If φ_K is in the form of $\varphi \vee \psi$, $\varphi \wedge \psi$, or $\mu X \varphi$, then $w_K = w_{K+1}$ and $x_{t(w)}(\varphi_K) \geq x_{t(w)}(\varphi_{K+1})$. Therefore the induction hypothesis leads to the conclusion. Since $k_0 \leq K$, φ_K cannot be in the form of $@n, \varphi$. Also, because $\varphi_K \in D$, it is neither an atomic formula or its negation.

The remaining cases are $\varphi_K = \langle m \rangle \varphi_{K+1}$ and $\varphi_K = [m]\varphi_{K+1}$, where $m \in \mathrm{Mod}$. In these cases, $w_{K+1} \neq w_K$ since W is a forest. Let M be the least index such that $K < M$ and $w_M = w_K$. Clearly $K + 1 \leq M -$

1 and $w_{K+1} = w_{M-1}$ since W is a forest. Let $w' = w_{K+1} = w_{M-1}$. By the induction hypothesis, we have (1) $x_{t(w')}(\varphi_{K+1}) \geq x_{t(w')}(\varphi_{M-1})$ and (2) $x_{t(w)}(\varphi_M) \geq x_{t(w)}(\varphi_L)$. Therefore if we show (3) $x_{t(w)}(\varphi_K) > x_{t(w)}(\varphi_M)$, we have $x_{t(w)}(\varphi_K) > x_{t(w)}(\varphi_L)$ as desired by combining (2) and (3). Note that either $\varphi_{M-1} = \langle m^{-1}\rangle\varphi_M$ or $\varphi_{M-1} = [m^{-1}]\varphi_M$ holds since w_{M-1} is different from w_M. In order to prove (3), it is sufficient to show either LoopFree$(t(w), t(w'), \varphi_K)$ or LoopFree$(t(w'), t(w), \varphi_{M-1})$ since we have (2).

First assume $\varphi_K = \langle m\rangle\varphi_{K+1}$. By Lemma 14 (3), $(t(w), t(w')) \in \mathrm{Tr}(\varphi_K)$. Therefore if $\varphi_{M-1} = [m^{-1}]\varphi_M$, LoopFree$(t(w), t(w'), \varphi_K)$ holds. In the other case, i.e. $\varphi_{M-1} = \langle m^{-1}\rangle\varphi_M$, if we assume $m \notin$ FMS, again by Lemma 14 (3), we have $(w, w') \in R^W$ and $(w', w) \in R^W$, which is impossible since W is a forest. Thus, we have $m \in$ FMS, and then, LoopFree$(t(w)$, $t(w'), \varphi_K)$ holds from the definition of $\mathrm{Tr}(m)$ and $\mathrm{Tr}(\langle m\rangle\varphi)$. The same argument can be done if we assume $\varphi_{M-1} = \langle m\rangle\varphi_M$.

The only case remained is $\varphi_K = [m]\varphi_{K+1}$ and $\varphi_{M-1} = [m^{-1}]\varphi_M$. Since $((\varphi_K, w_K), (\varphi_{K+1}, w_{K+1}))$ is a trace, $(w_K, w_{K+1}) \in R(m)$. Therefore either $(w_K, w_{K+1}) \in R^W$ and $l(w_K, w_{K+1}) = m$ or $(w_{K+1}, w_K) \in R^W$ and $l(w_{K+1}, w_K) = m^{-1}$. Then by Lemma 14 (4), $(t(w_K), t(w_{K+1})) \in \mathrm{Tr}(m)$ holds, and therefore we have LoopFree$(t(w)$, $t(w')$, $\varphi_K)$ as desired. ■

From this lemma, it follows that the set $\{k \in \omega \mid w = w_k\}$ is finite for any $w \in W$, We define $m(w)$ by $\max\{k \in \omega \mid w = w_k\}$, and a sequence $(L(i) \mid i \in \omega)$ of natural numbers by $L(0) = m(w_{k_0})$ and $L(i+1) = m(w_{L(i)+1})$. It is clear that $w_{L(i+1)} = w_{L(i)+1}$. Note that $\varphi_{L(i)} \in$ Lean since $w_{L(i)+1} \neq w_{L(i)}$.

CLAIM 19. There is a natural number $i_0 \in \omega$ such that for all $i \in \omega$, $i \geq i_0 \implies (w_{L(i)}, w_{L(i+1)}) \in R^W$.

Proof. Since $w_{L(i)} \neq w_{L(i)+1}$ and $L(i) \geq k_0$, $\varphi_{L(i)}$ is either $\langle m\rangle\varphi_{L(i)+1}$ or $[m]\varphi_{L(i)+1}$ for some $m \in$ Mod. Therefore $(w_{L(i)}, w_{L(i)+1}) \in R(m)$, that is, either $(w_{L(i)}, w_{L(i)+1}) \in R^W$ or $(w_{L(i)+1}, w_{L(i)}) \in R^W$ holds. Since W is a forest, the relation R^W is well-founded, so there is at least one $i_0 \in \omega$ that $(w_{L(i_0)}, w_{L(i_0)+1}) \in R^W$. We show by induction on i, $(w_{L(i)}, w_{L(i+1)}) \in R^W$ for all $i \geq i_0$. The case $i = i_0$ is trivial. The case $i+1$: recall that either $(w_{L(i+1)}, w_{L(i+1)+1}) \in R^W$ or $(w_{L(i+1)+1}, w_{L(i+1)}) \in R^W$ holds. If the latter is the case, since $(w_{L(i)}, w_{L(i)+1}) \in R^W$, $w_{L(i)+1} = w_{L(i+1)}$, and W is a forest, we have $w_{L(i)} = w_{L(i+1)+1}$, which contradicts the fact that $L(i)$ is the largest index since $L(i) < L(i+1)+1$. Therefore the former should be the case, i.e., $(w_{L(i+1)}, w_{L(i+1)+1}) \in R^W$. ■

The reflexive transitive closure of R^W is denoted by R^{W+}.

CLAIM 20. For any $j \in \omega$ and $k > L(i_j)$, $(w_{L(i_j)}, w_k) \in R^{W+}$ holds.

Proof. We prove the claim by induction on k. The base case $k = L(i_j)+1$ is clear from Claim 19. For the general case, assume $(w_{L(i_j)}, w_k) \in R^{W+}$. If we further assume $(w_{L(i_j)}, w_{k+1}) \notin R^{W+}$, w_{k+1} must be identical to $w_{L(i_j)}$

since $\varphi_k = \langle m \rangle \varphi_{k+1}$ or $\varphi_k = [m]\varphi_{k+1}$ for some $m \in \mathrm{Mod}$ because $k \geq k_0$. It again contradicts the fact that $L(i_j)$ is the maximum index. ■

By applying Lemma 15 (2) to the sequence $(w_{L(i)} \mid i_0 \leq i < \omega)$, we can take $i_1 \in \omega$ such that $i_1 \geq i_0$, $D(w_{L(i_1)}) = D$, and $E(w_{L(i_1)}) = \mathrm{sat}_D(t(w_{L(i_1)}))$. In particular, $\varphi_{L(i_1)} \in E_{L(i_1)}$.

CLAIM 21. For all $k \geq L(i_1)$, $D(w_k) = D$ and $t(w_k)^{E(w_k)} \Vdash \varphi_k$ holds.

Proof. We use induction on k. The base step $k = L(i_1)$ is trivial.

For the general step, we first consider the case $\varphi_k = \varphi_{k+1} \vee \psi$. Since $w_k = w_{k+1}$, $D(w_{k+1}) = D$. Note that $\psi \in D$, otherwise c should have chosen ψ. Let $(x, y) = t(w_k)$ and $E = E(w_k)$. By the definition of c, either $x(\varphi_{k+1}) < x(\psi)$ or $x(\varphi_{k+1}) = x(\psi)$ and $x^E(\varphi_{k+1}) \leq x^E(\psi)$ holds. In the former case, by Lemma 13, $x^E(\varphi_{k+1}) < x^E(\psi)$ or $x^E(\varphi_{k+1}) = x^E(\psi) = \infty$. Here, $x^E(\varphi_{k+1}) = x^E(\psi) = \infty$ implies $x^E(\varphi_k) = \infty$, which is impossible. Therefore, in both cases, $x^E(\varphi_{k+1}) \leq x^E(\psi)$, which implies $x^E(\varphi_{k+1}) = x^E(\varphi_k) < \infty$.

The cases $\varphi_k = \varphi_{k+1} \wedge \psi$ and $\mu X \psi$ are easy and we omit them.

The remaining cases are $\varphi_k = \langle m \rangle \varphi_{k+1}$ and $[m]\varphi_{k+1}$. In these two cases, either $(w_{k+1}, w_k) \in R^W$ or $(w_k, w_{k+1}) \in R^W$ holds. Assume first $(w_{k+1}, w_k) \in R^W$. Clearly $D(w_{k+1}) = D(w_k) = D$. Since for all j such that $L(i_1) \leq j < k$, one of $(w_{j+1}, w_j) \in R^W$, $(w_j, w_{j+1}) \in R^W$, or $w_j = w_{j+1}$ holds, so by Claim 20, there is i such that $L(i_1) \leq i < k+1$ such that $w_i = w_{k+1}$. Let $E = E(w_i)$ and $(x, y) = t(w_i)$. By induction hypothesis we have $x(\varphi_i) < \infty$. On the other hand we have $x(\varphi_i) > x(\varphi_{k+1})$ by Lemma 18. Therefore by Lemma 13, $x^E(\varphi_{k+1}) < x^E(\varphi_i) < \infty$.

Assume next $(w_k, w_{k+1}) \in R^W$ holds. In the case of $\varphi_k = \langle m \rangle \varphi_{k+1}$, we have $t(w_{k+1})^{E(w_{k+1})} \Vdash \varphi_{k+1}$ by the definition of c and the induction hypothesis. In the case of $\varphi_k = [m]\varphi_{k+1}$, since $\varphi_k \in E(w_k) \neq \varnothing$, $j(w_k)$ must be positive by Lemma 14 (5). Then again by Lemma 14 (5), $D(w_{k+1}) = D(w_k) = D$ and $t(w_{k+1})^{E(w_{k+1})} \Vdash \varphi_{k+1}$. ■

Claim 21 contradicts Lemma 15 (1), which establishes the condition (3) of Lemma 5. That completes the proof of Lemma 16. ■

THEOREM 22. *Assume that the decision procedure succeeds for φ_{I} in either $\mathcal{L}(\mu_{\mathrm{AF}}, \mathrm{nom}, \mathrm{back})$, $\mathcal{L}(\mu_{\mathrm{AF}}, \mathrm{back}, \mathrm{func})$, or $\mathcal{L}(\mu_{\mathrm{AF}}, \mathrm{nom}, \mathrm{func})$. Then, there is a Kripke structure and its state that satisfies φ_{I}.*

Proof. Lemma 12 guarantees that $\mathcal{K}$ is a Kripke structure. Since $t(w_{\mathrm{I}}) = t_{\mathrm{I}} \Vdash \varphi_{\mathrm{I}}$, $\mathcal{K}, w_{\mathrm{I}} \models \varphi$ by Lemma 16. ■

5 Complexity

Our decision procedure solves the satisfiability problems of three logics – $\mathcal{L}(\mu_{\mathrm{AF}}, \mathrm{nom}, \mathrm{back})$, $\mathcal{L}(\mu_{\mathrm{AF}}, \mathrm{nom}, \mathrm{func})$, and $\mathcal{L}(\mu_{\mathrm{AF}}, \mathrm{back}, \mathrm{func})$. Their complexities have already been known to be EXPTIME-complete [10]. More detailed calculation of the time complexity of our procedure is shown in the following proposition. Let n be the length of the formula φ_{I}.

PROPOSITION 23. *The time complexity of the decision procedure presented in Section 3 is* $2^{\mathcal{O}(n \log n)}$ *for formulas in* $\mathcal{L}(\mu_{\mathrm{AF}}, \mathrm{back}, \mathrm{func})$ *and* $2^{\mathcal{O}(n^2 \log n)}$ *for formulas in* $\mathcal{L}(\mu_{\mathrm{AF}}, \mathrm{nom}, \mathrm{back})$ *and* $\mathcal{L}(\mu_{\mathrm{AF}}, \mathrm{nom}, \mathrm{func})$.

Proof. Let us count the number of naming functions. The domain of a naming function is Nom and the range is Tab. Such functions exist at most $(n^n \cdot n^{2n})^n = 2^{3n^2 \log n}$. In the case of $\mathcal{L}(\mu_{\mathrm{AF}}, \mathrm{back}, \mathrm{func})$, the number of functions is 1 since there are no nominals. Therefore it is enough to show that the time complexity of the subprocedure for a naming function is $2^{\mathcal{O}(n \log n)}$.

Next we count the number A of nodes in the tableau. Component x can be regarded as a function from Lean to $\{0, 1, \ldots, n-1\}$. The number of such functions is at most n^n. Component y can be regarded as a function from Lean to $\{0, 1, \ldots, n^2 - 1\}$. The number of such functions is at most $(n^2)^n$. Therefore, $A \leq n^n \cdot (n^2)^n = 2^{\mathcal{O}(n \log n)}$.

The body of the subprocedure is a double-loop. In the outer loop, $T_0 \supset T_1 \supset \cdots \supset T_K$ are calculated. Since $|T_0| \leq A$, the number of repetition K does not exceed A. The inner loop calculates $V_0 \subset V_1 \subset \cdots \subset V_J$. Since each V_j is a subset of $T_0 \times \mathcal{P}(\mathrm{Lean})$, the number of repetition J does not exceed $B = A \cdot 2^n$.

In each repetition of the inner loop, V_{j+1} is calculated from V_j. This is done by verifying that each (t, E) is an element of V_{j+1}. The number of such (t, E) does not exceed B. Each (t, E) is checked against conditions (a), (b), and (c). They can be performed in polynomial time C once pair (t', E') is fixed and the number of candidates (t', E') does not exceed B.

Therefore, time required to perform the subprocedure is $A \cdot B \cdot B \cdot C \cdot B = C \cdot A^4 \cdot (2^n)^3 = 2^{\mathcal{O}(n \log n)}$. ■

Although we need to transform a given formula into a guarded PNF before applying the decision procedure, we can still presume that n in the proposition refers the length of the given formula, since the proof of Proposition 23 can be done by regarding n as the size of $\mathrm{cl}(\varphi_{\mathrm{I}})$.

6 Application

We apply the decision procedure to verify some properties of programs written in imperative languages that manipulate pointers.

We regard program heaps as Kripke structures. Thus, a property of a heap is expressed by a formula φ of the μ-calculus. Each operation σ on the heap can be regarded as a transformation of Kripke structures, and we compute the weakest precondition $\mathrm{wp}(\sigma, \varphi)$ as a formula. Then, we can determine whether a Kripke structure that satisfies φ can be transformed into a structure that satisfies ψ by verifying that formula $\varphi \wedge \mathrm{wp}(\sigma, \psi)$ is satisfiable. Refer to [19] for details. We built an experimental tool, called

id	len	# cl	# nom	time
regr56	35	11	3	461
listRevNoLeakB	92	25	5	610
listRevSwapA	148	37	6	1199
dswPopC2	188	46	7	2615
dswPushA1	372	54	7	9469

Table 1. Examples of formulas used in the analysis

MLAT [20], based on this method. It is programmed in Java, and the decision procedure is implemented using JavaBDD [7].

There are several versions of MLAT. By using one of the versions, we verified the partial correctness of the Deutsch-Schorr-Waite marking algorithm [21]. Details are reported in [19].

Using this version of the tool, the user constructs an abstract transition system by hand, and the tool checks whether the transition system is a correct abstraction of the concrete transition system corresponding to a given source code. As the logic for describing predicates, we employ $\mathcal{L}(\mu_{\mathrm{AF}}, \mathrm{nom}, \mathrm{back})$ for this version.

We simplify the procedure for efficiency; we do not use the component y at all. When the procedure with this simplification succeeds, a nominal may be satisfied with two different nodes s_1 and s_2 in the resulting "model"; however, such nodes are "indistinguishable" in the sense that $s_1 \models \varphi \iff s_2 \models \varphi$ holds for any $\varphi \in \mathrm{cl}(\varphi_{\mathrm{I}})$.

The modified procedure is still sound. In general, any sound decision procedure can be used in the predicate abstraction technique. The procedure should not necessarily be complete, but using incomplete procedures may affect the precision of the analysis. In our analysis, however, there is another reason for losing the precision: the selection of predicates. Our experiment shows that the second reason affects the results more than the first. We have several cases in which we failed to analyze the properties due to inappropriate selection of predicates, but there is no case in which the analysis failed due to incompleteness of the procedure. Apparently, "indistinguishability" is a good approximation for nominality in our application.

During the verification, a number of formulas were checked for satisfiability. Examples are shown in Table 1. The columns of the table are formula ID, the length of the formula, the size of the closure, the number of nominals in the formula, and elapsed time for the judgment in milliseconds. The machine used for the the measurement is Pentium 4 CPU 2.4GHz, 1GB memory running Microsoft Windows XP.

7 Future work

In the future we would like to implement the entire decision procedure. At the moment, we have only implemented modified versions of the procedure for particular applications, as described in Section 6.

Another possibility is to strengthen our decision procedures for applications to more powerful logics. The loosely guarded fragment (LGF) is a decidable sublogic of the first-order predicate logic, and the extension of the LGF with fixed-point operators is called μ-LGF, which can be regarded as a natural extension of the two-way modal μ-calculus. A natural question is whether our procedure can be extended to the alternation-free part of μ-LGF. A simple adaptation of our procedure to the logic does not work, and further investigations are required.

Acknowledgments. The authors are grateful to the anonymous reviewers for their careful reading. This research was partially supported by the research project "Solving the description explosion problem in verification by means of structure transformation" in CREST (Core Research for Evolution Science and Technology) program of the Japan Science and Technology Agency. This research was partially supported by Grant-in-Aid for Scientific Research by Ministry of Education, Culture, Science and Technology, Scientific Research(C) (2)18500003, "Abstraction of graphs to multisets using temporal logic".

BIBLIOGRAPHY

[1] Takahashi, K., Hagiya, M.: Abstraction of graph transformation using temporal formulas. In: Supplemental Volume of the 2003 International Conference on Dependable Systems and Networks (DNS-2003). (2003) W–65 to W–66

[2] Hagiya, M., Takahashi, K., Yamamoto, M., Sato, T.: Analysis of synchronous and asynchronous cellular automata using abstraction by temporal logic. In: FLOPS2004: The Seventh Functional and Logic Programming Symposium. Volume 2998 of Lecture Notes in Computer Science. (2004) 7–21

[3] Tanabe, Y., Takahashi, K., Yamamoto, M., Sato, T., Hagiya, M.: An implementation of a decision procedure for satisfiability of two-way CTL formulae using BDD (in Japanese). Computer Software **22**(3) (2005) 154–166

[4] Tanabe, Y., Takahashi, K., Yamamoto, M., Tozawa, A., Hagiya, M.: A decision procedure for the alternation-free two-way modal μ-calculus. In Beckert, B., ed.: TABLEAUX. Volume 3702 of Lecture Notes in Computer Science., Springer (2005) 277–291

[5] Sagiv, M., Reps, T., Wilhelm, R.: Parametric shape analysis via 3-valued logic. ACM Transactions on Programming Languages and Systems **24**(3) (2002) 217–298

[6] Bonatti, P.A., Peron, A.: On the undecidability of logics with converse, nominals, recursion and counting. Artificial Intelligence **158** (2004) 75–96

[7] JavaBDD: `http://javabdd.sourceforge.net/`

[8] Kozen, D.: Results on the propositional μ-calculus. Theoretical Computer Science **27**(3) (1983) 333–354

[9] Emerson, E.A., Jutla, C.S.: Tree automata, mu-calculus and determinacy (extended abstract). In: FOCS, IEEE (1991) 368–377

[10] Bonatti, P.A., Lutz, C., Murano, A., Vardi, M.Y.: The complexity of enriched μ-calculi. In Bugliesi, M., Preneel, B., Sassone, V., Wegener, I., eds.: ICALP (2). Volume 4052 of Lecture Notes in Computer Science., Springer (2006) 540–551

[11] Emerson, E.A.: Temporal and modal logic. In: Handbook of theoretical computer science (vol. B): formal models and semantics, Elsevier Science Publishers B.V. (1990) 995–1072

[12] Pan, G., Sattler, U., Vardi, M.Y.: BDD-based decision procedures for K. In: Proceedings of the 18th International Conference on Automated Deduction, Springer-Verlag (2002) 16–30

[13] Pan, G., Vardi, M.Y.: Optimizing a BDD-based modal solvar. In: 19th International Conference on Automated Deduction. Volume 2741 of Lecture Notes in Computer Science. (2003) 75–89

[14] Henriksen, J.G., Jensen, J.L., Jørgensen, M.E., Klarlund, N., Paige, R., Rauhe, T., Sandholm, A.: Mona: Monadic second-order logic in practice. In: Proceedings of the First International Workshop on Tools and Algorithms for Construction and Analysis of Systems. Volume 1019 of Lecture Notes in Computer Science., Springer-Verlag (1995) 89–110

[15] Kupferman, O., Vardi, M.Y.: Safraless decision procedures. In: FOCS, IEEE Computer Society (2005) 531–542

[16] van Eijck, J.: HyLoTab — tableau-based theorem proving for hybrid logics (2002) Manuscript, CWI, available from `http://www.cwi.nl/ jve/hylotab`.

[17] Walukiewicz, I.: Completeness of Kozens axiomatisation of the propositional mu-calculus. Information and Computation **157** (2000) 142–182

[18] Zappe, J.: Modal μ-calculus and alternating tree automata. In Grädel, E., Thomas, W., Wilke, T., eds.: Automata, Logics, and Infinite Games. Volume 2500 of Lecture Notes in Computer Science., Springer (2001) 171–184

[19] Tanabe, Y.: Satisfiability Judgment of Modal Logics and their Application to Verification Problems. PhD thesis, University of Tokyo (2008)

[20] Sekizawa, T., Tanabe, Y., Yuasa, Y., Takahashi, K.: MLAT: A tool for heap analysis based on predicate abstraction by modal logic. In: IASTED International Conference on Software Engineering (SE 2008). (2008) 310–317

[21] Schorr, H., Waite, W.M.: An efficient machine-independent procedure for garbage collection in various list structures. Commun. ACM **10**(8) (1967) 501–506

Yoshinori Tanabe
Graduate School of Information Science and Technology,
University of Tokyo.
Akihabara Daibiru 13F, 1-18-13, Sotokanda, Tokyo, 101-0021,
Japan.
`y-tanabe@ci.i.u-tokyo.ac.jp`

Koichi Takahashi
Research Center for Verification and Semantics (CVS),
National Institute of Advanced Industrial Science and Technology.
1-2-14, Shinsenri Nishimachi, Toyonaka, Osaka, 560-0083,
Japan.
`k.takahashi@aist.go.jp`

Masami Hagiya
Graduate School of Information Science and Technology,
University of Tokyo.
7-3-1, Hongo, Bunkyo-ku, Tokyo, 113-8656,
Japan.
`hagiya@is.s.u-tokyo.ac.jp`

Modal logic of time division

Tero Tulenheimo

ABSTRACT. A logic $\mathcal{L}_{TD}$ is defined, inspired by [37]. It is syntactically like basic modal logic with an additional unary operator but it has an interval-based semantics on structures with arbitrary linear frames. $\Box\psi$ is interpreted as meaning 'the current interval has a finite partition whose all members satisfy ψ.' $\mathcal{L}_{TD}$ is translatable into weak monadic second-order logic but not into first-order logic. The expressive power and the decidability properties of $\mathcal{L}_{TD}$ and its fragments are studied.

Keywords: decidability, expressive power, interval tense logic, linear order, negation, order type, von Wright, weak monadic second-order logic.

1 Introduction

G. H. von Wright suggested in his essay "Time, Change, and Contradiction" [37] an original approach to the logical investigation of time, where the basic objects of study are temporal intervals and their internal structure. In the present paper a formal semantics doing justice to von Wright's informally presented semantic ideas is formulated;[1] the resulting logic and its fragments are then studied for their expressive power and decidability properties.

Background. The guiding idea in von Wright's essay is to examine the relation between *time and change* on the one hand, and *time and contradiction* on the other. As von Wright saw it, in his paper — presented as the 22nd Eddington Memorial Lecture at Cambridge in 1968 — a new avenue in tense logic was opened up, leading to a study of the logic of the division of time into 'bits' of ever shorter duration [38, pp. xi–xii]. While Prior's tense logic [29] studies instants (time points) and their relationships, the basic relation being *succession* in time, in von Wright's approach one takes as the point of departure 'bits' of time and proceeds to the analysis of their internal structure; here the basic relation is *division* of time [39, p. 862].

Von Wright approaches the relation of time and contradiction by reference to the following modal-logical axioms: **(A1)** or $\Box(p \wedge q) \leftrightarrow (\Box p \wedge \Box q)$, **(A2)** or $\Box(p \vee q) \rightarrow (\Box p \vee \Box q)$, **(A3)** or $\Box(p \vee {\sim}p)$ and **(A4)** or ${\sim}\Box(p \wedge {\sim}p)$. Noting that one interpretation of $\Box$ satisfying the axioms is 'it will next be the case that,' he however proposes to use the axioms to speak of temporal occasions and read $\Box$ as 'it is completely or throughout the case that.' In this framework he formulates the notion of 'contradiction in nature,' which he

[1] For a different formal semantics based on von Wright's relevant logical ideas, see Dalla Chiara [3]; the logic she formulates is many-valued, actually a variant of Łukasiewicz logic.

 © *2008, the author(s).*

relates to the analysis of the notion of change. But what is a contradiction in nature? And how is von Wright led to recognize that type of contradiction?

Von Wright considers different cases arising from accepting some axioms and rejecting the others. He takes (A3) to state of an occasion that it can be divided into parts during each of which either p or else $\sim p$ holds throughout. He notes that hence not-(A3) is true of an occasion *any* partition of which has at least one member in which both p and $\sim p$ occur. Such an occasion is said to incorporate a *real contradiction* or a *contradiction in nature* — an idea that has, as von Wright notes, a rather Hegelian flavor. What is at stake, however, is not a *logical* contradiction, but the impossibility of an analysis of change without at least one of the building blocks of the analysis carrying logically contradictory but non-simultaneous constituents. Yet if ontological priority is given to extended occasions, occasions o making not-(A3) true deserve to be called contradictory in a sense: any analysis of o into sub-occasions has a component in which both p and $\sim p$ are present.

When discussing occasions not satisfying (A3), von Wright goes on to say that they positively satisfy $\sim\Box(p \vee \sim p)$, from which he further deduces that they satisfy $\Diamond(p \wedge \sim p)$. This reasoning is mistaken and illustrates problems to which one may be led when not being explicit about the semantics of the expressions involved. Negating (A3) cannot mean affirming $\sim\Box(p \vee \sim p)$ (unless the force of $\sim$ depends on its syntactic position vis-à-vis $\Box$). Namely, von Wright takes $\sim p$ to state at occasion o that not-p holds throughout o, whence $\sim\Box(p \vee \sim p)$ must mean that $\Box(p \vee \sim p)$ fails throughout o, while not-(A3) just says that $\Box(p \vee \sim p)$ fails at o. The temporal ontology of time intervals calls for a distinction between two negations:[2] $\neg\psi$ states at an occasion o that ψ fails at o, while $\sim\psi$ states that ψ fails at each time point in o. With this distinction at our disposal, it is seen that occasions incorporating a 'contradiction in nature' — i.e., occasions at which (A3) fails — satisfy $\neg\Box(p \vee \sim p)$. By the formal semantics to be given in the present paper, this formula is merely equivalent to $\Diamond(\neg\sim p \wedge \neg p)$, not to $\Diamond(p \wedge \sim p)$. (The formula $\neg\sim p$ is true at o iff p holds at least once during o, and $\neg p$ is true at o iff p fails at least once during o.) By contrast, $\Diamond(p \wedge \sim p)$ is logically contradictory (w.r.t. all occasions consisting of at least two instants). Von Wright overlooked the need for making the distinction between the two negations.[3] Also, he paid no attention to nested modal operators; and he did not note that the axioms may hold for *atomic* substitution instances of p and q without holding for arbitrary substitution instances.

Basic definitions. Let T be a set and $R \subseteq T^2$. R is *reflexive* (*irreflexive*) if $R(t,t)$ holds for all (no) t in T; *antisymmetric* if $R(s,t)$ and $R(t,s)$ implies $s = t$; *transitive* if $R(s,t)$ and $R(t,u)$ implies $R(s,u)$; *dichotomous* if $R(s,t)$ or $R(t,s)$ holds for every s,t; *trichotomous* if $R(s,t)$ or $R(t,s)$ or $s = t$ holds for every s,t; *linear order* if it is irreflexive, transitive and trichotomous. All

[2] Dalla Chiara [3] does not distinguish between two negations, but in a sense allows violations of the law of non-contradiction.

[3] Neither Prior [30] nor Smith [35] nor Mortensen [25] realizes that von Wright attaches two incompatible meanings to one and the same negation sign.

linear orders are antisymmetric and fail to be dichotomous. Somewhat confusingly, some authors use 'linear' as a synonym for 'trichotomous' [28]. By 'linear order' some authors mean antisymmetric, transitive and dichotomous — therefore reflexive — binary relation [22] (*reflexive linear orders*), while others mean what was above termed linear order [23]. When quoting other people's results, one must be careful about what they actually proved. A linear order $<$ is *dense* if $s < t$ implies the existence of u with $s < u < t$. Given a linear order $<$, write $\leq$ for its reflexive closure. It is assumed that the reader is familiar with propositional logic (**PL**), first-order logic (**FO**), basic modal logic (**ML**) and basic tense logic (**TL**) (see, e.g., [7, 1, 9]). The symbol $\top$ ($\bot$) denotes a propositional atom by stipulation true (false) under all valuations. The *quantifier rank* of an **FO**-formula is its maximum number of nested quantifiers. For the technique of using Ehrenfeucht-Fraïssé games to prove the elementary equivalence of two structures up to a given quantifier rank, see [6]. Recall that the future tense operators of **TL** are F and G. The reader is reminded of the logic **US** of *Until* and *Since* introduced by Kamp [19]. (For **US**, see [9].) *Weak monadic second-order logic* ($\mathcal{L}_{\mathrm{w}}^{\mathrm{mon}}$) [7, 21, 24] is obtained from **FO** by allowing atomic formulas $X(t)$ and complex formulas $\exists X\phi$, where X is a unary relation variable and t is a term. Crucially, the unary relation variables range over *finite* subsets of the domain. Allowing quantification over arbitrary subsets leads to monadic second-order logic ($\mathcal{L}^{\mathrm{mon}}$). $\mathcal{L}_{\mathrm{w}}^{\mathrm{mon}}$ is decidable over reflexive linear orders [22]. This implies the decidability of $\mathcal{L}_{\mathrm{w}}^{\mathrm{mon}}$ over linear orders, because the latter are (first-order) definable from the former.

We write $\phi \in L$ to indicate that ϕ is a formula of a logic L: we do not notationally distinguish a logic from its set of formulas. Henceforth, 'iff' abbreviates 'if and only if.' If S is a set of formulas, $\mathbf{Cl}_{\wedge,\vee}(S)$ is its closure under $\wedge$ and $\vee$: the set of formulas obtained from S by finitely many applications of $\wedge$ and $\vee$. Given a logic L, its satisfiability (validity) problem is denoted by L-SAT (L-VAL). If for every $\phi \in L$ there is $neg(\phi) \in L$ true precisely when ϕ is not true, and if $neg(\phi)$ is computed from ϕ in **PTIME**, L-SAT is decidable using an algorithm from complexity class C iff L-VAL is; for such L we may speak of decidability without specifying whether we mean L-SAT or L-VAL. **3-CNF** denotes the **NP**-complete problem of deciding whether a **PL**-formula in conjunctive normal form is satisfiable, given that each conjunct consists of just 3 disjuncts (each of which is a literal). In complexity results, time bounds are measured relative to the *length* of the input: its number of symbol tokens. If p_i is an atom ($i < \omega$), its length is $1 + b(i)$, where $b(i)$ is the number of digits of the numeral representing i in binary. If L, L' are (modal or abstract) logics defined over the same class of structures $\mathcal{K}$, a syntactic map $t : L \to L'$ is a *translation* of L into L' if for all $\phi \in L$ and $\mathfrak{M} \in \mathcal{K}$: ϕ is true in $\mathfrak{M}$ iff $t(\phi)$ is true therein. $L \leq L'$ means: a translation of L into L' exists; and $L < L'$ means: $L \leq L'$ but not $L' \leq L$. If $f : A \to B$ is a map, its *image* $Im(f)$ is the set $\{f(a) : a \in A\}$. Basic knowledge of order types and ordinals is assumed [8, 18, 34]. ω is the order type of natural numbers, ω^* is the dual of ω (having the order of ω

reversed), and η is the order type of the set of rational numbers (non-empty countable dense linear orders without end-points).

Plan of the paper. *Section* 2 introduces the *logic of time division* ($\mathcal{L}_{TD}$). Its expressive power is discussed in *Section* 3. Three fragments of $\mathcal{L}_{TD}$ are studied in some detail in *Sections* 4, 5 and 6. *Section* 7 concludes the paper by pointing out related work and questions for future research.

2 Logic of time division

Syntax. Let **prop** be a set of propositional atoms containing $\top$ and $\bot$. The syntax of the *logic of time division* (or $\mathcal{L}_{TD}$) is given by the grammar $\phi ::= p \mid \neg\phi \mid {\sim}\phi \mid (\phi \vee \phi) \mid (\phi \wedge \phi) \mid \Diamond\phi \mid \Box\phi$, with $p \in$ **prop**. Syntactically, $\mathcal{L}_{TD}$ is **ML** with an additional unary operator ($\sim$). $\Box$ and $\Diamond$ are *modal operators.* The *modal depth* $md(\phi)$ of ϕ is the maximum number of nested modal operators in ϕ. Formulas of the forms $p, \neg p, {\sim}p, \neg{\sim}p$ ($p \in$ **prop**) are termed *literals.* The notion of subformula is defined in the expected way.

Semantics. Only linear flows of time will be considered.[4] Von Wright [37] takes 'occasions' or intervals to be primary in relation to extensionless time points. These latter he views as 'idealizations'. The former he characterizes in the strict sense as 'bits' or 'stretches' of time during which no change takes place, but allows for a generalized sense in which an occasion is any interval offering a medium within which changes may occur. In the subsequent formal development this view on time results in letting a domain T consist of instants, but making evaluation, primarily, relative to extended intervals. (So the domain, as it is given, is a result of idealization, but the primary mode of evaluation reflects the conceptual priority of intervals over instants.) However, since we do not wish to outright exclude idealizations, evaluation relative to instants (or, singleton intervals) is admitted as well.

Frames are pairs $(T, <)$, where $T \neq \varnothing$ and $<$ is a linear order on T. For later purposes, we assume that for each frame an element $t^* \in T$ has been fixed. *Models* are triples $\mathcal{M} = (T, <, V)$ with $(T, <)$ a frame and $V : \textbf{prop} \to \mathcal{P}(T)$ a valuation. Always $V(\top) = T$ and $V(\bot) = \varnothing$. Models are, then, modal structures with a linear accessibility relation. However, formulas are evaluated relative to *certain kinds of subsets* of the domain (in **ML** all evaluation is relative to single elements of the domain). We define an *occasion* o in a frame $(T, <)$ to be a subset of T which either is of the form $]s, t] = \{x : s < x \leq t\}$ for some $s, t \in T$ or of the form $\{t\}$ for some $t \in T$. Occasions of the former kind are *occasions proper*; $\{t\}$ is an *idealized occasion* if t has no immediate predecessor in T (while if t has one, t', then $\{t\} =]t', t]$ is an occasion proper). The empty interval $\varnothing$ is an occasion proper: if $s \geq t$, then $]s, t] = \varnothing$.[5] The cardinality of o is denoted by $|o|$. If $o =]s, t]$ is non-empty, s is the *left bound* of o, denoted $l(o)$, and t its *right bound*, denoted $r(o)$. If o is empty, by stipulation $l(o) = t^* = r(o)$.

[4]This restriction reflects von Wright's interest in experienced time; certain time-related phenomena are better studied by reference to tree-like flows of time.

[5]We prefer not to preclude empty intervals at the outset. It will turn out that the emptiness and non-emptiness of an interval are properties definable in $\mathcal{L}_{TD}$.

(A stipulation is needed: we may have $]s,t] = \varnothing =]s',t']$ while $s \neq s'$ or $t \neq t'$.) Finally, $l(\{t\}) = t = r(\{t\})$. If $\mathcal{M}$ is a model and o is an occasion in its frame, the pair $(\mathcal{M}, o)$ is an *anchored model.*

We are led by aesthetic considerations when opting for intervals of the form $]s,t]$ and refraining to accommodate, e.g., intervals $\{x : s \leq x \leq t\}$ as well. The semantics of the modal operators $\Box$ and $\Diamond$ will be in terms of *divisions* of the current occasion. If for some $n > 0$ there are points $t_0 \notin o$ and $t_1, \ldots, t_{n+1} \in o$ such that $t_0 < t_1 < \ldots < t_{n+1}$ and $(]t_0,t_1], \ldots,]t_n, t_{n+1}])$ is a partition of the occasion o, this partition is termed the *division* of o *by the points* $t_1, \ldots, t_n$ and denoted $\mathbb{D}_o(t_1, \ldots, t_n)$. (Note that necessarily $t_0 = l(o)$ and $t_{n+1} = r(o)$.) It follows that an occasion o has a division iff $|o| \geq 2$. The members of a division are called its *cells.* If $o =]s,t]$ is finite and $|o| = n+1$, then o has $2^n - 1$ different divisions. While all divisions of an interval determine its partition, not all partitions are divisions; e.g., $(]1,2] \cup]3,4],]2,3])$ is a partition of the real interval $]1,4]$, but not its division. Among the *desiderata* guiding the definition of division is that all members of the appropriate partition will be non-empty and that whenever a division exists, it will be possible to choose the number of divisors so as to make the members of the partition to have pairwise the same cardinality. For these reasons it is convenient that the members of the partition are occasions of the same form as the occasion divided.

Given a model $\mathcal{M} = (T, <, V)$, define a binary relation $\mathcal{M}, o \models \psi$ among occasions o in T and formulas ψ of $\mathcal{L}_{TD}$ as the smallest set such that:

- $\mathcal{M}, o \models p$ if: $t \in V(p)$ for all $t \in o$
- $\mathcal{M}, o \models \sim\psi$ if: $\mathcal{M}, \{t\} \not\models \psi$ for all $t \in o$
- $\mathcal{M}, o \models \neg\psi$ if: $\mathcal{M}, o \not\models \psi$
- $\mathcal{M}, o \models (\psi \wedge \chi)$ if: $\mathcal{M}, o \models \psi$ and $\mathcal{M}, o \models \chi$
- $\mathcal{M}, o \models (\psi \vee \chi)$ if: $\mathcal{M}, o \models \psi$ or $M, o \models \chi$
- $\mathcal{M}, o \models \Box\psi$ if: for some positive integer n there are $t_1, \ldots, t_n$ with $l(o) < t_1 < \ldots < t_n < r(o)$ such that for each cell o' of the division $\mathbb{D}_o(t_1, \ldots, t_n)$, we have $\mathcal{M}, o' \models \psi$
- $\mathcal{M}, o \models \Diamond\psi$ if: for all positive integers n and all $t_1, \ldots, t_n$ with $l(o) < t_1 < \ldots < t_n < r(o)$, there is a cell o' of the division $\mathbb{D}_o(t_1, \ldots, t_n)$ such that $\mathcal{M}, o' \models \psi$.

It can be shown that all implications in the above definition can be reversed, i.e., that the relation $\{(o, \psi) : \mathcal{M}, o \models \psi\}$ is a fixed point of the inductive truth definition. If $\mathcal{M}, o \models \psi$, ψ is *true* in $\mathcal{M}$ at o; else *false* in $\mathcal{M}$ at o.

The symbols $\Box$ and $\Diamond$ have here a meaning very different from their meanings in **ML**. Seen as a generalized quantifier, $\Box$ ($\Diamond$) involves second-order existential (universal) and first-order universal (existential) quantification. $\Box\psi$ serves to assert at o that for all cells of some division of o, ψ holds. Dually, $\Diamond\psi$ asserts at o that for any division of o, some of its cells makes ψ true. $\Box$ is a kind of *chop-star* operator;[6] $\Diamond$ is its dual w.r.t. $\neg$ (the

[6]For a discussion on the relation of $\Box$ to *chop-star* as used in other logics, see *Sect.* 7.

literature seems not to have settled on any name for the dual of *chop-star*). Disjunction ($\vee$) and conjunction ($\wedge$) have their usual meanings and they are each other's duals w.r.t. $\neg$. While $\neg$ is the plain contradictory negation acting on occasions, $\sim$ acts on single points (technically, singleton intervals). $\neg\psi$ holds at o iff ψ does not hold at o, whereas $\sim\psi$ holds at o iff ψ fails separately at each point $t \in o$. In particular $\neg p$ holds at o if the atom p fails at some $t \in o$, while in order for $\sim p$ to hold at o, p must fail at each of its points. $\neg$ is termed the *contradictory negation* and $\sim$ the *universal negation*.[7] When speaking simply of negation, we mean $\neg$. The formula $\neg\sim p$ using both negations states that positively, p holds at some $t \in o$.

Note: Instead of $\sim$, we might consider the positive operator $\oplus$ with the following semantics: $\oplus\psi$ is true at o iff for every $t \in o$, ψ is true at $\{t\}$.[8] Employing $\oplus$, $\sim\psi$ would be definable as $\oplus\neg\psi$. (In the scope of $\oplus$, like in the scope of $\sim$, the difference between $\sim$ and $\neg$ vanishes.) Note that conversely, $\oplus\psi$ is definable as $\sim\sim\psi$, or, equivalently, as $\sim\neg\psi$. Using $\oplus$ could improve readability; e.g. the truth condition of $\sim(\sim p \wedge \sim q)$ may well be more accessible when referred to via $\oplus(p \vee q)$. For technical development it might be advisable to adopt the modified set of primitives. Yet von Wright's argument discussed in *Section* 1 seems to be best elucidated in terms of the two negations. To retain the connection to the proposed analysis of his argument, the syntax with $\neg$ and $\sim$ is kept in the present paper. $\dashv$

Let $\mathcal{K}$ be a class of anchored models and let $\phi, \psi \in \mathcal{L}_{TD}$. ϕ is *satisfiable (valid) over* $\mathcal{K}$ if $\mathcal{M}, o \models \phi$ for some (all) anchored models $(\mathcal{M}, o)$ in $\mathcal{K}$. A finite set of formulas is satisfiable (valid) if their conjunction is satisfiable (valid). ψ is a *logical consequence* of ϕ *over* $\mathcal{K}$, denoted $\phi \Rightarrow_{\mathcal{K}} \psi$, if $(\neg\phi \vee \psi)$ is valid over $\mathcal{K}$. Let $\mathcal{K}_0$ be the class of all anchored models. We write $\Rightarrow$ for the relation $\Rightarrow_{\mathcal{K}_0}$. Formulas ψ and ϕ are *logically equivalent*, denoted $\psi \equiv \phi$, if $\phi \Rightarrow \psi$ and $\psi \Rightarrow \phi$. The formula ϕ *characterizes* a property P on a class $\mathcal{K}$ if for all $(\mathcal{M}, o) \in \mathcal{K}$: $\mathcal{M}, o \models \phi$ iff o satisfies P. E.g., ϕ characterizes infinity on $\mathcal{K}_0$ if ϕ holds at all and only infinite occasions.

2.1 Some features of the semantics

Clearly all formulas of the forms p, $\top$, $\sim\psi$, $\Diamond\psi$ are true at the empty occasion, while no formula of the form $\Box\psi$ or $\bot$ is. Note that $\mathcal{M}, o \models \sim\top$ iff $o = \varnothing$ and thence $\mathcal{M}, o \models \neg\sim\top$ iff $o \neq \varnothing$. So emptiness and non-emptiness of an occasion are definable in $\mathcal{L}_{TD}$. The formula $(\Diamond\bot \wedge \neg\sim\top)$ is true at o iff ($o \neq \varnothing$ but o has no division) iff $|o| = 1$.

Considering $\Box$ and $\Diamond$ applied to literals, in 4 out of the total of 8 cases the resulting formula is definable in simpler terms. If $\theta \in \{p, \sim p\}$, $\Diamond\neg\theta \equiv (\neg\theta \vee \Diamond\bot)$ and $\Box\theta \equiv (\theta \wedge \Box\top)$. On the other hand, $\Diamond p$ says of o that p fails at most once during it, while $\Diamond\sim p$ states that p holds at most once. $\Box\neg p$ says that p fails at least twice and $\Box\neg\sim p$ that p holds at least twice. Interestingly, $\Diamond\Box\neg p$ asserts — of intervals of size at least 2 — that p fails infinitely often (any division has a cell that is further divisible into at least

[7] Relative to atoms, but not generally, $\neg$ could be termed the *existential negation*.

[8] I owe to an anonymous referee the suggestion to take $\oplus$ rather than $\sim$ as a primitive.

two cells so that in each p fails); while $\Box\Diamond p$ says that p fails only finitely many times (there is a division such that in each of its cells p fails at most once). Replacing in these examples p by an arbitrary formula does not, in general, serve to express an analogous property; e.g., $\Diamond\Box\neg p$ does not assert that $\Box\neg p$ fails at most once (actually, $\Box\neg p$ fails at each point). We may observe that there is a formula satisfiable exactly on occasions of even size: $|o|$ is even iff there is V such that $(T, <, V), o \models \Box(\neg{\sim}p \wedge \Diamond{\sim}p \wedge \neg p \wedge \Diamond p)$.

2.2 Negation normal form

Some peculiarities in the behavior of the universal negation (${\sim}$) are worth noting. First, ${\sim}$ and $\neg{\sim}$ both distribute over $\vee$: ${\sim}(\phi \vee \psi) \equiv ({\sim}\phi \wedge {\sim}\psi)$ and $\neg{\sim}(\phi \vee \psi) \equiv (\neg{\sim}\phi \vee \neg{\sim}\psi)$ for any $\phi, \psi \in \mathcal{L}_{TD}$. However, neither distributes over $\wedge$. E.g., ${\sim}({\sim}p \wedge {\sim}q)$ is true at a real interval $o =]s, t]$ whose every irrational point makes p true (but q false) and every rational point makes q true (but p false), while $(p \vee q)$ of course fails at o. Second, while ${\sim}{\sim}p \equiv p$ for any atom p, in general ${\sim}{\sim}\phi$ is not equivalent to ϕ. E.g., ${\sim}{\sim}(p \vee q) \equiv {\sim}({\sim}p \wedge {\sim}q) \not\equiv (p \vee q)$. Third, while ${\sim}\neg\phi \equiv {\sim}{\sim}\phi$ for any $\phi \in \mathcal{L}_{TD}$ (cf. the proof of Prop. 1), $\neg{\sim}\phi$ asserts of o that ϕ is true at some singleton interval $\{t\} \subseteq o$; it is equivalent neither to ϕ nor to ${\sim}{\sim}\phi$.

It will be convenient to deal with $\mathcal{L}_{TD}$ by reference to a normal form, where negation symbols $\neg$ are driven as deep as they go. Let the grammar

$$\psi \ ::= \ p \mid {\sim}p \mid (\psi \vee \psi) \mid (\psi \wedge \psi)$$

define the class of formulas $\mathcal{L}_{zero}$, and let the grammar

$$\chi \ ::= \ p \mid {\sim}p \mid \neg p \mid \neg{\sim}p \mid {\sim}(\mathbf{a} \wedge \mathbf{a}) \mid \neg{\sim}(\mathbf{a} \wedge \mathbf{a})$$

define the class of formulas $\mathcal{L}_{base}$, where $p \in \mathbf{prop}$ and $\mathbf{a} \in \mathcal{L}_{zero}$. The four tuples of negation signs (empty, $\neg$, ${\sim}$, $\neg{\sim}$) serve to represent the four basic modes of syllogistic assertions applied to time points in an interval: p is universal affirmative, ${\sim}p$ universal negative, $\neg{\sim}p$ particular affirmative and $\neg p$ particular negative. $\mathcal{L}_{zero}$ equals $\mathbf{Cl}_{\wedge,\vee}(\mathbf{prop} \cup \{{\sim}p : p \in \mathbf{prop}\})$; and $\mathcal{L}_{base}$ is obtained from $\mathbf{prop} \cup \{\neg p : p \in \mathbf{prop}\}$ by adding to it the universal negations of all $\mathcal{L}_{zero}$-formulas (except disjunctions), and the contradictory negations of the universal negations of all $\mathcal{L}_{zero}$-formulas (except disjunctions). Let $\mathcal{L}_{nnf}$ be the class of formulas produced by the grammar

$$\theta ::= \mathbf{b} \mid (\theta \vee \theta) \mid (\theta \wedge \theta) \mid \Diamond\theta \mid \Box\theta,$$

where $\mathbf{b} \in \mathcal{L}_{base}$. There is a truth-preserving map of type $\mathcal{L}_{TD} \to \mathcal{L}_{nnf}$.

PROPOSITION 1. *There is a map $t : \mathcal{L}_{TD} \to \mathcal{L}_{nnf}$ such that for all $\phi \in \mathcal{L}_{TD}$ and anchored models $(\mathcal{M}, o)$: $\mathcal{M}, o \models \phi$ iff $\mathcal{M}, o \models t(\phi)$.*

Proof. Think of formulas ${\sim}\psi$ first. If ψ' is the result of replacing $\neg$ by ${\sim}$ in ψ, then ${\sim}\psi \equiv {\sim}\psi'$: in the scope of ${\sim}$ all evaluation is w.r.t. single points, and $\mathcal{M}, \{t\} \models \neg\phi$ iff $\mathcal{M}, \{t\} \models {\sim}\phi$. Moreover, if ψ'' results from replacing in ψ' subformulas $\Box\phi$ by $\bot$ and $\Diamond\phi$ by $\top$, then ${\sim}\psi' \equiv {\sim}\psi''$. ($\Box\phi$ cannot hold at $\{t\}$: there is no s with $t < s < t$; dually, $\Diamond\phi$ is trivially true at $\{t\}$.) These observations motivate defining the following maps $[\,]^{\sim}$ and $[\,]^{zero}$, to be used when translating a formula of $\mathcal{L}_{TD}$ containing occurrences of $\sim$. First, if $\psi \in \mathcal{L}_{TD}$, let $[\psi]^{\sim}$ be the result of replacing all occurrences of $\neg$ in ψ by $\sim$, and putting the resulting formula in negation normal form (like in **ML**). So $[\psi]^{\sim} \in \mathcal{L}_{TD}$, $[\psi]^{\sim}$ contains no occurrences of $\neg$, and $\sim$ occurs in $[\psi]^{\sim}$ only prefixed to an atom. Second, define a map $[\,]^{zero}$ on the set $\{[\psi]^{\sim} : \psi \in \mathcal{L}_{TD}\}$ as follows: $[p]^{zero} = p$, $[\sim p]^{zero} = \sim p$, $[\Diamond\theta]^{zero} = \top$, $[\Box\theta]^{zero} = \bot$, $[(\theta_1 \circ \theta_2)]^{zero} = ([\theta_1]^{zero} \circ [\theta_2]^{zero})$ for $\circ \in \{\wedge, \vee\}$. Note that $Im([\,]^{zero}) \subseteq \mathcal{L}_{zero}$. The idea behind the maps $[\,]^{\sim}$ and $[\,]^{zero}$ is that if ψ appears in ψ' in the scope of $\sim$ (the maps will be applied in such cases), ψ can be replaced by $[[\psi]^{\sim}]^{zero}$ *salva veritate*. Let $[\,]^*$ be the composite map $([\,]^{zero} \circ [\,]^{\sim}) : \mathcal{L}_{TD} \to \mathcal{L}_{zero}$. Define a map $[\,]^{nnf} : \mathcal{L}_{TD} \to \mathcal{L}_{nnf}$ as follows:

$$
\begin{aligned}
[\theta]^{nnf} &= \theta \quad \text{for } \theta \in \{p, \neg p\}\\
[\neg\neg\phi]^{nnf} &= [\phi]^{nnf}\\
[\sim(\phi_1 \wedge \phi_2)]^{nnf} &= \sim[(\phi_1 \wedge \phi_2)]^*\\
[\neg\sim(\phi_1 \wedge \phi_2)]^{nnf} &= \neg\sim[(\phi_1 \wedge \phi_2)]^*\\
[\bigcirc\phi]^{nnf} &= \bigcirc[\phi^{nnf}] \quad \text{for } \bigcirc \in \{\Diamond, \Box\}\\
[(\phi_1 \circ \phi_2)]^{nnf} &= ([\phi_1]^{nnf} \circ [\phi_2]^{nnf}) \quad \text{for } \circ \in \{\vee, \wedge\}\\
[\neg\!\!\!\neg(\phi_1 \vee \phi_2)]^{nnf} &= ([\neg\!\!\!\neg\phi_1]^{nnf} \wedge [\neg\!\!\!\neg\phi_2]^{nnf}) \quad \text{for } \neg\!\!\!\neg \in \{\sim, \neg\}\\
[\neg\sim(\phi_1 \vee \phi_2)]^{nnf} &= ([\neg\sim\phi_1]^{nnf} \vee [\neg\sim\phi_2]^{nnf})\\
[\neg(\phi_1 \wedge \phi_2)]^{nnf} &= ([\neg\phi_1]^{nnf} \vee [\neg\phi_2]^{nnf})\\
[\neg\Diamond\phi]^{nnf} &= \Box[\neg\phi]^{nnf}\\
[\neg\Box\phi]^{nnf} &= \Diamond[\neg\phi]^{nnf}
\end{aligned}
$$

Any $\phi \in \mathcal{L}_{TD}$ can be thought of as being built from components of the forms $p, \sim\psi$ by applying $\neg, \vee, \wedge, \Box, \Diamond$. (A formula ψ prefixed by $\sim$ can be assumed to be an atom or a conjunction of $\mathcal{L}_{zero}$-formulas.) Relative to such 'atoms', $[\,]^{nnf}$ acts like a transformation producing negation normal form in **ML** (w.r.t. $\neg$). As the component formulas $\sim\psi$ have their inner structure, they will be processed further using the map $[\,]^*$ (applied to formulas of the forms $\sim\phi$, $\neg\sim\phi$). Doing so gets rid of all occurrences of $\neg$ in ψ, and drives the resulting formula in negation normal form (w.r.t. $\sim$). It is easy to show that $[\,]^*$ and $[\,]^{nnf}$ are truth-preserving. So we may take $t = [\,]^{nnf}$. ■

If $\phi \in \mathcal{L}_{TD}$, the *negation normal form* of ϕ is by definition the formula $[\phi]^{nnf}$, where $[\,]^{nnf}$ is the map defined in the proof of Proposition 1.

3 Expressive power

It turns out that $\mathcal{L}_{TD}$ is a very powerful logic.

EXAMPLE 2. $\mathcal{M}, o \models (\Box\top \wedge \Diamond\Box\top)$ iff $|o|$ is infinite. First assume $\mathcal{M}, o \models (\Box\top \wedge \Diamond\Box\top)$, supposing for contradiction that $|o| < \aleph_0$. Since $\Box\top$ holds at o, $|o| \geq 2$. Consider a division of o whose all cells are singletons. Then $\Box\top$ holds at one of these cells: a contradiction. For the converse, suppose $|o|$ is infinite. Then $\Box\top$ holds trivially at o. Choose any finite number $n > 0$ of points $t_1 < \ldots < t_n < r(o)$ in o, and consider the division $\mathbb{D}_o(t_1, \ldots, t_n)$.

At least one of its cells is infinite. At that cell, then, the formula $\Box\top$ holds. By what just established, $\mathcal{M}, o \models (\Diamond\bot \vee \Box\Diamond\bot)$ iff $|o|$ is finite. ⊣

A subset $S \subseteq T$ is *dense in* T if for all $x, y \in T$ with $x < y$, there is $z \in S$ such that $x < z < y$. The following fact will be used subsequently:

FACT 3. Let $(T, \prec)$ be of order type $\eta + 1$. There is a subset S of T such that both S and its complement $T \setminus S$ are dense in T.

Proof. W.l.o.g. consider the interval $T :=]0, 1] \cap \mathbb{Q}$. Put $S := \{\frac{m}{2^n} : n \geq 1$ and $m < 2^n$ and m is odd$\}$. Then both S and $T \setminus S$ are dense in T. ■

EXAMPLE 4. Consider the conjunction $\chi := (\neg{\sim}p \wedge \neg p)$, with $p \in$ **prop**. Clearly χ can only be true at an occasion of size at least two. Then the formula $(\Box\top \wedge \Diamond\chi)$ is satisfiable, but not true at any finite occasion. To see that there is an infinite occasion at which $(\Box\top \wedge \Diamond\chi)$ holds, let $o =]0, 1] \cap \mathbb{Q}$. By Fact 3 there is a subset S of o such that both S and its complement are dense in o. Define a model $\mathcal{M} = (T, <, V)$ with $T = [0, 1] \cap \mathbb{Q}$ by letting $<$ be the order of rationals in T by magnitude and $V(p) = S$. Now $\mathcal{M}, o \models (\Box\top \wedge \Diamond\chi)$. For, take any division of T by points $t_1 < \ldots < t_n$. Then the formula χ holds actually at any cell (while one would suffice). ⊣

The formulas calling for infinite models discussed in Examples 2 and 4 have extremely low modal depth: 2 *resp.* 1. Let us look at further examples that give some measure of the power of nesting modal operators.

EXAMPLE 5. If $\bigcirc \in \{\Box, \Diamond\}$, write $\bigcirc^n$ for the string consisting of n tokens of $\bigcirc$. Let n, m be positive integers.

(i) $\mathcal{M}, o \models \Box^n\top$ iff $|o| \geq 2^n$.

(i') $\mathcal{M}, o \models \Diamond^n\bot$ iff $|o| \leq 2^n - 1$.

(ii) $\mathcal{M}, o \models \Box^n\Diamond^m\bot$ iff $2^n \leq |o| < \aleph_0$.

(ii') $\mathcal{M}, o \models \Diamond^n\Box^m\top$ iff $2^n < |o|$ or $|o| \geq \aleph_0$.

(i') and (ii') are immediate from (i) *resp.* (ii). For (i), the estimate on $|o|$ is computed as $1 + \sum_{k=0}^{n-1} 2^k = 2^n$; it is the smallest number of time points allowing n iterated evaluations of $\Box$ in *each* minimal sub-occasion triggered by the previous evaluation step. For (ii), suppose first that $|o|$ is finite but at least 2^n. So we are sure to be able to evaluate $\Box^n$. Moreover, since $|o|$ is finite, we may choose the divisors of o so that the cells of the resulting division are all singletons. But then, by (i'), $\Diamond^m\bot$ is true at all those cells. Further, the value of the parameter $m > 0$ is irrelevant. Conversely, if the formula holds at o, it is possible to evaluate $\Box^n$, whence $|o| \geq 2^n$. Suppose for contradiction that $|o|$ is infinite. Then any division chosen to witness $\Box$ will have at least one infinite cell; at that cell, then, $\Diamond^m\bot$ cannot hold. ⊣

EXAMPLE 6. Let $\mathcal{O} = (\omega + 1, <, V)$, where V satisfies $\alpha \in V(p)$ iff $\alpha < \omega$ and α is odd. Let $o =]0, \kappa]$ with $\kappa \leq \omega$. Then $\mathcal{O}, o \models \Box(p \vee {\sim}p)$ iff $\kappa \neq \omega$, but $\mathcal{O}, o \models \Box(p \vee \neg p)$. If $\kappa = \omega$, $\Box(p \vee {\sim}p)$ fails at o, since whichever finite set of divisors is chosen to witness $\Box$, the rightmost of the cells of the corresponding division is infinite, and thus neither p nor $\sim p$ holds at it. On the other hand, $\Box(p \vee \neg p)$ indeed holds irrespective of the value of κ: if o' is a sub-occasion of o which does not satisfy p, there is at least one point in o' at which p does not hold; but then $\neg p$ is true at o'. For a further example, let $\mathcal{N} = (\omega, <, V)$, with V defined as follows: $n \in V(p)$ iff $n \equiv 1 \pmod 3$; $n \in V(q)$ iff $n \equiv 2 \pmod 3$; and $n \in V(r)$ iff $n \equiv 0 \pmod 3$. Let $o =]2, k]$ with $k < \omega$. Then $\mathcal{N}, o \models \Box(\neg{\sim}p \wedge \neg{\sim}q \wedge \neg{\sim}r \wedge \Diamond\Diamond\bot)$ iff $|o|$ is divisible by 3. The formula $\chi := (\neg{\sim}p \wedge \neg{\sim}q \wedge \neg{\sim}r \wedge \Diamond\Diamond\bot)$ states of an occasion o' that its size is at most 3 (cf. Ex. 5) and that each of the atoms p, q, r is true at o' at least once. Given the definition of the model, this only leaves open the possibility that $|o'| = 3$. But then, the formula $\Box\chi$ can only be true in $\mathcal{N}$ at an occasion whose size is divisible by 3. ⊣

EXAMPLE 7. Consider the model $\mathcal{Q} = (\mathbb{Q}, <, V)$ and the rational interval $o =]1, 2]$, given that $V(p) = \{t : t^2 < 2\}$. (Hence $\varnothing \neq V(p) \cap o \neq o$.) Then $\mathcal{Q}, o \models \neg\Box(p \vee {\sim}p)$. To see this, let $\mathbb{D}_o(t_1, \ldots, t_n)$ be any division of o. Then the divisors satisfy $1^2 < (t_1)^2 < \ldots < (t_n)^2 < 2^2$. Evidently, $(p \vee {\sim}p)$ can only be true at all cells of the division if for some $1 \leq i \leq n$, we have $(t_i)^2 = 2$. But this is impossible, as the t_i are rational. By contrast, if we move on to look at the model $\mathcal{R} = (\mathbb{R}, <, V)$ and the real interval $o =]1, 2]$, letting $V(p) = \{t : t^2 < 2\}$, we indeed have: $\mathcal{R}, o \models \Box(p \vee {\sim}p)$. There is exactly one witnessing division, namely $\mathbb{D}_o(\sqrt{2})$. ⊣

EXAMPLE 8. In Example 7, the truth of $\Box(p \vee {\sim}p)$ failed in a model based on rational numbers due to a gap — the non-existence of the supremum of the set of points making p true. This formula can fail in a model $\mathcal{Q}^* = (\mathbb{Q}, <, V^*)$ also for a different reason. By Fact 3 there is a subset S of the rational interval $o =]1, 2]$ such that both S and $]1, 2] \setminus S$ are dense in $]1, 2]$. Put $V^*(p) := S$. Then $\mathcal{Q}^*, o \models \neg\Box(p \vee {\sim}p)$. ⊣

Basic negative properties of $\mathcal{L}_{TD}$. Directly by Example 2, we have:

FACT 9. $\mathcal{L}_{TD}$ lacks the finite model property. ■

By contrast, **ML** has (strong) finite model property over the class of all pointed models [1, Thm. 2.34, Cor. 6.8]. Thinking of evaluation, in **ML** we climb up a tree (or more generally, a directed graph), while in $\mathcal{L}_{TD}$ we dig deeper into a given occasion. In view of the above examples, modal depth is a very bad measure of the structural requirements that an $\mathcal{L}_{TD}$-formula can impose on an occasion. A formula of modal depth 2 can characterize infinity (Ex. 2) and there are formulas of modal depth 1 true only at infinite occasions (Ex. 4). A simple formula such as $\Box^n\top$ forces its verifying occasion to be at least of size 2^n, while the same string of symbols, as a formula of **ML**, only requires of a pointed modal structure that its height be n. A property central to a great variety of modal languages $\mathcal{L}$ is having a *notion of finite degree* [1, Def. 7.58]: the existence of a function $f : \mathcal{L} \to \omega$ such that if ϕ is true in a pointed model at all, it is true in the result of removing from that pointed model everything that transcends the height $f(\phi)$. For **ML** we may take $f(\phi) = md(\phi)$. (Not all modal languages have a notion of finite degree, a much-studied case in point is modal mu-calculus.) $\mathcal{L}_{TD}$ fails to have a notion of finite degree, if by 'degree' of a formula of $\mathcal{L}_{TD}$ we understand a measure of the size of an occasion required to verify it.

If **prop** $= \{p_i : i < \omega\}$ and $\mathcal{M} = (T, <, V)$ is a model, let $\tau = \{\prec\} \cup \{P_i : i < \omega\}$ and define $\mathfrak{M}$ to be the τ-structure $(T, \prec^{\mathfrak{M}}, \langle P_i^{\mathfrak{M}} \rangle_{i<\omega})$, where $\prec^{\mathfrak{M}} = <$ and $P_i^{\mathfrak{M}} = V(p_i)$. ($\mathfrak{M}$ will be termed the *abstract structure induced by* $\mathcal{M}$.) If $\mathcal{L}$ is an abstract logic, say **FO** or $\mathcal{L}_{\mathrm{w}}^{\mathrm{mon}}$, $\mathcal{L}_{TD}$ is *translatable into* $\mathcal{L}$ if the following two conditions hold: (a) for every $\phi \in \mathcal{L}_{TD}$ there is a formula $\psi_\phi^1(x) \in \mathcal{L}[\tau]$ of one free variable such that for all $\mathcal{M}$ and $t \in T$: $\mathcal{M}, \{t\} \models \phi$ iff $\langle \mathfrak{M}, t \rangle \models \psi_\phi^1(x)$; and ($b$) for every $\phi \in \mathcal{L}_{TD}$ there is a formula $\psi_\phi^2(x, y) \in \mathcal{L}[\tau]$ of two free variables such that for all $\mathcal{M}$ and all occasions proper $o = \,]s, t] \subseteq T$: $\mathcal{M},]s, t] \models \phi$ iff $\langle \mathfrak{M}, s, t \rangle \models \psi_\phi^2(x, y)$. The formula ψ_ϕ^i will be called a *translation of* ϕ *of kind* i ($i := 1, 2$).

THEOREM 10. *$\mathcal{L}_{TD}$ is not translatable into* **FO**.

Proof. Let $\tau = \{<\}$. Given $n \in \mathbb{N}$, let $\mathfrak{M}_n$ be the τ-structure $(M_n, <^{\mathfrak{M}_n}, \varnothing)$ with $M_n = \{0, \ldots, 2^n\}$ and $<^{\mathfrak{M}_n}$ the order of $0, \ldots, 2^n$ by magnitude. Write $\prec$ for the order of integers by magnitude. Let $M = \mathbb{Z}$ and define a τ-structure $\mathfrak{M} = (M, <^{\mathfrak{M}}, \varnothing)$ by setting $z <^{\mathfrak{M}} z'$ iff $0 \preceq z \prec z'$ or $z \prec z' \prec 0$ or ($z \succ 0$ and $z' \prec 0$). So $<^{\mathfrak{M}}$ has the order type $\omega + \omega^*$. Note that $\min(<^{\mathfrak{M}}) = 0$ and $\max(<^{\mathfrak{M}}) = -1$. An easy Ehrenfeucht-Fraïssé game argument shows that for any $n \in \mathbb{N}$, the structures $\langle \mathfrak{M}, 0, -1 \rangle$ and $\langle \mathfrak{M}_n, 0, 2^n \rangle$ agree on all **FO**$[\tau]$-formulas of two free variables with quantifier rank at most n. Consider, then, the $\mathcal{L}_{TD}$-formula $\phi_{\inf} := (\Box\top \wedge \Diamond\Box\top)$, true in $\mathcal{M}$ at o iff $|o| \geq \aleph_0$. For contradiction, suppose $\phi_{\inf}$ has a translation $\chi(x, y)$ of kind 2 into **FO**$[\tau]$. Write q for the quantifier rank of χ. By what just observed, the structures $\langle \mathfrak{M}, 0, -1 \rangle$ and $\langle \mathfrak{M}_q, 0, 2^q \rangle$ are indistinguishable by $\chi(x, y)$. Note that the τ-structures $\mathfrak{M}_n$ and $\mathfrak{M}$ are, formally, also models for $\mathcal{L}_{TD}$. Now $\mathfrak{M}_q,]0, 2^q] \not\models \phi_{\inf}$ and $\mathfrak{M},]0, -1] \models \phi_{\inf}$. Since $\chi(x, y)$ is a translation of $\phi_{\inf}$, $\langle \mathfrak{M}_q, 0, 2^q \rangle \not\models \chi(x, y)$ and $\langle \mathfrak{M}, 0, -1 \rangle \models \chi(x, y)$, so after all $\chi(x, y)$ distinguishes the two structures: a contradiction. ■

Basic positive properties of $\mathcal{L}_{TD}$. As is immediate from the semantics, $\mathcal{L}_{TD} \leq \mathcal{L}_{\mathrm{w}}^{\mathrm{mon}}$. Write $ST_{x,y}$ for a translation (of kind 2) of $\mathcal{L}_{TD}$ into $\mathcal{L}_{\mathrm{w}}^{\mathrm{mon}}$. Note that **FO** $\not\leq \mathcal{L}_{TD}$; clearly e.g. the simple (**TL**-definable) formula $\exists z \exists v (x < z < v \leq y \wedge P(z) \wedge Q(v))$ has no translation into $\mathcal{L}_{TD}$.

PROPOSITION 11. $\mathcal{L}_{TD} < \mathcal{L}_{\mathrm{w}}^{\mathrm{mon}}$. ■

COROLLARY 12. *$\mathcal{L}_{TD}$ has downwards Löwenheim-Skolem property: any satisfiable formula of $\mathcal{L}_{TD}$ is true in a countable anchored model.*

Proof. Let $\phi \in \mathcal{L}_{TD}$. Suppose $\mathcal{M}, o \models \phi$, where $o = \,]s, t]$ and $s < t$. (If o is empty or a singleton there is nothing to prove.) Let ϕ_{lin} be an **FO**$[\{<\}]$-formula stating that the interpretation of $<$ is linear; then the sentence $\chi := (\exists x \exists y (x < y \wedge ST_{x,y}(\phi)) \wedge \phi_{\mathrm{lin}})$ of $\mathcal{L}_{\mathrm{w}}^{\mathrm{mon}}$ is true in the structure $\mathfrak{M}$ induced by $\mathcal{M}$. By the downwards Löwenheim-Skolem property of $\mathcal{L}_{\mathrm{w}}^{\mathrm{mon}}$ [24], there is a *countable* $\langle \mathfrak{M}', s', t' \rangle$ such that $\langle \mathfrak{M}', s', t' \rangle \models ST_{x,y}(\phi)$, the relation $<^{\mathfrak{M}'}$ is linear, and $s' <^{\mathfrak{M}'} t'$. Let $\mathcal{M}'$ be the model inducing $\mathfrak{M}'$. Then $\mathcal{M}',]s', t'] \models \phi$. ■

COROLLARY 13. *$\mathcal{L}_{TD}$ is decidable.*

Proof. Given $\phi \in \mathcal{L}_{TD}$, apply an algorithm solving $\mathcal{L}_{\mathrm{w}}^{\mathrm{mon}}$-SAT over linear orders [22] to the sentence $\exists x \exists y ST_{x,y}(\phi)$. (Note that ϕ is true at a singleton iff $ST_{x,y}(\phi)$ is satisfied in some $\langle \mathfrak{M}, t, t' \rangle$, where $]t, t']$ is a singleton.) ■

Further observations. Write ϕ_{inf} for the formula $(\Box\top \wedge \Diamond\Box\top)$ that was seen to hold at o iff $|o| \geq \aleph_0$ (Ex. 2). Using ϕ_{inf} we can build further formulas which will serve to capture classes of ordinals. (For any successor ordinal $\alpha = \beta + 1 = [0, \beta]$, the interval $]0, \beta]$ is an occasion proper in the class of all ordinals. However, we may consider α itself as an occasion proper.) E.g., $\Box\phi_{\mathrm{inf}}$ is true at an ordinal $\beta + 1$ iff $\beta \geq \omega \cdot 2$, while $(\Box\top \wedge \Diamond\Box\phi_{\mathrm{inf}})$ is true at $\beta + 1$ iff $\beta \geq \omega^2$. To see that the latter holds, suppose for contradiction that $\beta + 1 \models (\Box\phi_{\mathrm{inf}} \wedge \Diamond\Box\phi_{\mathrm{inf}})$ for some $\beta < \omega^2$. So $\beta = \omega \cdot n + k$ for some $n, k < \omega$. Consider a division of β by divisors $\omega, \omega \cdot 2, \ldots, \omega \cdot n$. Each cell of this division is of order type $\gamma_i \in \{k, \omega\}$ and yet one of them satisfies $\Box\phi_{\mathrm{inf}}$. This is impossible. More generally, we may define recursively formulas ϕ_{ω^n} by putting $\phi_{\omega^1} := (\Box\top \wedge \Diamond\Box\phi_{\mathrm{inf}})$ and $\phi_{\omega^{n+1}} := (\Box\top \wedge \Diamond\Box\phi_{\omega^n})$. Then we have for an ordinal β that $\beta + 1 \models \phi_{\omega^n}$ iff $\beta \geq \omega^n$. An ordinal $\beta + 1$ makes true *all* formulas of the set $\{\phi_{\omega^n} : n < \omega\}$ iff $\beta \geq \omega^\omega$.

Formula χ of $\mathcal{L}_{TD}$ is *idempotent* if $\Box\chi \Rightarrow \chi$: any occasion that can be cut into finitely many pieces so that each piece satisfies χ, itself satisfies χ. Not all formulas are idempotent. E.g., neither $(p \vee q)$ nor $(p \vee {\sim}p)$ nor $\Diamond\bot$ is. By contrast, all formulas p, $\neg p$, $\neg{\sim}p$, ${\sim}\phi$, $\Box\phi$, $(p \vee \neg p)$ are idempotent. Formula χ is *hyperconsistent* if $\Box\chi \Rightarrow \Diamond\chi$, i.e., if the truth of $\Box\chi$ excludes the truth of $\Box\neg\chi$. In other words, χ is hyperconsistent iff $(\Diamond\chi \vee \Diamond\neg\chi)$ is valid. Dually, formula χ is a *hypoantilogy* if $(\Box\chi \wedge \Box\neg\chi)$ is satisfiable. Hypoantilogies χ have the property that at least one occasion can be divided, in two ways, into at least two cells so that all cells of one division make χ true while all cells of the other division make $\neg\chi$ true. Hypoantilogies are formulas which can, by multiplication so to say, be fitted into one and the same interval together with their contradictories. A formula is hyperconsistent (hypoantilogy) iff its negation is. Directly by definitions, if a formula χ and its negation both are idempotent, χ is hyperconsistent. Hyperconsistency is a very restrictive condition. E.g., $(p \vee q)$ is not hyperconsistent, as witnessed by a rational interval $]1, 4]$ where q holds throughout $]1, 2]$ and $]3, 4]$, p holds throughout $]2, 3]$, and p fails at $1\frac{1}{2}$ and $3\frac{1}{2}$ while q fails at $2\frac{1}{4}$ and $2\frac{3}{4}$. Then $\Box(p \vee q)$ holds at $]1, 4]$ as witnessed by the division $\mathbb{D}_{]1,4]}(2, 3)$. Yet also $\Box\neg(p \vee q)$ holds at $]1, 4]$. This is witnessed by the division $\mathbb{D}_{]1,4]}(2\frac{1}{2})$. In both cells $]1, 2\frac{1}{2}]$ and $]2\frac{1}{2}, 4]$, both literals $\neg p$ and $\neg q$ hold. Other examples of hypoantilogies are $(r \vee (q \wedge s))$, $\Diamond\bot$ and $\Box\top$. E.g., in order for $\Diamond\bot$ to be hyperconsistent, the formula $(\Diamond\Diamond\bot \vee \Diamond\neg\Diamond\bot)$ must be valid; however, this formula is false at all finite occasions of size at least 4. The existence of hypoantilogies shows immediately that the distribution laws $(\Box\chi \wedge \Box\theta) \equiv \Box(\chi \wedge \theta)$ and $(\Diamond\chi \vee \Diamond\theta) \equiv \Diamond(\chi \vee \theta)$ *fail* in $\mathcal{L}_{TD}$.

Write $\phi \to \psi$ for $(\neg\phi \vee \psi)$. All of the following modal-logical axiom schemata fail in $\mathcal{L}_{TD}$: **4** or $\Box\theta \to \Box\Box\theta$, **T** or $\Box\theta \to \theta$, **B** or $\theta \to \Box\Diamond\theta$, **D** or $\Box\theta \to \Diamond\theta$, **.3** or $(\Diamond\theta \wedge \Diamond\chi) \to \Diamond(\theta \wedge \Diamond\chi) \vee \Diamond(\theta \wedge \chi) \vee \Diamond(\chi \wedge \Diamond\theta)$, **L** or $\Box(\Diamond\neg p \vee p) \to \Box p$ and **K** or $\Box(\neg\theta \vee \chi) \to (\Diamond\neg\theta \vee \Box\chi)$. The failure of **D** follows from the existence of hypoantilogies: it fails in $\mathcal{L}_{TD}$ for a rather 'substantial' reason. (In **ML** an instance of this schema can only fail at a state having no successor states.) The schema **.3** holds for *atomic* θ, χ — in particular $(\Diamond p \wedge \Diamond q) \Rightarrow \Diamond(p \wedge \Diamond q)$ — but fails already for negated atomic formulas. On the positive side, the axiom schema that in basic modal logic corresponds to the density of the accessibility relation, viz. $\Box\Box\theta \to \Box\theta$, actually holds for $\mathcal{L}_{TD}$. Here it has nothing to do with density, but is a simple consequence of the semantics of $\Box$. Since **K** fails in $\mathcal{L}_{TD}$, nominally $\mathcal{L}_{TD}$ is not a normal modal logic. Whether it would do justice to $\mathcal{L}_{TD}$ to call it a non-normal modal logic, or a modal logic at all, is debatable; what is clear is that $\mathcal{L}_{TD}$ is some sort of modal-like temporal logic.

4 Prenex formulas

A reasonable way to get to grips with characteristic features of $\mathcal{L}_{TD}$ is to distinguish its fragments and study them in isolation. What is learned from such case studies can, then, hopefully shed light on the general features of the logic. In this section we take up the study of 'prenex formulas'; in *Sections* 5 and 6 the 'propositional fragment' *resp.* the 'simple fragment' are considered. Already attempts to reach an overview of such relatively straightforward fragments of $\mathcal{L}_{TD}$ lead to rather involved considerations; there would be no realistic hope of understanding the details of the semantic behavior of $\mathcal{L}_{TD}$ from scratch. E.g., should anyone wish to design a decision algorithm specifically for $\mathcal{L}_{TD}$, such a detailed understanding would be needed. From this perspective, the question whether the chosen fragments are 'natural' in a more global setting is immaterial. (The 'propositional fragment' actually turns out to be a natural fragment of **TL**.)

A *prenex formula* is any $\mathcal{L}_{TD}$-formula of one of the forms

$$(\Box\Diamond)^n\ell,\ (\Diamond\Box)^n\ell,\ \Diamond(\Box\Diamond)^n\ell \text{ and } \Box(\Diamond\Box)^n\ell,$$

with $n < \omega$ and ℓ a literal. Prenex formulas with ℓ in their matrix are *ℓ-formulas*. E.g. $\Box\Diamond\Box p$, $\Diamond\Box\neg{\sim}p$ and ${\sim}p$ are prenex formulas; the first is p-formula, the second $\neg{\sim}p$-formula and the third ${\sim}p$-formula. Write $\mathcal{L}_{PR}$ for the class of all prenex formulas, and let $\mathcal{L}_{BPR} := \mathbf{Cl}_{\wedge,\vee}(\mathcal{L}_{PR})$. E.g., $(\Box\Diamond p \vee (\Diamond{\sim}q \wedge \neg r))$ is in $\mathcal{L}_{BPR}$ but $\Box(p \vee \Diamond q)$ is not. Note that semantically, $\mathcal{L}_{BPR}$ is closed under $\neg$. We will study $\mathcal{L}_{BPR}$-SAT w.r.t. the class $\mathcal{D}$ of all anchored models whose linear order is dense and whose occasion is proper. By Corollary 13 it is already known that this problem is decidable (density is first-order definable), but here a more fine-grained analysis is attempted. Dropping the assumption of density would complicate the details of the proofs; we leave the study of the more general case for future research.

Over $\mathcal{D}$, we have the equivalences $\Diamond\neg p \equiv \neg p$, $\Diamond\neg{\sim}p \equiv \neg{\sim}p$, $\Box p \equiv p$ and $\Box{\sim}p \equiv {\sim}p$. (For these it suffices to exclude occasions o with $0 \neq |o| \neq 1$, cf. *Subsect.* 2.1.) So we may, w.l.o.g., restrict attention to prenex formulas of the forms $(\Box\Diamond)^n p$, $(\Box\Diamond)^n{\sim}p$, $\Diamond(\Box\Diamond)^n p$, $\Diamond(\Box\Diamond)^n{\sim}p$ and $(\Diamond\Box)^n\neg p$, $(\Diamond\Box)^n\neg{\sim}p$, $\Box(\Diamond\Box)^n\neg p$, $\Box(\Diamond\Box)^n\neg{\sim}p$. Recall that $\mathcal{M}, o \models \Box\Diamond p$ iff $\neg p$ occurs only finitely often in o; and $\mathcal{M}, o \models \Diamond\Box\neg p$ iff $\neg p$ occurs infinitely often in o. Similarly, $\Box\Diamond{\sim}p$ asserts of an occasion that p holds only finitely often in it and $\Diamond\Box\neg{\sim}p$ that p holds infinitely often in it. We prove some auxiliary results and then show that $\mathcal{L}_{BPR}$-SAT over $\mathcal{D}$ is **NP**-complete. If $\theta \in \mathcal{L}_{PR}$, write $|\theta|$ for the number of modal operator tokens in its prefix.

LEMMA 14. **(a)** *If θ_1 and θ_2 are p-formulas (${\sim}p$-formulas) with $|\theta_1| \leq |\theta_2|$, then $\theta_1 \Rightarrow_{\mathcal{D}} \theta_2$.* **(b)** *If θ_1 and θ_2 are $\neg p$-formulas ($\neg{\sim}p$-formulas) with $|\theta_1| \leq |\theta_2|$, then $\theta_2 \Rightarrow_{\mathcal{D}} \theta_1$.*

Proof. For (a), consider how the complement of $V(p)$ looks like in anchored models $\langle (T, <, V), o\rangle \in \mathcal{D}$ satisfying a given p-formula. It is not difficult to see that the situation is as summarized by Table 1. Note that if θ_1 and θ_2 are p-formulas with $|\theta_1| \leq |\theta_2|$, any occasion satisfying θ_1 is a special case of an occasion satisfying θ_2. Therefore $\theta_1 \Rightarrow_{\mathcal{D}} \theta_2$. The case of ${\sim}p$-formulas is analogous. Item (b) is immediate from (a). Namely, if χ_1 and χ_2 are $\neg p$-formulas with $|\chi_1| \leq |\chi_2|$, then $\neg\chi_i$ is logically equivalent to a p-formula θ_i ($i := 1, 2$) such that $|\chi_i| = |\theta_i|$. Therefore by (a), $\theta_1 \Rightarrow_{\mathcal{D}} \theta_2$. But this means that $\chi_2 \Rightarrow_{\mathcal{D}} \chi_1$. The case of $\neg{\sim}p$-formulas is analogous. ■

Formula	*Order type of $T \setminus V(p)$ in o*
$\Diamond p$	≤ 1
$\Box\Diamond p$	n_0
$\Diamond\Box\Diamond p$	α_1
$\Box\Diamond\Box\Diamond p$	$\alpha_1 \cdot n_1$
$\Diamond\Box\Diamond\Box\Diamond p$	$\alpha_1 \cdot n_1 \cdot \alpha_2$
$\Box\Diamond\Box\Diamond\Box\Diamond p$	$\alpha_1 \cdot n_1 \cdot \alpha_2 \cdot n_2$
...	...
$(\Box\Diamond)^m p$	$\prod_{i=1}^m \alpha_1 \cdot n_i$
$\Diamond(\Box\Diamond)^m p$	$\prod_{i=1}^m \alpha_1 \cdot n_i \cdot \alpha_{m+1}$
...	...

$1 < n_i < \omega$; α_i finite or $\alpha_i \in \{\omega, \omega^*\}$

Table 1

Formula	*Order type of $T \setminus V(p)$ in o*
$\Box\neg p$	α_0
$\Diamond\Box\neg p$	∞_1
$\Box\Diamond\Box\neg p$	$\infty_1 \cdot \alpha_1$
$\Diamond\Box\Diamond\Box\neg p$	$\infty_1 \cdot \alpha_1 \cdot \infty_2$
$\Box\Diamond\Box\Diamond\Box\neg p$	$\infty_1 \cdot \alpha_1 \cdot \infty_2 \cdot \alpha_2$
$\Diamond\Box\Diamond\Box\Diamond\Box\neg p$	$\infty_1 \cdot \alpha_1 \cdot \infty_2 \cdot \alpha_2 \cdot \infty_3$
...	...
$(\Diamond\Box)^m\neg p$	$(\prod_{i=1}^{m-1} \infty_i \cdot \alpha_i) \cdot \infty_m$
$\Box(\Diamond\Box)^m\neg p$	$\prod_{i=1}^m \infty_i \cdot \alpha_i$
...	...

$0 \neq \alpha_i \neq 1$; ∞_i is infinite

Table 2

Table 2 shows how the complement of $V(p)$ looks like in occasions satisfying a given $\neg p$-formula. Indeed if χ_1 and χ_2 are $\neg p$-formulas with $|\chi_1| \leq |\chi_2|$, any occasion satisfying χ_2 is a special case of an occasion satisfying χ_1.

LEMMA 15. **(a)** *Suppose $\langle (T, <, V), o\rangle \in \mathcal{D}$ and $(T, <, V), o \models \theta$. If θ is p-formula, the set $V(p) \cap o$ is dense but the set $o \setminus V(p)$ is not. Similarly, if θ is ${\sim}p$-formula, the set $o \setminus V(p)$ is and the set $V(p) \cap o$ is not dense.* **(b)** *If θ is p-formula and χ is ${\sim}p$-formula, $(\theta \wedge \chi)$ is not satisfiable over $\mathcal{D}$.* **(c)** *If θ is p-formula (${\sim}p$-formula) and χ is $\neg p$-formula ($\neg{\sim}p$-formula), $(\theta \wedge \chi)$ is satisfiable iff $|\theta| > |\chi|$.* **(d)** *If θ is p-formula (${\sim}p$-formula) and Σ is the set of all $\neg{\sim}p$-formulas (all $\neg p$-formulas), $\theta \Rightarrow_{\mathcal{D}} \psi$ for all $\psi \in \Sigma$.*

Proof. (a) is immediate from Table 1. For (b), suppose for contradiction that $(T, <, V), o \models (\theta \wedge \chi)$, with $<$ dense and o proper. By (a), the set $V(p) \cap o$ is and is not dense. Item (c) is immediate from Tables 1 and 2. For (d), suppose θ is p-formula and $\mathcal{M}, o \models \theta$ with $(\mathcal{M}, o) \in \mathcal{D}$. From Table 1 it is seen that o can be partitioned into ω cells in each of which p occurs infinitely often, whence all $\neg\sim p$-formulas hold at o. ∎

THEOREM 16. *Let $\Sigma(p)$ be any set of p-formulas, $\Sigma(\neg p)$ any set of $\neg p$-formulas, $\Sigma(\sim p)$ be any set of $\sim p$-formulas and $\Sigma(\neg\sim p)$ any set of $\neg\sim p$-formulas. The union $\Sigma(p) \cup \Sigma(\neg p) \cup \Sigma(\sim p) \cup \Sigma(\neg\sim p)$ is satisfiable iff one of the following two conditions holds: (i) $\Sigma(\sim p) = \varnothing$ and $\Sigma(\neg p)$ is finite and* $\max\{|\theta| : \theta \in \Sigma(\neg p)\} < \min\{|\chi| : \chi \in \Sigma(p)\}$, *($ii$) $\Sigma(p) = \varnothing$ and $\Sigma(\neg\sim p)$ is finite and* $\max\{|\theta| : \theta \in \Sigma(\neg\sim p)\} < \min\{|\chi| : \chi \in \Sigma(\sim p)\}$.

Proof. By Lemma 14(a), the set $\Sigma(p)$ alone is satisfiable. By Lemmas 14 and 15(c), $\Sigma(p) \cup \Sigma(\neg p)$ is satisfiable iff $\Sigma(\neg p)$ is finite and $\max\{|\theta| : \theta \in \Sigma(\neg p)\} < \min\{|\chi| : \chi \in \Sigma(p)\}$. By Lemma 15($d$), again, $\Sigma(p) \cup \Sigma(\neg p)$ is satisfiable iff $\Sigma(p) \cup \Sigma(\neg p) \cup \Sigma(\neg\sim p)$ is satisfiable. So if $\Sigma(\sim p) = \varnothing$, the union $\Sigma(p) \cup \Sigma(\neg p) \cup \Sigma(\sim p) \cup \Sigma(\neg\sim p)$ is satisfiable iff condition (i) holds. Suppose, then, $\Sigma(\sim p) \neq \varnothing$. By Lemma 15($b$), no occasion satisfying $\Sigma(\sim p)$ can satisfy $\Sigma(p)$ unless $\Sigma(p) = \varnothing$. Dually to what noted above, $\Sigma(\sim p) \cup \Sigma(\neg\sim p) \cup \Sigma(\neg p)$ is satisfiable iff $\Sigma(\neg\sim p)$ is finite and $\max\{|\theta| : \theta \in \Sigma(\neg\sim p)\} < \min\{|\chi| : \chi \in \Sigma(\sim p)\}$. That is, if $\Sigma(\sim p) \neq \varnothing$, $\Sigma(p) \cup \Sigma(\neg p) \cup \Sigma(\sim p) \cup \Sigma(\neg\sim p)$ is satisfiable iff condition (ii) holds. ∎

THEOREM 17. *Over $\mathcal{D}$, $\mathcal{L}_{BPR}$-SAT ($\mathcal{L}_{BPR}$-VAL) is* **NP***-complete.*

Proof. It suffices to consider satisfiability ($\mathcal{L}_{BPR}$ is semantically closed under $\neg$ and a formula expressing the negation of a given formula is computed in constant time). **Inclusion:** Given $\theta \in \mathcal{L}_{BPR} \setminus \mathcal{L}_{PR}$, non-deterministically guess a map $d : \cup_{i \leq n}\{0,1\}^i \to \{0,1\}$, with $n+1$ the maximum number of nested disjunctions and conjunctions in θ; the map d can be used to determine for every disjunctive subformula of θ one of the disjuncts in an obvious way. Starting with $S_0 := \{\theta\}$, generate a set $S_n \subseteq \mathcal{L}_{PR}$ by letting S_{i+1} contain $S_i \cap \mathcal{L}_{PR}$, both conjuncts of every conjunction in S_i, for every disjunction in S_i, the disjunct determined by d, and no other formulas. (If $\theta \in \mathcal{L}_{PR}$, proceed with $S_0 := \{\theta\}$.) Using Theorem 16, we determine whether S_n is satisfiable over $\mathcal{D}$: first see if S_n contains both a p-formula and a $\sim p$-formula. If so, S_n is not satisfiable over $\mathcal{D}$. Else, if S_n contains no $\sim p$-formula, check if S_n contains a $\neg p$-formula whose prefix is not shorter than the prefixes of all p-formulas in S_n; if it does, S_n is not satisfiable over $\mathcal{D}$, otherwise it is. If, again, S_n contains no p-formulas, similarly check if S_n contains a $\neg\sim p$-formula whose prefix is not shorter than the prefixes of all $\sim p$-formulas in S_n; if so, S_n is not satisfiable over $\mathcal{D}$, else it is. The induced algorithm runs in **NP**: the non-deterministically guessed map d was employed to generate in constant time the set S_n, whose satisfiability over $\mathcal{D}$ was then checked in polynomial time. **Hardness:** The **NP**-complete problem **3-CNF** can be simulated in $\mathcal{L}_{BPR}$ w.r.t. $\mathcal{D}$: the **PL**-formula $\bigwedge_{i<n}(\ell_{i1} \vee \ell_{i2} \vee \ell_{i3})$ is satisfiable iff the formula $(\bigwedge_{i<n}(\ell_{i1} \vee \ell_{i2} \vee \ell_{i3})) \wedge \neg{\sim}\top$ of $\mathcal{L}_{PR}$ is satisfiable over $\mathcal{D}$, given that each ℓ_{ij} equals p or $\neg p$ for some atom p. (The conjunct $\neg{\sim}\top$ serves to exclude the empty occasion). ■

5 Propositional fragment

Define the *propositional fragment* $\mathcal{L}_{prop}$ of $\mathcal{L}_{TD}$ to be $\mathbf{Cl}_{\wedge,\vee}(\mathcal{L}_{base})$. Now $\mathcal{L}_{prop}$ has genuine tense-logical content — despite its 'propositional' nature. Use $\sim$ as the negation symbol in **TL**; let **SLF** or *simple logic of future* be the fragment of **TL** whose syntax is generated thus:

$$\gamma \ ::= Fp \mid F{\sim}p \mid Gp \mid G{\sim}p \mid F(\beta \wedge \beta) \mid G(\beta \vee \beta) \mid (\gamma \wedge \gamma) \mid (\gamma \vee \gamma),$$

with $p \in \mathbf{prop}$ and $\beta \in \mathcal{L}_{zero}$. (Syntactically, $\mathcal{L}_{zero} \subset \mathbf{TL}$.) Semantically, **SLF** coincides with the fragment of **TL** whose formulas have modal depth at most one and make no use of the past tense operators. $\mathcal{L}_{prop}$ and **SLF** are intertranslatable, in the sense expressed by Fact 18. If $\mathcal{M} = (T, <, V)$ is a model, $t, t' \in T$ and $t < t'$, let $\mathcal{M}_{t,t'}$ be the model $([t, t'], <_{t,t'}, V_{t,t'})$, where $<_{t,t'}$ ($V_{t,t'}$) is the restriction of $<$ (*resp.* V) to the interval $[t, t']$.

FACT 18. There are **PTIME**-computable functions $T : \mathcal{L}_{prop} \to \mathbf{SLF}$ and $S : \mathbf{SLF} \to \mathcal{L}_{prop}$ such that for all $\chi \in \mathcal{L}_{prop}$, models $\mathcal{M} = (T, <, V)$ and non-empty $]t, t'] \subseteq T$, we have: $\mathcal{M},]t, t'] \models \chi$ iff $\mathcal{M}_{t,t'}, t \models T[\chi]$. Conversely, for all $\theta \in \mathbf{SLF}$, models $\mathcal{N} = (N, <, V)$ and non-empty $]t, t'] \subseteq N$, we have: $\mathfrak{N}_{t,t'}, t \models \theta$ iff $\mathcal{N},]t, t'] \models S[\theta]$.

Proof. If $\psi \in \mathcal{L}_{TD}$, let $[\psi]^{nnf}$ be its negation normal form. Define a map $T : \mathcal{L}_{prop} \to \mathbf{SLF}$ as follows: $T[p] = Gp$, $T[{\sim}p] = G{\sim}p$, $T[\neg p] = F{\sim}p$, $T[\neg{\sim}p] = Fp$, $T[\alpha] = \alpha$, $T[{\sim}(\alpha \wedge \alpha')] = G(T[[{\sim}\alpha]^{nnf}] \vee T[[{\sim}\alpha']^{nnf}])$, $T[\neg{\sim}(\alpha \wedge \alpha')] = F(T[\alpha] \wedge T[\alpha'])$ and $T[(\chi \circ \chi')] = (T[\chi] \circ T[\chi'])$ for $\circ \in \{\wedge, \vee\}$, where $\alpha, \alpha' \in \mathcal{L}_{zero}$ and $\chi, \chi' \in \mathcal{L}_{prop}$. Conversely, define a map $S : \mathbf{SLF} \to \mathcal{L}_{prop}$ by putting $S[Fp] = \neg{\sim}p$, $S[F{\sim}p] = \neg p$, $S[Gp] = p$, $S[G{\sim}p] = {\sim}p$, $S[\beta] = \beta$, $S[F(\beta \wedge \beta')] = \neg{\sim}(S[\beta] \wedge S[\beta'])$, $S[G(\beta \vee \beta')] = {\sim}([S[{\sim}\beta]]^{nnf} \wedge [S[{\sim}\beta']]^{nnf})$ and $S[(\gamma \circ \gamma')] = (S[\gamma] \circ S[\gamma'])$ for $\circ \in \{\wedge, \vee\}$, where $\beta, \beta' \in \mathcal{L}_{zero}$ and $\gamma, \gamma' \in \mathbf{SLF}$. Clearly T and S are **PTIME**-computable and translations in the sense required. ■

TL-SAT over linear orders is **NP**-complete. (In [28] **TL**-SAT was proven **NP**-complete over transitive trichotomous orders; this settles also the case of linear orders: a transitive trichotomous order can be turned by 'bulldozing' into an irreflexive transitive trichotomous order, cf. [1, Thm. 4.56].) By Fact 18, then, $\mathcal{L}_{prop}$-SAT is in **NP**. However, let us look at the situation in detail. Let $\mathcal{C}_0 = \{(\mathcal{M}, o) : \mathcal{M} \text{ is a model and } o \text{ is non-empty and proper}\}$.

LEMMA 19. *The satisfiability and validity problems of* $\mathcal{L}_{prop}$ *are* **NP**-*complete, both over* $\mathcal{C}_0$ *and over* $\mathcal{D}_0 = \{(\mathcal{M}, o) \in \mathcal{D} : o \neq \varnothing\}$.

Proof. We show that the satisfiability problem of **SLF** is **NP**-complete over all (dense) models $\mathcal{M}_{t,t'}$ with $t < t'$. The claims about $\mathcal{L}_{prop}$ follow due to Fact 18. **Inclusion:** Any $\gamma \in \mathbf{SLF}$ is obtained from formulas of the forms $F\beta$ or $G\beta$ (for suitable $\beta \in \mathcal{L}_{zero}$) by finitely many applications of $\wedge$ and $\vee$. Non-deterministically guess a map choosing a disjunct for each disjunctive subformula of γ, and use it to produce a set X of formulas of the forms F and G such that γ is satisfiable iff the set X is. There are, then, formulas $\alpha_i, \beta_j \in \mathcal{L}_{zero}$ and numbers x, y such that $X = \{G\alpha_1, \ldots, G\alpha_x, F\beta_1, \ldots, F\beta_y\}$. Evidently X is satisfiable iff all sets $\{\alpha_1, \ldots, \alpha_x, \beta_j\}$ are satisfiable ($j := 1, \ldots, y$). This, again, is the case iff for a non-deterministically chosen number l, the set $X^l := \{\alpha_1, \ldots, \alpha_x, \beta_l\}$ is satisfiable. To decide whether it is, non-deterministically resolve the disjunctions in the formulas $\alpha_1, \ldots, \alpha_x, \beta_l$ and obtain a set $A_1^l \cup \ldots \cup A_x^l \cup B^l$ of literals. If this set contains an atom and its negation, X^l is not satisfiable, otherwise it is. The algorithm runs in **NcoP** = **NP**. If it accepts the input γ, the induced model $\mathcal{M}_{t,t'}$ satisfies: $|\{z : t \leq z \leq t'\}| = y+1$, where y is the number of formulas of the form F in the non-deterministically guessed set X. $\mathcal{M}_{t,t'}$ can then be turned into a model $\mathcal{M}'_{t,t'}$ with a dense linear order, e.g. by adding an isomorphic copy of the rational interval $]0,1[$ between each point in the domain of $\mathcal{M}_{t,t'}$ and making all α_i with $G\alpha_i \in X$ true throughout those intervals. So $\mathcal{L}_{prop}$-SAT is **NP**-decidable also over $\mathcal{D}_0$. **Hardness: 3-CNF** can be simulated in **SLF**: the **PL**-formula $\bigwedge_{i<n}(\ell_{i1} \vee \ell_{i2} \vee \ell_{i3})$ is satisfiable iff the formula $(\bigwedge_{i<n}(\ell'_{i1} \vee \ell'_{i2} \vee \ell'_{i3}))$ of **SLF** is satisfiable over arbitrary (*resp.* dense) models $\mathcal{M}_{t,t'}$, given that each ℓ_{ij} equals p or $\neg p$ for some atom p, and $\ell'_{ij} = Gp$ if $\ell_{ij} = p$ and $\ell'_{ij} = F{\sim}p$ if $\ell_{ij} = \neg p$. ∎

6 Simple fragment

The syntax of the *simple fragment* $\mathcal{L}^1_{TD}$ of $\mathcal{L}_{TD}$ is generated by the grammar

$$\chi ::= \mathbf{b} \mid \Box\mathbf{b} \mid \Diamond\mathbf{b} \mid (\chi \wedge \chi) \mid (\chi \vee \chi),$$

with $\mathbf{b} \in \mathcal{L}_{prop}$. The syntax excludes nested modal operators; semantically $\mathcal{L}^1_{TD}$ equals $\{\phi \in \mathcal{L}_{TD} : md(\phi) \leq 1\}$. We observe some facts about $\mathcal{L}^1_{TD}$.

Diamond formulas. Let us call $\mathcal{L}^1_{TD}$-formulas of the form $\Diamond\theta$ *diamond formulas.* They can impose quite strong requirements. E.g., $\Diamond(\neg p \wedge \neg{\sim}p)$ has only infinite models. Restricting attention to anchored models from $\mathcal{D}$, consider the kinds of statements that can be made using formulas $\Diamond\theta$.

Fix some auxiliary notation. If the χ_i are $\mathcal{L}_{base}$-formulas, let $\mathcal{M},]s, s'] \models \mathsf{low}(\chi_1, \ldots, \chi_n)$ state that for any given $t \in]s, s']$ and any i, there is $t'_i \in]s, t[$ such that χ_i holds at t'_i. Similarly, $\mathsf{high}(\chi_1, \ldots, \chi_n)$ states that above any given point each χ_i is true. Let $\mathcal{M}, o \models \mathsf{close}(\chi_1, \ldots, \chi_n)$ mean that there is $t \in o$ such that either every χ_i is true arbitrarily close to t above t or every χ_i is true arbitrarily close to t below t. If $\psi \in \mathcal{L}_{prop}$, let $]s, s'[\models \psi$ be otherwise the same statement as $]s, s'] \models \psi$ except that the metalanguage first-order quantifiers are not allowed to range over the right bound s' of the occasion $]s, s']$. Then if $\psi, \chi, \phi \in \mathcal{L}_{prop}$, let $\sharp(\psi, \chi, \phi)$ state at $]t, t']$ that there is a point $s \in]t, t'[$ such that $]t, s[\models \psi$ and $\{s\} \models \chi$ and $]s, t'] \models \phi$. Finally, define $\S(\psi)$ to be true at $]t, t']$ if there is s such that $]t, s] \models \psi$ or $]s, t'] \models \psi$. The abbreviated statements are not themselves definable in

$\mathcal{L}_{TD}$; they will be used in analyzing what can be stated in terms of diamond formulas relative to $\mathcal{D}$. (Incidentally, all statements are **US**-definable.)

EXAMPLE 20. The following equivalences are relative to $\mathcal{D}$; they are straightforward consequences of the semantics of $\Diamond$.

(*a*) $\Diamond(p \vee q)$ iff $\sharp(p \vee q, \top, p \vee q)$.

(*b*) $\Diamond(\neg p \vee \neg q)$ iff $(\neg p \vee \neg q)$.

(*c*) $\Diamond(p \wedge q)$ iff $\sharp(p \wedge q, \top, p \wedge q)$.

(*d*) $\Diamond(\neg p \wedge \neg q)$ iff $\mathsf{close}(\neg p, \neg q) \vee \mathsf{low}(\neg p, \neg q) \vee \mathsf{high}(\neg p, \neg q)$.

(*e*) $\Diamond(p \vee \neg q)$ iff $\Diamond p \vee \Diamond \neg q$ iff $\Diamond p \vee \neg q$.

(*f*) $\Diamond(p \vee [\neg q \wedge \neg r])$ iff $\Diamond p \vee \Diamond(\neg q \wedge \neg r) \vee \S(p \wedge \neg q \wedge \neg r)$.

(*g*) $\Diamond(p \vee \neg q \vee \neg r)$ iff $\Diamond p \vee \Diamond(\neg q \wedge \neg r) \vee \S(p \wedge [\neg q \vee \neg r])$.

(*h*) $\Diamond(p \wedge \neg q)$ iff $(p \wedge \neg q)$ or $[\sharp(p, \neg p, p) \wedge \mathsf{low}(\neg q) \wedge \mathsf{high}(\neg q)]$.

(*i*) $\Diamond[(p \wedge \neg q) \vee (r \wedge \neg s)]$ iff $\Diamond(p \wedge \neg q)$ or $\Diamond(r \wedge \neg s)$ or $[\sharp(p, \neg p, r) \wedge \mathsf{low}(\neg q) \wedge \mathsf{high}(\neg s)]$ or $[\sharp(r, \neg r, p) \wedge \mathsf{low}(\neg s) \wedge \mathsf{high}(\neg q)]$.

(*j*) $\Diamond[(p \wedge \neg q) \vee (r \wedge \neg s) \vee (t \wedge \neg u)]$ iff $\Diamond[(p \wedge \neg q) \vee (r \wedge \neg s)] \vee \Diamond[(p \wedge \neg q) \vee (t \wedge \neg u)] \vee \Diamond[(r \wedge \neg s) \vee (t \wedge \neg u)]$. ⊣

An $\mathcal{L}_{base}$-formula is *universal* if it is of the form p, $\sim p$ or $\sim(\chi \wedge \chi')$, and *existential* if of the form $\neg p$, $\neg\sim p$ or $\neg\sim(\chi \wedge \chi')$. By their semantics, universal formulas make a universal statement about an occasion, while existential formulas are witnessed by a single point in an occasion.

LEMMA 21. *Let the* u_i (resp. e_i) *range over universal (existential) formulas of* $\mathcal{L}_{base}$. *The following equivalences hold relative to* $\mathcal{D}$.

(*a*) $\Diamond(\bigvee_i u_i \vee \bigvee_j e_j)$ *iff* $\bigvee_j e_j \vee \sharp(\bigvee_i u_i, \top, \bigvee_i u_i)$.

(*b*) *Suppose that* $n \geq 1$ *and that there are* i, j *with* $e_i \neq e_j$. *Then* $\Diamond(\bigwedge_{i<n+1} e_i)$ *iff* $\mathsf{close}(e_0, \ldots, e_n) \vee \mathsf{low}(e_0, \ldots, e_n) \vee \mathsf{high}(e_0, \ldots, e_n)$.

(*c*) $\Diamond[\bigwedge_i u_i \wedge \bigwedge_j e_j]$ *iff* $\bigwedge_i u_i \wedge \bigwedge_j e_j \vee [\sharp(\bigwedge_i u_i, \top, \bigwedge_i u_i) \wedge \mathsf{low}(e_0, \ldots, e_n) \wedge \mathsf{high}(e_0, \ldots, e_n)]$.

(*d*) $\Diamond(\bigwedge_i u_i \vee \bigwedge_j e_j)$ *iff* $\Diamond(\bigwedge_i u_i) \vee \Diamond(\bigwedge_j e_j) \vee \S(\bigwedge_i u_i \wedge \bigwedge_j e_j)$.

(*e*) $\Diamond([\bigwedge_i u_i \wedge \bigwedge_j e_j] \vee [\bigwedge_k u'_k \wedge \bigwedge_l e'_l])$ *iff* $\Diamond[\bigwedge_i u_i \wedge \bigwedge_j e_j] \vee \Diamond[\bigwedge_k u'_k \wedge \bigwedge_l e'_l] \vee$ $[\sharp(\bigwedge_i u_i, \top, \bigwedge_k u'_k) \wedge \mathsf{low}(e_0, \ldots, e_n) \wedge \mathsf{high}(e'_0, \ldots, e'_m)] \vee$ $[\sharp(\bigwedge_k u'_k, \top, \bigwedge_i u_i) \wedge \mathsf{low}(e'_0, \ldots, e'_m) \wedge \mathsf{high}(e_0, \ldots, e_n)]$.

(*f*) $\Diamond(\bigwedge_{i_1} \theta^1_{i_1} \vee \ldots \vee \bigwedge_{i_n} \theta^n_{i_n})$ *iff* $\bigvee_{(k,l) \in \{1,\ldots,n\}^2} \Diamond(\bigwedge_{i_k} \theta^k_{i_k} \vee \bigwedge_{i_l} \theta^l_{i_l})$.

Proof. Generalizing the observations incorporated in Example 20. ■

Formula θ of $\mathcal{L}_{prop}$ is in *disjunctive normal form* (DNF) if $\theta = \bigvee_i \bigwedge_j \theta_{ij}$, where the θ_{ij} are literals. Formula $\Diamond\theta$ of $\mathcal{L}^1_{TD}$ is in DNF if θ is; and $\Diamond\theta$ is *non-degenerate* if it is in DNF and each disjunct of θ is of the form $(\bigwedge_i u_i \wedge \bigwedge_j e_j)$ for positive i, j. Observe the following decidability result.

THEOREM 22. *The satisfiability (and validity) of finite sets of non-degenerate diamond formulas over $\mathcal{D}$ is decidable in* **NP**.

Proof. Let $\Theta := \{\Diamond\theta_i : 1 \leq i \leq n\}$ with the $\Diamond\theta_i$ non-degenerate. By Lemma 21(f), Θ is satisfiable iff for every θ_i there are disjuncts θ_i^1 and θ_i^2 (having both universal and existential conjuncts) such that the set $\Theta' := \{\Diamond(\theta_i^1 \vee \theta_i^2) : 1 \leq i \leq n\}$ is satisfiable. By Lemma 21(c, e), $\Diamond([\bigwedge_i u_i \wedge \bigwedge_j e_j] \vee [\bigwedge_k u'_k \wedge \bigwedge_l e'_l])$ holds at a dense occasion o iff one of the following six conditions holds at o:

(1) $\bigwedge_i u_i \wedge \bigwedge_j e_j$;
(2) $\bigwedge_k u'_k \wedge \bigwedge_l e'_l$;
(3) $[\sharp(\bigwedge_i u_i, \top, \bigwedge_i u_i) \wedge \mathsf{low}(e_0, \ldots, e_n) \wedge \mathsf{high}(e_0, \ldots, e_n)]$;
(4) $[\sharp(\bigwedge_k u'_k, \top, \bigwedge_k u'_k) \wedge \mathsf{low}(e'_0, \ldots, e'_m) \wedge \mathsf{high}(e'_0, \ldots, e'_m)]$;
(5) $[\sharp(\bigwedge_i u_i, \top, \bigwedge_k u'_k) \wedge \mathsf{low}(e_0, \ldots, e_n) \wedge \mathsf{high}(e'_0, \ldots, e'_m)]$;
(6) $[\sharp(\bigwedge_k u'_k, \top, \bigwedge_i u_i) \wedge \mathsf{low}(e'_0, \ldots, e'_m) \wedge \mathsf{high}(e_0, \ldots, e_n)]$.

Enumerate these options as $1, \ldots, 6$ in the above order and guess an assignment $f : \{1, \ldots, n\} \to \{1, \ldots, 6\}$. If it so happens that $\{1, 2\} \cap Im(f) = \varnothing$, it is easy to decide whether Θ' is satisfiable. Let $\mathfrak{L}_u$ ($\mathfrak{R}_u$) be the set of all universal formulas that appear as conjuncts in the first (third) argument of the metalanguage connective $\sharp$ in a condition $f(x)$ for some $1 \leq x \leq n$. Let $\mathfrak{L}_e$ ($\mathfrak{R}_e$) be the set of all existential formulas that appear as arguments of the metalanguage connective low (high) in a condition $f(x)$ with $1 \leq x \leq n$. Then check, applying the **NP**-algorithm provided by the proof of Lemma 19, whether for all $\epsilon \in \mathfrak{L}_e$ and all $\epsilon' \in \mathfrak{R}_e$, both sets $\{\epsilon\} \cup \mathfrak{L}_u$ and $\{\epsilon'\} \cup \mathfrak{R}_u$ are satisfiable. If they are, then so is Θ', otherwise Θ' is not satisfiable.

Under the assumption $\{1, 2\} \cap Im(f) = \varnothing$, all existential formulas from $\mathfrak{L}_e$ ($\mathfrak{R}_e$) must be compatible with *all* universal formulas from $\mathfrak{L}_u$ ($\mathfrak{R}_u$). By contrast, there is some more room in accommodating $\bigwedge_i u_i \wedge \bigwedge_j e_j$ with formulas of the forms $3, 4, 5, 6$. Perhaps an occasion has two points of division z and v, and A, B, C, D are conjunctions of universal formulas such that A (C) holds to the left of z (v) and B (D) to the right of z (v). Then for satisfying $\bigwedge_j e_j$ simultaneously with A it is perfectly sufficient that some conjuncts are compatible with C and the rest with D — provided that A and D are mutually compatible. For the general case, let $1 \leq x \leq n$. If $1 \leq f(x) \leq 2$, let $\mathfrak{L}_u^x = \mathfrak{R}_u^x$ contain all universal conjuncts of $f(x)$, otherwise let $\mathfrak{L}_u^x$ ($\mathfrak{R}_u^x$) be the set of all universal formulas that appear as conjuncts in the first (third) argument of the metalanguage connective $\sharp$ in the condition $f(x)$. Similarly, if $1 \leq f(x) \leq 2$, let $\mathfrak{L}_e^x = \mathfrak{R}_e^x$ contain all existential conjuncts of $f(x)$, otherwise let $\mathfrak{L}_e^x$ ($\mathfrak{R}_e^x$) be the set of all existential formulas that appear as arguments of the metalanguage connective low (high) in the condition $f(x)$. Using the **NP**-algorithm employed to prove Lemma 19, check if for every $\epsilon \in \bigcup_{f(x) \notin \{1,2\}} \mathfrak{L}_e^x$ and every $\epsilon' \in \bigcup_{f(x) \notin \{1,2\}} \mathfrak{R}_e^x$, the sets

$\{\epsilon\} \cup \bigcup_x \mathfrak{L}_u^x$ and $\{\epsilon'\} \cup \bigcup_x \mathfrak{R}_u^x$ are satisfiable. If not, Θ' is not satisfiable. Otherwise proceed to check if the existential formulas brought in by formulas $f(x) \in \{1, 2\}$ can be accommodated in a model. To this end, think of each formula $f(x)$ as inducing a point dividing an attempted model into a left side and a right side. Distinct formulas may but need not induce the same division point. Let $1 \leq k \leq n$ and let $g : \{1, \ldots, n\} \to \{1, \ldots, k\}$ be a surjection. Intuitively, g determines a linear order for the division points corresponding to the formulas $f(x)$, allowing the identity of points induced by several formulas. Introduce sets S_y $(1 \leq y \leq k + 1)$ as follows: S_y includes those $\mathfrak{L}_u^x$ for which $y \leq g(x)$ and those $\mathfrak{R}_u^x$ for which $y > g(x)$; further, for every x guess a partition of $\mathfrak{L}_e^x$, including each member of the partition into one of the sets S_y with $y \leq g(x)$, and guess a partition of $\mathfrak{R}_e^x$, including each member of the partition into one of the sets S_y with $y > g(x)$. Do not include any other elements into the S_y. Then decide using the **NP**-algorithm of Lemma 19 whether all sets S_y are satisfiable. If they are, then so is Θ', otherwise Θ' is not satisfiable. ■

Theorem 22 is clearly generalizable to finite sets of arbitrary diamond formulas in DNF; however, a proof of this fact is left to another occasion.

Box formulas. Any $\mathcal{L}_{TD}^1$-formula $\Box\theta$ is a *box formula.* We content ourselves with a couple of observations concerning finite sets of box formulas. A box formula $\Box\theta$ is in DNF if the $\mathcal{L}_{prop}$-formula θ is; and it is *purely universal* if it contains no existential $\mathcal{L}_{base}$-formulas as components.

FACT 23. Let $\theta_1, \ldots, \theta_n$ be purely universal $\mathcal{L}_{prop}$-formulas in DNF. The set $\{\Box\theta_1, \ldots, \Box\theta_n\}$ is satisfiable iff there are formulas χ_i $(i := 1, \ldots, n)$ such that χ_i is a disjunct of θ_i and $(\chi_1 \wedge \ldots \wedge \chi_n)$ is satisfiable.

Proof. *Right to left:* If $(\chi_1 \wedge \ldots \wedge \chi_n)$ is satisfiable, so is $(\theta_1 \wedge \ldots \wedge \theta_n)$. Thus $\Box(\theta_1 \wedge \ldots \wedge \theta_n)$ is satisfiable and, *a fortiori*, so is $(\Box\theta_1 \wedge \ldots \wedge \Box\theta_n)$. *Left to right:* Write $\theta_i = \bigvee_{j<n_i} \bigwedge_{l<m_{ji}} \theta_{jl}^i$. Suppose $\bigwedge_{1\leq i\leq n} \Box(\bigvee_{j<n_i} \bigwedge_{l<m_{ji}} \theta_{jl}^i)$ is true in $\mathcal{M}$ at o for some $(\mathcal{M}, o)$. Each conjunct $\Box(\bigvee_{j<n_i} \bigwedge_{l<m_{ji}} \theta_{jl}^i)$ determines a finite division D_i of o into $N_i \geq 1$ cells such that each cell makes true one of the formulas $\bigwedge_{l<m_{ji}} \theta_{jl}^i$ $(j < n_i)$. Jointly the conjuncts determine, therefore, a partition of o into $(\sum_{i=1}^n (N_i - 1)) + 1$ sets. Each member S_r' of this partition is an intersection $S_1 \cap \ldots \cap S_n$, where S_i is a cell of the division D_i; hence S_r' is an occasion and satisfies a conjunction $C_r := (\chi_1^r \wedge \ldots \wedge \chi_n^r)$, where χ_i^r is a disjunct of θ_i true at S_i. ■

The case of arbitrary box formulas is considerably more involved. Let $\{\Box\theta_1, \ldots, \Box\theta_n\}$ be any set of box formulas in DNF. To check if it is satisfiable, we should be able to operate with the formulas θ_i as follows. Let Θ_i be the set whose members are the sets of conjuncts of the disjuncts of θ_i. Suppose the attempted model we are constructing has a sub-occasion satisfying $S_j^i \in \Theta_i$. At the moment we introduce S_j^i, we will have to check, separately for every existential formula $e \in S_j^i$, that from each of the remaining sets Θ_l with $l \neq i$ we can find a member $S_{k_l}^l$ such that $\{e\} \cup \bigcup_l \forall(S_{k_l}^l) \cup \forall(S_j^i)$ is satisfiable. (If $S \subseteq \mathcal{L}_{base}$, $\forall(S)$ denotes the set of universal formulas of S.) Checking this for just one existential formula $e \in S^i_j$, we get committed to checking the analogous condition for the total number of existential formulas in the sets $S^l_{k_l}$. In order for this process to terminate, we must, evidently, be able to exhibit finite numbers $N_1, \ldots, N_n$ such that $\{\Box\theta_1, \ldots, \Box\theta_n\}$ is satisfiable iff it is satisfiable when attention is restricted to divisions triggered by $\Box\theta_i$ into at most N_i cells. Without such bounds, we can only show that the set X of finite satisfiable sets of box formulas is recursively enumerable. By Corollary 13, we know that actually X is recursive. We leave it as an open question how to determine the numbers $N_1, \ldots, N_n$ (depending on the sizes of the input formulas). Note that it *can* indeed happen that the model-seeking process described above in general terms never terminates while the relevant set of box formulas is *not* satisfiable.

EXAMPLE 24. Let $\theta_1 := [p \wedge (\neg r \wedge \neg s)] \vee [q \wedge (\neg r \wedge \neg s)]$ and $\theta_2 := [r \wedge (\neg p \wedge \neg q)] \vee [s \wedge (\neg p \wedge \neg q)]$. It is easy to see that $\{\Box\theta_1, \Box\theta_2\}$ is not satisfiable. Yet there *is* $(\mathcal{M}, o)$ such that o has a division into ω cells each of which satisfies θ_1 and another division into ω cells each of which satisfies θ_2. ⊣

7 Concluding remarks

Related work. In the literature various temporal logics using intervals in their semantics have been formulated; see [14] for a discussion. (On the whole, interval-based logics remain, however, less studied than point-based ones.) Among different modal logics of intervals, $\mathcal{L}_{TD}$ bears the closest resemblance to the *propositional interval temporal logic* or **PITL** [13, 26, 27] and (the propositional fragment of) the *duration calculus with iteration* or **DC*** [5, 10, 15]. These logics have a strong motivation deriving from computer science, with applications e.g. to hardware description and verification. It is of some interest to note that von Wright's philosophically driven considerations dating from 1968 led largely to the same logical conceptualizations as those that some 15 years later emerged in **PITL** — which, again, has an unmistakable connection to regular expressions, the formulation of which goes back to Kleene's work [20] around 15 years before von Wright's paper. On finite intervals, the operators *chop* and *chop-star* of **PITL** are related to catenation *resp.* catenation closure (Kleene star). (Regular expressions as such have, though, nothing to do with temporal logic.)

PITL uses natural numbers to model time, while **DC*** employs real numbers. Formulas of **PITL** are evaluated relative to finite or infinite maps $\sigma : I \rightarrow 2^{\mathbf{prop}}$ ($I \subseteq \mathbb{N}$) called *intervals*. For each $i \in I$, the object $\sigma(i)$ is termed a *state*; hence states are truth-value distributions over a fixed set **prop** of atoms. An atom p holds at σ iff $\sigma(\mathbf{min})(p) = 1$, where **min** is the smallest element in $dom(\sigma)$. Disjunction ($\vee$), negation ($\neg$) and *verum* ($\top$) have their expectable semantics. The modal operators *eventually* ($\Diamond$) and *next* ($\bigcirc$) satisfy: $\sigma \models \Diamond\chi$ iff $\sigma' \models \chi$ for some suffix σ' of σ; and $\sigma \models \bigcirc\chi$ iff $\sigma' \models \chi$, given that $\sigma = \sigma(\mathbf{min})^\frown\sigma'$. The dual of $\Diamond$ is denoted by $\Box$. Characteristic of **PITL** are the binary operator *chop* (written ;) for sequential composition and the unary operator *chop-star* (written *) for the closure of sequential composition. If σ is finite, by definition $\sigma \models \psi;\chi$ iff there is $k \in dom(\sigma)$ such that $\sigma(\mathbf{min})\ldots\sigma(k) \models \psi$ and $\sigma(k)\ldots\sigma(\mathbf{max}) \models \chi$. And $\sigma \models \psi^*$ iff there are $m < \omega$ and $k_0,\ldots,k_m \in dom(\sigma)$ such that $k_0 = \mathbf{min}$ and $k_m = \mathbf{max}$ and $\sigma(k_i)\ldots\sigma(k_{i+1}) \models \psi$ for all $0 \leq i < m$. The clauses for *chop* and *chop-star* are similar in $\mathbf{DC}^*$; the evaluation in $\mathbf{DC}^*$ is relative to *closed real intervals*.[9] The semantics of *chop-star* is obviously very close to the semantics of the operator $\Box$ of $\mathcal{L}_{TD}$. The main difference is that when the latter is evaluated, the relevant interval is divided into finitely many *disjoint* pieces, while when evaluating *chop-star*, adjacent subintervals have exactly one state in common. So it is $\Box$ rather than *chop-star* that has a really straightforward connection to regular expressions: if the 'extension' of ψ is denoted by a regular expression r, the 'extension' of $\Box\psi$ is denoted by rrr^*. (Recall that $\Box\psi$ is only true of intervals divisible into at least two cells satisfying ψ.) On finite intervals, the following map t translates $\mathcal{L}_{nnf}$ into $\mathbf{PITL}$, whereas by Proposition 1, $\mathcal{L}_{TD} \leq \mathcal{L}_{nnf}$. Define a map $s : \mathcal{L}_{zero} \to \mathbf{PITL}$ by stipulating that $s(\psi)$ is the result of replacing all occurrences of $\sim$ in ψ by $\neg$. Then put: $t(p) = \Box p$, $t({\sim}\phi) = \Box\neg s(\phi)$, $t(\neg\phi) = \neg t(\phi)$, $t(\phi \circ \chi) = t(\phi) \circ t(\chi)$ for $\circ \in \{\vee, \wedge\}$, $t(\Box\phi) = t(\phi); \bigcirc t(\phi); [\bigcirc t(\phi)]^*$, $t(\Diamond\phi) = \neg(\neg t(\phi); \bigcirc\neg t(\phi); [\bigcirc\neg t(\phi)]^*)$. It is not immediately clear whether $\mathcal{L}_{TD} \leq \mathbf{DC}^*$; relative to real intervals the *next* operator is not available to help expressing the semantics of $\Box$. Both $\mathbf{PITL}$ and $\mathbf{DC}^*$ have been extensively studied from the proof-theoretic viewpoint. $\mathbf{PITL}$ is known to admit of a complete proof system both over finite and over infinite time; see the bibliography of [27]. A complete proof system for $\mathbf{DC}^*$ relative to so-called abstract-time semantics has likewise been presented [10]. For more information about duration calculus, see [15, 16]. Unlike $\mathbf{PITL}$ and $\mathbf{DC}^*$, $\mathcal{L}_{TD}$ is not for its semantics restricted to any particular class of linear orders. On the other hand, relative to the appropriate classes of linear orders, $\mathcal{L}_{TD}$ is less expressive than either $\mathbf{PITL}$ or $\mathbf{DC}^*$: in particular the operator *chop* is neither syntactically given nor definable in $\mathcal{L}_{TD}$.

Questions for future research. The present paper leaves it for future research to estimate the complexity of an optimal algorithm solving $\mathcal{L}_{TD}$-SAT. Läuchli's proof [22] does not yield an explicit upper bound on the time complexity of $\mathcal{L}^{\text{mon}}_{\text{w}}$-SAT, while Rabin's proof [31] provides a non-elementary upper bound (i.e., the time complexity is not bounded by any stack of exponentials of a fixed height). Another obvious task is to provide a complete proof system for $\mathcal{L}_{TD}$. One might also attempt relating fragments of $\mathcal{L}_{TD}$ to independently interesting fragments of $\mathcal{L}^{\text{mon}}_{\text{w}}$.

The study of $\mathcal{L}_{TD}$ can be pursued in different directions. One option is to study its finite models; technically, these are *word models*. By Büchi's theorem, cf. [6], $\mathcal{L}^{\text{mon}}$ can define over word models precisely the denotations of regular expressions. (Over finite models, of course $\mathcal{L}^{\text{mon}}_{\text{w}} = \mathcal{L}^{\text{mon}}$.) One can ask in which precise way $\mathcal{L}_{TD}$ falls short of capturing regular languages;

[9] For *chop* as a modal operator employing a ternary accessibility relation, see [36].

cf. [17] for related research. The question of whether $\mathcal{L}_{TD}$ has 0-1 law could also be studied. (Evenness is, apparently, not definable.) It might be possible to link $\mathcal{L}_{TD}$ to certain logics of trees with a *yield* operator [4].[10]

$\mathcal{L}_{TD}$ could be modified by adding an operator $\Box'$ otherwise like $\Box$ but involving a division into a fixed finite number of cells, say 2; this might be further strengthened into a binary *chop* modality: ϕ holds first and then ψ. The semantics of $\Box$ itself can also be varied. We might, e.g., allow the number of divisors in the semantic clause for $\Box$ to be countably infinite or arbitrary. Without further restrictions, however, the resulting logic would be very expressive indeed and the whole idea of division would take a somewhat unintended form: e.g. a rational interval would have a division into singleton cells. A more interesting variant is obtained by imposing a condition on the *induced order* of the divisors. We might, say, restrict attention to sets of divisors whose induced order is of type ω. Actually von Wright did not forbid infinite divisions but still took occasions to be always divided into discrete 'bits' or 'stretches' [37, p. 127]. If $\mathcal{C}$ is a class of *discrete* order types, we might study the logic $\mathcal{L}^{\mathcal{C}}_{TD}$ with $\Box$ strengthened so as to allow any sets of divisors the induced order of which has a type $\alpha \in \mathcal{C}$. The semantics of $\Box$ in $\mathcal{L}_{TD}$ results from letting $\mathcal{C}$ consist of all finite order types $n \geq 1$. On arbitrary linear orders, $\mathcal{L}^{\mathcal{C}}_{TD}$ can merely be translated into $\mathcal{L}^{\text{mon}}$ if $\mathcal{C}$ admits of infinite order types; more specifically, into the fragment $\mathcal{L}^{\text{mon}}_{\mathcal{C}}$ of $\mathcal{L}^{\text{mon}}$ whose second-order quantifiers range over subsets meeting the appropriate order type requirement. There is no immediate way of settling whether $\mathcal{L}^{\mathcal{C}}_{TD}$ is decidable: $\mathcal{L}^{\text{mon}}$ is undecidable over arbitrary linear orders — notably over the real line [33, 12]. Over some classes of linear orders $\mathcal{L}^{\text{mon}}$ is decidable, however, e.g., countable linear orders [31], $\{\omega_1\}$ and $\{\alpha : \alpha < \omega_2\}$ [2]. As to $\{\omega_2\}$, ZFC does not determine which sentences of $\mathcal{L}^{\text{mon}}$ are true of ω_2 [11, 33]. Evidently $\mathcal{L}^{\mathcal{C}}_{TD} < \mathcal{L}^{\text{mon}}_{\mathcal{C}} < \mathcal{L}^{\text{mon}}$; it might happen that $\mathcal{L}^{\mathcal{C}}_{TD}$ is decidable over arbitrary linear orders.

Finally, one might experiment with studying $\mathcal{L}_{TD}$ relative to a larger class of models, say tree structures. This would require finding a suitable interpretation to the idea of division relative to trees. Conceivably divisor points might be replaced by *bars* (i.e., sets B such that every branch of the tree intersects B exactly once). Or by non-comparable nodes such that every maximal branch belongs to the 'neighborhood' determined by one of the nodes. Or the cells of divisions might be taken to be subtrees of a tree.

Acknowledgments. The author wishes to thank the three anonymous referees for their detailed, useful comments. The research was carried out within the project "Modalities, Games and Independence in Logic" funded by the Academy of Finland.

BIBLIOGRAPHY

[1] P. Blackburn, M. de Rijke & Y. Venema. *Modal Logic*. CUP, 2002.

[10] I am indebted to an anonymous referee for this suggestion.

[2] J. R. Büchi. The monadic second-order theory of ω_1. In *Lecture Notes in Mathematics* Vol. 328, 1–127, Springer, 1973.
[3] M. L. Dalla Chiara. Von Wright on time, change, and contradiction. In [32], 637–45.
[4] H. Comon *et al.* *Tree Automata Techniques and Applications.* Available at `http://tata.gforge.inria.fr/`, released October 12, 2007.
[5] H. Dang Van & J. Wang. On the design of hybrid control systems using automata models. In LNCS 1180, Springer, 415–38, 1996.
[6] H.-D. Ebbinghaus & J. Flum. *Finite Model Theory.* Springer, 1999.
[7] H.-D. Ebbinghaus, J. Flum & W. Thomas. *Mathematical Logic.* Springer, 1984.
[8] A. Fraenkel. *Abstract Set Theory.* North-Holland, 1961.
[9] D. M. Gabbay, I. Hodkinson & M. Reynolds. *Temporal Logic*, Vol. 1. OUP, 1994.
[10] D. P. Guelev & H. Dang Van. On the completeness and decidability of duration calculus with iteration. *Theor. Comp. Science* 337, 278–304, 2005.
[11] Y. Gurevich, M. Magidor & S. Shelah. The monadic theory of ω_2. *J. Symb. Log.* 48(2), 387–98, 1983.
[12] Y. Gurevich & S. Shelah. Monadic theory of order and topology in ZFC. *Annals of Math. Logic* 23, 179–98, 1982.
[13] J. Y. Halpern, Z. Manna & B. Moszkowski. A hardware semantics based on temporal intervals. In LNCS 154, Springer, 278–91, 1983.
[14] J. Y. Halpern & Y. Shoham. A propositional modal logic of time intervals. *J. ACM* 38(4), 935–62, 1991.
[15] M. R. Hansen & H. Dang Van. A theory of duration calculus with application. In LNCS 4710, Springer, 119–76, 2007.
[16] M. R. Hansen & C. Zhou. Duration calculus: logical foundations. *Formal Aspects of Computing* 9, 283–330, 1997.
[17] L. Hella & T. Tulenheimo. On the existence of a modal-logical basis for monadic second-order logic. Unpublished manuscript, 2008.
[18] K. Hrbacek & T. Jech. *Introduction to Set Theory.* Marcel Dekker AG, 1999.
[19] H. Kamp. *Tense Logic and the Theory of Linear Order*, Ph.D. thesis, UCLA, 1968.
[20] S. C. Kleene. Representation of events in nerve nets and finite automata. In C. E. Shannon & J. McCarthy (eds.): *Automata Studies*, 3–42, Princeton UP, 1956. (First published as RAND research memorandum RM-704, 15 December 1951, 101 pages.)
[21] D. Leivant. Higher order logic. In *Handbook of Logic in Artificial Intelligence and Logic Programming*, 229–321, OUP, 1994.
[22] H. Läuchli. A decision procedure for the weak second order theory of linear order. In *Proc. of the Logic Colloquium, Hannover, 1966*, North-Holland, 189–97, 1968.
[23] H. Läuchli & J. Leonard. On the elementary theory of linear order. *Fund. Math.* 59, 109–16, 1966.
[24] J. D. Monk. *Mathematical Logic.* Springer, 1976.
[25] C. Mortensen. Change. *The Stanford Encyclopedia of Philosophy* (Winter 2006), E. N. Zalta (ed.), `http://plato.stanford.edu/archives/win2006/entries/change/`.
[26] B. Moszkowski. *Executing Temporal Logic Programs.* CUP, 1986.
[27] B. Moszkowski. Using temporal logic to analyse temporal logic: a hierarchical approach based on intervals. *J. Log. Comp.* 17(2), 333–409, 2007.
[28] H. Ono & A. Nakamura. On the size of refutation Kripke models for some linear modal and tense logics. *Studia Logica* 39(4), 325–33, 1980.
[29] A. N. Prior. *Past, Present and Future.* OUP, 1967.
[30] A. N. Prior. Review of [37]. *Brit. J. Phil. Sci.* 20, 372–4.
[31] M. Rabin. Decidability of second-order theories and automata on infinite trees. *Trans. Amer. Math. Soc.* 141, 1–35, 1969.
[32] P. A. Schilpp & L. E. Hahn (eds.): *The Philosophy of Georg Henrik von Wright.* Library of Living Philosophers Vol. XIX, Open Court, 1989.
[33] S. Shelah. The monadic theory of order. *Ann. of Math.* 102(3), 379–419, 1975.
[34] W. Sierpiński. *Cardinal and Ordinal Numbers.* Monografie Matematyczne, Vol. 34. Państwowe Wydawnictwo Naukowe, 1958.

[35] J. W. Smith. Time, change and contradiction. *Australas. J. Phil.* 68(2), 178–88, 1990.
[36] Y. Venema. A modal logic for chopping intervals. *J. Log. Comp.* 1(4), 453–76, 1991.
[37] G. H. von Wright. Time, change, and contradiction. CUP, 1969. Reprinted in [38], 115–31. (References in the body of the paper are to [38].)
[38] G. H. von Wright. *Philosophical Logic. Philosophical Papers* Vol. 2, Blackwell, 1983.
[39] G. H. von Wright. Dalla Chiara on time, change, and contradiction. In [32], 862–4.

Tero Tulenheimo
Academy of Finland
Department of Philosophy,
University of Helsinki
P.O. Box 9, 00014 University of Helsinki,
Finland.
tero.tulenheimo@helsinki.fi

Three 13th-century views of quantified modal logic

SARA L. UCKELMAN

ABSTRACT. There are two reasons why medieval logic is of interest to modern logician: One is to see how similar it is to modern logic and the other is to see how different it is. We study three 13th-century works on modal logic and give two examples of how their views of modal logic differ from modern views of the same: the nature of modality and the truth conditions for modal sentences. Because of the different goals of the medieval logicians, modern logicians must take care in arguing for or against the correctness of the medieval logical theories.

Keywords: 13th century, modal proposition, modal syllogism, modal square of opposition, William of Sherwood.

1 Two reasons to study medieval logic

There are two reasons why the study of medieval logic is of interest to the modern logician. The first is to see how closely logical theories in different branches (modal logic, temporal logic, quantifier logic, etc.) resemble modern logical theories in these same branches. The second is to see how much they differ. Investigating a topic in medieval logic for either of these reasons will result in something informative and illuminating. If the medieval theory is similar to the modern theory, one can ask to what extent we can shed new light on the medieval theory by modeling it with modern formal tools. If the medieval theory differs from the modern theory, one can ask what the causes of these differences are, whether they are purely historical, accidental, or whether they reflect conscious differences in goals and aims, and, if the latter, what we can learn from these differences.

In this paper we compare contemporary philosophical (as opposed to mathematical) modal logic with three 13th-century views of modal logic. What we discover falls under the heading of the second reason: The comparison demonstrates that there is a fundamental difference between how these 13th-century logicians approached and used modal logic and how philosophical logicians of the 21st-century approach and use modal logic. This gives us cause to be careful that we do not discount medieval modal logic as being narrow or unfruitful: Because its aims are different from ours, we should not expect it to be applicable in the same circumstances.

The three 13th-century texts that we consider are William of Sherwood's *Introductiones in logicam* [11] (translated into English with commentary in

[10]), the short text *De propositionibus modalibus* [9], and Pseudo-Aquinas's *Summa totius logicae Aristotelis* [8].[1] Of the three, the provenance of the *Introductiones* is best known; the author can be ascribed with confidence, and while a definitive date of the text is not known, it is quite likely that the text was compiled between 1240 and 1248, a period in which Sherwood was a master in the Arts Faculty at the University of Paris [10, p. 8]. The other two texts are both connected to St. Thomas Aquinas. Aquinas was long considered to be the author of the *Summa*, though current thought is that this is highly unlikely. Conversely, the authorship of the *De modalibus* text was considered doubtful until the early 20th century when Grabmann attributed it to Aquinas; if he is the author of the *De modalibus*, it is a juvenile and early work [7, p. 13]. We shall follow Grabmann in attributing *De modalibus* to Aquinas, but reflecting our uncertainty about the authorship of the *Summa*, we shall refer to the author of that text as Pseudo-Aquinas. Despite questions about the authorship of the two text, it is clear from their content that they date from the same period as Sherwood's *Introductiones* or slightly later [2].

Before we can discuss the views of these three authors, and how they compare to modern approaches to modal logic, we must first address the question of what we mean by the phrase "modal logic". Broadly speaking, the phrase can apply to any type of logic to which new operators expressing different modes, such as modes of belief, knowledge, time, necessity, agency, etc., are added. It is not uncommon, today, to speak of deontic logic, epistemic logic, temporal logic, etc. all as "modal" logics. However, we take the phrase in a more narrow sense, using it to refer only to the logic of necessity and possibility. The term "modal" comes from Latin *modus* 'mode, mood', and when medieval authors speak of adding a *modus* to a sentence, they generally specify that it is one of the following six modes: *verum, falsum, necessarium, impossibile, possibile, contingens.*[2] By restricting our attention to just statements of necessity and possibility in this paper, we are following their customary usage of the phrase "modal proposition".

A note about references: Citations from William of Sherwood refer to page number unless a section number is explicitly indicated. The Aquinas text is referenced by sentence number, and Pseudo-Aquinas by tract, chapter, and sentence number.

2 Modes and modal propositions

All three of the 13th-century authors define modal propositions as being constructed from categorical propositions (recall that a categorical proposition or statement is, à la Sherwood, *cuius substantia consistit ex subiecto et praedicato* [11, p. 12][3]). The class of modal propositions is defined in a jointly semantic-syntactic fashion. First, on the syntactic side, a modal proposition is a categorical proposition to which a mode has been added.

[1] Translations of quotes from these two sources are by the present author.

[2] True, false, necessary, impossible, possible, and contingent, respectively.

[3] "one whose substance consists of a subject and a predicate" [10, p. 27].

The three authors all give slightly different definitions of *modus* 'mode'. Aquinas says that a mode is a *determinatio adiacens rei, quae quidem fit per adiectionem nominis adiectivi, quod determinat substantivum... vel per adverbium, quod determinat verbum* [9, 2][4], that is, both adverbs and adjectives are modes. Pseudo-Aquinas says a mode is an *adjacens rei determinatio; idest, determinatio facta per adjectivum* [8, tract. 6, cap. 11, 2][5], that is, modes are adjectives. And Sherwood takes the other route; his definition of mode includes only adverbs: *Modus igitur dicitur communiter et proprie. Communiter sic: Modus est determinatio alicuius actus, et secundum hoc convenit omni adverbio* [11, p. 32].[6]

But not all categorical sentences to which adverbs or adjectives have been added are, strictly speaking, modal. The second part of the definition, which the three authors all include, is the semantic side: It is only those categorical statements where the adverb determines or modifies the composition of the subject and the predicate that are correctly called modal.[7] This *determinatio* is a semantic concept, as it modifies the *significatio* ('signification', roughly, the meaning) of the sentence. The six modes which can determine the inherence expressed in a categorical sentence are *verum*, *falsum*, *necessarium*, *impossibile*, *possibile*, and *contingens*. However, because the addition of "true" and "false" to a categorical proposition does not change its signification (because *nihil addunt supra significationes propositionum de inesse* [9, 9][8]) these two modes will be omitted from consideration and the focus will be on the four modes *necessarium*, *impossibile*, *possibile*, and *contingens*.

At this point in his presentation of modality, Sherwood makes a distinction which the other two authors do not. He notes that there are two ways that *impossibile* and *necessarium* can be used. Both ways can be expressed in terms of temporal notions:

> *uno modo, quod non potest nec poterit nec potuit esse verum, et est impossible per se... alio modo, quod non potest nec poterit esse verum, potuit tamen ... et est impossibile per accidens. Et similiter dicitur necessarium per se, quod non potest nec potuit nec poterit esse falsum... Necessarium autem per accidens est,*

[4] "a determining attribute of a thing, which is made by an addition of an adjective word, which determines a substantive... or by an adverb, which determines a verb".

[5] "an adjoining determination of a thing; that is, a determination made through an adjective".

[6] "The word 'mode' is used both broadly and strictly. Broadly speaking, a mode is the determination of an act, and in this respect it goes together with every adverb" [10, p. 40].

[7] *Quidam determinat compositionem ipsam praedicati ad subiectum... et ab hoc solo modo dicitur propositio modalis*, "some determine the composition itself of the predicate with the subject... and by this mode alone is a proposition called modal" [9, 6]; *modalis vero in qua inhaerentia praedicati ad subjectum modificatur*, "a modal [is that] in which the inherence of the predicate to the subject is modified" [8, tract. 6, cap. 7, 4]; *proprie sic: modus est determinatio praedicati in subiecto* [11, p. 32], "strictly speaking, a mode is the determination of [the inherence of] the predicate in the subject" [10, p. 40].

[8] "they attach nothing above the significations of the assertoric propositions".

$$\begin{array}{rcl}\Box_{ps}\varphi & := & \varphi \wedge G\varphi \wedge H\varphi \\ \Box_{pa}\varphi & := & \varphi \wedge G\varphi \wedge \Diamond\neg H\varphi\end{array}$$

Figure 1. Sherwood's necessity operators

> *quod non potest nec poterit esse falsum, potuit tamen* [11, p. 34].[9]

Essentially, Sherwood is defining the necessity operators found in the table in Figure 1, where we list them translated into the familiar notation of temporal logic. (We discuss the correct interpretation of the $\Diamond$ in the definition of necessity *per accidens* in §4.) As we'll see in §3, we can define the impossibility operators from the necessity operators by negation, so we do not need to list them separately. According to Sherwood, *possibile* and *contingens* also have twofold usage. On the one hand, they can be used of statements which can both be true and be false, and so are neither impossible or necessary; this is the sense which is generally ascribed to "contingent" in modern usage. On the other hand, they can be used of statements which can be true, even if they cannot be false; this is the sense which is generally ascribed to "possibility" in modern usage, under the assumption that the axiom $\Box\varphi \rightarrow \Diamond\varphi$ is valid. While some medieval authors follow this distinction, using *possibile* for things which can be true, even if they cannot be false, and *contingens* in the stricter fashion for things which can be true or false, the two terms were regularly conflated, and as they were in the three texts we're considering, we'll follow their lead.

2.1 Construction

Once the relevant modes have been identified, the syntactic ways that they can be added to a categorical proposition must be distinguished. There are two ways that a mode can determine the composition of a categorical proposition. The three authors each make the distinction, but in slightly different ways and with different labels.

Aquinas's text divides modal propositions into those which are modal *de dicto* and those which are modal *de re*. This text is generally credited as being the source of the use of this distinction in modern philosophy and modal logic.[10] He makes the distinction this way:

[9] [impossible] is used in one way of whatever cannot be true now or in the future or in the past; and this is "impossible *per se*"... It is used in the other way of whatever cannot be true now or in the future although it could have been true in the past... and this is "impossible *per accidens*". Similarly, in the case something cannot be false now or in the future or in the past it is said to be "necessary *per se*"... But it is "necessary *per accidens*" in case something cannot be false now or in the future although it could have been [false] in the past [10, p. 41].

[10] See [12, p. 1], where the terms are first introduced in modern contexts. Von Wright credits Aquinas with this distinction, probably in reference to *De modalibus*, as this was attributed to Aquinas by the 1950's. Dutilh Novaes in [1, fn. 9] notes that von Wright was introduced to the distinction by Peter Geach.

> *Modalis de dicto est, in qua totum dictum subiicitur et modus praedicatur, ut Socrates currere est possibile; modalis de re est, in qua modus interponitur dicto, ut Socratem possibile est currere* [9, 16].[11]

The *dictum* of a sentence is what the sentence expresses; a categorical proposition's *dictum* can be formed, as the Aquinas tells us, by substituting the infinitive form for the indicative verb, and the accusative case for the nominative subject.[12] This same distinction is found in Pseudo-Aquinas but in a more elaborate fashion:

> *Ad sciendum autem earum quantitatem, notandum quod quaedam sunt propositiones modales de dicto, ut, Socratem currere est necesse; in quibus scilicet dictum subjicitur, et modus praedicatur: et istae sunt vere modales, quia modus hic determinat verbum ratione compositionis, ut supra dictum est. Quaedam autem sunt modales de re, in quibus videlicet modus interponitur dicto, ut, Socratem necesse est currere: non enim modo est sensus, quod hoc dictum sit necessarium, scilicet Socratem currere; sed hujus sensus est, quod in Socrate sit necessitas ad currendum* [8, tract. 6, cap. 11, 14–15].[13]

Sherwood makes the same distinction but does not use the *de dicto*/*de re* terminology. Instead he distinguishes between adverbial modes and nominal modes; categorical propositions with adverbial modes correspond to the class of *de re* modal sentences, and those with nominal modes correspond to the class of *de dicto* modal sentences [11, pp. 34–38].

2.2 Quantity

The type of modal sentence (that is, whether it is *de re* (or adverbial) or *de dicto* (or nominal)) must be established before the further properties of the sentence can be determined. Modal propositions, like categorical propositions, have both quantity and quality, and the authors give rules by which the quantity and quality of modal propositions can be recognized. The quantity of a categorical proposition can be one of four types: singular,

---

[11] "Modality is *de dicto* in which the whole *dictum* is made the subject and the mode is predicated, as in 'that Socrates runs is possible'; modality is *de re* in which the mode is inserted into the *dictum*, as in 'Socrates is possibly running'."

[12] *quod quidem fit si pro verbo indicativo propositionis sumatur infinitivus, et pro nominativo accusativus*, "because indeed it happens if an infinitive verb is assumed for the indicative verb of the proposition, and the accusative case [is assumed] for the nominative the accusative case" [9, 12].

[13] "However for knowing the quantity of them, it must be noted that certain ones are modal propositions *de dicto*, as in 'that Socrates runs is necessary', in which clearly the *dictum* is made subject, and a mode is predicated: and these are truly modals, because a mode here determines the verb by reason of composition, as in what's said above. However certain others are modal [propositions] *de re*, in which a mode is interposed in the dictum, as in 'Socrates is necessarily running': indeed by this mode the sense is not that this dictum is necessary, namely 'that Socrates runs', but of this the sense is that in Socrates is necessity for running."

particular, universal, or indefinite. A categorical proposition is singular when the subject term picks out only one object, e.g., because it is either a proper name or because it is modified by a definite article such as *hoc* or *illud*. It is particular when the subject term picks out more than one object, because it is modified by a particular quantifier such as *quoddam* or *aliquid*. It is universal when the subject term picks out all objects of which the term can be truly predicated, because is modified by a universal quantifier such as *omnem* or *nullum*. Finally, a categorical is indefinite; this is when the subject term refers to some object or objects, but no particular object or objects, because no quantifier or definite article is used, and the subject is not a proper name.

The division into modal statements *de dicto* and *de re* is motivated partly by the differences in how the quantity of the two types of statements is determined. Modal *de re* statements have the same quantity as their underlying categorical sentences. But this is not the case for modal *de dicto* statements. According to both Aquinas and Pseudo-Aquinas, *de dicto* statements always have singular quantity, even though they may contain universal or particular quantifiers within them.[14] This is because the subject of a *de dictum* sentence is a *dictum*, and a *dictum* is essentially a proper name; it has a unique referent. Because Sherwood doesn't use the *de dicto*/*de re* distinction, his identification of the quantity is phrased somewhat differently, but with the same end result: When a categorical statement with a nominal mode is interpreted as if it had an adverbial mode, then the quantity of the sentence is determined by the quantity of the underlying categorical claim. But when it is not interpreted this way, then the *dictum* of the sentence is the subject, and this is singular.

2.3 Quality

The quality of a proposition (categorical or otherwise) is determined by the presence or absence of a negation: For categorical sentences, it is the negation of the composition between the subject and the predicate, for modal sentences it is the negation of the mode. If the composition or the mode is affirmed, then the sentence is affirmative, and if it is denied, then it is negative. In this way, a categorical proposition which is negative can become positive when made into a modal proposition, and similarly a positive categorical proposition can become negative when made modal, because, as Aquinas notes, *propositio modalis dicitur affirmativa vel negativa secundum affirmationem vel negationem modi, et non dicti* [9, 19].[15] As an exam-

[14] *Sciendum quod omnes enunciationes modales de dicto sunt singulares, quantumcumque sit in eis signum universale*, "it must be known that all modal *de dicto* assertions are singulars, although in it may be a universal sign" [8, tract. 6, cap. 11, 21]; *sciendum est autem quod omnes modales de dicto sunt singulares, eo quod modus praedicatur de hoc vel de illo sicut de quodam singulari*, "However it must be understand that all modals *de dicto* are singulars, because a mode is predicated of this or that in the same way as of a certain singular" [9, 17].

[15] "a modal proposition is called affirmative or negative according to the affirmation or negation of the mode, and not of the *dictum*".

ple, *Socrates non currit* is a negative categorical proposition, but *Socrates non currere est possibile* is an affirmative modal proposition. Note that the quantity of a proposition is a syntactic property, because it depends on the presence or absence of the term *non*, whereas the quality of a categorical proposition is semantic, because it does not depend on the addition of a specific term but rather on the truth conditions of various predications of the subject term on different objects.

The importance of being able to determine the quality and quantity of a modal proposition is grounded in the importance which is ascribed to the inferential relations of modal propositions, as it is the quality and the quantity that determines which propositions can be inferred from which others. We discuss these next.

3 Inferential relations

The inferential relations which are discussed in the three treatises can be divided into two groups: implications and conversions. The implications considered are the relations of contradiction, contrariety, subcontrariety, subalternation, and superalternation (the relations which make up the square of opposition). The conversions considered are the traditional Aristotelian ones, conversion *per accidens* and conversion *simplex*, along with equivalences which can be generated through the square of opposition. These implications and conversions are used to develop a modal syllogistic.

3.1 Implications

Sherwood notes that modes can be combined with negation in one of the following four ways [10, p. 48]:

A without negation

B with more than one negation

C with one negation, before the mode

D with one negation, after the mode

Since we have four modes, and four ways that a mode can be combined with negation, this gives us sixteen syntactically different modes; these modes can occur both adverbially and nominally (or, to say the same thing, in *de dicto* and in *de re* statements). The question is whether these sixteen syntactically different modes are all semantically distinct, or whether there are any pairs which are equipollent (to call them by their standard medieval name *equipollens*). The answer is that each of the sixteen can be placed into one of four groups, called *ordines* 'orders' or 'series' (see Figure 2). An order is essentially an equivalence class, since *omnes propositiones, quae sunt in eodem ordine, aequipollent* [9, 23].[16] The four orders make up the corners of a square of opposition illustrating the inferential relationships (see Figure 3). This square of opposition can be found in the manuscripts of each of

[16] "[A]ll propositions which are in the same order are equipollent".

ordo 1	*possibile* *contingens* *non impossible* *non necessarium non*	*non possibile* *non contingens* *impossibile* *necessarium non*	*ordo* 3
ordo 2	*possibile non* *contingens non* *non impossibile non* *non necessarium*	*non possibile non* *non contingens non* *impossibile non* *necessarium*	*ordo* 4

Figure 2. The four *ordines*

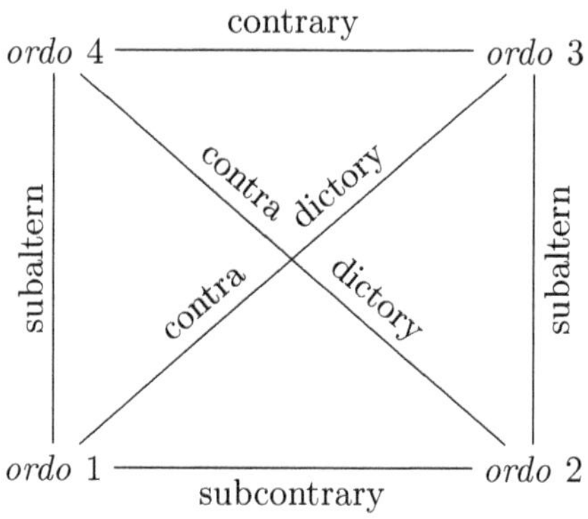

Figure 3. Modal square of opposition

the three treatises. After the square of opposition is presented, Aquinas's treatise makes reference to the mnemonic poem for constructing the square, whereupon the text ends.[17]

[17]The mnemonic poem shows up in various forms in 13th-century texts. Sherwood gives the text as follows: *Sit tibi linea subcontraria prima secunde. / Tertius est quarto semper contrarius ordo. / Tertius est primo contradictorius ordo. / Pugnat cum quarto contradicendo secundus. / Prima subest quarte vice particularis habens se. / Hac habet ad seriem se lege secunda sequentem* [10, fn. 91]. Aquinas rearranges the lines, adds a few of his own, and omit some of Sherwood's: *Tertius est quarto semper contrarius ordo. Pugnat cum quarto contradicendo secundus. Sit subcontraria linea tibi prima secundae. Tertius est primo contradictorius ordo. Prima subest quartae vicem particularis habens. Sed habet ad seriem se lege secunda sequentem. Vel ordo subalternus sit primus sive secundus. Primus amabimus, edentulique secundus. Tertius illiace, purpurea reliquus. Destruit u totum sed a confirmat utrumque, destruit e dictum, destruit i que modum* [9, 35–4]. Pseudo-Aquinas reduces the poem to just the names for each of the corners of the square: *Amabimus, edentuli, illiace, purpurea* [8, tract. 6, cap. 13, 22].

3.2 Conversions

Both Sherwood and Pseudo-Aquinas discuss how modal propositions can be converted from one form to another. By "conversion" both authors mean the two types of conversion which Aristotle presents in giving rules for the proving of syllogisms, conversion *simplex* or *per se* and conversion *per accidens*. Simple conversion of a categorical exchanges the subject and predicate terms, leaving the quality and the quantity of the sentence unchanged; accidental conversion swaps the subject and predicate, but also changes the quantity, from universal to particular or vice versa.[18] Sherwood also mentions a third type of conversion, conversion *per contrapositionem*, where the subject and predicate are swapped and replaced with their infinite counterparts (e.g., 'man' is replaced with 'non-man'; *infinitum* is the standard medieval name for such terms.)

In tract. 7, cap. 3, Pseudo-Aquinas tells us that *propositiones de necessario et impossibili eodem modo convertuntur sicut propositiones de inesse, et per idem principium probantur* [8, 2]. Though he does not say so explicitly, it is clear from all of his examples that he is discussing conversion principles for *de dicto* statements; all of his examples use nominal modes, not adverbial ones. Because it is not obvious that necessary and impossible propositions can be converted in the same way that assertoric (that is, categorical) propositions can be, he gives proofs for various conversions. We give the first, because it exemplifies the techniques used in the rest. It is a proof that

(1) *necesse est nullum b esse a*

can be simply converted into

(2) *necesse est nullum a esse b*

First, Pseudo-Aquinas notes that the opposite of (2) implies the opposite of (1). But the opposite of (2),

(3) *non necesse est nullum a esse b*

is equipollent to

(4) *possibile est aliquod a esse b*

The equipollence between (3) and (4) holds because *impossibile* and *non necessarium non* are equipollent (as we saw in the previous section), and this latter equipollence holds because *non nullus* and *aliquis* are equipollent. Next, he notes that from (4) the following can be proved through an expository syllogism (an expository syllogism is one which one premise is a singular proposition. Pseudo-Aquinas discusses these in tract. 7, cap. 2.):

[18] *Dicitur autem conversio simplex, quando de praedicato fit subjectum, et de subjecto praedicatum, manente secunda propositione in eadem qualitate et quantitate cum prima. Per accidens vero dicitur, quando de subjecto fit praedicatum, et e converso, manente eadem qualitate propositionis, sed mutata quantitate* [8, tract. 7, cap. 2, 4–5]; a similar definition can be found in [11, cap. 3, §2].

(5) *possibile est aliquod b esse a*

But (5) is the contradictory of (1). Since we were able to prove the contradictory of the antecedent from the contradictory of the consequent, we can conclude that (1) can be converted into (2). That (2) can be converted back into (1) by similar reasoning is obvious.

The other proofs are similar and so will not be discussed further here.

3.3 Modal syllogisms

Sherwood tells the reader, before he even gives the definition of a mode, that the reason it is important to separate modal propositions from assertoric ones is that

> *Cum intentio sit de enuntiatione propter syllogismum, consideranda est sub differentiis, in quibus differentiam facit in syllogismo. Quales sunt haec: ... modale, de inesse et aliae huiusmodi. Differt enim syllogismus a syllogismo per has differentias* [11, p. 30].[19]

Kretzmann points out that "in spite of this remark, which seems to promise a consideration of the modal/assertoric difference as it relates to the syllogism, there is no treatment of modal syllogisms in any of the works that have been ascribed to Sherwood" [10, fn. 58].

This leaves us with the *Summa*. Pseudo-Aquinas discusses modal syllogisms in tract. 7, caps. 13–15. Unfortunately, in many cases, his presentation is less than clear. The three chapters are devoted to the different combinations of necessary, impossible, and contingent premises with assertoric premises in syllogisms. Each combination is considered, and if it is valid, no argument is given, and if it is invalid, a counterexample is given. The result is an unfortunate tangle of case-by-case examples and rules with limited applicability.

Additionally, in giving the various examples of valid and invalid syllogisms, Pseudo-Aquinas moves between *de dicto* and *de re* formulations indiscriminately. For example, when he says that a syllogism in any mood or figure (for the technical details and terminology of Aristotelian syllogisms, see the Appendix) which has two necessary premises will have a necessary conclusion, the example that he gives is the following [tract. 7, cap. 13, 7–9]:

Necesse est omnem hominem esse animal.
Necesse est omne risibile esse hominem.
Ergo necesse est omne risibile esse animal.

But when he gives an example to show that a necessary conclusion does not follow from an assertoric major and a necessary minor, he uses *de re* modalities [tract. 7, cap. 13, 21–23]:

[19] [s]ince our treatment is oriented toward syllogism, we have to consider them under those differences that make a difference in syllogism. These are such differences as... modal, assertoric; and others of that sort. For one syllogism differs from another as a result of those differences [10, p. 39].

Omnis homo est albus.
Omne risibile necessario est homo.
Ergo omne risibile necessario est album.

The unclarity which results from his indiscriminate use of *de dicto* and *de re* statements in his examples is compounded by the fact that very few explicit rules for resolving the validity of classes of syllogisms are given. In assertoric syllogisms, the two rules commonly discussed are the *dici de omni* and the *dici de nullo*:

> *Est autem dici de omni, quando nihil est sumere sub subjecto, de quo non dicatur praedicatum; dici vero de nullo est, quando nihil est sumere sub subjecto, a quo non removeatur praedicatum* [8, tract. 7, cap. 1, 36].

Pseudo-Aquinas often appeals to these two rules when he gives arguments for the invalidity of certain syllogisms with one modal and one assertoric premise. It is only when he considers syllogisms which have one necessary premise and one contingent or possible premise that he formulates a new rule. The rule is:

> *si aliquod subjectum sit essentialiter sub aliquo praedicato, quicquid contingit sub subjecto, contingit sub praedicato* [8, tract. 7, cap. 15, 10].

Clearly this rule is an attempt to make a modal variant of the *dici de omni.*

It is at this point in the treatise that the modern logician could be forgiven for finding himself frustrated. The lack of both precision and perspicuity make one wonder whether there is anything to be gained in further study. If one is interested solely in developing a reliable modal syllogistic, there are other authors where this material is more easily accessible. But if one is interested in understanding the parts of the modal theory which are difficult not just because they are unclear but because they are fundamentally different from modern modal theories, then there are a number of things that can be said; we turn to these in the next section.

4 Contrasts with modern views of modal logic

We are now in the position to note two places where the medieval conception of modality and modal reasoning diverge from the modern conception of the same, with interesting consequences for our understanding of medieval modal logic.

4.1 The nature of modality

The first is that in modern propositional modal logic, the modality being expressed is the *de dicto* modality. A modal operator is an operator at the level of *formulas*. A formula of the form $\Box\varphi$ is read "it is necessary that φ", where the addition of "that" before "φ" is the syntactic construct in English for forming the *dictum* of a sentence. It isn't even clear that *de re* modality, with its emphasis on the inherence of the subject in the predicate, can be interpreted in a propositional context in a coherent fashion. Because of the subject-predicate nature of the *de re* sentences, it is clear that we are working with some type of first-order logic, not a propositional logic. But in the context of predicate logic, there is some temptation to say that *de re* statements aren't *really* about modality; they're just about a (perhaps special) type of predicates which we could call, e.g., possibly-P. But syntactically, these are just like any other predicate, and semantically, we would be perfectly within our bounds to give the truth conditions to predicates like possibly-P in the same way that we do predicates like P, through an assignment function. Then we could use $\Box$ and $\Diamond$ to express *real* modality, modality applying at the level of entire formulas.

This approach to modality is in direct contrast with that of William of Sherwood. Sherwood is reluctant to accept categorical statements with nominal modes (that is, *de dicto* modals) as modal statements [11, p. 36]. Recall that in §2 when we presented the different definitions of 'mode', all three authors agreed that under the most strict interpretation, only those categorical sentences where the mode determines the inherence of the subject and predicate are really modal. Both Aquinas and Pseudo-Aquinas are willing to let sentences such as *possibile est aliquod a esse b* to count as being determinations of the subject a in the predicate b, without really spelling out how we are to understand this determination, but Sherwood will only call such sentences modal when they are interpreted in the *de re* fashion. Under this interpretation:

> *Si enim dicam 'Socratem currere est contingens', idem est secundum rem ac si dicerem 'Socrates contingenter currit'* [11, p. 38].[20]

Can modifications in the inherence of a subject in a predicate even be represented in first-order modal logic? If the underlying categorical statement is universal or particular, then the distinction between the nominal and adverbial modes is easy: It is just the distinction between, e.g., $\Box\forall x F(x)$ and $\forall x \Box F(x)$ (see, e.g., [4, §4.3]). But this will not work for singular or indefinite statements, where there is no quantifier.

In [3, p. 108], Fitting gives two different ways that the formula $\Diamond P(c)$ could be read. If we take as models 5-tuples $\mathcal{M} = \langle W, R, D, I, V\rangle$, where W is the set of worlds, R the accessibility relation, D the domain function assigning a non-empty set of objects to each world, I the interpretation function which assigns each constant to an object in each world and each n-ary predicate to a set of n-tuples of objects in each world, and V is a valuation function assigning values to free variables, and we stipulate that every object in a world has a constant which is interpreted as that object, then the two possibilities for $\mathcal{M}, w \models \Diamond P(c)$ can be represented as:

[20] if I say 'that Socrates is running is contingent', it is just the same, with respect to what is signified, as if I were to say 'Socrates is contingently running' [10, p. 45].

1 There is a world x such that wRx and $\mathcal{M}, x \models_V Py$ where $V(y) = I(c, x)$.

2 There is a world x such that wRx and $\mathcal{M}, x \models_V Py$ where $V(y) = I(c, w)$.

The first reading can be interpreted as modality *de dicto*: The most natural reading of "it is possible that c is P" is "there is a possible world where the interpretation of c at that world is in the interpretation of P". The second is a plausible reading of modality *de re*, namely that what c actually is in the current world, that very thing is in the interpretation of P in another possible world.

This means that sentences of the form $\Diamond P(c)$ are essentially ambiguous: Their syntactic structure gives no clues as to whether they should be interpreted in the first or the second way. But from the point of view of the medieval logicians, this is precisely what they want: Natural language sentences such as *Socrates est possibile currere* are ambiguous, and we, as users of natural language, must make a choice in the interpretation of the sentence (perhaps based on context) when we wish to reason about it in a formal setting. The choice of interpretation will, naturally, affect the validity of the syllogisms in which these premises are found.[21]

This distinction is given in terms of simple predications, but its analysis easily extends to more complicated sentences such as *Omnis homo est possibile currere*. If we formalize this as $\forall y(Hy \rightarrow \Diamond_{dr} Cy)$[22] to show that we are interested in the *de re* analysis, then $\mathcal{M}, w \models \forall y(Hy \rightarrow \Diamond_{dr} Cy)$ is true if and only if for arbitrary m

$$\text{if } I(m) \in I(H, w), \text{ then } \exists x, wRx \text{ and } \mathcal{M}, x \models_v C(y) \text{ where } y \in I(m, w)$$

Note that x can be different for different m; this is exactly what we want, for if we required that it be the same world where all the currently existing men are running, then the sentence would collapse into the *de dicto* reading.

4.2 The truth conditions of modal sentences

The second discrepancy between modern modal logic and medieval logic as presented in these three texts comes from the emphasis. In modern modal logic, emphasis is placed on the truth conditions of the modal propositions considered in and of themselves; when working with Kripke semantics, this emphasis manifests itself in the choice of the R relation or a restriction on the valuation functions for the propositions. This is in contrast to the three texts that we've seen, where the emphasis is placed on the inferential relations between modal propositions, e.g., the relations which form the Square of Opposition, conversions and of modal propositions, and classes of valid syllogisms. (Speaking anachronistically, we could say that the medieval logicians were more interested in proof theory than in model theory.) Pseudo-Aquinas does not provide any explicit truth conditions for modal

---

[21] Since this is not an acceptable solution for many contemporary logicians, Fitting in [3, §3], Fitting and Mendelsohn in [4, ch. 9], and Garson in [5, ch. 19] all introduce lambda abstraction to solve the problem.

[22] For present purposes it does no harm to omit consideration of existential import.

propositions *considered in themselves* (as opposed to *considered with respect to other modal propositions*). This is most surprising when considered in conjunction with the stated goal of the entire treatise. *Omnes homines natura scire desiderant*, the text opens [8, Prologue, 1]. But, he goes on to say, knowledge only comes as a result of demonstration, and a demonstration is a valid syllogism with necessarily true premises. Because this is the only route to knowledge (valid syllogisms which have merely, but not necessarily, true premises can only lead to probable knowledge; these syllogisms are subsumed under 'dialectic', which our author says he will not consider in this treatise [8, Prologue, 11]), it is quite surprising that nothing is said about how to determine whether a premise is necessarily true, or (a slightly different question) whether a necessary premise is true.

Aquinas devotes two sentences to the truth conditions of modal propositions, when he draws a conceptual parallel between the four modes and the four combinations of quality and quantity in categorical propositions. He says:

> *Attendendum est autem quod necessarium habet similitudinem cum signo universali affirmativo, quia quod necesse est esse, semper est; impossibile cum signo universali negativo, quia quod est impossibile esse, nunquam est. Contingens vero et possibile similitudinem habent cum signo particulari: quia quod est contingens et possibile, quandoque est, quandoque non est* [9, 20–21].[23]

This interpretation of necessity and impossibility corresponds to Sherwood's definition of necessity and impossibility *per se* that we saw in §2. And as we saw in §3 that impossibility can be defined from necessity using negation, so too can possibility; so the type of possibility that Aquinas is discussing here is possibility *per se*, meaning that we can also formalize it with temporal notions, as

(6) $\Diamond_{ps}\varphi := (\varphi \vee F\varphi \vee P\varphi) \wedge (\neg\varphi \vee F\neg\varphi \vee P\neg\varphi)$

But if this temporal formula expresses the truth conditions of sentences of possibility and contingency, and there little reason to think that Sherwood would reject this definition while accepting the other, then we are left with the question of what exactly Sherwood means when he says that a statement which is necessary *per accidens* "could have been false in the past". That is, we must ask what type of possibility is being expressed by the $\Diamond$ in

(7) $\Box_{pa}\varphi := \varphi \wedge G\varphi \wedge \Diamond\neg H\varphi$

[23] "However, it must be understood that 'necessary' has a likeness with a universal affirmative sign, because what is necessary, always is; [and] 'impossible' [has a likeness] with a universal negative sign, because what is impossible, never is. But 'contingent' and 'possible' have a likeness with a particular sign: because what is contingent and possible, sometimes is, sometimes isn't."

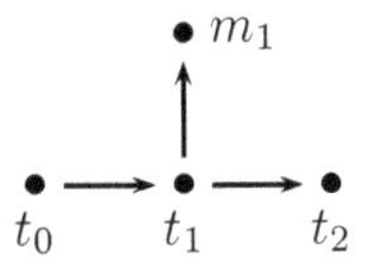

Figure 4. $\Box_{ps}$ is evaluated w.r.t. t_n, $\Diamond_{dr}$ w.r.t m_1

We can prove easily that $\Diamond\varphi$ here cannot be a short-hand for $(\varphi \vee F\varphi \vee P\varphi) \wedge (\neg\varphi \vee F\neg\varphi \vee P\neg\varphi)$. Let w be an arbitrary point where $\Box_{pa}\varphi$ is true. We know then that $w \models \varphi$ and $w \models G\varphi$, and

(8) $w \models (\neg H\varphi \vee F\neg H\varphi \vee P\neg H\varphi) \wedge (H\varphi \vee FH\varphi \vee PH\varphi)$

The problem is the second conjunct. If $\Box_{pa}\varphi$ is to be distinguished from $\Box_{ps}\varphi$, we know that neither $H\varphi$ nor $FH\varphi$ can be true, for then the two would be equivalent. Thus there is some $t \ll w$ such that $t \models H\varphi$; t cannot be an immediate predecessor of w or otherwise $PH\varphi$ would be equivalent with $H\varphi$, if we assume reflexivity. But this with the first conjunct forces there to be some t', $t < t' < w$ where $t' \models \neg\varphi$. And in this case, the interpretation of $\Box_{pa}\varphi$ would be that φ is true now and always in the future, but was false at some point in the past, and not that φ is true now and always in the future but *could have been* false in the past (even if it never was). If we take seriously Sherwood's counterfactual truth conditions for necessity *per accidens*, then the possibility involved cannot be temporal possibility.

There is a natural solution to the problem, though it is not one explicitly endorsed by Sherwood. If we remember that φ is not just a simple propositional construct, but a subject-predicate sentence like *Socrates est necessario currere*, then we can solve the question of the interpretation of $\Diamond$ by using the formal distinction between the *de re* and *de dicto* readings that we presented in the previous section. Then if *Socrates est necessario currere* is interpreted with necessity *per accidens*, it can be rewritten as

(9) $C(s) \wedge \Box_{ps}FC(s) \wedge \Diamond_{dr}\neg C(s)$

that is, Socrates is running now, it is necessary *per se* that he is running in the future, but he is possibly (*de re*) not running. The reason that this explication doesn't collapse the same way that the other one did is that the *de re* possibility here is not defined with respect to past, present, or future times, but to possible worlds; i.e., this type of possibility is in a sense perpendicular to the temporal notion of possibility (see Figure 4). And thus we see how Sherwood's insistence that it is the adverbial modal sentences which are the real modal sentences, and not the nominal ones, can be used to explain how, under a temporal notion of modality, the distinction between necessity *per accidens* and necessity *per se* can be maintained in the way that he has defined them.

5 Concluding remarks

Section 4.4 of [4] addresses the question "is quantified modal logic possible?" Fitting and Mendelsohn note that

> for much of the latter half of the twentieth century, there has been considerable antipathy toward the development of modal logic in certain quarters. Many of the philosophical objectors find their inspiration in the work of W.V.O. Quine, who as early as (Quine, 1943), expressed doubts about the coherence of the project... Quine does not believe that quantified modal logic can be done coherently... [4, p. 89]

These philosophical doubts are cited as the cause for the lack of development of quantified modal logic in modern times; Garson in his introduction says

> The problem is that quantified modal logic is not as well developed... Philosophical worries about whether quantification is coherent or advisable in certain modal settings partly explains this lack of attention [5, p. xiii]

This suspicion of quantified modal logic is deep-seated and pervasive among contemporary philosophical logicians; skim through any article which discusses quantified modal logic from a philosophical (as opposed to mathematical) point of view, and you will find at least one disparaging remark about it. In this paper we have demonstrated a lesson worth learning from the medieval logicians: quantified modal logic does not have to be a scary, intractable field of study, but in fact can be developed in a systematic fashion from the logic of simple categorical statements. Not only is this development conceptually quite natural, it is in some sense more natural than a modal logic for unanalyzed propositions.

Appendix

This appendix is a brief refresher course on basic (non-modal) Aristotelian syllogisms. Aristotelian syllogisms can be divided into four figures; the figure determines the order of the terms in the premises and conclusion (see Figure 5). The predicate term of the conclusion is called the *major term*;

1st figure	2nd figure
A_C, B_C: A_C	B_A, B_C: A_C
3rd figure	**4th figure**
A_B, C_B: A_C	B_A, C_B: A_C

Figure 5. The four figures

the subject term of the conclusion is called the *minor term*. The term which occurs in both premises but not in the conclusion is the *middle term*. The premise containing the major term is called the *major premise* and the premise containing the minor term is called the *minor premise*. Moods are created from the figures by inserting one of the four copulae

a "Every __ is a __" (universal affirmative)

e "No __ is a __" (universal negative)

i "Some __ is a __" (particular affirmative)

o "Some __ is not a __" (particular negative)

Since each figure has three slots and there are four different copulae, this means there are 64 moods. Only 24 of these moods are valid. The medievals gave mnemonic names to 19 of the 24 valid moods, where the vowels indicate the copulae of the major premise, the minor premise, and the conclusion (in that order), and the consonants indicate which of the four basic syllogism moods it is to be converted into, and by which conversion methods. This list has been extended in modern times to include names for all 24 of the valid moods. These are [6, §1]:

1st figure Barbara, Celarent, Darii, Ferio, Barbari, Celaront

2nd figure Cesare, Camestres, Festino, Baroco, Cesaro, Camestrop

3rd figure Darapti, Disamis, Datisi, Felapton, Bocardo, Ferison

4th figure Bramantip, Camenes, Dimaris, Fesapo, Fresison, Camenop

BIBLIOGRAPHY

[1] Dutilh Novaes, C. 2004. "A medieval reformulation of the de dicto / de re distinction", in Libor Běhounek, *Logica Yearbook 2003*, (Prague: Filosofia): 111–124.

[2] Eschmann, I.T. 1956. "A catalogue of St. Thomas' works: bibliographical notes" appendix to E. Gilson, *The Christian philosophy of Saint Thomas Aquinas* (New York: Random House): 381–439.

[3] Fitting, M. 1999. "On quantified modal logic", *Fundamenta informaticae* 39, no. 1: 105–121.

[4] Fitting, M. & R.L. Mendelsohn. 1998. *First-order modal logic*, Synthese Historical Library vol. 277 (Dordrecht: Kluwer Academic Publishers).

[5] Garson, J.W. 2006. *Modal logic for philosophers* (New York: Cambridge University Press).

[6] Lagerlund, H. 2004. "Medieval theories of the syllogism", in Edward N. Zalta, ed., *The Stanford encyclopedia of philosophy* (Spring 2004 edition). `http://plato.stanford.edu/archives/spr2004/entries/medieval-syllogism/`.

[7] O'Grady, P. 1997. "Aquinas on modal propositions: introduction, text, and translation", *International philosophical quarterly* 37, no. 1, issue no. 145: 13–27.

[8] Pseudo-Thomas Aquinas. 2006. *Summa totius Logicae Aristotelis* (WWW: Fundación Tomás de Aquino quoad hanc editionem Iura omnia asservantur). `http://www.corpusthomisticum.org/xpl.html`

[9] Thomas Aquinas. 2006. *De propositionibus modalibus* (WWW: Fundación Tomás de Aquino quoad hanc editionem Iura omnia asservantur). `http://www.corpusthomisticum.org/dpp.html`

[10] William of Sherwood. 1966. *William of Sherwood's introduction to logic*, trans. by N. Kretzmann, (Minneapolis, MN: University of Minnesota Press).

[11] William of Sherwood. 1995. *Introductiones in logicam / Einführung in die Logik*, ed. and trans. by H. Brands & Ch. Kann (Hamburg: Felix Meiner Verlag).
[12] von Wright, G.H. 1951. *An essay in modal logic* (Amsterdam: North-Holland).

Sara L. Uckelman
Institute for Logic, Language, and Computation,
Plantage Muidergracht 24, 1018TV Amsterdam,
The Netherlands
S.L.Uckelman@uva.nl

www.ingramcontent.com/pod-product-compliance
Ingram Content Group UK Ltd.
Pitfield, Milton Keynes, MK11 3LW, UK
UKHW021316180426
11947UKWH00015B/1259